U.S. Environmental Protection Agency

FY 2012 Annual Performance Report
FY 2014 Annual Plan

DATA QUALITY RECORDS

This document represents the verification and validation component of EPA's annual *Justification of Appropriation Estimates for the Committee on Appropriations.* It contains the following sections:

- Background information about EPA's performance data quality procedures;
- Data Quality Records (DQRs) for selected performance measures; and
- A DQR for the Agency's Budget Formulation System (BFS).

NOTE ABOUT SUPPORTING ATTACHMENTS NOT INCLUDED IN THIS DOCUMENT:
Some individual DQRs reference supporting attachments, indicated at the end of a DQR field under "Attached Documents."

These attachments are not accessible through this PDF, but are available upon request by sending an email to OCFOINFO@epa.gov. The email should indicate the measure number and text associated with the DQR, and the filename shown underneath the icon for the attachment.

Table of Contents

OFFICE OF INDIAN AND TRIBAL AFFAIRS (OITA) RECORD(S)   217

OFFICE OF CHEMICAL STRATEGIES AND POLLUTION PREVENTION (OCSPP) RECORD(S)   223

Office of the Inspector General (OIG) Record(s)

Measure Code: 35A - Environmental and business actions taken for improved performance or risk reduction.

Office of the Inspector General (OIG)

Goal Number and Title:
0 -

Objective Number and Title:
0 -

Sub-Objective Number and Title:
0 -

Strategic Target Code and Title:
0 -

Managing Office:
Chief of Staff in the Immediate Office of the Inspector General

1a. Performance Measure Term Definitions:
Number of environmental and business actions taken for improvements made or risks reduced in response to or influenced by OIG recommendations. OIG performance results are a chain of linked events, starting with OIG outputs (e.g., recommendations, reports of best practices, and identification of risks). The subsequent actions taken by EPA or its stakeholders/partners, as a result of OIG's outputs, to improve operational efficiency and environmental program delivery are reported as intermediate outcomes. The resulting improvements in operational efficiency, risks reduced/eliminated, and conditions of environmental and human health are reported as outcomes. By using common categories of performance measures, quantitative results can be summed and reported. Each outcome is also qualitatively described, supported, and linked to an OIG product or output. The OIG can only control its outputs and has no authority, beyond its influence, to implement its recommendations that lead to environmental and management outcomes. # Environmental/Health Improvements: Identifiable and documented environmental or human health improvements resulting from, or influenced by, any OIG work. Measured by the number and types of improvements. Narrative should describe the type of improvement and results in better environmental or human health conditions. The significance in improvements or impacts can be described in terms of physical characteristics, numbers of people affected, health and behavioral changes, and compliance with standards, including a percent change in a recognized environmental/health performance measure or indicator. Example: Faster cleanup of toxic waste dumps resulted from a process improvement that was recommended by the OIG and implemented by EPA reducing cases of illness. # Best Practices Implemented: Environmental program or business/operational best practices that were disseminated through OIG work and implemented by Agency offices, States, or other government agencies. Describe each best practice implemented and its implication for efficiency, effectiveness or economy.

Example 1: An OIG audit finds that one Region has improved its grants process through a best practice using a data control check system, resulting in better data accuracy and tracking of grant funds. OIG auditors recommend that another Region use the same system, and the best practice is successfully implemented to improve the Region's grants program. Example 2: An audit report describes a successful new method, developed by one EPA Region, to track and pursue fines for violators of waste manifest regulations. As a result of the report, several other EPA Regions decide to use the new method.

Risks Reduced or Eliminated: Environmental or business risks reduced or eliminated as a result of any OIG work. Measured in terms of the number of types (not occurrences) of risks reduced or eliminated. Narrative should describe the risk by type of environmental or human health exposure, incidence, financial, integrity or security or threat. Agency actions, which were influenced by OIG recommendations or advice, taken to resolve management challenges, Agency level or material weaknesses. Describe FMFIA weakness or management challenge addressed, and the action taken and implications. Example: Indictment/conviction regarding illegal dumping, or closure of fraudulent asbestos removal company, reduces the risk of exposure to harmful pollutants.

Additional Information:

U.S. EPA, Office of Inspector General, Audits, Evaluations, and Other Publications;   Available on the Internet at www.epa.gov/oig , last updatedAugust 2011.

Federal Government Inspector General Quality Standards.
Except for justified exceptions, OIG adheres to the following standards, which apply across the federal government:
·   Overall Governance: Quality Standards for Federal Offices of Inspector General. (President's Council on Integrity and Efficiency (PCIE) and Executive Council on Integrity and Efficiency (ECIE), October 2003). (http://www.ignet.gov/pande/standards/igstds.pdf This document contains quality standards for the management, operation and conduct of the Federal Offices of Inspector General (OIG). This document specifies that each federal OIG shall conduct, supervise, and coordinate its audits, investigations, inspections, and evaluations in compliance with the applicable professional standards listed below:
·   For Investigations: Quality Standards for Investigations. (President's Council on Integrity and Efficiency (PCIE) and Executive Council on Integrity and Efficiency (ECIE), December 2003). http://www.ignet.gov/pande/standards/invstds.pdf Consistent with appropriate Department of Justice Directives.
·   For Inspections and Evaluations: Quality Standards for Inspections. (President's Council on Integrity and Efficiency (PCIE) and Executive Council on Integrity and Efficiency (ECIE), January 2005). http://www.ignet.gov/pande/standards/oeistds.pdf
·   For Audits: Government Auditing Standards, issued by the US General Accounting Office (GAO). The professional standards and guidance in the Yellow Book are commonly referred to as generally accepted government auditing standards (GAGAS). These standards and guidance provide a framework for conducting high quality government audits and attestation engagements with competence, integrity, objectivity, and independence. The current version of the Yellow Book (July 2007) can be located in its entirety at the

following Website: www.gao.gov/govaud/d07162g.pdf

EPA OIG-Specific Operating Standards. The Project Management Handbook is the Office of Inspector General (OIG) policy document for conducting audit, program evaluation, public liaison, follow-up, and related projects. The Handbook describes the processes and standards the OIG uses to conduct the various phases of its work and helps ensure the quality, consistency, and timeliness of its products. Each OIG office may issue, upon approval by the Inspector General, supplemental guidance over assignments for which that office has responsibility.... This Handbook describes the audit, evaluation, public liaison, and follow-up processes and phases; it does not address OIG investigative processes although it does apply to audits/evaluations performed by the Office of Investigations (OI) [within EPA OIG]....OIG audit, program evaluation, public liaison, and follow-up reviews are normally conducted in accordance with appropriate Government Auditing Standards, as issued by the Comptroller General of the United States, commonly known as the Yellow Book.

Staff may use GAGAS in conjunction with other sets of professional standards. OIG reports may cite the use of other standards as appropriate. Teams should use GAGAS as the prevailing standard for conducting a review and reporting results should inconsistencies exist between GAGAS and other professional standards.

For some projects, adherence to all of the GAGAS may not be feasible or necessary. For these projects, the Product Line Director (PLD) will provide a rationale, the applicable standards not followed, and the impact on project results. The PLD's decision should be made during the design meeting, documented in the working papers, and described in the Scope and Methodology section of the report. [Source: Project Management Handbook].

Product Line Directors. Product Line Directors oversee one or more particular work areas and multiple project teams. The OIG product lines are as follows: Air/Research and Development; Water; Superfund/Land; Cross Media; Public Liaison and Special Reviews; Assistance Agreements; Contracts; Forensic Audits; Financial Management; Risk Assessment and Program Performance; Information Resources Management; Investigations; US Chemical Safety and Hazard Investigation Board; Legal Reviews; Briefings; OIG Enabling Support Programs; and Other Activities.

For more information on the PLD responsibilities, see Chapter 5 of the OIG Project Management Handbook, attached to this record.

2a. Original Data Source:

Data track EPA programs' environmental and business actions taken or improvements made and risks reduced or avoided as a result of OIG performance evaluations, audits, inspections and investigations. OIG collects such data from EPA programs and from EPA's contractors, partners and stakeholders.

2b. Source Data Collection:

Collection mode of information supporting this measure can vary.

OIG must determine whether the Agency's/auditee's corrective actions have adequately addressed and corrected the problems identified in the report. (Additional information on OIG's follow-up process can be found at

at http://oigintra.epa.gov/policy/policies/documents/OIG-04Follow-upPolicy.pdf

Project Managers (PMs) may make and document periodic inquiries concerning the Agency's/auditee's progress in implementing corrective actions resulting from OIG work. As part of this process, OIG may also request documentation supporting the progress or completion of actions taken to implement the Agency's corrective actions plan. OIG may also request the Agency's views and concurrence on the actual benefits resulting from the report. When a report is closed upon issuance, the transmittal memorandum should state that OIG will make periodic inquiries of the Agency's/auditee's progress in implementing corrective actions resulting from OIG work.

EPA Manual 2750 provides policy and direction for program managers to report and coordinate their corrective action plans with the OIG. (EPA's Audit Management Process, 2750 Change 2, December 3, 1988, Website: http://intranet.epa.gov/rmpolicy/ads/manuals/2750_2_t.pdf. This document requires OIG, as part of an effective system of internal controls, to evaluate the adequacy of such efforts before the recommendations can be closed out in the Agency's follow-up database. Evaluation of the corrective actions taken will allow the OIG to measure performance and accountability against OIG's performance targets and strategic goals. On an annual basis, a portion of OIG resources will be devoted to conducting follow-up reviews on specific significant reports. Each Assistant Inspector General (AIG), in consultation with his or her Product Line Director (PLD), will identify such work during the annual planning process.

2c. Source Data Reporting:

Data comes from OIG audit, evaluations and investigations that are performed under strict compliance with professional standard of the US Government Accountability Office and the US Department of Justice and subject to independent peer review. Data in the form of activities, output, and outcomes is entered by designated staff into the Inspector General Enterprise Management System. All original data is quality controlled for compliance with professional standard and data entered is quality reviewed for accuracy, completeness, timeliness and adequately supported.

3a. Relevant Information Systems:

OIG Performance Measurement and Results System (PMRS). PMRS captures and aggregates information on an array of OIG measures in a logic model format, linking immediate outputs with long-term intermediate outcomes and results. (The logic model can be found in OIG's Annual Performance Report at http://www.epa.gov/oig/planning.htm. PMRS is the OIG official system for collecting performance results data, in relation to its strategic and annual goals. All outputs (recommendations, best practices, risks identified) and outcome results (actions taken, changes in policies, procedures, practices, regulations, legislation, risks reduced, certifications for decisions, environmental improvements) influenced by OIG's current or prior work, and recognized during FY 2010 and beyond, should be entered into PMRS.

PMRS was developed as a prototype in FY 2001. Since then, there have been system improvements for ease of use. For example, during FY 2009 the PMRS was converted to a relational database directly linked to the new Inspector General Enterprise Management System (IGEMS).

IGEMS is an OIG employee time-tracking and project cost-tracking database that generates management reports. IGEMS is used to generate a project tracking number and a work product number. This system also tracks project progress and stores all related cost information.

AutoAudit and Teammate. These are repositories for all project working papers.

3b. Data Quality Procedures:

Data quality assurance and control are performed as an extension of OIG products and services, subject to rigorous compliance with the Government Auditing Standards of the Comptroller General, and are regularly reviewed by OIG management, an independent OIG Management Assessment Review Team, and external independent peer reviews (e.g., by accountancies qualified to evaluate OIG procedures against Government Auditing Standards). Each Assistant Inspector General certifies the completeness and accuracy of performance data.

All data reported are audited internally for accuracy and consistency.

OIG processes, including data processes, are governed by the quality standards described in "Additional Information" under the Performance Term Definition field. Notably, the Project Management Handbook (which governs audits) provides a QA checklist (see Appendix 4, of the 2008 Project Management Handbook, attached to this record). The Project Manager (PM) is responsible for completing the Quality Assurance (QA) checklist throughout the project. The PM prepares the checklist and submits it to the Product Line Director (PLD) upon completion of the Post Reporting Phase of the Project. The Checklist should be completed for all projects, recognizing that some steps in the checklist may not be applicable to all projects. The QA Checklist asks teams to ensure the integrity of data that resides in all of the OIG data systems. [Source: Project Management Handbook].

During FY 2008, OIG implemented an Audit Follow-up Policy to independently verify the status of Agency actions on OIG recommendations, which serve as the basis for OIG intermediate outcome results reported in the OIG PMRS.

(Additional information on the OIG's follow-up process can be found at http://oigintra.epa.gov/policy/policies/documents/OIG-04Follow-upPolicy.pdf

Attached Documents:
Policy101.PMH.Final.05.08.08.pdf

3c. Data Oversight:

There are three levels of PMRS access: View Only, Edit and Administrator. Everyone with IGEMS access has view only privileges. Individuals tasked with adding or editing PMRS entries must be granted PMRS Edit privileges. Contact a PMRS administrator to request Edit privileges.

Each Product Line Director (PLD), each of whom oversees one or more OIG work areas (e.g., Superfund, Contracts, etc.) and multiple project management teams, is responsible for ensuring that teams maintain proper integrity, accessibility, and retrievability of working papers in accordance with OIG policies. Likewise, they must ensure that information in OIG's automated systems is updated regularly by the team. (See field 2i, Additional Information, for more information about PLDs.)

3d. Calculation Methodology:

Database measures include numbers of: 1) recommendations for environmental and management improvement; 2) legislative, regulatory policy, directive, or process changes; 3) environmental, program management, security and resource integrity risks identified, reduced, or eliminated; 4) best practices identified and implemented; 5) examples of environmental and management actions taken and improvements made; 6) monetary value of funds questioned, saved, fined, or recovered; 7) criminal, civil, and administrative actions taken, 8) public or congressional inquiries resolved; and 9) certifications, allegations disproved, and cost corrections.

Because intermediate and long-term results may not be realized over a period of several years, only verifiable results are reported in the year completed.

Unit of measurement: Individual outcomes/actions

4a. Oversight and Timing of Final Results Reporting:

Data comes from OIG audit, evaluations and investigations that are performed under strict compliance with professional standard of the US Government Accountability Office and the US Department of Justice and subject to independent peer review. Data in the form of activities, output, and outcomes is entered by designated staff into the Inspector General Enterprise Management System. All original data is quality controlled for compliance with professional standard and data entered is quality reviewed for accuracy, completeness, timeliness and adequately supported. All data entered is carefully reviewed several times a years as it is entered and subsequently reported on a quarterly baThe OIG Assistant Inspectors General oversee the quality of the data used to generate reports of performance. The Office of the Chief of Staff oversee the data quality and the IG reviews the documents and date use for external consumption. Data is audited and quality test on a continuous basis through several steps from origin to final use.

4b. Data Limitations/Qualifications:

Because intermediate and long-term results may not be realized over a period of several years, only verifiable results are reported in the year completed.

Although all OIG staff are responsible for data accuracy in their products and services, there is a possibility of incomplete, miscoded, or missing data in the system due to human error or time lags. Data supporting achievement of results are often from indirect or external sources, with their own methods or standards for data verification/validation. Such data are reviewed according to the appropriate OIG quality standards (see "Additional Information"), and any questions about the quality of such data are documented in OIG reports and/or the PMRS.

The error rate for outputs is estimated at +/-2%, while the error rate for reported long-term outcomes is presumably greater because of the longer period needed for tracking results and difficulty in verifying a nexus between our work and subsequent actions and impacts beyond OIG's control. (The OIG logic model in the Annual Performance Report clarifies the kinds of measures that are output-oriented, like risks identified, versus outcome-oriented, like risks reduced.) Errors tend to be those of omission. Some errors may result from duplication as well.

4c. Third-Party Audits:

There have not been any previous audit findings or reports by external groups on data or database weaknesses in PMRS.

A December 2008 independent audit (www.epa.gov/oig/reports/2009/QualityReviewofEPAOIG-20081216.pdf found the following with regard to general OIG processes:

"We determined that the EPA OIG audits methodology, policies and procedures adequately complied with the Government Auditing Standards. The EPA OIG quality control system adequately documented compliance with professional and auditing standards for : Independence; Professional Judgment; Competence; Audit Planning; Supervision; Evidence and Audit Documentation; Reports on Performance Audits; Nonaudit Services; and the Quality Control Process. The auditors documented, before the audit report was issued, evidence of supervisory review of the work performed that supports findings, conclusions, and recommendations contained in the audit report.

"We determined that EPA OIG adequately followed the quality control policies established in the EPA OIG Project Management Handbook for conducting audit, program evaluation, and related projects. The audit documentation adequately includes evidence of work performed in the major three phases: Preliminary Research, Field Work and Reporting.

"We determined that EPA OIG adequately followed the standards and principles set forth in the PCIE and Executive Council on Integrity and Efficiency Quality Standards for Investigations, as applicable. The investigation adequately documented compliance with the guidelines applicable to the investigation efforts of criminal investigators working for the EPA OIG."

The audit also identified two minor conditions, related working paper review/approval and completion/update status. OIG agreed with the auditor recommendations related to the conditions and adapted its Project Management Handbook to address the concerns.

A June 2010 internal OIG review of OIG report quality (which included a review of reporting procedures) found no substantial issues (see http://www.epa.gov/oig/reports/2010/20100602-10-N-0134.pdf

Measure Code: 35B - Environmental and business recommendations or risks identified for corrective action.

Office of the Inspector General (OIG)

Goal Number and Title:
0 -

Objective Number and Title:
0 -

Sub-Objective Number and Title:
0 -

Strategic Target Code and Title:
0 -

Managing Office:
Chief of Staff in the Immediate Office of the Inspector General

1a. Performance Measure Term Definitions:
This is a measure of the number of OIG recommendations or risks identified for action, correction or improvement. OIG performance results are a chain of linked events, starting with OIG outputs (e.g., recommendations, reports of best practices, and identification of risks). The subsequent actions taken by EPA or its stakeholders/partners, as a result of OIG's outputs, to improve operational efficiency and environmental program delivery are reported as intermediate outcomes. The resulting improvements in operational efficiency, risks reduced/eliminated, and conditions of environmental and human health are reported as outcomes. By using common categories of performance measures, quantitative results can be summed and reported. Each outcome is also qualitatively described, supported, and linked to an OIG product or output. The OIG can only control its outputs and has no authority, beyond its influence, to implement its recommendations that lead to environmental and management outcomes. # Recommendations for Improvement: Number of recommendations for action in OIG reports, formal presentations or analyses. When the final product is issued, the number of report recommendations should be recorded in PMRS whether or not the Agency has concurred with or implemented the recommendations. (Do not count observations, suggestions, or editorial comments.) Describe each recommendation and its implications for environmental or management action and improvement. # Best Practices Identified: Best practices identified by OIG work for environmental or management program implementation to resolve a problem or risk, or improve a condition, process or result (from any source: EPA, State, other agency, etc.). Results are measured by the number of best practices identified. Narrative should explain the significance by describing the potential environmental or management change, action or impact. Example 1: In reviewing several States' partnership roles for an audit issue, we found that one State had developed very efficient and cost-effective water quality measures that could be applicable to other States or nationwide. Example 2: An audit determines that a Region has improved its management of a grant program

because of a workgroup the Region set up to coordinate grant and cooperative agreement functions.

Environmental or Business/ Operational/ Control Risks Identified (including noncompliance): Actual or potential environmental, health or operational risks identified by any OIG work. Measured in terms of the number of risks by type including the number of FMFIA disclosed program assurance issues, EPA management challenges and specific risks or internal control weaknesses. Includes issues presented in EPA financial statement audits and internal OIG reviews. Narrative should describe the risks and potential/actual environmental, health, and safety vulnerabilities, behaviors or conditions, risk of financial or resource loss or internal control weakness and their implications. Example 1: An OIG report on hog farm waste identifies environmental risks for drinking water contamination in nearby wells. Example 2: An OIG report identified that grants were given to grantees without specific performance objectives or verification that the grantees had acceptable financial accountability systems or controls.

Additional Information:

U.S. EPA, Office of Inspector General, Audits, Evaluations, and Other Publications;   Available on the Internet at www.epa.gov/oig , last updatedAugust 2011.

Federal Government Inspector General Quality Standards.
Except for justified exceptions, OIG adheres to the following standards, which apply across the federal government:
·   Overall Governance: Quality Standards for Federal Offices of Inspector General. (President's Council on Integrity and Efficiency (PCIE) and Executive Council on Integrity and Efficiency (ECIE), October 2003). (http://www.ignet.gov/pande/standards/igstds.pdf This document contains quality standards for the management, operation and conduct of the Federal Offices of Inspector General (OIG). This document specifies that each federal OIG shall conduct, supervise, and coordinate its audits, investigations, inspections, and evaluations in compliance with the applicable professional standards listed below:
·   For Investigations: Quality Standards for Investigations. (President's Council on Integrity and Efficiency (PCIE) and Executive Council on Integrity and Efficiency (ECIE), December 2003). http://www.ignet.gov/pande/standards/invstds.pdf Consistent with appropriate Department of Justice Directives.
·   For Inspections and Evaluations: Quality Standards for Inspections. (President's Council on Integrity and Efficiency (PCIE) and Executive Council on Integrity and Efficiency (ECIE), January 2005). http://www.ignet.gov/pande/standards/oeistds.pdf
·   For Audits: Government Auditing Standards, issued by the US General Accounting Office (GAO). The professional standards and guidance in the Yellow Book are commonly referred to as generally accepted government auditing standards (GAGAS). These standards and guidance provide a framework for conducting high quality government audits and attestation engagements with competence, integrity, objectivity, and independence. The current version of the Yellow Book (July 2007) can be located in its entirety at the following Website: www.gao.gov/govaud/d07162g.pdf

EPA OIG-Specific Operating Standards. The Project Management Handbook is the Office of Inspector General

(OIG) policy document for conducting audit, program evaluation, public liaison, follow-up, and related projects. The Handbook describes the processes and standards the OIG uses to conduct the various phases of its work and helps ensure the quality, consistency, and timeliness of its products. Each OIG office may issue, upon approval by the Inspector General, supplemental guidance over assignments for which that office has responsibility.... This Handbook describes the audit, evaluation, public liaison, and follow-up processes and phases; it does not address OIG investigative processes although it does apply to audits/evaluations performed by the Office of Investigations (OI) [within EPA OIG]....OIG audit, program evaluation, public liaison, and follow-up reviews are normally conducted in accordance with appropriate Government Auditing Standards, as issued by the Comptroller General of the United States, commonly known as the Yellow Book.

Staff may use GAGAS in conjunction with other sets of professional standards. OIG reports may cite the use of other standards as appropriate. Teams should use GAGAS as the prevailing standard for conducting a review and reporting results should inconsistencies exist between GAGAS and other professional standards.

For some projects, adherence to all of the GAGAS may not be feasible or necessary. For these projects, the Product Line Director (PLD) will provide a rationale, the applicable standards not followed, and the impact on project results. The PLD's decision should be made during the design meeting, documented in the working papers, and described in the Scope and Methodology section of the report. [Source: Project Management Handbook].

Product Line Directors.   Product Line Directors oversee one or more particular work areas and multiple project teams. The OIG product lines are as follows: Air/Research and Development; Water; Superfund/Land; Cross Media; Public Liaison and Special Reviews; Assistance Agreements; Contracts; Forensic Audits; Financial Management; Risk Assessment and Program Performance; Information Resources Management; Investigations; US Chemical Safety and Hazard Investigation Board; Legal Reviews; Briefings; OIG Enabling Support Programs; and Other Activities.

For more information on the PLD responsibilities, see Chapter 5 of the OIG Project Management Handbook, attached to this record.

2a. Original Data Source:
Data track environmental and business recommendations or risks identified for corrective action as a result of OIG performance evaluations, audits, inspections and investigations. OIG collects such data from EPA programs and from EPA's contractors, partners and stakeholders.

2b. Source Data Collection:
Collection mode of information supporting this measure can vary. OIG must determine whether the Agency's/auditee's corrective actions have adequately addressed and corrected the problems identified in the report. (Additional information on OIG's follow-up process can be found at at http://oigintra.epa.gov/policy/policies/documents/OIG-04Follow-upPolicy.pdf

Project Managers (PMs) may make and document periodic inquiries concerning the Agency's/auditee's progress in implementing corrective actions resulting from OIG work. As part of this process, OIG may also request documentation supporting the progress or completion of actions taken to implement the Agency's corrective actions plan. OIG may also request the Agency's views and concurrence on the actual benefits resulting from the report. When a report is closed upon issuance, the transmittal memorandum should state that OIG will make periodic inquiries of the Agency's/auditee's progress in implementing corrective actions resulting from OIG work.

EPA Manual 2750 provides policy and direction for program managers to report and coordinate their corrective action plans with the OIG. (EPA's Audit Management Process, 2750 Change 2, December 3, 1988, Website: http://intranet.epa.gov/rmpolicy/ads/manuals/2750_2_t.pdf. This document requires OIG, as part of an effective system of internal controls, to evaluate the adequacy of such efforts before the recommendations can be closed out in the Agency's follow-up database. Evaluation of the corrective actions taken will allow the OIG to measure performance and accountability against OIG's performance targets and strategic goals. On an annual basis, a portion of OIG resources will be devoted to conducting follow-up reviews on specific significant reports. Each Assistant Inspector General (AIG), in consultation with his or her Product Line Director (PLD), will identify such work during the annual planning process.

2c. Source Data Reporting:

Data comes from OIG audit, evaluations and investigations that are performed under strict compliance with professional standard of the US Government Accountability Office and the US Department of Justice and subject to independent peer review. Data in the form of activities, output, and outcomes is entered by designated staff into the Inspector General Enterprise Management System. All original data is quality controlled for compliance with professional standard and data entered is quality reviewed for accuracy, completeness, timeliness and adequately supported.

3a. Relevant Information Systems:

OIG Performance Measurement and Results System (PMRS). PMRS captures and aggregates information on an array of OIG measures in a logic model format, linking immediate outputs with long-term intermediate outcomes and results. (The logic model can be found in OIG's Annual Performance Report at http://www.epa.gov/oig/planning.htm. PMRS is the OIG official system for collecting performance results data, in relation to its strategic and annual goals. All outputs (recommendations, best practices, risks identified) and outcome results (actions taken, changes in policies, procedures, practices, regulations, legislation, risks reduced, certifications for decisions, environmental improvements) influenced by OIG's current or prior work, and recognized during FY 2010 and beyond, should be entered into PMRS.

PMRS was developed as a prototype in FY 2001. Since then, there have been system improvements for ease of use. For example, during FY 2009 the PMRS was converted to a relational database directly linked to the new Inspector General Enterprise Management System (IGEMS).

IGEMS is an OIG employee time-tracking and project cost-tracking database that generates management reports. IGEMS is used to generate a project tracking number and a work product number. This system also tracks project progress and stores all related cost information.

AutoAudit and Teammate. These are repositories for all project working papers.

3b. Data Quality Procedures:

Data quality assurance and control are performed as an extension of OIG products and services, subject to rigorous compliance with the Government Auditing Standards of the Comptroller General, and are regularly reviewed by OIG management, an independent OIG Management Assessment Review Team, and external independent peer reviews (e.g., by accountancies qualified to evaluate OIG procedures against Government Auditing Standards). Each Assistant Inspector General certifies the completeness and accuracy of performance data.

All data reported are audited internally for accuracy and consistency.

OIG processes, including data processes, are governed by the quality standards described in "Additional Information" under the Performance Term Definition field. Notably, the Project Management Handbook (which governs audits) provides a QA checklist (see Appendix 4, of the 2008 Project Management Handbook, attached to this record). The Project Manager (PM) is responsible for completing the Quality Assurance (QA) checklist throughout the project. The PM prepares the checklist and submits it to the Product Line Director (PLD) upon completion of the Post Reporting Phase of the Project. The Checklist should be completed for all projects, recognizing that some steps in the checklist may not be applicable to all projects. The QA Checklist asks teams to ensure the integrity of data that resides in all of the OIG data systems. [Source: Project Management Handbook].

During FY 2008, OIG implemented an Audit Follow-up Policy to independently verify the status of Agency actions on OIG recommendations, which serve as the basis for OIG intermediate outcome results reported in the OIG PMRS.

(Additional information on the OIG's follow-up process can be found at
http://oigintra.epa.gov/policy/policies/documents/OIG-04Follow-upPolicy.pdf

Attached Documents:
Policy101.PMH.Final.05.08.08.pdf

3c. Data Oversight:

There are three levels of PMRS access: View Only, Edit and Administrator. Everyone with IGEMS access has view only privileges. Individuals tasked with adding or editing PMRS entries must be granted PMRS Edit privileges. Contact a PMRS administrator to request Edit privileges.

Each Product Line Director (PLD), each of whom oversees one or more OIG work areas (e.g., Superfund, Contracts, etc.) and multiple project management teams, is responsible for ensuring that teams maintain proper integrity, accessibility, and retrievability of working papers in accordance with OIG policies. Likewise, they must ensure that information in OIG's automated systems is updated regularly by the team. (See field 2i, Additional Information, for more information about PLDs.)

3d. Calculation Methodology:

Database measures include numbers of: 1) recommendations for environmental and management improvement; 2) legislative, regulatory policy, directive, or process changes; 3) environmental, program management, security and resource integrity risks identified, reduced, or eliminated; 4) best practices identified and implemented; 5) examples of environmental and management actions taken and improvements made; 6) monetary value of funds questioned, saved, fined, or recovered; 7) criminal, civil, and administrative actions taken, 8) public or congressional inquiries resolved; and 9) certifications, allegations disproved, and cost corrections.

Because intermediate and long-term results may not be realized over a period of several years, only verifiable results are reported in the year completed.

Unit of measurement: Individual recommendations/risks

4a. Oversight and Timing of Final Results Reporting:

The OIG Assistant Inspectors General oversee the quality of the data used to generate reports of performance. The Office of the Chief of Staff oversee the data quality and the IG reviews the documents and date use for external consumption. Data is audited and quality test on a continuous basis through several steps from origin to final use.

4b. Data Limitations/Qualifications:

Because intermediate and long-term results may not be realized over a period of several years, only verifiable results are reported in the year completed.

Although all OIG staff are responsible for data accuracy in their products and services, there is a possibility of incomplete, miscoded, or missing data in the system due to human error or time lags. Data supporting achievement of results are often from indirect or external sources, with their own methods or standards for data verification/validation. Such data are reviewed according to the appropriate OIG quality standards (see "Additional Information"), and any questions about the quality of such data are documented in OIG reports and/or the PMRS.

The error rate for outputs is estimated at +/-2%, while the error rate for reported long-term outcomes is presumably greater because of the longer period needed for tracking results and difficulty in verifying a nexus between our work and subsequent actions and impacts beyond OIG's control. (The OIG logic model in the Annual Performance Report clarifies the kinds of measures that are output-oriented, like risks identified, versus outcome-oriented, like risks reduced.) Errors tend to be those of omission. Some errors may result

from duplication as well.

4c. Third-Party Audits:

There have not been any previous audit findings or reports by external groups on data or database weaknesses in PMRS.

A December 2008 independent audit (www.epa.gov/oig/reports/2009/QualityReviewofEPAOIG-20081216.pdf found the following with regard to general OIG processes:

"We determined that the EPA OIG audits methodology, policies and procedures adequately complied with the Government Auditing Standards. The EPA OIG quality control system adequately documented compliance with professional and auditing standards for : Independence; Professional Judgment; Competence; Audit Planning; Supervision; Evidence and Audit Documentation; Reports on Performance Audits; Nonaudit Services; and the Quality Control Process. The auditors documented, before the audit report was issued, evidence of supervisory review of the work performed that supports findings, conclusions, and recommendations contained in the audit report.

"We determined that EPA OIG adequately followed the quality control policies established in the EPA OIG Project Management Handbook for conducting audit, program evaluation, and related projects. The audit documentation adequately includes evidence of work performed in the major three phases: Preliminary Research, Field Work and Reporting.

"We determined that EPA OIG adequately followed the standards and principles set forth in the PCIE and Executive Council on Integrity and Efficiency Quality Standards for Investigations, as applicable. The investigation adequately documented compliance with the guidelines applicable to the investigation efforts of criminal investigators working for the EPA OIG."

The audit also identified two minor conditions, related working paper review/approval and completion/update status. OIG agreed with the auditor recommendations related to the conditions and adapted its Project Management Handbook to address the concerns.

A June 2010 internal OIG review of OIG report quality (which included a review of reporting procedures) found no substantial issues (see http://www.epa.gov/oig/reports/2010/20100602-10-N-0134.pdf

Measure Code: 35C - Return on the annual dollar investment, as a percentage of the OIG budget, from audits and investigations.

Office of the Inspector General (OIG)

Goal Number and Title:
0 -

Objective Number and Title:
0 -

Sub-Objective Number and Title:
0 -

Strategic Target Code and Title:
0 -

Managing Office:
Chief of Staff in the Immediate Office of the Inspector General

1a. Performance Measure Term Definitions:
This is a measure of the total dollar amount of questioned costs, cost efficiencies, civil settlements, fines and recoveries from OIG audits and investigations compared to annual budget investments in the OIG. OIG performance results are a chain of linked events, starting with OIG outputs (e.g., recommendations, reports of best practices, and identification of risks). The subsequent actions taken by EPA or its stakeholders/partners, as a result of OIG's outputs, to improve operational efficiency and environmental program delivery are reported as intermediate outcomes. The resulting improvements in operational efficiency, risks reduced/eliminated, and conditions of environmental and human health are reported as outcomes. By using common categories of performance measures, quantitative results can be summed and reported. Each outcome is also qualitatively described, supported, and linked to an OIG product or output. The OIG can only control its outputs and has no authority, beyond its influence, to implement its recommendations that lead to environmental and management outcomes. $s Questioned Costs Sustained: Dollar amount of questioned costs accepted or agreed to by the Agency or other action official. Describe the EPA total amount questioned and its nature. $s Efficiencies or Adjustments Sustained: Dollar amount of efficiencies or cost adjustments, accepted or agreed to by the Agency or other action official. Describe the total amount identified as an efficiency/adjustment and its nature. Actual Costs Recovered: Questioned costs or cost efficiencies that are recovered. $ Questioned Costs: (actual dollars) The dollar value of questioned costs as defined by the IG Act. Describe nature of costs questioned. The IG Act defines a questioned cost as "a cost that is questioned by the Office because of 1) an alleged violation or provision of law, regulation, contract, grant, or cooperative agreement, or other agreement or document governing the expenditure of funds; 2) a finding that at the time of the audit,

such cost is not supported by adequate documentation; or 3) a finding that the expenditure of funds for the intended purpose is unnecessary or unreasonable."

It is the amounts paid by EPA for which the OIG recommends EPA pursue recovery, including Government property, services or benefits provided to ineligible recipients; recommended collections of money inadvertently or erroneously paid out; and recommended collections or offsets for overcharges or ineligible claims.

For contract/grant reports, it is contractor or grantee costs the "auditor" recommends be disallowed by the contracting officer, grant official, or other management official on an EPA portion of a contract or grant. Costs normally result from a finding that expenditures were not made in accordance with applicable laws, regulations, contracts, grants, or other agreements; or a finding that the expenditure of funds for the intended purpose was unnecessary or unreasonable.

$ Recommended Efficiencies, Costs Saved or Avoided: (monetized results) The immediate and near future monetary benefit of savings or funds put to better use on an EPA project as a result of OIG work:
1) Savings from eliminating work products or office functions, which were no longer of use or too costly; and
2) The savings from new or streamlined processes or work products, instituted to save time and/or money. Describe the nature of the savings including monetary value of time saved.

For cost efficiencies, the IG Act defines a recommendation that funds be put to better use as "a recommendation by the Office that funds could be used more efficiently if management of an establishment took actions to implement and complete the recommendation, including: 1) Reductions in outlays; 2) Deobligations of funds from programs or operations; 3) Withdrawal of interest subsidy costs on loans or loan guarantees, insurance, or bonds; 4) Costs not incurred by implementing recommended improvements related to the operations of the establishment, a contractor, or grantee; 5) Avoidance of unnecessary expenditures noted in preaward reviews of contract or grants; or 6) Other savings which are specifically identified.

Cost efficiencies, funds put to better use, represent a quantity of funds that could be used more efficiently if management took actions to complete recommendations pertaining to deobligation of funds, costs not incurred by implementing recommended improvements, and other savings identified.

$ Cost Adjustments (Savings, Questioned) Made During the Audit, But Not Reported for Resolution: During the conduct of an audit or evaluation, costs may be questioned or opportunities for savings and adjustments may be identified which are acknowledged and acted upon/resolved prior to the report being issued. These costs may not be reported to the Agency since they are resolved prior to issuance and therefore do not go into the Agency Audit Resolution Process. These $ costs/savings or adjustments should be reported in PMRS as Valued Added results by the OIG or its surrogates as long as they can be substantiated. Also, report adjustments know as "Cost Realism", where a contract is adjusted to reflect accurate costs that may change a decision, or impact future funding of a contract or project. Describe the action taken and anticipated or actual impact.

$ Fines, Recoveries, Restitutions, Collections: Dollar value of investigative recoveries, meaning: 1) Recoveries during the course of an investigation before any criminal or civil prosecution; 2) criminal or civil court-ordered fines, penalties, and restitutions; 3) out-of-court settlements, including non-court settlements resulting from administrative actions. Describe nature of amounts and reason.

Additional Information:

U.S. EPA, Office of Inspector General, Audits, Evaluations, and Other Publications;   Available on the Internet at www.epa.gov/oig , last updatedAugust 2011.

Federal Government Inspector General Quality Standards.
Except for justified exceptions, OIG adheres to the following standards, which apply across the federal government:
·   Overall Governance: Quality Standards for Federal Offices of Inspector General. (President's Council on Integrity and Efficiency (PCIE) and Executive Council on Integrity and Efficiency (ECIE), October 2003). (http://www.ignet.gov/pande/standards/igstds.pdf This document contains quality standards for the management, operation and conduct of the Federal Offices of Inspector General (OIG). This document specifies that each federal OIG shall conduct, supervise, and coordinate its audits, investigations, inspections, and evaluations in compliance with the applicable professional standards listed below:
·   For Investigations: Quality Standards for Investigations. (President's Council on Integrity and Efficiency (PCIE) and Executive Council on Integrity and Efficiency (ECIE), December 2003). http://www.ignet.gov/pande/standards/invstds.pdf Consistent with appropriate Department of Justice Directives.
·   For Inspections and Evaluations: Quality Standards for Inspections. (President's Council on Integrity and Efficiency (PCIE) and Executive Council on Integrity and Efficiency (ECIE), January 2005). http://www.ignet.gov/pande/standards/oeistds.pdf
·   For Audits: Government Auditing Standards, issued by the US General Accounting Office (GAO). The professional standards and guidance in the Yellow Book are commonly referred to as generally accepted government auditing standards (GAGAS). These standards and guidance provide a framework for conducting high quality government audits and attestation engagements with competence, integrity, objectivity, and independence. The current version of the Yellow Book (July 2007) can be located in its entirety at the following Website: www.gao.gov/govaud/d07162g.pdf

EPA OIG-Specific Operating Standards. The Project Management Handbook is the Office of Inspector General (OIG) policy document for conducting audit, program evaluation, public liaison, follow-up, and related projects. The Handbook describes the processes and standards the OIG uses to conduct the various phases of its work and helps ensure the quality, consistency, and timeliness of its products. Each OIG office may issue, upon approval by the Inspector General, supplemental guidance over assignments for which that office has responsibility.... This Handbook describes the audit, evaluation, public liaison, and follow-up processes and phases; it does not address OIG investigative processes although it does apply to audits/evaluations performed by the Office of Investigations (OI) [within EPA OIG]....OIG audit, program evaluation, public liaison, and follow-up reviews are normally conducted in accordance with appropriate Government Auditing Standards, as issued by the Comptroller General of the United States, commonly known as the Yellow Book.

Staff may use GAGAS in conjunction with other sets of professional standards. OIG reports may cite the use of other standards as appropriate. Teams should use GAGAS as the prevailing standard for conducting a review and reporting results should inconsistencies exist between GAGAS and other professional standards.

For some projects, adherence to all of the GAGAS may not be feasible or necessary. For these projects, the Product Line Director (PLD) will provide a rationale, the applicable standards not followed, and the impact on project results. The PLD's decision should be made during the design meeting, documented in the working papers, and described in the Scope and Methodology section of the report. [Source: Project Management Handbook].

Product Line Directors.   Product Line Directors oversee one or more particular work areas and multiple project teams. The OIG product lines are as follows: Air/Research and Development; Water; Superfund/Land; Cross Media; Public Liaison and Special Reviews; Assistance Agreements; Contracts; Forensic Audits; Financial Management; Risk Assessment and Program Performance; Information Resources Management; Investigations; US Chemical Safety and Hazard Investigation Board; Legal Reviews; Briefings; OIG Enabling Support Programs; and Other Activities.

For more information on the PLD responsibilities, see Chapter 5 of the OIG Project Management Handbook, attached to this record.

2a. Original Data Source:

Data is collected and reported by designated OIG staff members in OIG Performance Measurement Databases as a result of OIG performance evaluations, audits, inspections and investigations and other analysis of proposed and existing Agency Policies, regulations and laws. OIG collects such data from the activities, outputs, intermediate outcomes and long-term outcome results of OIG operations. OIG collects such data from EPA programs and from court and other public data sources.

2b. Source Data Collection:

Performance information is entered by designated staff into the Inspector General Enterprise Management System from OIG audits, evaluations and investigations performed under strict compliance with applicable professional standards. All OIG products go through a rigorous quality assurance process and are subject to independent peer review.

2c. Source Data Reporting:

Data is derived from the results of audits, evaluations, investigations and special analysis that are performed in accordance with Professional Standards of the US Government Accountability Office or the Us Department of Justice. All OIG products are quality controlled and subject to independent peer review for compliance with a all professional standards. Data is entered, in compliance with EPA and OIG data quality standards into the Inspector General Enterprise Management System and which is further reviewed for quality and consistency by the OIG performance quality staff members.

3a. Relevant Information Systems:

OIG Performance Measurement and Results System (PMRS). PMRS captures and aggregates information on an array of OIG measures in a logic model format, linking immediate outputs with long-term intermediate outcomes and results. (The logic model can be found in OIG's Annual Performance Report at http://www.epa.gov/oig/planning.htm. PMRS is the OIG official system for collecting performance results data, in relation to its strategic and annual goals. All outputs (recommendations, best practices, risks identified) and outcome results (actions taken, changes in policies, procedures, practices, regulations, legislation, risks reduced, certifications for decisions, environmental improvements) influenced by OIG's current or prior work, and recognized during FY 2010 and beyond, should be entered into PMRS.

PMRS was developed as a prototype in FY 2001. Since then, there have been system improvements for ease of use. For example, during FY 2009 the PMRS was converted to a relational database directly linked to the new Inspector General Enterprise Management System (IGEMS).

IGEMS is an OIG employee time-tracking and project cost-tracking database that generates management reports. IGEMS is used to generate a project tracking number and a work product number. This system also tracks project progress and stores all related cost information.

AutoAudit and Teammate. These are repositories for all project working papers.

3b. Data Quality Procedures:

Data quality assurance and control are performed as an extension of OIG products and services, subject to rigorous compliance with the Government Auditing Standards of the Comptroller General, and are regularly reviewed by OIG management, an independent OIG Management Assessment Review Team, and external independent peer reviews (e.g., by accountancies qualified to evaluate OIG procedures against Government Auditing Standards). Each Assistant Inspector General certifies the completeness and accuracy of performance data.

All data reported are audited internally for accuracy and consistency.

OIG processes, including data processes, are governed by the quality standards described in "Additional Information" under the Performance Term Definition field. Notably, the Project Management Handbook (which governs audits) provides a QA checklist (see Appendix 4, of the 2008 Project Management Handbook, attached to this record). The Project Manager (PM) is responsible for completing the Quality Assurance (QA) checklist throughout the project. The PM prepares the checklist and submits it to the Product Line Director (PLD) upon completion of the Post Reporting Phase of the Project. The Checklist should be completed for all projects, recognizing that some steps in the checklist may not be applicable to all projects. The QA Checklist asks teams to ensure the integrity of data that resides in all of the OIG data systems. [Source: Project Management Handbook].

During FY 2008, OIG implemented an Audit Follow-up Policy to independently verify the status of Agency actions on OIG recommendations, which serve as the basis for OIG intermediate outcome results reported in the OIG PMRS.

(Additional information on the OIG's follow-up process can be found at http://oigintra.epa.gov/policy/policies/documents/OIG-04Follow-upPolicy.pdf

Attached Documents:
Policy101.PMH.Final.05.08.08.pdf

3c. Data Oversight:

There are three levels of PMRS access: View Only, Edit and Administrator. Everyone with IGEMS access has view only privileges. Individuals tasked with adding or editing PMRS entries must be granted PMRS Edit privileges. Contact a PMRS administrator to request Edit privileges.

Each Product Line Director (PLD), each of whom oversees one or more OIG work areas (e.g., Superfund, Contracts, etc.) and multiple project management teams, is responsible for ensuring that teams maintain proper integrity, accessibility, and retrievability of working papers in accordance with OIG policies. Likewise, they must ensure that information in OIG's automated systems is updated regularly by the team. (See field 2i, Additional Information, for more information about PLDs.)

3d. Calculation Methodology:

Database measures include numbers of: 1) recommendations for environmental and management improvement; 2) legislative, regulatory policy, directive, or process changes; 3) environmental, program management, security and resource integrity risks identified, reduced, or eliminated; 4) best practices identified and implemented; 5) examples of environmental and management actions taken and improvements made; 6) monetary value of funds questioned, saved, fined, or recovered; 7) criminal, civil, and administrative actions taken, 8) public or congressional inquiries resolved; and 9) certifications, allegations disproved, and cost corrections.

Because intermediate and long-term results may not be realized over a period of several years, only verifiable results are reported in the year completed.

Unit of measurement: Individual outcomes/actions

Unit of Measurement: Percentage (of the OIG budget)

4a. Oversight and Timing of Final Results Reporting:

Data comes from OIG audit, evaluations and investigations that are performed under strict compliance with professional standard of the US Government Accountability Office and the US Department of Justice and subject to independent peer review. Data in the form of activities, output, and outcomes is entered by designated staff into the Inspector General Enterprise Management System. All original data is quality controlled for compliance with professional standard and data entered is quality reviewed for accuracy, completeness, timeliness and adequately supported. All data entered is carefully reviewed several times a years as it is entered and subsequently reported on a quarterly basis. The OIG Assistant Inspectors General oversee the quality of the data used to generate reports of performance. The Office of the Chief of Staff oversee the data quality and the IG reviews the documents and date use for external consumption. Data is audited and quality test on a continuous basis through several steps from origin to final public consumption

4b. Data Limitations/Qualifications:

Because intermediate and long-term results may not be realized over a period of several years, only verifiable results are reported in the year completed.

Although all OIG staff are responsible for data accuracy in their products and services, there is a possibility of incomplete, miscoded, or missing data in the system due to human error or time lags. Data supporting achievement of results are often from indirect or external sources, with their own methods or standards for data verification/validation. Such data are reviewed according to the appropriate OIG quality standards (see "Additional Information"), and any questions about the quality of such data are documented in OIG reports and/or the PMRS.

The error rate for outputs is estimated at +/-2%, while the error rate for reported long-term outcomes is presumably greater because of the longer period needed for tracking results and difficulty in verifying a nexus between our work and subsequent actions and impacts beyond OIG's control. (The OIG logic model in the Annual Performance Report clarifies the kinds of measures that are output-oriented, like risks identified, versus outcome-oriented, like risks reduced.) Errors tend to be those of omission. Some errors may result from duplication as well.

4c. Third-Party Audits:

There have not been any previous audit findings or reports by external groups on data or database weaknesses in PMRS.

A December 2008 independent audit (www.epa.gov/oig/reports/2009/QualityReviewofEPAOIG-20081216.pdf found the following with regard to general OIG processes:

"We determined that the EPA OIG audits methodology, policies and procedures adequately complied with the Government Auditing Standards. The EPA OIG quality control system adequately documented compliance with professional and auditing standards for : Independence; Professional Judgment; Competence; Audit Planning; Supervision; Evidence and Audit Documentation; Reports on Performance Audits; Nonaudit Services; and the Quality Control Process. The auditors documented, before the audit report was issued, evidence of supervisory review of the work performed that supports findings, conclusions, and recommendations contained in the audit report.

"We determined that EPA OIG adequately followed the quality control policies established in the EPA OIG Project Management Handbook for conducting audit, program evaluation, and related projects. The audit documentation adequately includes evidence of work performed in the major three phases: Preliminary

Research, Field Work and Reporting.

"We determined that EPA OIG adequately followed the standards and principles set forth in the PCIE and Executive Council on Integrity and Efficiency Quality Standards for Investigations, as applicable. The investigation adequately documented compliance with the guidelines applicable to the investigation efforts of criminal investigators working for the EPA OIG."

The audit also identified two minor conditions, related working paper review/approval and completion/update status. OIG agreed with the auditor recommendations related to the conditions and adapted its Project Management Handbook to address the concerns.

A June 2010 internal OIG review of OIG report quality (which included a review of reporting procedures) found no substantial issues (see http://www.epa.gov/oig/reports/2010/20100602-10-N-0134.pdf

Measure Code: 35D - Criminal, civil, administrative, and fraud prevention actions.

Office of the Inspector General (OIG)

Goal Number and Title:
0 -
Objective Number and Title:
0 -
Sub-Objective Number and Title:
0 -
Strategic Target Code and Title:
0 -
Managing Office:
Chief of Staff in the Immediate Office of the Inspector General
1a. Performance Measure Term Definitions:
This is a measure of the total number of convictions, indictments, civil and administrative actions from OIG investigations. OIG performance results are a chain of linked events, starting with OIG outputs (e.g., recommendations, reports of best practices, and identification of risks). The subsequent actions taken by EPA or its stakeholders/partners, as a result of OIG's outputs, to improve operational efficiency and environmental program delivery are reported as intermediate outcomes. The resulting improvements in operational efficiency, risks reduced/eliminated, and conditions of environmental and human health are reported as outcomes. By using common categories of performance measures, quantitative results can be summed and reported. Each outcome is also qualitatively described, supported, and linked to an OIG product or output. The OIG can only control its outputs and has no authority, beyond its influence, to implement its recommendations that lead to environmental and management outcomes. # Criminal/Civil/Administrative Actions: Measured by the number of: 1) Indictments or informations where there is preliminary evidence of a violation of law; 2) convictions, guilty pleas, pre-trial diversion agreements, and based on the proof of evidence as decided by a judicial body affecting EPA operations and environmental programs; 3) Civil actions arising from OIG work. Civil actions include civil judgments and civil settlements from law suits for recovery; and 4) Administrative actions as a result of OIG work, which include: a) Personnel actions, such as reprimands, suspensions, demotions, or terminations of Federal, State, and local employees (including Federal contractor/grantee employees); b) Contractor or grantee (individual and entity) suspensions and/or debarments from doing business with the Federal government; and c) Compliance agreements. Additional Information: U.S. EPA, Office of Inspector General, Audits, Evaluations, and Other Publications;   Available on the Internet at www.epa.gov/oig , last updatedAugust 2011.

Federal Government Inspector General Quality Standards.

Except for justified exceptions, OIG adheres to the following standards, which apply across the federal government:

· Overall Governance: Quality Standards for Federal Offices of Inspector General. (President's Council on Integrity and Efficiency (PCIE) and Executive Council on Integrity and Efficiency (ECIE), October 2003). (http://www.ignet.gov/pande/standards/igstds.pdf This document contains quality standards for the management, operation and conduct of the Federal Offices of Inspector General (OIG). This document specifies that each federal OIG shall conduct, supervise, and coordinate its audits, investigations, inspections, and evaluations in compliance with the applicable professional standards listed below:

· For Investigations: Quality Standards for Investigations. (President's Council on Integrity and Efficiency (PCIE) and Executive Council on Integrity and Efficiency (ECIE), December 2003). http://www.ignet.gov/pande/standards/invstds.pdf Consistent with appropriate Department of Justice Directives.

· For Inspections and Evaluations: Quality Standards for Inspections. (President's Council on Integrity and Efficiency (PCIE) and Executive Council on Integrity and Efficiency (ECIE), January 2005). http://www.ignet.gov/pande/standards/oeistds.pdf

· For Audits: Government Auditing Standards, issued by the US General Accounting Office (GAO). The professional standards and guidance in the Yellow Book are commonly referred to as generally accepted government auditing standards (GAGAS). These standards and guidance provide a framework for conducting high quality government audits and attestation engagements with competence, integrity, objectivity, and independence. The current version of the Yellow Book (July 2007) can be located in its entirety at the following Website: www.gao.gov/govaud/d07162g.pdf

EPA OIG-Specific Operating Standards. The Project Management Handbook is the Office of Inspector General (OIG) policy document for conducting audit, program evaluation, public liaison, follow-up, and related projects. The Handbook describes the processes and standards the OIG uses to conduct the various phases of its work and helps ensure the quality, consistency, and timeliness of its products. Each OIG office may issue, upon approval by the Inspector General, supplemental guidance over assignments for which that office has responsibility.... This Handbook describes the audit, evaluation, public liaison, and follow-up processes and phases; it does not address OIG investigative processes although it does apply to audits/evaluations performed by the Office of Investigations (OI) [within EPA OIG]....OIG audit, program evaluation, public liaison, and follow-up reviews are normally conducted in accordance with appropriate Government Auditing Standards, as issued by the Comptroller General of the United States, commonly known as the Yellow Book.

Staff may use GAGAS in conjunction with other sets of professional standards. OIG reports may cite the use of other standards as appropriate. Teams should use GAGAS as the prevailing standard for conducting a review and reporting results should inconsistencies exist between GAGAS and other professional standards.

For some projects, adherence to all of the GAGAS may not be feasible or necessary. For these projects, the Product Line Director (PLD) will provide a rationale, the applicable standards not followed, and the impact on project results. The PLD's decision should be made during the design meeting, documented in the working papers, and described in the Scope and Methodology section of the report. [Source: Project Management Handbook].

Product Line Directors.   Product Line Directors oversee one or more particular work areas and multiple project teams. The OIG product lines are as follows: Air/Research and Development; Water; Superfund/Land; Cross Media; Public Liaison and Special Reviews; Assistance Agreements; Contracts; Forensic Audits; Financial Management; Risk Assessment and Program Performance; Information Resources Management; Investigations; US Chemical Safety and Hazard Investigation Board; Legal Reviews; Briefings; OIG Enabling Support Programs; and Other Activities.

For more information on the PLD responsibilities, see Chapter 5 of the OIG Project Management Handbook, attached to this record.

2a. Original Data Source:
Data is collected and reported by designated OIG staff members in OIG Performance Measurement Databases as a result of OIG performance evaluations, audits, inspections and investigations and other analysis of proposed and existing Agency Policies, regulations and laws. OIG collects such data from the activities, outputs, intermediate outcomes and long-term outcome results of OIG operations.
2b. Source Data Collection:
Performance information is entered by designated staff into the Inspector General Enterprise Management System from OIG audits, evaluations and investigations performed under strict compliance with applicable professional standards. All OIG products go through a rigorous quality assurance process and are subject to independent peer review.
2c. Source Data Reporting:
Data is derived from the results of audits, evaluations, investigations and special analysis that are performed in accordance with Professional Standards of the US Government Accountability Office or the Us Department of Justice. All OIG products are quality controlled and subject to independent peer review for compliance with a all professional standards. Data is entered, in compliance with EPA and OIG data quality standards into the Inspector General Enterprise Management System and which is further reviewed for quality and consistency by the OIG performance quality staff members.
3a. Relevant Information Systems:
OIG Performance Measurement and Results System (PMRS). PMRS captures and aggregates information on an array of OIG measures in a logic model format, linking immediate outputs with long-term intermediate outcomes and results. (The logic model can be found in OIG's Annual Performance Report at http://www.epa.gov/oig/planning.htm. PMRS is the OIG official system for collecting performance results data, in relation to its strategic and annual goals. All outputs (recommendations, best practices, risks identified) and outcome results (actions taken, changes in policies, procedures, practices, regulations, legislation, risks reduced, certifications for decisions, environmental improvements) influenced by OIG's current or prior work, and recognized during FY 2010 and beyond, should be entered into PMRS.

PMRS was developed as a prototype in FY 2001. Since then, there have been system improvements for ease of use. For example, during FY 2009 the PMRS was converted to a relational database directly linked to the new Inspector General Enterprise Management System (IGEMS).

IGEMS is an OIG employee time-tracking and project cost-tracking database that generates management reports. IGEMS is used to generate a project tracking number and a work product number. This system also tracks project progress and stores all related cost information.

AutoAudit and Teammate. These are repositories for all project working papers.

3b. Data Quality Procedures:

Data quality assurance and control are performed as an extension of OIG products and services, subject to rigorous compliance with the Government Auditing Standards of the Comptroller General, and are regularly reviewed by OIG management, an independent OIG Management Assessment Review Team, and external independent peer reviews (e.g., by accountancies qualified to evaluate OIG procedures against Government Auditing Standards). Each Assistant Inspector General certifies the completeness and accuracy of performance data.

All data reported are audited internally for accuracy and consistency.

OIG processes, including data processes, are governed by the quality standards described in "Additional Information" under the Performance Term Definition field. Notably, the Project Management Handbook (which governs audits) provides a QA checklist (see Appendix 4, of the 2008 Project Management Handbook, attached to this record). The Project Manager (PM) is responsible for completing the Quality Assurance (QA) checklist throughout the project. The PM prepares the checklist and submits it to the Product Line Director (PLD) upon completion of the Post Reporting Phase of the Project. The Checklist should be completed for all projects, recognizing that some steps in the checklist may not be applicable to all projects. The QA Checklist asks teams to ensure the integrity of data that resides in all of the OIG data systems. [Source: Project Management Handbook].

During FY 2008, OIG implemented an Audit Follow-up Policy to independently verify the status of Agency actions on OIG recommendations, which serve as the basis for OIG intermediate outcome results reported in the OIG PMRS.

(Additional information on the OIG's follow-up process can be found at
http://oigintra.epa.gov/policy/policies/documents/OIG-04Follow-upPolicy.pdf

Attached Documents:
Policy101.PMH.Final.05.08.08.pdf

3c. Data Oversight:

There are three levels of PMRS access: View Only, Edit and Administrator. Everyone with IGEMS access has view only privileges. Individuals tasked with adding or editing PMRS entries must be granted PMRS Edit privileges. Contact a PMRS administrator to request Edit privileges.

Each Product Line Director (PLD), each of whom oversees one or more OIG work areas (e.g., Superfund, Contracts, etc.) and multiple project management teams, is responsible for ensuring that teams maintain proper integrity, accessibility, and retrievability of working papers in accordance with OIG policies. Likewise, they must ensure that information in OIG's automated systems is updated regularly by the team. (See field 2i, Additional Information, for more information about PLDs.)

3d. Calculation Methodology:

Database measures include numbers of: 1) recommendations for environmental and management improvement; 2) legislative, regulatory policy, directive, or process changes; 3) environmental, program management, security and resource integrity risks identified, reduced, or eliminated; 4) best practices identified and implemented; 5) examples of environmental and management actions taken and improvements made; 6) monetary value of funds questioned, saved, fined, or recovered; 7) criminal, civil, and administrative actions taken, 8) public or congressional inquiries resolved; and 9) certifications, allegations disproved, and cost corrections.

Because intermediate and long-term results may not be realized over a period of several years, only verifiable results are reported in the year completed.
Unit of measurement: Individual actions

4a. Oversight and Timing of Final Results Reporting:

Data comes from OIG audit, evaluations and investigations that are performed under strict compliance with professional standard of the US Government Accountability Office and the US Department of Justice and subject to independent peer review. Data in the form of activities, output, and outcomes is entered by designated staff into the Inspector General Enterprise Management System. All original data is quality controlled for compliance with professional standard and data entered is quality reviewed for accuracy, completeness, timeliness and adequately supported. All data entered is carefully reviewed several times a years as it is entered and subsequently reported on a quarterly baThe OIG Assistant Inspectors General oversee the quality of the data used to generate reports of performance. The Office of the Chief of Staff oversee the data quality and the IG reviews the documents and date use for external consumption. Data is audited and quality test on a continuous basis through several steps from origin to final use.

4b. Data Limitations/Qualifications:

Because intermediate and long-term results may not be realized over a period of several years, only verifiable results are reported in the year completed.

Although all OIG staff are responsible for data accuracy in their products and services, there is a possibility of incomplete, miscoded, or missing data in the system due to human error or time lags. Data supporting achievement of results are often from indirect or external sources, with their own methods or standards for data verification/validation. Such data are reviewed according to the appropriate OIG quality standards (see "Additional Information"), and any questions about the quality of such data are documented in OIG reports and/or the PMRS.

The error rate for outputs is estimated at +/-2%, while the error rate for reported long-term outcomes is presumably greater because of the longer period needed for tracking results and difficulty in verifying a nexus between our work and subsequent actions and impacts beyond OIG's control. (The OIG logic model in the Annual Performance Report clarifies the kinds of measures that are output-oriented, like risks identified, versus outcome-oriented, like risks reduced.) Errors tend to be those of omission. Some errors may result from duplication as well.

4c. Third-Party Audits:

There have not been any previous audit findings or reports by external groups on data or database weaknesses in PMRS.

A December 2008 independent audit
(www.epa.gov/oig/reports/2009/QualityReviewofEPAOIG-20081216.pdf found the following with regard to general OIG processes:
"We determined that the EPA OIG audits methodology, policies and procedures adequately complied with the Government Auditing Standards. The EPA OIG quality control system adequately documented compliance with professional and auditing standards for : Independence; Professional Judgment; Competence; Audit Planning; Supervision; Evidence and Audit Documentation; Reports on Performance Audits; Nonaudit Services; and the Quality Control Process. The auditors documented, before the audit report was issued, evidence of supervisory review of the work performed that supports findings, conclusions, and recommendations contained in the audit report.
"We determined that EPA OIG adequately followed the quality control policies established in the EPA OIG Project Management Handbook for conducting audit, program evaluation, and related projects. The audit documentation adequately includes evidence of work performed in the major three phases: Preliminary Research, Field Work and Reporting.
"We determined that EPA OIG adequately followed the standards and principles set forth in the PCIE and Executive Council on Integrity and Efficiency Quality Standards for Investigations, as applicable. The investigation adequately documented compliance with the guidelines applicable to the investigation efforts of criminal investigators working for the EPA OIG."
The audit also identified two minor conditions, related working paper review/approval and completion/update status. OIG agreed with the auditor recommendations related to the conditions and adapted its Project Management Handbook to address the concerns.

A June 2010 internal OIG review of OIG report quality (which included a review of reporting procedures) found no substantial issues (see http://www.epa.gov/oig/reports/2010/20100602-10-N-0134.pdf

Office of the Administrator (OA) Record(s)

Measure Code: AD3 - Cumulative number of major grant, loan, contract, or technical assistance agreement programs that integrate climate science data into climate sensitive projects that have an environmental outcome

Office of the Administrator (OA)

Goal Number and Title:	
1 - Taking Action on Climate Change and Improving Air Quality	
Objective Number and Title:	
1 - Address Climate Change	
Sub-Objective Number and Title:	
1 - Address Climate Change	
Strategic Target Code and Title:	
5 - EPA will build resilience to climate change by integrating considerations of climate change impacts	
Managing Office:	
Office of Policy	
1a. Performance Measure Term Definitions:	
EPA will measure the amount of grants, loans, contracts, or technical assistance agreements. The term project is defined as an individual funding agreement and a program is defined as multiple projects. For example, the Great Lakes Restoration Initiative (GLRI) is a program that includes funding for grants. This EPA-led interagency initiative targets the most significant problems in the region, including invasive aquatic species, non-point source pollution, and contaminated sediment. It has outcome-oriented performance goals and measures, many of which are climate-sensitive. To ensure the overall success of the initiative, it is imperative that consideration of climate change and climate adaptation be integrated into GLRI grants and projects. Aside from GLRI, other climate-sensitive programs across the Agency include those for land revitalization and cleanup, air quality monitoring and protection, wetlands and water protection and restoration to name a few. Greenhouse gas mitigation programs and projects would not be included in this total. Climate change data needs to be integrated into climate-sensitive projects funded through EPA grants, loans, contracts, or technical assistance agreements. The 2011-2015 Strategic Plan is the driver for this annual measure Here is the adaptation website: http://www.epa.gov/climatechange/effects/adaptation.html	
2a. Original Data Source:	
Data will be submitted to the Office of Policy (OP) from environmental and research programs across the Agency. The data originate from each of the National Program Offices and Regional Offices; they collect the information from their program contacts.	
2b. Source Data Collection:	
The data are submitted to the Senior Advisor for Climate Adaptation in the Office of Policy. The data are entered into a spreadsheet. The climate change adaptation advisor will determine whether the result meets	

the criteria.
2c. Source Data Reporting:
The Program Offices (OAR, OW, OCSPP, OSWER, OITA) and Regional Offices will contact the climate change adaptation advisor to report this information. Tracked in a spreadsheet and maintained by the Office of Policy (OP).
3a. Relevant Information Systems:
Performance data are tracked in a spreadsheet and maintained by the Office of Policy (OP). This is source data from the Program Offices and Regional Offices, and is summed to be entered into PERS. Information system integrity standards don't apply. The Budget Automation System (BAS) is the final step for data entry.
3b. Data Quality Procedures:
The climate change adaptation advisor verifies the information with his climate change adaptation team through conversations with the programs and then has one of his staff enter the data into BAS.
3c. Data Oversight:
EPA Senior Advisor for Climate Adaptation
3d. Calculation Methodology:
The "program" measure is calculated by assigning a numeric value of one (1) to any major programs that integrate climate change data. This is an annual, not cumulative measure A program may only be counted once.
4a. Oversight and Timing of Final Results Reporting:
Climate Change Adaptation Science Advisor
4b. Data Limitations/Qualifications:
It is difficult to firmly define when climate change data have been adequately integrated into the grants, loans, contracts, or technical assistance agreements used in an environmental management program. Whether this has adequately been done requires verification by the climate change adaptation advisor. Some programs might not be captured in this measure. The final tabulation is a conservative count of the work completed. There is no data lag.A program may only be counted once.
4c. Third-Party Audits:
Not applicable

Measure Code: AD2 - Cumulative number of major rulemakings with climate sensitive, environmental impacts, and within existing authorities, that integrate climate change science data

Office of the Administrator (OA)

Goal Number and Title:	
1 - Taking Action on Climate Change and Improving Air Quality	
Objective Number and Title:	
1 - Address Climate Change	
Sub-Objective Number and Title:	
1 - Address Climate Change	
Strategic Target Code and Title:	
4 - EPA will account for climate change by integrating climate change science trend and scenario infor	
Managing Office:	
Office of Policy	
1a. Performance Measure Term Definitions:	
EPA is defining a "major" rule based upon guidelines published by the Office of Management and Budget. Specifically, a major rule is one that has an annual effect on the economy of $100 million or more. Also, the term "rule" refers to a proposed rule. Climate change data needs to be considered and integrated into the rulemaking process. The 2011-2015 Strategic Plan is the driver for this annual measure Here is the adaptation website: http://www.epa.gov/climatechange/effects/adaptation.html	
2a. Original Data Source:	
Data will be submitted to the Office of Policy (OP) from environmental and research programs across the Agency. The data originate from each of the National Program Offices; they collect the information from their program contacts.	
2b. Source Data Collection:	
The data are submitted to the Senior Advisor for Climate Adaptation in the Office of Policy. The climate change advisor will determine whether the result meets the criteria.	
2c. Source Data Reporting:	
The programs (OAR, OW, OCSPP, OSWER) will contact the climate change adaptation advisor to report this information. The information is maintained by the Office of Policy (OP)	
3a. Relevant Information Systems:	
Performance data are tracked in a spreadsheet and maintained by the Office of Policy (OP). This is source data from the programs and is summed to be entered into PERS. Information system integrity standards don't apply. The Budget Automation System (BAS) is the final step for data entry.	
3b. Data Quality Procedures:	

The climate change adaptation advisor verifies the information with his climate change adaptation team through conversations with the programs and then has one of his staff enter the data into BAS.
3c. Data Oversight:
EPA Senior Advisor on Climate Adaptation
3d. Calculation Methodology:
The "proposed rule making" measure is calculated by assigning a numeric value of one (1) to any major rule proposed. This is an annual, not cumulative measure A rule may only be counted once.
4a. Oversight and Timing of Final Results Reporting:
Climate Change Adaptation Science Advisor
4b. Data Limitations/Qualifications:
There are different ways for accounting for climate change in a rule making process (e.g., in the rule itself; in guidance issued for implementing the rule). Where climate change has adequately been accounted for in a rule making process requires verification by the climate change adaptation advisor. Some programs might not be captured in this measure. The final tabulation is a conservative count of the work completed. There is no data lag. A rule may only be counted once.
4c. Third-Party Audits:
Not applicable

Measure Code: AD1 - Cumulative number of major scientific models and decision support tools used in implementing environmental management programs that integrate climate change science data

Office of the Administrator (OA)

Goal Number and Title:
1 - Taking Action on Climate Change and Improving Air Quality

Objective Number and Title:
1 - Address Climate Change

Sub-Objective Number and Title:
1 - Address Climate Change

Strategic Target Code and Title:
3 - EPA will integrate climate change science trend and scenario information into five major scientific

Managing Office:
Office of Policy

1a. Performance Measure Term Definitions:
Consistent with this approach, EPA is defining a major scientific model and/or decision support tool as one that may influence a major agency rule or action. For example, the BASINS CAT model is a decision support tool that enhances the ability of U.S. cities and communities with combined sewer systems to meet the requirements of EPA's Combined Sewer Overflow (CSO) Control Policy [1]. In 1996, EPA estimated the cost of CSO control, consistent with the CSO Control Policy, to be $44.7 billion (1996 dollars). For this reason, the BASIN CAT model is an appropriate decision support tool to include. A program is defined as multiple projects. For example, the Great Lakes Restoration Initiative (GLRI) is a program that includes funding for grants. This EPA-led interagency initiative targets the most significant problems in the region, including invasive aquatic species, non-point source pollution, and contaminated sediment. It has outcome-oriented performance goals and measures, many of which are climate-sensitive. To ensure the overall success of the initiative, it is imperative that consideration of climate change and climate adaptation be integrated into GLRI grants and projects. Aside from GLRI, other climate-sensitive programs across the Agency include those for land revitalization and cleanup, air quality monitoring and protection, wetlands and water protection and restoration to name a few. Greenhouse gas mitigation programs and projects would not be included in this total. Climate change data needs to be integrated into the tool or model. The 2011-2015 Strategic Plan is the driver for this annual measure Here is the adaptation website: http://www.epa.gov/climatechange/effects/adaptation.html

2a. Original Data Source:
Data will be submitted to the Office of Policy (OP) from environmental and research programs across the Agency. The data originate from each of the National Program Offices and Regional Offices; they collect the information from their program contacts.

2b. Source Data Collection:
The data are submitted to the Senior Advisor for Climate Adaptation in the Office of Policy. The climate adaptation advisor will determine whether the result meets the criteria.

2c. Source Data Reporting:
The Program Offices (OAR, OW, OCSPP, OSWER, OITA) and Regional Offices will contact the climate change adaptation advisor to report this information. Tracked in a spreadsheet and maintained by the Office of Policy (OP).

3a. Relevant Information Systems:
Performance data are tracked in a spreadsheet and maintained by the Office of Policy (OP). This is source data from the Program Offices and Regional Offices, and is summed to be entered into PERS. Information system integrity standards don't apply. The Budget Automation System (BAS) is the final step for data entry.

3b. Data Quality Procedures:
The climate adaptation advisor verifies the information with his climate change adaptation team through conversations with the Program and Regional Offices, and then has one of his staff enter the data into BAS.

3c. Data Oversight:
EPA Senior Advisor for Climate Adaptation

3d. Calculation Methodology:
The "scientific models/decisions support tools" measure is calculated by assigning a numeric value of one (1) to any major scientific model or decision support tool. This is an annual, not cumulative measure. A model/tool may only be counted once.

4a. Oversight and Timing of Final Results Reporting:
Climate Change Adaptation Science Advisor

4b. Data Limitations/Qualifications:
It is difficult to firmly define when a particular scientific model or decision-support tool has been adequately integrated into an environmental management program. Whether this has adequately been done requires verification by the climate change adaptation advisor. Some programs might not be captured in this measure. The final tabulation is a conservative count of the work completed. There is no data lag. A model/tool may only be counted once.

4c. Third-Party Audits:
Not applicable

Office of Air and Radiation (OAR) Record(s)

Measure Code: 001 - Cumulative percentage reduction in tons of toxicity-weighted (for cancer risk) emissions of air toxics from 1993 baseline.

Office of Air and Radiation (OAR)

Goal Number and Title:
1 - Taking Action on Climate Change and Improving Air Quality
Objective Number and Title:
2 - Improve Air Quality
Sub-Objective Number and Title:
2 - Reduce Air Toxics
Strategic Target Code and Title:
1 - By 2015, reduce toxicity-weighted (for cancer) emissions of air toxics
Managing Office:
Office of Air Quality Planning and Standards
1a. Performance Measure Term Definitions:
Toxicity-weighted emissions: Toxicity-weighted emissions are an approach to normalize the mass of the HAP release (in tons per year) by a toxicity factor. The toxicity factors are based on either the HAPs cancer potency or noncancer potency. The more toxic the HAP the more "weight" it receives. Air toxics: Air toxics, also known as hazardous air pollutants, are those pollutants emitted into the air that are known or suspected to cause cancer or other serious health effects, such as reproductive effects or birth defects, or adverse environmental effects. As defined by the Section 112 of the Clean Air Act; the EPA currently regulates 187 air toxics released into the environment   Cancer risk: The probability of contracting cancer over the course of a lifetime (assumed to be 70 years for the purposes of most risk characterization). A risk level of "N" in a million implies a likelihood that up to "N" people, out of one million equally exposed people would contract cancer if exposed continuously (24 hours per day) to the specific concentration over 70 years (an assumed lifetime). This risk would be an excess cancer risk that is in addition to any cancer risk borne by a person not exposed to these air toxics.
2a. Original Data Source:
Emissions inventories are from many primary sources. The baseline National Toxics Inventory (for base years 1990 - 1993) is based on data collected during the development of Maximum Achievable Control Technology (MACT) standards, state and local data, Toxics Release Inventory (TRI) data, and emissions estimates using accepted emission inventory methodologies.

The primary source of data in the 1996 and 1999 toxics emissions inventories are state and local air pollution control agencies and Tribes. These data vary in completeness, format, and quality. EPA evaluates these data and supplements them with data gathered while developing Maximum Achievable Control Technology (MACT) and residual risk standards, industry data, and Toxics Release Inventory data.

The health risk data were obtained from various data sources including EPA, the U.S. Agency for Toxic Substances and Disease Registry, California Environmental Protection Agency, and the International Agency for Research on Cancer. The numbers from the health risk database are used for estimating the risk of contracting cancer and the level of hazard associated with adverse health effects other than cancer.

2b. Source Data Collection:

Source Data Collection Methods: Field monitoring; estimation

Date/time Intervals Covered by Source Data: Each inventory year provides an annual emissions sum for that year

EPA QA requirements/guidance governing collection: The overarching QA requirements and guidance are covered in the OAQPS Quality Assurance Project Plan [insert reference].

EPA's uniform data standards relevant to the NEI for HAPs are the: SIC/NAICS, Latitude/Longitude, Chemical Identification, Facility Identification, Date, Tribal and Contact Data Standards.

For more information on compliance of the NEI for HAPs with EPA's Information Quality Guidelines and new EPA data standards, please refer to the following web site for a paper presented at the 2003 Emission Inventory Conference in San Diego. "The Challenge of Meeting New EPA Data Standards and Information Quality Guidelines in the Development of the 2002 NEI Point Source Data for HAPs", Anne Pope, et al. www.epa.gov/ttn/chief/conference/ei12/dm/pope.pdf

Geographical Extent of Source Data: National

Spatial Detail Covered By the Source Data: 2002 and2005 NEI data—by facility address. Earlier—by county

Emissions Data: The National Emissions Inventory (NEI) for Hazardous Air Pollutants (HAPs) includes emissions from large and small industrial sources inventoried as point sources, smaller stationary area and other sources, such as fires inventoried as non-point sources, and mobile sources.

Prior to the 1999 NEI for HAPs, there was the National Toxics Inventory (NTI). The baseline NTI (for base years 1990 - 1993) includes emissions information for 188 hazardous air pollutants from more than 900 stationary sources and from mobile sources. The baseline NTI contains county level emissions data and cannot be used for modeling because it does not contain facility specific data.

The 2002 NEI and a slightly modified/updated 2005 NEI for HAPs contain stationary and mobile source estimates. These inventories also contain estimates of facility-specific HAP emissions and their source specific parameters such as location (latitude and longitude) and facility characteristics (stack height, exit velocity, temperature, etc.). Furthermore for 2005, a 2005 inventory was developed for the National Air Toxics Assessment (NATA) http://www.epa.gov/nata2005/ which provides the most updated source of air toxics emissions for 2005.

The 2008 NEI contains HAP emissions reported by state, local, and tribal agencies as well as data from the 2008 TRI and EPA data developed as part of MACT regulation development. Detailed documentation including QA procedures is underdevelopment as of January, 2012.

Information on EPA's Health Criteria Data for Risk Characterization:
http://www.epa.gov/ttn/atw/toxsource/summary.html
Contents: Tabulated dose response values for long-term (chronic) inhalation and oral
exposures; and values for short term (acute) inhalation exposure

EPA's Health Criteria Data for Risk Characterization is a compendium of cancer and noncancer health risk criteria used to develop a risk metric. This compendium includes tabulated values for long-term (chronic) inhalation for many of the 188 hazardous air pollutants.

Audience: Public

2c. Source Data Reporting:
Form/mechanism for receiving data and entering into EPA system: During the development of the 1999 National Emission Inventory (NEI) for Hazardous Air Pollutants (HAPs), all primary data submitters and reviewers were required to submit their data and revisions to EPA in a standardized format using the Agency's Central Data Exchange (CDX). For more information on CDX, please go the following web site: www.epa.gov/ttn/chief/nif/cdx.html

This approach was also used for the 2002 and 2005 NEI. Starting with the 2008 NEI, a new CDX-based mechanism was used called the Emissions Inventory System (EIS). http://www.epa.gov/ttn/chief/eis/gateway/index.html The data are transmitted automatically through CDX into the EIS data system.

Timing and frequency of reporting: Other [NEI data are calculated every 3 years] |

3a. Relevant Information Systems:
The NEI data and documentation are available at the following sites:

Emissions Inventory System (EIS): http://www.epa.gov/ttn/chief/eis/gateway/index.html
Available inventories: 2002 NEI, 2005 NEI, 2008 NEI
Contents: Detailed raw final inventories
Audience: EPA staff and state/local/tribal reporting agencies |

The EIS is the interface for state, local, and tribal agencies to upload their emissions inventory data. It works using the Central Data Exchange (CDX) network to directly transfer data from external agencies to EPA. EIS also allows EPA inventory development staff to upload data to augment inventories, particularly for HAP emissions, which the states are not required to submit to EPA. EIS includes a "Quality Assurance Environment" that allows states to quality assure their data before submitting to EPA. During this phase of use, EIS runs hundreds of quality assurance checks on the data to ensure that the format (e.g., required data fields) and content (e.g., data codes, range checks) of the data are valid. After using the QA environment, states submit using the production environment, which also runs the QA checks. EIS further allows reporting agencies to make changes as needed to correct any data that passed the QA checks but is not correct. EIS allows both data submitters and all EPA staff to view the data. EIS reports facilitate the QA and augmentation of the data by EPA inventory preparation staff. EIS facilitates EPA's automatic compilation of all agency data and EPA data using a hierarchical selection process, but which EPA staff define the order of precedence for using datasets when multiple emissions values exist from more than one group (for example, state data versus EPA estimated data).

Clearinghouse for Inventories and Emission Factors (CHIEF):
- Contents: Modeling data files for each state, summary data files for the nation, documentation, and README file
- Audience: State/local/Tribal agencies, industry, EPA, and the public.
- 1999 NEI: http://www.epa.gov/ttn/chief/net/1999inventory.html

Contents: 1999 NEI for HAPs data development materials;
1999 Data Incorporation Plan - describes how EPA compiled the 1999 NEI for HAPs; QC tool for data submitters; Data Augmentation Memo describes procedures EPA will use to augment data; 99 NTI Q's and A's provides answers to frequently asked questions; NIF (Input Format) files and descriptions; CDX Data Submittal Procedures - instructions on how to submit data using CDX; Training materials on development of HAP emission inventories; and Emission factor documents, databases, and models.
- 2002 NEI: http://www.epa.gov/ttn/chief/net/2002inventory.html - inventorydata
- 2005 NEI: http://www.epa.gov/ttn/chief/net/2005inventory.html - inventorydata
- 2005 NATA: http://www.epa.gov/ttn/atw/nata2005/methods.html - emissions
- 2008 NEI: http://www.epa.gov/ttn/chief/net/2008inventory.html

Additional information:

3b. Data Quality Procedures:

Starting with the 2008 NEI, EPA has used the Emissions Inventory System (EIS) for collecting and compiling the National Emissions Inventory (NEI). EIS includes a "Quality Assurance Environment" that allows states to quality assure their data before submitting to EPA. During this phase of use, EIS runs hundreds of quality assurance checks (~650 as of January 2012) on the data to ensure that the format (e.g., required data fields) and content (e.g., data codes, emissions range checks, duplicate prevention) of the data are valid. After using the QA environment, states submit using the production environment, which also runs the QA checks. QA checks are partly documented in Appendix 5 of the 2008 NEI Implementation Plan available at http://www.epa.gov/ttn/chief/net/neip/index.html and fully documented on the EIS gateway at

Data submitters are given feedback reports containing errors for missed requirements and warnings for non-required checks, such as emissions range checks. After data are compiled, EPA inventory preparation staff perform numerous procedures on the data that are not yet automated. In many cases, EPA further consulted with the data external data providers to obtain revised data submissions to correct issues identified. These checks and data improvements included:

· Comparison to past inventories including 2005 NATA to identify missing data (facilities, pollutants), particularly for facilities identified in past efforts as high risk

· Comparison of latitude longitude locations to county boundaries

· Augmentation of HAP emissions data with TRI

· Augmentation of HAP emissions data using emission factor ratios

· Augmentation of HAP emissions with EPA data developed for MACT and RTR standards

· Outlier analysis

Detailed documentation including QA procedures is underdevelopment as of January, 2012.

Prior to 2008, EIS was unavailable and so many of the data techniques used by EIS were done in a more manual fashion. The EPA performed extensive quality assurance/quality control (QA/QC) activities, including checking data provided by other organizations to improve the quality of the emission inventory. Some of these activities include: (1) the use of an automated format QC tool to identify potential errors of data integrity, code values, and range checks; (2) use of geographical information system (GIS) tools to verify facility locations; and (3) automated content analysis by pollutant, source category and facility to identify potential problems with emission estimates such as outliers, duplicate sites, duplicate emissions, coverage of a source category, etc. The content analysis includes a variety of comparative and statistical analyses. The comparative analyses help reviewers prioritize which source categories and pollutants to review in more detail based on comparisons using current inventory data and prior inventories. The statistical analyses help reviewers identify potential outliers by providing the minimum, maximum, average, standard deviation, and selected percentile values based on current data. Documentation on procedures used prior to 2008 is most readily available in the documentation for the 2002 NEI, available at http://www.epa.gov/ttn/chief/net/2002inventory.html

The NTI database contains data fields that indicate if a field has been augmented and identifies the augmentation method. After performing the content analysis, the EPA contacts data providers to reconcile potential errors. The draft NTI is posted for external review and includes a README file, with instructions on review of data and submission of revisions, state-by-state modeling files with all modeled data fields, and summary files to assist in the review of the data. One of the summary files includes a comparison of point source data submitted by different organizations. During the external review of the data, state and local agencies, Tribes, and industry provide external QA of the inventory. The EPA evaluates proposed revisions from external reviewers and prepares memos for individual reviewers documenting incorporation of revisions and explanations if revisions were not incorporated. All revisions are tracked in the database with the source of original data and sources of subsequent revision.

The external QA and the internal QC of the inventory have resulted in significant changes in the initial emission estimates, as seen by comparison of the initial draft NEI for HAPs and its final version. For more information on QA/QC of the NEI for HAPs, please refer to the following web site for a paper presented at the 2002 Emission Inventory Conference in Atlanta: "QA/QC - An Integral Step in the Development of the 1999 National Emission Inventory for HAPs", Anne Pope, et al. www.epa.gov/ttn/chief/conference/ei11/qa/pope.pdf

The tables used in the EPA's Health Criteria Data for Risk Characterization (found at www.epa.gov/ttn/atw/toxsource/summary.html are compiled assessments from various sources for many of the 188 substances listed as hazardous air pollutants under the Clean Air Act of 1990. The data are reviewed to make sure they support hazard identification and dose-response assessment for chronic exposures as defined in the National Academy of Sciences (NAS) risk assessment paradigm (www.epa.gov/ttn/atw/toxsource/paradigm.html Because the health criteria data were obtained from various sources they are prioritized for use (in developing the performance measure, for example) according to 1) conceptual consistency with EPA risk assessment guidelines and 2) various levels of scientific peer review. The prioritization process is aimed at incorporating the best available scientific data.

3c. Data Oversight:

Source Data: Air Quality Assessment Division, Emissions Inventory Assessment Group
Information Systems: Health & Environmental Impacts Division, Air Toxics Assessment Group

3d. Calculation Methodology:

Explanation of the Calculations: As the NEI is only developed every three years, EPA utilizes an emissions modeling system to project inventories for "off-years" and to project the inventory into the future. This model, the EMS-HAP (Emissions Modeling System for Hazardous Air Pollutants), can project future emissions, by adjusting stationary source emission data to account for growth and emission reductions resulting from emission reduction scenarios such as the implementation of the Maximum Achievable Control Technology (MACT) standards.

Information on the Emissions Modeling System for Hazardous Air Pollutants (EMS-HAP):
http://www.epa.gov/scram001/userg/other/emshapv3ug.pdf
http://www.epa.gov/ttn/chief/emch/projection/emshap.html
Contents: 1996 NTI and 1999 NEI for HAPs Audience: public

Explanation of Assumptions: Once the EMS-HAP process has been performed, the EPA would tox-weight the inventory by "weighting" the emissions for each pollutant with the appropriate health risk criteria. This would be accomplished through a multi-step process. Initially, pollutant by pollutant values would be obtained from the NEI for the current year and the baseline year (1990/93). Conversion of actual tons for each pollutant for the current year and the baseline year to "toxicity-weighted" tons would be accomplished by multiplying the appropriate values from the health criteria database such as the unit risk estimate (URE) or lifetime cancer risk (defined at http://www.epa.gov/ttn/atw/toxsource/summary.html to get the noncancer tons. These toxicity-weighted values act as a surrogate for risk and allow EPA to compare the toxicity-weighted values against a 1990/1993 baseline of toxicity-weighted values to determine the percentage reduction in risk on an annual basis.

Information on EPA's Health Criteria Data for Risk Characterization (Health Criteria Data):

http://www.epa.gov/ttn/atw/toxsource/summary.html

Contents: Tabulated dose
response values for
long-term (chronic) inhalation
and oral exposures; and values for
short-term (acute) inhalation
exposure.
Audience: Public

Identification of Unit of Measure and Timeframe: Cumulative percentage reduction in tons of toxicity-weighted emissions as a surrogate for actual risks reduction to the public.

4a. Oversight and Timing of Final Results Reporting:

Oversight of Final Reporting: OAQPS will update the actual toxicity-weighted emissions approximately every three years to coincide with updated toxic inventories.

Timing of Results Reporting: Annually. NEI data are calculated every three years; in years when NEI data are not calculated, the annual measure is reported based upon modeled results.

4b. Data Limitations/Qualifications:

While emissions estimating techniques have improved over the years, broad assumptions about the behavior of sources and serious data limitations still exist. The NTI and the NEI for HAPs contain data from other primary references. Because of the different data sources, not all information in the NTI and the NEI for HAPs has been developed using identical methods. Also, for the same reason, there are likely some geographic areas with more detail and accuracy than others.

The 1996 NTI and 1999 NEI for HAPs are a significant improvement over the baseline NTI because of the added facility-level detail (e.g., stack heights, latitude/longitude locations), making it more useful for dispersion model input.

For further discussion of the data limitations and the error estimates in the 1999 NEI for HAPs, please refer to the discussion of Information Quality Guidelines in the documentation at: www.epa.gov/ttn/chief/net/index.html - haps99

The tables used in the EPA's Health Criteria Data for Risk Characterization (found at www.epa.gov/ttn/atw/toxsource/summary.html are compiled assessments from various sources for many of the 188 substances listed as hazardous air pollutants under the Clean Air Act of 1990. Because different sources developed these assessments at different times for purposes that were similar but not identical, results are not totally consistent. To resolve these discrepancies and ensure the validity of the data, EPA applied a consistent priority scheme consistent with EPA risk assessment guidelines and various levels of scientific peer review. These risk assessment guidelines can be found at http://www.epa.gov/risk/guidance.htm

While the Agency has made every effort to utilize the best available science in selecting appropriate health criteria data for toxicity-weighting calculations, there are inherent limitations and errors (uncertainties) associated with this type of data. Most of the agencies health criteria are derived from response models and laboratory experiments involving animals. The parameter used to convert from exposure to cancer risk (i.e. the Unit Risk Estimate or URE) is based on default science policy processes used routinely in EPA assessments. First, some air toxics are known to be carcinogens in animals but lack data in humans. These have been assumed to be human carcinogens. Second, all the air toxics in this assessment were assumed to have linear relationships between exposure and the probability of cancer (i.e. effects at low exposures were extrapolated from higher, measurable, exposures by a straight line). Third, the URE used for some air toxics compounds represents a maximum likelihood estimate, which might be taken to mean the best scientific estimate. For other air toxics compounds, however, the URE used was an "upper bound" estimate, meaning that it probably leads to an overestimation of risk if it is incorrect. For these upper bound estimates, it is assumed that the URE continues to apply even at low exposures. It is likely, therefore, that this linear model over-predicts the risk at exposures encountered in the environment. The cancer weighting-values for this approach should be considered "upper bound" in the science policy sense.

All of the noncancer risk estimates have a built-in margin of safety. All of the Reference Concentrations (RfCs) used in toxicity-weighting of noncancer are conservative, meaning that they represent exposures which probably do not result in any health effects, with a margin of safety built into the RfC to account for sources of uncertainty and variability. Like the URE used in cancer weighting the values are, therefore, considered "upper bound" in the science policy sense. Further details on limitations and uncertainties associated with the agencies health data can be found at: www.epa.gov/ttn/atw/nata/roy/page9.html - L10

4c. Third-Party Audits:

In 2004, the Office of the Inspector General (OIG) released a final evaluation report on "EPA's Method for Calculating Air Toxics Emissions for Reporting Results Needs Improvement" (report can be found at www.epa.gov/oig/reports/2004/20040331-2004-p-00012.pdf The report stated that although the methods used have improved substantially, unvalidated assumptions and other limitations underlying the NTI continue to impact its use as a GPRA performance measure. As a result of this evaluation and the OIG recommendations for improvement, EPA prepared an action plan and is looking at ways to improve the accuracy and reliability of the data. EPA will meet bi-annually with OIG to report on its progress in completing the activities as outlined in the action plan.

EPA staff, state and local agencies, Tribes, industry and the public review the NTI and the NEI for HAPs. To assist in the review of the 1999 NEI for HAPs, the EPA provided a comparison of data from the three data sources (MACT/residual risk data, TRI, and state, local and Tribal inventories) for each facility. For the 1999 NEI for HAPs, two periods were available for external review - October 2001 - February 2002 and October 2002 - March 2003. The final 1999 NEI was completed and posted on the Agency website in the fall of 2003.

The EMS-HAP has been subjected to the scrutiny of leading scientists throughout the country in a process called "scientific peer review". This ensures that EPA uses the best available scientific methods and information. In 2001, EPA's Science Advisory Board (SAB) reviewed the EMS-HAP model as part of the 1996 national-scale assessment. The review was generally supportive of the assessment purpose, methods, and presentation; the committee considers this an important step toward a better understanding of air toxics. Additional information is available on the Internet: www.epa.gov/ttn/atw/nata/peer.html

Measure Code: A01 - Annual emissions of sulfur dioxide (SO2) from electric power generation sources.

Office of Air and Radiation (OAR)

Goal Number and Title:
1 - Taking Action on Climate Change and Improving Air Quality

Objective Number and Title:
2 - Improve Air Quality

Sub-Objective Number and Title:
1 - Reduce Criteria Pollutants and Regional Haze

Strategic Target Code and Title:
1 - By 2015, concentrations of ozone (smog) in monitored counties will decrease to .073 ppm

Managing Office:
Office of Atmospheric Programs

1a. Performance Measure Term Definitions:
Emissions of SO2: Sulfur dioxide (also sulphur dioxide) is the chemical compound with the formula SO2. Electric power generation sources: The Acid Rain Program, established under Title IV of the Clean Air Act Amendments of 1990, requires major reductions in sulfur dioxide (SO2) and nitrogen oxide (NOx) emissions from the U.S. electric power generation industry.   The program implements Title IV by continuing to measure, quality assure, and track emissions for SO2 and/or NOx from Continuous Emissions Monitoring Systems (CEMS) or equivalent direct measurement methods at over 3,600 affected electric generation units in the U.S.

2a. Original Data Source:
More than 3,400 fossil fuel-fired utility units affected under the Title IV Acid Rain Program collect hourly measurements of SO2, NOx, volumetric flow, CO2, and other emission-related parameters using certified continuous emission monitoring systems (CEMS) or equivalent continuous monitoring methods. For a description of EPA's Acid Rain Program, see the program's website at http://www.epa.gov/acidrain/index.html and the electronic Code of Federal Regulations at http://www.epa.gov/docs/epacfr40/chapt-I.info/subch-C.html (40 CFR parts 72-78.)

2b. Source Data Collection:
Source Data Collection Methods: Field monitoring using certified continuous emission monitoring systems (CEMS) or equivalent continuous monitoring methods, collected hourly. EPA QA requirements/guidance governing collection: Promulgated QA/QC requirements dictate performing a series of quality assurance tests of CEMS performance. For these tests, emissions data are collected under highly structured, carefully designed testing conditions, which involve either high quality standard reference materials or multiple instruments performing simultaneous emission measurements. The resulting data are screened and analyzed using a battery of statistical procedures, including one that tests for systematic bias. If a CEM fails the bias test, indicating a potential for systematic underestimation of emissions, the source of the error must be identified and corrected or the data are adjusted to minimize the bias. Each affected plant is

required to maintain a written QA plan documenting performance of these procedures and tests.

The ETS provides instant feedback to sources on data reporting problems, format errors, and inconsistencies. The electronic data file QA checks are described at http://www.epa.gov/airmarkets/business/report-emissions.html

Geographical Extent of Source Data: National

Spatial Detail Covered By the Source Data: Spatial detail for SO2 emissions can be obtained at the following website: http://camddataandmaps.epa.gov/gdm/index.cfm?fuseaction=emissions.wizard This website allows access to current and historical emissions data via Quick Reports. Annual, quarterly, monthly, daily and hourly data are available at the unit level and the monitoring location level.

2c. Source Data Reporting:

Form/mechanism for receiving data and entering into EPA system: Beginning with the first quarter of 2009, and quarterly thereafter, all industry sources regulated under the Acid Rain and Clean Air Interstate Rule (CAIR) programs are required use the Emissions Collection and Monitoring Plan System (ECMPS) to submit their monitoring plan, QA/cert test, and emissions data to the EPA.

The new XML file format allows the data to be organized based on dates and hours instead of pollutant type.

See also the ECMPS Reporting Instructions Emissions document: http://www.epa.gov/airmarkets/business/ecmps/docs/ECMPSEMRI2009Q2.pdf

Timing and frequency of reporting: Emissions data are submitted to the ECMPS and represent hourly values for measured parameters, calculated hourly emissions values, instrument calibration data, and aggregated summary data. An emissions file contains one calendar quarter of hourly and aggregate emissions measurements for a specified unit or group of related units, including stacks and pipes.

Each unit that is required to submit emissions data for a particular calendar quarter must be included in one and only one emissions file for that quarter. Each emissions file should contain all relevant operating, daily quality assurance, and emissions data for all units, common stacks, multiple stacks, or common pipes that were in a common monitoring configuration for any part of the quarter.

You must submit an emissions file for each quarter or, for ozone season only reporters, for the second and third calendar quarters of each year.

3a. Relevant Information Systems:

Emissions Tracking System (ETS) /
Emissions Collection and Monitoring Plan System (ECMPS)

Additional information:
EPA's Clean Air Markets Division (CAMD) has undertaken a project to re-engineer the process and data systems associated with emissions, monitoring plan, and certification data. As part of the project, CAMD

reviewed how monitoring plan information, certification/ recertification applications, on-going quality assurance data, and emissions data are maintained, quality assured and submitted. CAMD also reviewed the tools available for checking and submitting data on a quarterly and ozone season basis. Once the review was complete, CAMD developed a number of goals for the ECMPS project. They include:

· Creating a single client tool for all users to check and submit data.
· Providing users with the ability to quality assure data prior to submission.
· Providing users with one set of feedback.
· Allowing for seamless updates to the client tool.
· Providing direct access to EPA's database through the client tool.
· Maintaining select data outside of the electronic data report.
· Creating new XML file format.
· Developing new security requirements.

Adding flexibility to the process is one of the main reasons for changing how monitoring and emissions data are quality assured and submitted. There are several changes to the process that will involve adding flexibility:

· Monitoring plans will no longer be required as part of the quarterly file.
· On-going quality assurance test data may be submitted after the tests are performed—users will not have to wait to submit the data as part of a quarterly report.

[Source: http://www.epa.gov/airmarkets/business/ecmps/index.html

The ECMPS contain source data.

The ECMPS meets relevant EPA standards for information system integrity.

3b. Data Quality Procedures:

EPA analyzes all quarterly reports to detect deficiencies and to identify reports that must be resubmitted to correct problems. EPA also identifies reports that were not submitted by the appropriate reporting deadline. Revised quarterly reports, with corrected deficiencies found during the data review process, must be obtained from sources by a specified deadline. All data are reviewed, and preliminary and final emissions data reports are prepared for public release and compliance determination.

For a review of the ETS data audit process, see: http://www.epa.gov/airmarkets/ presentations/docs/epri06/epri_electronic_audit_revised.ppt.

3c. Data Oversight:

Branch Chief, Emissions Monitoring Branch is responsible for source data reporting.
Branch Chief, Market Operations Branch is responsible for the information systems utilized in producing the performance result.

3d. Calculation Methodology:

Definition of variables: The ECMPS Reporting Instructions Emissions document at http://www.epa.gov/airmarkets/business/ecmps/docs/ECMPSEMRI2009Q2.pdf is the data dictionary for the ECMPS.

Explanation of Calculations: Promulgated methods are used to aggregate emissions data across all United States' utilities for each pollutant and related source operating parameters such as heat inputs.The ECMPS Reporting Instructions Emissions document at

http://www.epa.gov/airmarkets/business/ecmps/docs/ECMPSEMRI2009Q2.pdf provides the methods used to aggregate emissions data across all United States' utilities.

Unit of analysis: Tons of emission

4a. Oversight and Timing of Final Results Reporting:

Branch Chief, Assessment And Communications Branch, oversees final reporting by the National Program Office.

4b. Data Limitations/Qualifications:

None

4c. Third-Party Audits:

In July of 2010, the Quality Staff of the Office of Environmental Information completed a Quality System Assessment (QSA) for the Office of Atmospheric Programs. The results of the assessment were summarized as follows: "Please note that there are no findings requiring corrective action. Review of QA requirements and interviews with management and staff revealed no weaknesses in the overall Quality System management for OAP. Controls appear to be in place, the QA structure appears effective, there is project-level planning QA documentation (QAPPs, QARFs) in place as well as the appropriate training and records management practices".

Measure Code: M91 - Cumulative percentage reduction in population-weighted ambient concentration of fine particulate matter (PM-2.5) in all monitored counties from 2003 baseline.

Office of Air and Radiation (OAR)

Goal Number and Title:
1 - Taking Action on Climate Change and Improving Air Quality
Objective Number and Title:
2 - Improve Air Quality
Sub-Objective Number and Title:
1 - Reduce Criteria Pollutants and Regional Haze
Strategic Target Code and Title:
2 - By 2015,concentrations of inhalable fine particles in monitored counties will decrease to 10.5 μg/m3
Managing Office:
Office of Air Quality Planning and Standards
1a. Performance Measure Term Definitions:
Population-weighted: The ambient concentration multiplied by total county population, using constant population values for all years. Ambient concentration: The highest reported site-level annual standard design value; i.e., the 3-year average annual mean 24-hour average concentration of PM-2.5. Fine particulate matter (PM 2.5): Particles with a diameter of 5 microns or less. Monitored counties: The counties in the current time-frame with at least one site meeting completeness criteria that also were present in the base period (i.e., contained at least one complete site in the period 2001-2003).
2a. Original Data Source:
State and local agency data are from State and Local Air Monitoring Stations (SLAMS). Population data are from the Census Bureau/Department of Commerce (2000 Census)
2b. Source Data Collection:
Source Data Collection Methods: Field monitoring; survey (2000 Census) Date/Time Intervals Covered by Source Data: 2003 to present (for air pollution data). 2000 (for census data) EPA QA Requirements/Guidance Governing Collection: To ensure quality data, the SLAMS are required to meet the following: 1) each site must meet network design and site criteria; 2) each site must provide adequate QA assessment, control, and corrective action functions according to minimum program requirements; 3) all sampling methods and equipment must meet EPA reference or equivalent requirements; 4) acceptable data validation and record keeping procedures must be followed; and 5) data from SLAMS must be summarized and reported annually to EPA. Finally, there are system audits that regularly review the overall

air quality data collection activity for any needed changes or corrections. Further information is available on the Internet at http://www.epa.gov/cludygxb/programs/namslam.html and through United States EPA's Quality Assurance Handbook (EPA-454/R-98-004 Section 15).

Geographical Extent of Source Data: National

Spatial Detail Covered By the Source Data: 437 counties in the 48 continental States plus D.C.

2c. Source Data Reporting:

Agencies submit air quality data to AQS thru the Agency's Central Data Exchange (CDX).

3a. Relevant Information Systems:

The Air Quality Subsystem (AQS) stores ambient air quality data used to evaluate an area's air quality levels relative to the National Ambient Air Quality Standards (NAAQS).

AQS has been enhanced to comply with the Agency's data standards (e.g., latitude/longitude, chemical nomenclature).

All annual mean concentration data used in the performance analysis were extracted from the AQS. Population data were obtained from the Bureau of the Census.

Additional information:

In January 2002, EPA completed the reengineering of AQS to make it a more user friendly, Windows-based system. As a result, air quality data are more easily accessible via the Internet.

Beginning in July 2003, agencies submitted air quality data to AQS thru the Agency's Central Data Exchange (CDX). CDX is intended to be the portal through which all environmental data coming to or leaving the Agency will pass.

3b. Data Quality Procedures:

The AQS QA/QC process also involves participation in the EPA's National Performance Audit Program (NPAP), system audits, and network reviews. Please see www.epa.gov/ttn/amtic/npaplist.html for more information. Under NPAP, all agencies required to report gaseous criteria pollutant data from their ambient air monitoring stations to EPA's Air Quality System (AQS) for comparison to the National Ambient Air Quality Standard (NAAQS) are required to participate in EPA's NPAP TTP program. Guidance for participating in this program requires NPAP audits of at least 20% of a Primary Quality Assurance Organization's (PQAO's) sites each year; and all sites in 5 years.

3c. Data Oversight:

National Air Data Group [Outreach and Information Division, OAQPS] oversees operations of the Air Quality System, the database used to store and deliver the source data.

Air Quality Monitoring Group [Air Quality Assessment Division (AQAD), OAQPS] oversees the monitoring and quality assurance of the source data.

Air Quality Analysis Group (AQAG) [AQAD, OAQPS] oversees the transformation and data reporting aspects associated with the Calculation of this performance measure.

3d. Calculation Methodology:

Explanation of Calculations: Air quality levels are evaluated relative to the baseline level and the design value. The change in air quality concentrations is then multiplied by the number of people living in the county.

Explanation of Assumptions: Design values are calculated for every county with adequate monitoring data. The design value is the mathematically determined pollutant concentration at a particular site that must be reduced to, or maintained at or below the National Ambient Air Quality Standards (NAAQS) in order to assure attainment. The design value may be calculated based on ambient measurements observed at a local monitor in a 3-year period or on model estimates. The design value varies from year to year due to both the pollutant emissions and natural variability such as meteorological conditions, wildfires, dust storms, volcanic activities etc. For more information on design values, including a definition, see www.epa.gov/ttn/oarpg/t1/memoranda/cdv.pdf This analysis assumes that the populations of the areas are held constant at 2000 Census levels. Data comparisons over several years allow assessment of the air program's success.

Unit of analysis: Cumulative percent reduction in population-weighted ambient concentration

4a. Oversight and Timing of Final Results Reporting:

Air Quality Assessment Group, OAQPS, OAR is directly responsible for the calculations associated with this performance measure.

4b. Data Limitations/Qualifications:

There is uncertainty in the projections and near term variations in air quality (due to meteorological conditions, for example).

4c. Third-Party Audits:

Design Values used in this performance measure are vetted with the State and Local data reporting agencies.

Measure Code: M9 - Cumulative percentage reduction in population-weighted ambient concentration of ozone in monitored counties from 2003 baseline.

Office of Air and Radiation (OAR)

Goal Number and Title:
1 - Taking Action on Climate Change and Improving Air Quality
Objective Number and Title:
2 - Improve Air Quality
Sub-Objective Number and Title:
1 - Reduce Criteria Pollutants and Regional Haze
Strategic Target Code and Title:
1 - By 2015, concentrations of ozone (smog) in monitored counties will decrease to .073 ppm
Managing Office:
Office of Air Quality Planning and Standards
1a. Performance Measure Term Definitions:
Population-weighted: Multiply (or weight) these concentrations by the number of people living in the county where the monitor is located. The population estimates are from the U.S. Census Bureau (2000 decennial census). Ambient concentration: EPA tracks improvements in air quality on an annual basis by measuring the change in ambient air quality concentrations of 8-hour ozone in counties with monitoring data weighted by the number of people living in these counties. This measure makes use of actual, observed changes in ambient ozone levels over time to determine NAAQS program effectiveness. Three year averages of the 4th highest daily maximum ozone values (i.e., design values) are used to help mitigate the influence of meteorology which would otherwise confound measurement of actual program progress. Other than this that I pulled from the attached, I could add that ambient air is the air we breathe vs emitted air from a pollution source, and a concentration is measured at a monitor. Ozone: Ozone (O_3) is a gas composed of three oxygen atoms. It is not usually emitted directly into the air, but at ground-level is created by a chemical reaction between oxides of nitrogen (NOx) and volatile organic compounds (VOC) in the presence of sunlight. Ozone has the same chemical structure whether it occurs miles above the earth or at ground-level and can be "good" or "bad," depending on its location in the atmosphere. Monitored counties: Calculate 8-hour ozone design values for 2001-2003 for every county with adequate monitoring data. A monitoring site's design value for 8-hour ozone is expressed as the average of the fourth-highest daily maximum 8-hour average ozone concentration for each of three consecutive years. A county's design value is the highest of these site-level design values. The national ozone monitoring network conforms to uniform criteria for monitor siting, instrumentation, and quality assurance.
2a. Original Data Source:
State and local agency data are from State and Local Air Monitoring Stations (SLAMS). Population data are from the Census Bureau/Department of Commerce (2000 Census)

2b. Source Data Collection:
Source Data Collection Methods: Field monitoring; survey (2000 Census)

Date/time intervals covered by source data: 2003 to present (for air pollution data). 2000 (for census data)

EPA QA requirements/guidance governing collection: To ensure quality data, the SLAMS are required to meet the following: 1) each site must meet network design and site criteria; 2) each site must provide adequate QA assessment, control, and corrective action functions according to minimum program requirements; 3) all sampling methods and equipment must meet EPA reference or equivalent requirements; 4) acceptable data validation and record keeping procedures must be followed; and 5) data from SLAMS must be summarized and reported annually to EPA. Finally, there are system audits that regularly review the overall air quality data collection activity for any needed changes or corrections. Further information is available on the Internet at http://www.epa.gov/cludygxb/programs/namslam.html and through United States EPA's Quality Assurance Handbook (EPA-454/R-98-004 Section 15).

Geographical Extent of Source Data: National

Spatial Detail Covered By the Source Data: State, Local and Tribal air pollution control agencies

2c. Source Data Reporting:
State, Local and Tribal air pollution control agencies submit data within 30 days after the end of each calendar quarter. The data can be submitted in one of three different formats, and is submitted using an Exchange Network Node or the agency's Central Data Exchange web interface. The submitted data are then quality assured and loaded into the AQS database.

3a. Relevant Information Systems:
The Air Quality Subsystem (AQS) stores ambient air quality data used to evaluate an area's air quality levels relative to the National Ambient Air Quality Standards (NAAQS).

AQS has been enhanced to comply with the Agency's data standards (e.g., latitude/longitude, chemical nomenclature).

AQS stores the as-submitted source data and data that are aggregated to the daily, monthly, quarterly and annual values by the system.

3b. Data Quality Procedures:
AQS: The QA/QC of the national air monitoring program has several major components: the Data Quality Objective (DQO) process, reference and equivalent methods program, EPA's National Performance Audit Program (NPAP), system audits, and network reviews. Please see www.epa.gov/ttn/amtic/npaplist.html for more information.

The AQS QA/QC process also involves participation in the EPA's National Performance Audit Program (NPAP), system audits, and network reviews. Please see www.epa.gov/ttn/amtic/npaplist.html for more information. Under NPAP, all agencies required to report gaseous criteria pollutant data from their ambient air monitoring stations to EPA's Air Quality System (AQS) for comparison to the National Ambient Air Quality Standard (NAAQS) are required to participate in EPA's NPAP TTP program. Guidance for participating in this program requires NPAP audits of at least 20% of a Primary Quality Assurance Organization's (PQAO's) sites each year; and all sites in 5 years.

3c. Data Oversight:

Team Member, Central Operations and Resources Staff, OAQPS

3d. Calculation Methodology:

Decision Rules for Selecting Data:

All available air quality measurement data is included in the Design Value calculations except as indicated below:.

1.   Individual measurements that are flagged as being exceedances caused by "Exceptional Events" (as defined in 40 CFR Part 50.14) and that are concurred by the EPA Regional Office are excluded.

Definitions of Variables:

For each AQS monitor, the following variables are calculated:

8-Hour Average: Arithmetic mean of eight consecutive hourly measurements, with the time for the average defined to be the begin hour. (There will be 24 8-hour averages for each day.) Missing values (measurements for a specifc hour) are handled as follows: If there are less than 6 measurements in the 8-hour period, ½ of the Method Detection Limit for the method is used in place of the missing value.

Daily Maximum: The maximum 8-hour average for the calendar day.

Annual 4th Maximum: The fourth highest daily maximum for the year.

Three-Year Design Value: The average of the annual 4th maxima for the three year period.

Explanation of Calculations: Air quality levels are evaluated relative to the baseline level and the design value. The change in air quality concentrations is then multiplied by the number of people living in the county.

Explanation of Assumptions: Design values are calculated for every county with adequate monitoring data. The design value is the mathematically determined pollutant concentration at a particular site that must be reduced to, or maintained at or below the National Ambient Air Quality Standards (NAAQS) in order to assure attainment. The design value may be calculated based on ambient measurements observed at a local monitor in a 3-year period or on model estimates. The design value varies from year to year due to both the pollutant emissions and natural variability such as meteorological conditions, wildfires, dust storms, volcanic activities etc. For more information on design values, including a definition, see

www.epa.gov/ttn/oarpg/t1/memoranda/cdv.pdf This analysis assumes that the populations of the areas are held constant at 2000 Census levels. Data comparisons over several years allow assessment of the air program's success.
Unit of analysis: Cumulative percent reduction in population-weighted ambient concentration
4a. Oversight and Timing of Final Results Reporting:
Director, Central Operations and Resources Staff, OAQPS
4b. Data Limitations/Qualifications:
There is uncertainty in the projections and near term variations in air quality (due to meteorological conditions, for example).
4c. Third-Party Audits:
2008 OIG system audit 2010 System Risk Assessment

Measure Code: G16 - Million metric tons of carbon equivalent (mmtco2e) of greenhouse gas reductions in the industry sector.

Office of Air and Radiation (OAR)

Goal Number and Title:
1 - Taking Action on Climate Change and Improving Air Quality

Objective Number and Title:
1 - Address Climate Change

Sub-Objective Number and Title:
1 - Address Climate Change

Strategic Target Code and Title:
2 - Additional programs from across EPA will promote practices to help Americans save energy and conserv

Managing Office:
Office of Atmospheric Programs

1a. Performance Measure Term Definitions:
Carbon equivalent of Greenhouse Gas Emissions: Carbon dioxide (CO_2) is the base of the global warming potential (GWP) system and has a GWP of 1. All other greenhouse gases' ability to increase global warming is expressed in terms of CO_2. The CO_2e for a gas is derived by multiplying the tons of the gas by that gas's GWP. Commonly expressed as "million metric tons of carbon dioxide equivalents" (MMTCO2e).   Industry Sector: The industrial sector is an important part of the U.S. economy: manufacturing goods valued at nearly $5.5 trillion, contributing over 11 percent to the U.S. GDP, and providing more than 12.7 million jobs paying an average of $47,500 annually. The industrial sector also generates more than a quarter of the nation's annual GHG emissions. Through EPA's voluntary programs, EPA enables the industrial sector to cost-effectively reduce GHG emissions.

2a. Original Data Source:
Carbon emissions related to baseline energy use (e.g., business-as-usual" without the impact of EPA's voluntary climate programs) comes from the Energy Information Agency (EIA) and from EPA's Integrated Planning Model (IPM) of the U.S. electric power sector. Baseline data for non-carbon dioxide (CO_2) emissions, including nitrous oxide and other high global warming potential gases, are maintained by EPA. The non-CO_2 data are compiled with input from industry and also independently from partners' information.   Data collected by EPA's voluntary programs include partner reports on facility- specific improvements (e.g. space upgraded, kilowatt-hours (kWh) reduced), national market data on shipments of efficient products, and engineering measurements of equipment power levels and usage patterns.   Additional Information: The accomplishments of many of EPA's voluntary programs are documented in the Climate Protection Partnerships Division Annual Report. The most recent version is ENERGY STAR and Other Climate Protection Partnerships 2008 Annual Report. http://www.energystar.gov/ia/partners/annualreports/annual_report_2008.pdf

2b. Source Data Collection:

See Section 3b

2c. Source Data Reporting:

See Section 3b

3a. Relevant Information Systems:

Climate Protection Partnerships Division Tracking System. The tracking system's primary purpose is to maintain a record of the annual greenhouse gas emissions reduction goals and accomplishments for the voluntary climate program using information from partners and other sources.

The Climate Protection Partnerships Division Tracking System contains transformed data.

The Climate Protection Partnerships Division Tracking System meets relevant EPA standards for information system integrity.

3b. Data Quality Procedures:

The Industry sector includes a variety of programs. Data Quality procedures vary by program as follows:

The Combined Heat and Power (CHP) Partnership Partnership dismantles the market barriers stifling investment in environmentally beneficial CHP projects. Program partners such as project owners voluntarily provide project-specific information on newly operational CHP projects to EPA. These data are screened and any issues resolved. Energy savings are determined on a project-by-project basis, based on fuel type, system capacity, and operational profile. Estimates of the use of fossil and renewable fuels are developed, as well as the efficiency of thermal and electrical use or generation, as appropriate. Emissions reductions are calculated on a project-by-project basis to reflect the greater efficiency of onsite CHP. Avoided emissions of GHGs from more efficient energy generation are determined using marginal emissions factors derived from energy efficiency scenario runs of IPM, and displaced emissions from boiler-produced thermal energy are developed through engineering estimates. In addition, emissions reductions may include avoided transmission and distribution losses, as appropriate. Only the emissions reductions from projects that meet the assistance criteria for the program are included in the program benefit estimates. EPA also addresses the potential for double counting benefits between this and other partnerships by having program staff meet annually to identify and resolve any overlap issues.

The Green Power Partnership boosts supply of clean energy by helping U.S. organizations purchase electricity from eligible renewable generation sources. As a condition of partnership, program partners submit data annually on their purchases of qualifying green power products. These data are screened and any issues resolved. Avoided emissions of GHGs are determined using marginal emissions factors for CO_2 derived from scenario runs of IPM. The potential for double counting, such as counting green power purchases that may be required as part of a renewable portfolio standard or may rely on resources that are already part of the system mix, is addressed through a partnership requirement that green power purchases be incremental to what is already required. EPA estimates that the vast majority of the green power purchases made by program partners are due to the partnership, as partners comply with aggressive green power procurement requirements (usually at incremental cost) to remain in the program. Further, EPA estimates that its efforts to foster a growing voluntary green power market have likely led to additional voluntary green power purchases that have not been reported through the program.

EPA's methane programs facilitate recovering methane from landfills, natural gas extraction systems, agriculture, and coal mines, as well as using methane as a clean energy resource. The expenditures used in the program analyses include the capital costs agreed to by partners to bring projects into compliance with program specifications and any additional operating costs engendered by program participation.

Within the Natural Gas STAR Program, as a condition of partnership, program partners submit implementation plans to EPA describing the emissions reduction practices they plan to implement and evaluate. In addition, partners submit progress reports detailing specific emissions reduction activities and accomplishments each year. EPA does not attribute all reported emissions reductions to Natural Gas STAR. Partners may only include actions that were undertaken voluntarily, not those reductions attributable to compliance with existing regulations. Emissions reductions are estimated by the partners either from direct before-and-after measurements or by applying peer-reviewed emissions reduction factors.

Within the Landfill Methane Outreach Program, EPA maintains a comprehensive database of the operational data on landfills and landfill gas energy projects in the United States. The data are updated frequently based on information submitted by industry, the Landfill Methane Outreach Program's (LMOP's) outreach efforts, and other sources. Reductions of methane that are the result of compliance with EPA's air regulations are not included in the program estimates. In addition, only the emissions reductions from projects that meet the LMOP assistance criteria are included in the program benefit estimates. EPA uses emissions factors that are appropriate to the project. The factors are based on research, discussions with experts in the landfill gas industry, and published references.

Within the Coalbed Methane Outreach Program, through collaboration with the U.S. Mine Safety & Health Administration, state oil and gas commissions, and the mining companies themselves, EPA collects mine-specific data annually and estimates the total methane emitted from the mines and the quantity of gas recovered and used. There are no regulatory requirements for recovering and using CMM; such efforts are entirely voluntary. EPA estimates CMM recovery attributable to its program activities on a mine-specific basis, based on the program's interaction with each mine.

Within the Voluntary Aluminum Industry Partnership program, VAIP partners agree to report aluminum production and anode effect frequency and duration in order to estimate annual FGHG emissions. Reductions are calculated by comparing current emissions to a BAU baseline that uses the industry's 1990 emissions rate. Changes in the emissions rate (per ton production) are used to estimate the annual GHG emissions and reductions that are a result of the program. The aluminum industry began making significant efforts to reduce FGHG emissions as a direct result of EPA's climate partnership program. Therefore, all reductions achieved by partners are assumed to be the result of the program.

Within the HFC-23 Emission Reduction Program, program partners report HCFC-22 production and HFC-23 emissions to a third party that aggregates the estimates and submits the total estimates for the previous year to EPA. Reductions are calculated by comparing current emissions to a BAU baseline that uses the industry's 1990 emissions rate. Changes in the emissions rate are used to estimate the annual GHG emissions and reductions that are a consequence of the program. Subsequent to a series of meetings with EPA, industry began making significant efforts to reduce HFC-23 emissions. All U.S. producers participate in the program; therefore, all reductions achieved by manufacturers are assumed to be the result of the program.

EPA's Environmental Stewardship Programs include the FGHG Partnership for the Semiconductor Industry and the SF6 Partnerships for Electric Power Systems and Magnesium Industries. Partners report emissions and emissions reductions based on jointly developed estimation methods and reporting protocols. Data collection methods are sector specific, and data are submitted to EPA either directly or through a designated third party. Reductions are calculated by comparing current emissions to a BAU baseline, using industry-wide or company-specific emissions rates in a base year. The reductions in emissions rates are used to calculate the overall GHG emissions reductions from the program. The share of the reductions attributable to EPA's programs is identified based on a detailed review of program activities and industry-specific information.

Within the Responsible Appliance Disposal (RAD) Program, as a condition of partnership, RAD partners submit annual data to EPA on their achievements. Submitted data includes the number and type of appliances collected and processed as well as the quantity and fate of the individual components. GHG reductions are calculated by measuring the emissions avoided by recovering refrigerant, foam blowing agents, and recycling durable components in addition to the energy savings from early appliance retirement from utility programs.

Within the GreenChill Partnership, partner emissions reductions are calculated both year-to-year and aggregate. Partners set annual refrigerant emissions reduction goals and submit refrigerant management plans to detail their reduction initiatives.

Peer-reviewed carbon-conversion factors are used to ensure consistency with generally accepted measures of greenhouse gas (GHG) emissions, and peer-reviewed methodologies are used to calculate GHG reductions from these programs.

3c. Data Oversight:

The Non-CO2 Program Branch is responsible for overseeing (1) source data reporting and (2) the information systems utilized in producing the performance result for Methane Programs and the Voluntary Aluminum Industry Partnership program.

The Energy Supply & Industry Branch is responsible for overseeing (1) source data reporting and (2) the information systems utilized in producing the performance result for the Combined Heat and Power and Green Power Partnership programs.

The Alternatives and Emissions Reduction Branch is responsible for overseeing (1) source data reporting and (2) the information systems utilized in producing the performance result for the GreenChill Partnership, the Responsible Appliance Disposal, and the HFC-23 Emission Reduction Program.

3d. Calculation Methodology:

Explanation of Assumptions: Most of the voluntary climate programs' focus is on energy efficiency. For these programs, EPA estimates the expected reduction in electricity consumption in kilowatt-hours (kWh). Emissions prevented are calculated as the product of the kWh of electricity saved and an annual emission factor (e.g., metric tons carbon equivalent (MMTCE) prevented per kWh). Other programs focus on directly lowering greenhouse gas emissions (e.g., non-CO2 Partnership programs, Landfill Methane Outreach, and Coalbed Methane Outreach); for these, greenhouse gas emission reductions are estimated on a project-by-project basis.

Explanation of the Calculations: The Integrated Planning Model, used to develop baseline data for carbon emissions, is an important analytical tool for evaluating emission scenarios affecting the U.S. power sector.

Baseline information is discussed at length in the U.S. Climate Action Report 2002. The report includes a complete chapter dedicated to the U.S. greenhouse gas inventory (sources, industries, emissions, volumes, changes, trends, etc.). A second chapter addresses projected greenhouse gases in the future (model assumptions, growth, sources, gases, sectors, etc.) Please see http://www.gcrio.org/CAR2002 and www.epa.gov/globalwarming/publications/car/index.html

Unit of Measure: Million metric tons of carbon equivalent (MMTE) of greenhouse gas emissions

Additional information:

The IPM has an approved quality assurance project plan that is available from EPA's program office.

Background information on the IPM can be found on the website for EPA's Council for Regulatory Environmental Modeling: http://cfpub.epa.gov/crem/knowledge_base/crem_report.cfm?deid=74919

4a. Oversight and Timing of Final Results Reporting:

Branch Chief, Non-CO2 Program Branch is responsible for overseeing final reporting for Methane Programs and the Voluntary Aluminum Industry Partnership program.
Branch Chief, Energy Supply & Industry Branch is responsible for overseeing final reporting for the Combined Heat and Power and Green Power Partnership programs.
Branch Chief, Alternatives and Emissions Reduction Branch is responsible for overseeing final reporting for the GreenChill Partnership, the Responsible Appliance Disposal, and the HFC-23 Emission Reduction Program.

4b. Data Limitations/Qualifications:

These are indirect measures of GHG emissions (carbon conversion factors and methods to convert material-specific reductions to GHG emissions reductions). Although EPA devotes considerable effort to obtaining the best possible information on which to evaluate emissions reductions from its voluntary programs, errors in the performance data could be introduced through uncertainties in carbon conversion factors, engineering analyses, and econometric analyses. Comprehensive documentation regarding the IPM and uncertainties associated with it can be found at the IPM website: http://www.epa.gov/airmarkets/progsregs/epa-ipm/ Also,

the voluntary nature of the programs may affect reporting.
4c. Third-Party Audits:
The Administration regularly evaluates the effectiveness of its climate programs through interagency evaluations. The second such interagency evaluation, led by the White House Council on Environmental Quality, examined the status of U.S. climate change programs. The review included participants from EPA and the Departments of State, Energy, Commerce, Transportation, and Agriculture. The results were published in the U.S. Climate Action Report-2002 as part of the United States' submission to the Framework Convention on Climate Change (FCCC). The previous evaluation was published in the U.S. Climate Action Report-1997. A 1997 audit by EPA's Office of the Inspector General concluded that the climate programs examined "used good management practices" and "effectively estimated the impact their activities had on reducing risks to health and the environment..."

Measure Code: G06 - Million metric tons of carbon equivalent (mmtco2e) of greenhouse gas reductions in the transportation sector.

Office of Air and Radiation (OAR)

Goal Number and Title:
1 - Taking Action on Climate Change and Improving Air Quality

Objective Number and Title:
1 - Address Climate Change

Sub-Objective Number and Title:
1 - Address Climate Change

Strategic Target Code and Title:
2 - Additional programs from across EPA will promote practices to help Americans save energy and conserv

Managing Office:
Office of Transportation and Air Quality

1a. Performance Measure Term Definitions:
Carbon equivalent of Greenhouse Gas Emissions: Carbon dioxide (CO2) is the base of the global warming potential (GWP) system and has a GWP of 1. All other greenhouse gases' ability to increase global warming is expressed in terms of CO2. The CO2e for a gas is derived by multiplying the tons of the gas by that gas's GWP. Commonly expressed as "million metric tons of carbon dioxide equivalents" (MMTCO2e) Transportation Sector: Mobile Sources

2a. Original Data Source:
Carbon emissions related to baseline energy use (e.g., business-as-usual" without the impact of EPA's voluntary climate programs) comes from the Energy Information Agency (EIA) and from EPA's Integrated Planning Model (IPM) of the U.S. electric power sector. Baseline data for non-carbon dioxide (CO2) emissions, including nitrous oxide and other high global warming potential gases, are maintained by EPA. The non-CO2 data are compiled with input from industry and also independently from partners' information. Data on the effects of EPA regulatory programs on baseline transportation emissions are obtained from EPA's Motor Vehicle Emissions Simulator (MOVES) model. Data collected by EPA's voluntary programs include partner reports on facility- specific improvements (e.g. space upgraded, kilowatt-hours (kWh) reduced), national market data on shipments of efficient products, and engineering measurements of equipment power levels and usage patterns. Additional Information: The accomplishments of many of EPA's voluntary programs are documented in the Climate Protection Partnerships Division Annual Report. The most recent version is ENERGY STAR and Other Climate Protection Partnerships 2008 Annual Report. http://www.energystar.gov/ia/partners/annualreports/annual_report_2008.pdf

2b. Source Data Collection:
Partners provide information annually on freight transportation activity. Data Collection is ongoing, as new partners join and existing partners are retained.

2c. Source Data Reporting:
Data is submitted through the use of EPA-provided assessment tools. Data is submitted annually and entered into a program data base.

3a. Relevant Information Systems:
Climate Protection Partnerships Division Tracking System. The tracking system's primary purpose is to maintain a record of the annual greenhouse gas emissions reduction goals and accomplishments for the voluntary climate program using information from partners and other sources.   The data base contains source data from partners.   Partners submit data to SmartWay in a spreadsheet-based reporting tool, which is uploaded in XML format to an Oracle database containing all partner data, including prior year data and data from companies that have left the program.

3b. Data Quality Procedures:
Partners do contribute actual emissions data biannually after their facility-specific improvements but these emissions data are not used in tracking the performance measure. EPA, however, validates the estimates of greenhouse gas reductions based on the actual emissions data received.   For transportation emissions, data is calculated from operation activity (fuel use, miles driven, etc). Partner activity metrics were developed and peer reviewed according to EPA peer review requirements. Peer-reviewed carbon-conversion factors are used to ensure consistency with generally accepted measures of greenhouse gas (GHG) emissions, and peer-reviewed methodologies are used to calculate GHG reductions from these programs.

3c. Data Oversight:
Supervisory EPS, Transportation and Climate Division (TCD) is program manager, with overall oversight responsibility.   Environmental Scientist, TCD is responsible for maintaining data results and program goals and results.   Environmental Engineer, TCD is responsible for maintaining the information systems (partner forms and data base.)

3d. Calculation Methodology:
Explanation of Assumptions: Most of the voluntary climate programs' focus is on energy efficiency. For these programs, EPA estimates the expected reduction in electricity consumption in kilowatt-hours (kWh). Emissions prevented are calculated as the product of the kWh of electricity saved and an annual emission factor (e.g., metric tons carbon equivalent (MMTCE) prevented per kWh). Other programs focus on directly lowering greenhouse gas emissions (e.g., non-CO2 Partnership programs, Landfill Methane Outreach, and Coalbed Methane Outreach); for these, greenhouse gas emission reductions are estimated on a project-by-project basis. Other programs focused on transportation (e.g., SmartWay) calculate emissions reductions as the product of fuel saved and an annual emission factor (e.g., metric tons carbon equivalent (MMTCE) prevented per gallon of fuel saved).

Explanation of the Calculations: The Integrated Planning Model, used to develop baseline data for carbon emissions, is an important analytical tool for evaluating emission scenarios affecting the U.S. power sector.

Baseline information is discussed at length in the U.S. Climate Action Report 2002. The report includes a complete chapter dedicated to the U.S. greenhouse gas inventory (sources, industries, emissions, volumes, changes, trends, etc.). A second chapter addresses projected greenhouse gases in the future (model assumptions, growth, sources, gases, sectors, etc.) Please see http://www.gcrio.org/CAR2002 and www.epa.gov/globalwarming/publications/car/index.html

Unit of Measure: Million metric tons of carbon equivalent (MMTE) of greenhouse gas emissions

Additional information:

The IPM has an approved quality assurance project plan that is available from EPA's program office.

Background information on the IPM can be found on the website for EPA's Council for Regulatory Environmental Modeling: http://cfpub.epa.gov/crem/knowledge_base/crem_report.cfm?deid=74919

4a. Oversight and Timing of Final Results Reporting:

Program Analyst, Planning & Budget Office, Office of Transportation and Air Quality oversees the reporting process.

4b. Data Limitations/Qualifications:

These are indirect measures of GHG emissions (carbon conversion factors and methods to convert material-specific reductions to GHG emissions reductions). Although EPA devotes considerable effort to obtaining the best possible information on which to evaluate emissions reductions from its voluntary programs, errors in the performance data could be introduced through uncertainties in carbon conversion factors, engineering analyses, and econometric analyses. Comprehensive documentation regarding the IPM and uncertainties associated with it can be found at the IPM website: http://www.epa.gov/airmarkets/progsregs/epa-ipm/ Also, the voluntary nature of the programs may affect reporting.

4c. Third-Party Audits:

The Administration regularly evaluates the effectiveness of its climate programs through interagency evaluations. The second such interagency evaluation, led by the White House Council on Environmental Quality, examined the status of U.S. climate change programs. The review included participants from EPA and the Departments of State, Energy, Commerce, Transportation, and Agriculture. The results were published in the U.S. Climate Action Report-2002 as part of the United States' submission to the Framework Convention on Climate Change (FCCC). The previous evaluation was published in the U.S. Climate Action Report-1997. A 1997 audit by EPA's Office of the Inspector General concluded that the climate programs examined "used good management practices" and "effectively estimated the impact their activities had on reducing risks to health and the environment..."

An August 30, 2012 report by EPA's Office of Inspector General (OIG) found that while SmartWay "performs some checks of data provided by industry … there is no direct verification by EPA of data submitted by SmartWay participants." OIG recommended that EPA "protect the integrity of its program by implementing some form of direct verification or other measures to deter companies from submitting data that result in overstated scores." EPA has instituted an annual Data Verification Program where EPA staff visit selected partners each year and review data management practices. EPA has also published an industry-reviewed guidance on best practices in data management for SmartWay partners.

Measure Code: G02 - Million metric tons of carbon equivalent (mmtco2e) of greenhouse gas reductions in the buildings sector.

Office of Air and Radiation (OAR)

Goal Number and Title:
1 - Taking Action on Climate Change and Improving Air Quality
Objective Number and Title:
1 - Address Climate Change
Sub-Objective Number and Title:
1 - Address Climate Change
Strategic Target Code and Title:
2 - Additional programs from across EPA will promote practices to help Americans save energy and conserv
Managing Office:
Office of Atmospheric Programs
1a. Performance Measure Term Definitions:
Carbon equivalent of Greenhouse Gas Emissions: Carbon equivalent of Greenhouse Gas Emissions: Carbon dioxide (CO_2) is the base of the global warming potential (GWP) system and has a GWP of 1. All other greenhouse gases' ability to increase global warming is expressed in terms of CO_2. The CO_2e for a gas is derived by multiplying the tons of the gas by that gas's GWP. Commonly expressed as "million metric tons of carbon dioxide equivalents" ($MMTCO_2e$).   Buildings Sector: The Buildings Sector includes the following Energy Star partnerships: Energy Star Labeling, Energy Star Homes, and the Energy Star Buildings programs. In the Energy Star Labeling program, the American public continues to look to ENERGY STAR as the national symbol for energy efficiency to inform purchasing choices, save money on utility bills, and protect the environment. In 2010, Americans purchased about 200 million products that had earned the ENERGY STAR across more than 60 product categories for a cumulative total of about 3.5 billion ENERGY STAR qualified products purchased since 2000. Qualified products—including appliances, heating and cooling equipment, consumer electronics, office equipment, lighting, and more—offer consumers savings of as much as 65 percent relative to standard models while providing the features and functionality consumers expect. In the Energy Star Homes program we focus on the 17 percent of the GHGs emitted in the United States that are attributed to the energy we use to heat, cool, and light our homes, as well as power the appliances and electronics in them. By making energy-efficient choices in the construction of new homes and the improvement of existing homes, American homeowners, renters, homebuilders, and home remodelers can lower household utility bills while helping to protect the environment. Through ENERGY STAR, EPA offers an array of useful tools and resources to households and the housing industry to increase the energy efficiency of the nation's housing stock. In the the Energy Star Buildings program we focus on efforts to improve energy efficiency in commercial buildings across the country by 20 percent over the next decade. Through the ENERGY STAR program, EPA is already helping the commercial building sector improve energy efficiency in the places where consumers work, play, and learn. In turn, these efforts will help create jobs, save money, reduce dependence on foreign oil, and contribute to cleaner air and the protection of people's health. These and future efficiency efforts are of critical importance,

as commercial buildings are responsible for approximately 20 percent of all energy consumption in the United States.

2a. Original Data Source:

Carbon emissions related to baseline energy use (e.g., business-as-usual" without the impact of EPA's voluntary climate programs) comes from the Energy Information Agency (EIA) and from EPA's Integrated Planning Model (IPM) of the U.S. electric power sector. Baseline data for non-carbon dioxide (CO_2) emissions, including nitrous oxide and other high global warming potential gases, are maintained by EPA. The non-CO_2 data are compiled with input from industry and also independently from partners' information.

Data collected by EPA's voluntary programs include partner reports on facility- specific improvements (e.g. space upgraded, kilowatt-hours (kWh) reduced), national market data on shipments of efficient products, and engineering measurements of equipment power levels and usage patterns.

Additional Information:

The accomplishments of many of EPA's voluntary programs are documented in the Climate Protection Partnerships Division Annual Report. The most recent version is ENERGY STAR and Other Climate Protection Partnerships 2008 Annual Report. http://www.energystar.gov/ia/partners/annualreports/annual_report_2008.pdf

2b. Source Data Collection:

Avoided emissions of GHGs are determined using marginal emissions factors for CO_2 equivalency based on factors established as part of the U.S. government's reporting process to the UN Framework Convention on Climate Change, as well as historical emissions data from EPA's eGRID database. For future years, EPA uses factors derived from energy efficiency scenario runs of the integrated utility dispatch model, Integrated Planning Model (IPM®).

2c. Source Data Reporting:

Carbon emissions related to baseline energy use (e.g., business-as-usual" without the impact of EPA's voluntary climate programs) comes from the Energy Information Agency (EIA) and from EPA's Integrated Planning Model (IPM) of the U.S. electric power sector. Baseline data for non-carbon dioxide (CO_2) emissions, including nitrous oxide and other high global warming potential gases, are maintained by EPA. The non-CO_2 data are compiled with input from industry and also independently from partners' information.

Data collected by EPA's voluntary programs include partner reports on facility- specific improvements (e.g. space upgraded, kilowatt-hours (kWh) reduced), national market data on shipments of efficient products, and engineering measurements of equipment power levels and usage patterns.

3a. Relevant Information Systems:

Climate Protection Partnerships Division Tracking System. The tracking system's primary purpose is to maintain a record of the annual greenhouse gas emissions reduction goals and accomplishments for the voluntary climate program using information from partners and other sources.

The Climate Protection Partnerships Division Tracking System contains transformed data.

The Climate Protection Partnerships Division Tracking System meets relevant EPA standards for information system integrity.

3b. Data Quality Procedures:

ENERGY STAR program procedures for oversight, review and quality assurance include the following. To participate, product manufacturers and retailers enter into formal partnership agreements with the government and agree to adhere to the ENERGY STAR Identity Guidelines, which describe how the ENERGY STAR name and mark may be used. EPA continually monitors the use of the brand in trade media, advertisements, and stores and on the Internet. The Agency also conducts biannual onsite store-level assessments of ENERGY STAR qualified products on the stores' shelves to ensure the products are presented properly to consumers. To ensure that ENERGY STAR remains a trusted symbol for environmental protection through superior efficiency, EPA completed comprehensive enhancements of the product qualification and verification processes. Third-party certification of ENERGY STAR products went into effect, as scheduled, on January 1, 2011. Before a product can be labeled with the ENERGY STAR under the new requirements, its performance must be certified by an EPA-recognized third party based on testing in an EPA-recognized lab. In addition, ENERGY STAR manufacturer partners must participate in verification testing programs run by the approved certification bodies. By the end of 2010, EPA had recognized 21 accreditation bodies, 132 laboratories, and 15 certification bodies.

Enforcing proper use of the ENERGY STAR mark is essential to maintaining the integrity of the program. As the result of multiple off-the-shelf testing efforts, EPA disqualified 17 products from the ENERGY STAR program in 2010 for failure to meet performance standards. Manufacturers of those products were required to discontinue use of the label and take additional steps to limit product exposure in the market. In an effort to ensure fair and consistent commitment among ENERGY STAR partners, EPA also took steps this year to suspend the partner status of manufacturers failing to comply with program requirements.

Peer-reviewed carbon-conversion factors are used to ensure consistency with generally accepted measures of greenhouse gas (GHG) emissions, and peer-reviewed methodologies are used to calculate GHG reductions from these programs.

3c. Data Oversight:

The Energy Star Labeling Branch is responsible for overseeing (1) source data reporting and (2) the information systems utilized in producing the performance result for the Energy Star Labeling program. The Energy Star Residential Branch is responsible for overseeing (1) source data reporting and (2) the information systems utilized in producing the performance result for the Energy Star Homes program. The Energy Star Commercial & Industrial Branch is responsible for overseeing (1) source data reporting and (2) the information systems utilized in producing the performance result for the Energy Star Commercial Buildings program.

3d. Calculation Methodology:

Explanation of Assumptions: Most of the voluntary climate programs' focus is on energy efficiency. For these programs, EPA estimates the expected reduction in electricity consumption in kilowatt-hours (kWh). Emissions prevented are calculated as the product of the kWh of electricity saved and an annual emission factor (e.g., metric tons carbon equivalent (MMTCE) prevented per kWh). Other programs focus on directly lowering greenhouse gas emissions (e.g., non-CO2 Partnership programs, Landfill Methane Outreach, and Coalbed

Methane Outreach); for these, greenhouse gas emission reductions are estimated on a project-by-project basis.

Explanation of the Calculations: The Integrated Planning Model, used to develop baseline data for carbon emissions, is an important analytical tool for evaluating emission scenarios affecting the U.S. power sector.

Baseline information is discussed at length in the U.S. Climate Action Report 2002. The report includes a complete chapter dedicated to the U.S. greenhouse gas inventory (sources, industries, emissions, volumes, changes, trends, etc.). A second chapter addresses projected greenhouse gases in the future (model assumptions, growth, sources, gases, sectors, etc.) Please see http://www.gcrio.org/CAR2002 and www.epa.gov/globalwarming/publications/car/index.html

Unit of Measure: Million metric tons of carbon equivalent (MMTE) of greenhouse gas emissions

Additional information:

The IPM has an approved quality assurance project plan that is available from EPA's program office.

Background information on the IPM can be found on the website for EPA's Council for Regulatory Environmental Modeling: http://cfpub.epa.gov/crem/knowledge_base/crem_report.cfm?deid=74919

4a. Oversight and Timing of Final Results Reporting:
Branch Chief, Energy Star Labeling Branch is responsible for the Energy Star Labeling program. Branch Chief, Energy Star Residential Branch is responsible for the Energy Star Homes program. Branch Chief, Energy Star Commercial & Industrial Branch is responsible for the Energy Star Commercial Buildings program.
4b. Data Limitations/Qualifications:
These are indirect measures of GHG emissions (carbon conversion factors and methods to convert material-specific reductions to GHG emissions reductions). Although EPA devotes considerable effort to obtaining the best possible information on which to evaluate emissions reductions from its voluntary programs, errors in the performance data could be introduced through uncertainties in carbon conversion factors, engineering analyses, and econometric analyses. Comprehensive documentation regarding the IPM and uncertainties associated with it can be found at the IPM website: http://www.epa.gov/airmarkets/progsregs/epa-ipm/ Also, the voluntary nature of the programs may affect reporting.
4c. Third-Party Audits:
The Administration regularly evaluates the effectiveness of its climate programs through interagency evaluations. The second such interagency evaluation, led by the White House Council on Environmental Quality, examined the status of U.S. climate change programs. The review included participants from EPA and the Departments of State, Energy, Commerce, Transportation, and Agriculture. The results were published in

the U.S. Climate Action Report-2002 as part of the United States' submission to the Framework Convention on Climate Change (FCCC). The previous evaluation was published in the U.S. Climate Action Report-1997. A 1997 audit by EPA's Office of the Inspector General concluded that the climate programs examined "used good management practices" and "effectively estimated the impact their activities had on reducing risks to health and the environment…"

Office of Air and Radiation (OAR)

Goal Number and Title:
1 - Taking Action on Climate Change and Improving Air Quality
Objective Number and Title:
2 - Improve Air Quality
Sub-Objective Number and Title:
1 - Reduce Criteria Pollutants and Regional Haze
Strategic Target Code and Title:
5 - By 2015, reduce emissions of direct particulate matter (PM)
Managing Office:
Office of Transportation and Air Quality
1a. Performance Measure Term Definitions:
Mobile sources: Includes onroad cars/trucks, nonroad engines such as farms/construction, locomotives, commercial marine, and aircraft. Particulate matter (PM-2.5): Solid material 2.5 microns or smaller as defined by the EPA National Ambient Air Quality Standard and measurement methods.
2a. Original Data Source:
Estimates for on-road and off-road mobile source emissions are built from inventories fed into the relevant models. Data for the models are from many sources, including Vehicle Miles Traveled (VMT) estimates by state (Federal Highway Administration), the mix of VMT by type of vehicle (Federal Highway Administration), temperature, gasoline properties, and the designs of Inspection/Maintenance (I/M) programs. Usage data for nonroad comes largely from fuel consumption information from DOE.
2b. Source Data Collection:
Source Data Collection Methods: Emission tests for engines/vehicles come from EPA, other government agencies (including state/local governments). academic institutions, and industry. The data come from actual emission tests measuring HC, CO, NOx , and PM emissions. Usage surveys for vehicle miles traveled are obtained from DOT surveys and fuel usage for nonroad vehicles/engines are obtained from a variety of sources such as DOE. Geographical Extent of Source Data: National and state level Spatial Detail Covered By the Source Data: County level data
2c. Source Data Reporting:
Form/mechanism for receiving data and entering into EPA system: EPA develops and receives emission data in a g/mile or g/unit work (or unit fuel consumed) basis.

Timing and frequency of reporting: The inputs to MOVES/MOBILE 6 and NONROAD 2008 and other models are reviewed and updated, sometimes on an annual basis for some parameters. Generally, Vehicle Miles Traveled (VMT), the mix of VMT by type of vehicle (Federal Highway Administration (FHWA)-types), temperature, gasoline properties, and the designs of Inspection/Maintenance (I/M) programs are updated each year.

Emission factors for all mobile sources and activity estimates for non-road sources are revised at the time EPA's Office of Transportation and Air Quality provides new information.

Updates to the inputs to the models means the emissions inventories will change.

3a. Relevant Information Systems:

National Emissions Inventory Database. Obtained by modeling runs using MOBILE/MOVES, NONROAD, and other models.

Please see: http://www.epa.gov/ttn/chief/trends/ for a summary of national emission inventories and how the numbers are obtained in general.

The emission inventory contains source test data as well as usage information compiled from other sources. Also, for consistency from year to year and to provide a baseline over time, the emission inventories are updated for these performance measure only when it is essential to do so. The source data (emissions and usage) are "transformed" into emission inventories.

The models and input undergo peer review receiving scientific input from a variety of sources including academic institutions and public comments.

3b. Data Quality Procedures:

The emissions inventories are reviewed by both internal and external parties, including the states, locals and industries. EPA works with all of these parties in these reviews. Also, EPA reviews the inventories comparing them to other derived in earlier years to assure that changes in inputs provide reasonable changes in the inventories themselves

3c. Data Oversight:

EPA emission inventories for the performance measure are reviewed by various OTAQ Center Directors in the Assessment and Standards Division. The Center Directors are responsible for vehicle, engine, fuel, and modeling data used in various EPA programs.

3d. Calculation Methodology:

Explanation of the Calculations:

EPA uses models to estimate mobile source emissions, for both past and future years. The emission inventory estimate is detailed down to the county level and with over 30 line items representing mobile sources.

The MOVES (Motor Vehicle Emission Simulator) model replacing the earlier MOBILE6 vehicle emission factor model is a software tool for predicting gram per mile emissions of hydrocarbons, carbon monoxide, oxides of nitrogen, carbon dioxide, particulate matter, and toxics from cars, trucks, and motorcycles under various conditions. Inputs to the model include fleet composition, activity, temporal information, and control program characteristics. For more information on the MOBILE6 model, please visit http://www.epa.gov/otaq/m6.htm

The NONROAD 2008 emission inventory model replacing earlier versions of NONROAD is a software tool for predicting emissions of hydrocarbons, carbon monoxide, oxides of nitrogen, particulate matter, and sulfur dioxides from small and large off road vehicles, equipment, and engines. Inputs to the model include fleet composition, activity and temporal information. For more information on the NONROAD model, please visit http://www.epa.gov/oms/nonrdmdl.htm

Additional information:

To keep pace with new analysis needs, new modeling approaches, and new data, EPA is currently working on a new modeling system termed the Multi-scale Motor Vehicles and Equipment Emission System (MOVES). This new system will estimate emissions for on road and off road sources, cover a broad range of pollutants, and allow multiple scale analysis, from fine scale analysis to national inventory estimation. When fully implemented, MOVES will serve as the replacement for MOBILE6 and NONROAD. The new system will not necessarily be a single piece of software, but instead will encompass the necessary tools, algorithms, underlying data and guidance necessary for use in all official analyses associated with regulatory development, compliance with statutory requirements, and national/regional inventory projections. Additional information is available on the Internet at http://www.epa.gov/otaq/ngm.htm

Unit of analysis: tons of emissions, vehicle miles traveled, and hours (or fuel) used]

4a. Oversight and Timing of Final Results Reporting:
Team Member, Planning and Budget Office, OTAQ
4b. Data Limitations/Qualifications:

The limitations of the inventory estimates for mobile sources come from limitations in the modeled emission factors (based on emission factor testing and models predicting overall fleet emission factors in g/mile) and also in the estimated vehicle miles traveled for each vehicle class (derived from Department of Transportation data)..

For nonroad emissions, the estimates come from a model using equipment populations, emission factors per hour or unit of work, and an estimate of usage. This nonroad emissions model accounts for over 200 types of nonroad equipment. Any limitations in the input data will carry over into limitations in the emission inventory estimates.

Additional information about data integrity for the MOVES/MOBILE6 and NONROAD models is available on the Internet at http://www.epa.gov/otaq/m6.htm and http://www.epa.gov/oms/nonrdmdl.htm respectively.

When the method for estimating emissions changes significantly, older estimates of emissions in years prior to the most recent year may be revised to be consistent with the new methodology when possible.

Methods for estimating emission inventories are frequently updated to reflect the most up-to-date inputs and assumptions. Past emission estimates that inform our performance measure frequently do not keep pace with the changing inventories associated with more measures in 2002, making both current and future year projections for on-road and nonroad. The emission estimates have been updated numerous times since then for rulemaking packages and will be updated for these performance measures.

4c. Third-Party Audits:

All of the inputs for the models, the models themselves, and the resultant emission inventories are reviewed as appropriate by academic experts and, also, by state/local governments which use some of this information for their State Implementation Plans to meet the National Ambient Air Quality Standards.

Measure Code: S01 - Remaining US Consumption of hydrochlorofluorocarbons (HCFCs), chemicals that deplete the Earth's protective ozone layer, measured in tons of Ozone Depleting Potential (ODP).

Office of Air and Radiation (OAR)

Goal Number and Title:
1 - Taking Action on Climate Change and Improving Air Quality
Objective Number and Title:
3 - Restore the Ozone Layer
Sub-Objective Number and Title:
1 - Reduce Consumption of Ozone-depleting Substances
Strategic Target Code and Title:
1 - By 2015, U.S. reduce consumption of hydrochlorofluorocarbons (HCFCs), chemicals
Managing Office:
Office of Atmospheric Programs
1a. Performance Measure Term Definitions:
Remaining: The term "Remaining" is defined as that which remains, especially after something else has been removed.

US consumption: Class II controlled substances are compounds that have an ozone depletion potential (ODP) less than 0.2, and are all hydrochlorofluorocarbons (HCFCs). HCFCs were developed as transitional substitutes for Class I substances and are subject to a later phaseout schedule than Class I substances.

Although there are currently 34 controlled HCFCs, only a few are commonly used. The most widely used have been HCFC-22 (usually a refrigerant), HCFC-141b (a solvent and foam-blowing agent), and HCFC-142b (a foam-blowing agent and component in refrigerant blends).

As a Party to the Montreal Protocol, the U.S. must incrementally decrease HCFC consumption and production, culminating in a complete HCFC phaseout in 2030. The major milestones that are upcoming for developed countries are a reduction in 2010 to at least 75 percent below baseline HCFC levels and a reduction in 2015 to at least 90 percent below baseline.

Section 605 of the Clean Air Act sets the U.S. phaseout targets for Class II substances. In 1993, the EPA established the phaseout framework and the "worst-first" approach that focused first on HCFC-22, HCFC-141b, and HCFC-142b because these three HCFCs have the highest ODPs of all HCFCs. To meet the required 2004 reduction, the EPA phased out HCFC-141b in 2003 and froze the production and consumption of HCFC-22 and HCFC-142b. In 2009, EPA reduced the production and import of virgin HCFC-22 and HCFC-142b and limited the use of those compounds to meet the Montreal Protocol's 2010 milestones.

EPA ensures that HCFC consumption in the U.S. is 75% below the U.S. baseline (as required under the Montreal Protocol) by issuing allowances to producers and importers of HCFCs. The "2010 HCFC Allocation

Rule" allocated allowances for each year between 2010 and 2014. To meet the stepdown, the number of allowances for HCFC-22 and HCFC-142b were less than for the 2003-2009 control periods. EPA also issued allowances for HCFC-123, HCFC-124, HCFC-225ca, and HCFC-225cb. The rules also limited the use of virgin HCFC-22 and HCFC-142b to existing refrigeration and air-conditioning equipment. The "Pre-Charted Appliances Rule" banned the sale or distribution of air-conditioning and refrigeration products containing HCFC-22, HCFC-142b, or blends containing one or both of these substances, beginning January 1, 2010.

The "2010 HCFC Allocation Rule" was challenged in the U.S. Court of Appeals for the D.C. Circuit in Arkema v EPA. In August, 2010, the court decided against EPA. EPA interprets the Court's decision as vacating the portion of the rule that establishes company-by-company production and consumption baselines and calendar-year allowances for HCFC-22 and HCFC-142b. All other aspects of the rule are intact. On August 5, 2011, EPA issued an interim final rule that establishes new company-by-company HCFC-22 and HCFC-142b baselines and allocates production and consumption allowances for 2011.

EPA is developing regulations that will issue allowances for the 2012-2014 control periods in response to the court's decision in Arkema v EPA.

Hydrochlorofluorocarbon (HCFC): a compound consisting of hydrogen, chlorine, fluorine, and carbon
The HCFCs are one class of chemicals being used to replace the chlorofluorocarbons (CFCs). They contain chlorine and thus deplete stratospheric ozone, but to a much lesser extent than CFCs. HCFCs have ozone depletion potentials (ODPs) ranging from 0.01 to 0.1.

Class II Ozone-Depleting Substance (ODS): a chemical with an ozone-depletion potential of less than 0.2
Currently, all of the HCFCs are class II substances, and the only Class II substances are HCFCs.

Ozone Depletion Potential (ODP): a number that refers to the amount of ozone depletion caused by a substance
The ODP is the ratio of the impact on ozone of a chemical compared to the impact of a similar mass of CFC-11. Thus, the ODP of CFC-11 is defined to be 1.0. Other CFCs and HCFCs have ODPs that range from 0.01 to 1.0.

Tons of Ozone Depleting Potential: metric tons of ODS weighted by their Ozone Depletion Potential (ODP), otherwise referred to as ODP tons.

See http://www.epa.gov/ozone/desc.html for additional information on ODSs. See http://www.epa.gov.ozone/intpol/index.html for additional information about the Montreal Protocol. See http://www.unmfs.org/ for more information about the Multilateral Fund.

2a. Original Data Source:

US Companies Producing, Importing and Exporting ODS. Progress on restricting domestic exempted consumption of Class II HCFCs is tracked by monitoring industry reports of compliance with EPA's phase-out regulations. Data are provided by U.S. companies producing, importing, and exporting ODS. Corporate data are typically submitted as quarterly reports. Specific requirements, as outlined in the Clean Air Act, are available on the Internet at: http://www.epa.gov/ozone/title6/index.html

The International Trade Commission also provides monthly information on US production, imports, and exports.

2b. Source Data Collection:

Source Data Collection Methods: § 82.24   Recordkeeping and reporting requirements for class II controlled substances.

a) Recordkeeping and reporting. Any person who produces, imports, exports, transforms, or destroys class II controlled substances must comply with the following recordkeeping and reporting requirements:

(1) Reports required by this section must be mailed to the Administrator within 30 days of the end of the applicable reporting period, unless otherwise specified.

(2) Revisions of reports that are required by this section must be mailed to the Administrator within 180 days of the end of the applicable reporting period, unless otherwise specified.

(3) Records and copies of reports required by this section must be retained for three years.

(4) Quantities of class II controlled substances must be stated in terms of kilograms in reports required by this section.

(5) Reports and records required by this section may be used for purposes of compliance determinations. These requirements are not intended as a limitation on the use of other evidence admissible under the Federal Rules of Evidence. Failure to provide the reports, petitions and records required by this section and to certify the accuracy of the information in the reports, petitions and records required by this section, will be considered a violation of this subpart. False statements made in reports, petitions and records will be considered violations of Section 113 of the Clean Air Act and under 18 U.S.C. 1001.

(b) Producers. Persons ("producers") who produce class II controlled substances during a control period must comply with the following recordkeeping and reporting requirements:

(1) Reporting—Producers. For each quarter, each producer of a class II controlled substance must provide the Administrator with a report containing the following information:

(i) The quantity (in kilograms) of production of each class II controlled substance used in processes resulting in their transformation by the producer and the quantity (in kilograms) intended for transformation by a second party;

(ii) The quantity (in kilograms) of production of each class II controlled substance used in processes resulting in their destruction by the producer and the quantity (in kilograms) intended for destruction by a second party;

(iii) The expended allowances for each class II controlled substance;

(iv) The producer's total of expended and unexpended production allowances, consumption allowances, export production allowances, and Article 5 allowances at the end of that quarter;

(v) The quantity (in kilograms) of class II controlled substances sold or transferred during the quarter to a person other than the producer for use in processes resulting in their transformation or eventual destruction;

(vi) A list of the quantities and names of class II controlled substances, exported by the producer to a Party to the Protocol, that will be transformed or destroyed and therefore were not produced expending production or consumption allowances;

(vii) For transformation in the U.S. or by a person of another Party, one copy of a transformation verification from the transformer for a specific class II controlled substance and a list of additional quantities shipped to that same transformer for the quarter;

(viii) For destruction in the U.S. or by a person of another Party, one copy of a destruction verification as required in paragraph (e) of this section for a particular destroyer, destroying the same class II controlled substance, and a list of additional quantities shipped to that same destroyer for the quarter;

(ix) In cases where the producer produced class II controlled substances using export production allowances, a list of U.S. entities that purchased those class II controlled substances and exported them to a Party to the Protocol;

(x) In cases where the producer produced class II controlled substances using Article 5 allowances, a list of U.S. entities that purchased those class II controlled substances and exported them to Article 5 countries; and

(xi) A list of the HCFC 141b-exemption allowance holders from whom orders were received and the quantity (in kilograms) of HCFC-141b requested and produced.

(2) Recordkeeping—Producers. Every producer of a class II controlled substance during a control period must maintain the following records:

(i) Dated records of the quantity (in kilograms) of each class II controlled substance produced at each facility;

(ii) Dated records of the quantity (in kilograms) of class II controlled substances produced for use in processes that result in their transformation or for use in processes that result in their destruction;

(iii) Dated records of the quantity (in kilograms) of class II controlled substances sold for use in processes that result in their transformation or for use in processes that result in their destruction;

(iv) Dated records of the quantity (in kilograms) of class II controlled substances produced with export production allowances or Article 5 allowances;

(v) Copies of invoices or receipts documenting sale of class II controlled substances for use in processes that result in their transformation or for use in processes that result in their destruction;

(vi) Dated records of the quantity (in kilograms) of each class II controlled substance used at each facility as feedstocks or destroyed in the manufacture of a class II controlled substance or in the manufacture of any other substance, and any class II controlled substance introduced into the production process of the same class II controlled substance at each facility;

(vii) Dated records of the quantity (in kilograms) of raw materials and feedstock chemicals used at each facility for the production of class II controlled substances;

(viii) Dated records of the shipments of each class II controlled substance produced at each plant;

(ix) The quantity (in kilograms) of class II controlled substances, the date received, and names and addresses of the source of used materials containing class II controlled substances which are recycled or reclaimed at each plant;

(x) Records of the date, the class II controlled substance, and the estimated quantity of any spill or release of a class II controlled substance that equals or exceeds 100 pounds;

(xi) Transformation verification in the case of transformation, or the destruction verification in the case of destruction as required in paragraph (e) of this section showing that the purchaser or recipient of a class II controlled substance, in the U.S. or in another country that is a Party, certifies the intent to either transform or destroy the class II controlled substance, or sell the class II controlled substance for transformation or destruction in cases when allowances were not expended;

(xii) Written verifications from a U.S. purchaser that the class II controlled substance was exported to a Party in accordance with the requirements in this section, in cases where export production allowances were expended to produce the class II controlled substance;

(xiii) Written verifications from a U.S. purchaser that the class II controlled substance was exported to an Article 5 country in cases where Article 5 allowances were expended to produce the class II controlled substance;

(xiv) Written verifications from a U.S. purchaser that HCFC-141b was manufactured for the express purpose of meeting HCFC-141b exemption needs in accordance with information submitted under §82.16(h), in cases where HCFC-141b exemption allowances were expended to produce the HCFC-141b.

(3) For any person who fails to maintain the records required by this paragraph, or to submit the report required by this paragraph, the Administrator may assume that the person has produced at full capacity during the period for which records were not kept, for purposes of determining whether the person has violated the prohibitions at §82.15.

(c) Importers. Persons ("importers") who import class II controlled substances during a control period must comply with the following recordkeeping and reporting requirements:

(1) Reporting—Importers. For each quarter, an importer of a class II controlled substance (including importers of used class II controlled substances) must submit to the Administrator a report containing the following information:

(i) Summaries of the records required in paragraphs (c)(2)(i) through (xvi) of this section for the previous quarter;

(ii) The total quantity (in kilograms) imported of each class II controlled substance for that quarter;

(iii) The commodity code for the class II controlled substances imported, which must be one of those listed in Appendix K to this subpart;

(iv) The quantity (in kilograms) of those class II controlled substances imported that are used class II controlled substances;

(v) The quantity (in kilograms) of class II controlled substances imported for that quarter and totaled by chemical for the control period to date;

(vi) For substances for which EPA has apportioned baseline production and consumption allowances, the importer's total sum of expended and unexpended consumption allowances by chemical as of the end of that quarter;

(vii) The quantity (in kilograms) of class II controlled substances imported for use in processes resulting in their transformation or destruction;

(viii) The quantity (in kilograms) of class II controlled substances sold or transferred during that quarter to each person for use in processes resulting in their transformation or eventual destruction; and

(ix) Transformation verifications showing that the purchaser or recipient of imported class II controlled substances intends to transform those substances or destruction verifications showing that the purchaser or recipient intends to destroy the class II controlled substances (as provided in paragraph (e) of this section).

(x) [Reserved]

(xi) A list of the HCFC 141b-exemption allowance holders from whom orders were received and the quantity (in kilograms) of HCFC-141b requested and imported.

(2) Recordkeeping—Importers. An importer of a class II controlled substance (including used class II controlled substances) must maintain the following records:

(i) The quantity (in kilograms) of each class II controlled substance imported, either alone or in mixtures, including the percentage of each mixture which consists of a class II controlled substance;

(ii) The quantity (in kilograms) of those class II controlled substances imported that are used and the information provided with the petition where a petition is required under paragraph (c)(3) of this section;

(iii) The quantity (in kilograms) of class II controlled substances other than transhipments or used substances imported for use in processes resulting in their transformation or destruction;

(iv) The quantity (in kilograms) of class II controlled substances other than transhipments or used substances imported and sold for use in processes that result in their destruction or transformation;

(v) The date on which the class II controlled substances were imported;

(vi) The port of entry through which the class II controlled substances passed;

(vii) The country from which the imported class II controlled substances were imported;

(viii) The commodity code for the class II controlled substances shipped, which must be one of those listed in Appendix K to this subpart;

(ix) The importer number for the shipment;

(x) A copy of the bill of lading for the import;

(xi) The invoice for the import;

(xii) The quantity (in kilograms) of imports of used class II controlled substances;

(xiii) The U.S. Customs entry form;

(xiv) Dated records documenting the sale or transfer of class II controlled substances for use in processes resulting in their transformation or destruction;

(xv) Copies of transformation verifications or destruction verifications indicating that the class II controlled substances will be transformed or destroyed (as provided in paragraph (e) of this section).

(xvi) Written verifications from a U.S. purchaser that HCFC-141b was imported for the express purpose of meeting HCFC-141b exemption needs in accordance with information submitted under §82.16(h), and that the quantity will not be resold, in cases where HCFC-141b exemption allowances were expended to import the HCFC-141b.

(3) Petition to import used class II controlled substances and transhipment-Importers. For each individual shipment over 5 pounds of a used class II controlled substance as defined in §82.3 for which EPA has apportioned baseline production and consumption allowances, an importer must submit directly to the Administrator, at least 40 working days before the shipment is to leave the foreign port of export, the following information in a petition:

(i) The name and quantity (in kilograms) of the used class II controlled substance to be imported;

(ii) The name and address of the importer, the importer ID number, the contact person, and the phone and fax numbers;

(iii) Name, address, contact person, phone number and fax number of all previous source facilities from which the used class II controlled substance was recovered;

(iv) A detailed description of the previous use of the class II controlled substance at each source facility and a best estimate of when the specific controlled substance was put into the equipment at each source facility, and, when possible, documents indicating the date the material was put into the equipment;

(v) A list of the name, make and model number of the equipment from which the material was recovered at each source facility;

(vi) Name, address, contact person, phone number and fax number of the exporter and of all persons to whom the material was transferred or sold after it was recovered from the source facility;

(vii) The U.S. port of entry for the import, the expected date of shipment and the vessel transporting the chemical. If at the time of submitting a petition the importer does not know the U.S. port of entry, the expected date of shipment and the vessel transporting the chemical, and the importer receives a non-objection notice for the individual shipment in the petition, the importer is required to notify the Administrator of this information prior to the actual U.S. Customs entry of the individual shipment;

(viii) A description of the intended use of the used class II controlled substance, and, when possible, the name, address, contact person, phone number and fax number of the ultimate purchaser in the United States;

(ix) The name, address, contact person, phone number and fax number of the U.S. reclamation facility, where applicable;

(x) If someone at the source facility recovered the class II controlled substance from the equipment, the name and phone and fax numbers of that person;

(xi) If the imported class II controlled substance was reclaimed in a foreign Party, the name, address, contact person, phone number and fax number of any or all foreign reclamation facility(ies) responsible for reclaiming the cited shipment;

(xii) An export license from the appropriate government agency in the country of export and, if recovered in another country, the export license from the appropriate government agency in that country;

(xiii) If the imported used class II controlled substance is intended to be sold as a refrigerant in the U.S., the name and address of the U.S. reclaimer who will bring the material to the standard required under subpart F of this part, if not already reclaimed to those specifications; and

(xiv) A certification of accuracy of the information submitted in the petition.

(4) Review of petition to import used class II controlled substances and transhipments—Importers. Starting on the first working day following receipt by the Administrator of a petition to import a used class II controlled substance, the Administrator will initiate a review of the information submitted under paragraph (c)(3) of this section and take action within 40 working days to issue either an objection-notice or a non-objection notice for the individual shipment to the person who submitted the petition to import the used class II controlled substance.

(i) The Administrator may issue an objection notice to a petition for the following reasons:

(A) If the Administrator determines that the information is insufficient, that is, if the petition lacks or appears to lack any of the information required under paragraph (c)(3) of this section;

(B) If the Administrator determines that any portion of the petition contains false or misleading information, or the Administrator has information from other U.S. or foreign government agencies indicating that the petition contains false or misleading information;

(C) If the transaction appears to be contrary to provisions of the Vienna Convention on Substances that Deplete the Ozone Layer, the Montreal Protocol and Decisions by the Parties, or the non-compliance procedures outlined and instituted by the Implementation Committee of the Montreal Protocol;

(D) If the appropriate government agency in the exporting country has not agreed to issue an export license for the cited individual shipment of used class II controlled substance;

(E) If reclamation capacity is installed or is being installed for that specific class II controlled substance in the country of recovery or country of export and the capacity is funded in full or in part through the Multilateral Fund.

(ii) Within ten (10) working days after receipt of the objection notice, the importer may re-petition the Administrator, only if the Administrator indicated "insufficient information" as the basis for the objection notice. If no appeal is taken by the tenth working day after the date on the objection notice, the objection shall become final. Only one re-petition will be accepted for any original petition received by EPA.

(iii) Any information contained in the re-petition which is inconsistent with the original petition must be identified and a description of the reason for the inconsistency must accompany the re-petition.

(iv) In cases where the Administrator does not object to the petition based on the criteria listed in paragraph (c)(4)(i) of this section, the Administrator will issue a non-objection notice.

(v) To pass the approved used class II controlled substances through U.S. Customs, the petition and the non-objection notice issued by EPA must accompany the shipment through U.S. Customs.

(vi) If for some reason, following EPA's issuance of a non-objection notice, new information is brought to EPA's attention which shows that the non-objection notice was issued based on false information, then EPA has the right to:

(A) Revoke the non-objection notice;

(B) Pursue all means to ensure that the class II controlled substance is not imported into the U.S.; and

(C) Take appropriate enforcement actions.

(vii) Once the Administrator issues a non-objection notice, the person receiving the non-objection notice is permitted to import the individual shipment of used class II controlled substance only within the same control period as the date stamped on the non-objection notice.

(viii) A person receiving a non-objection notice from the Administrator for a petition to import used class II controlled substances must maintain the following records:

(A) A copy of the petition;

(B) The EPA non-objection notice;

(C) The bill of lading for the import; and

(D) U.S. Customs entry documents for the import that must include one of the commodity codes from Appendix K to this subpart.

(5) Recordkeeping for transhipments—Importers. Any person who tranships a class II controlled substance must maintain records that indicate:

(i) That the class II controlled substance shipment originated in a foreign country;

(ii) That the class II controlled substance shipment is destined for another foreign country; and

(iii) That the class II controlled substance shipment will not enter interstate commerce within the U.S.

(d) Exporters. Persons ("exporters") who export class II controlled substances during a control period must comply with the following reporting requirements:

(1) Reporting—Exporters. For any exports of class II controlled substances not reported under §82.20 (additional consumption allowances), or under paragraph (b)(2) of this section (reporting for producers of class II controlled substances), each exporter who exported a class II controlled substance must submit to the Administrator the following information within 30 days after the end of each quarter in which the unreported exports left the U.S.:

(i) The names and addresses of the exporter and the recipient of the exports;

(ii) The exporter's Employer Identification Number;

(iii) The type and quantity (in kilograms) of each class II controlled substance exported and what percentage, if any of the class II controlled substance is used;

(iv) The date on which, and the port from which, the class II controlled substances were exported from the U.S. or its territories;

(v) The country to which the class II controlled substances were exported;

(vi) The quantity (in kilograms) exported to each Article 5 country;

(vii) The commodity code for the class II controlled substances shipped, which must be one of those listed in Appendix K to this subpart;

(viii) For persons reporting transformation or destruction, the invoice or sales agreement containing language similar to the transformation verifications that the purchaser or recipient of imported class II controlled substances intends to transform those substances, or destruction verifications showing that the purchaser or recipient intends to destroy the class II controlled substances (as provided in paragraph (e) of this section).

(2) Reporting export production allowances—Exporters. In addition to the information required in paragraph (d)(1) of this section, any exporter using export production allowances must also provide the following to the Administrator:

(i) The Employer Identification Number on the Shipper's Export Declaration Form or Employer Identification Number of the shipping agent shown on the U.S. Customs Form 7525;

(ii) The exporting vessel on which the class II controlled substances were shipped; and

(iii) The quantity (in kilograms) exported to each Party.

(3) Reporting Article 5 allowances—Exporters. In addition to the information required in paragraph (d)(1) of this section, any exporter using Article 5 allowances must also provide the following to the Administrator:

(i) The Employer Identification Number on the Shipper's Export Declaration Form or Employer Identification Number of the shipping agent shown on the U.S. Customs Form 7525; and

(ii) The exporting vessel on which the class II controlled substances were shipped.

(4) Reporting used class II controlled substances—Exporters. Any exporter of used class II controlled substances must indicate on the bill of lading or invoice that the class II controlled substance is used, as defined in §82.3.

(e) Transformation and destruction. Any person who transforms or destroys class II controlled substances must comply with the following recordkeeping and reporting requirements:

(1) Recordkeeping—Transformation and destruction. Any person who transforms or destroys class II controlled substances produced or imported by another person must maintain the following:

(i) Copies of the invoices or receipts documenting the sale or transfer of the class II controlled substances to the person;

(ii) Records identifying the producer or importer of the class II controlled substances received by the person;

(iii) Dated records of inventories of class II controlled substances at each plant on the first day of each quarter;

(iv) Dated records of the quantity (in kilograms) of each class II controlled substance transformed or destroyed;

(v) In the case where class II controlled substances were purchased or transferred for transformation purposes, a copy of the person's transformation verification as provided under paragraph (e)(3)of this section.

(vi) Dated records of the names, commercial use, and quantities (in kilograms) of the resulting chemical(s) when the class II controlled substances are transformed; and

(vii) Dated records of shipments to purchasers of the resulting chemical(s) when the class II controlled substances are transformed.

(viii) In the case where class II controlled substances were purchased or transferred for destruction purposes, a copy of the person's destruction verification, as provided under paragraph (e)(5) of this section.

(2) Reporting—Transformation and destruction. Any person who transforms or destroys class II controlled substances and who has submitted a transformation verification ((paragraph (e)(3) of this section) or a destruction verification (paragraph (e)(5) of this section) to the producer or importer of the class II controlled substances, must report the following:

(i) The names and quantities (in kilograms) of the class II controlled substances transformed for each control period within 45 days of the end of such control period; and

(ii) The names and quantities (in kilograms) of the class II controlled substances destroyed for each control period within 45 days of the end of such control period.

(3) Reporting—Transformation. Any person who purchases class II controlled substances for purposes of transformation must provide the producer or importer with a transformation verification that the class II controlled substances are to be used in processes that result in their transformation.

(i) The transformation verification shall include the following:

(A) Identity and address of the person intending to transform the class II controlled substances;

(B) The quantity (in kilograms) of class II controlled substances intended for transformation;

(C) Identity of shipments by purchase order number(s), purchaser account number(s), by location(s), or other means of identification;

(D) Period of time over which the person intends to transform the class II controlled substances; and

(E) Signature of the verifying person.

(ii) [Reserved]

(4) Reporting—Destruction. Any person who destroys class II controlled substances shall provide EPA with a one-time report containing the following information:

(i) The destruction unit's destruction efficiency;

(ii) The methods used to record the volume destroyed;

(iii) The methods used to determine destruction efficiency;

(iv) The name of other relevant federal or state regulations that may apply to the destruction process;

(v) Any changes to the information in paragraphs (e)(4)(i), (ii), and (iii) of this section must be reflected in a revision to be submitted to EPA within 60 days of the change(s).

(5) Reporting—Destruction. Any person who purchases or receives and subsequently destroys class II controlled substances that were originally produced without expending allowances shall provide the producer or importer from whom it purchased or received the class II controlled substances with a verification that the class II controlled substances will be used in processes that result in their destruction.

(i) The destruction verification shall include the following:

(A) Identity and address of the person intending to destroy class II controlled substances;

(B) Indication of whether those class II controlled substances will be completely destroyed, as defined in §82.3, or less than completely destroyed, in which case the destruction efficiency at which such substances will be destroyed must be included;

(C) Period of time over which the person intends to destroy class II controlled substances; and

(D) Signature of the verifying person.

(ii) [Reserved]

(f) Heels-Recordkeeping and reporting. Any person who brings into the U.S. a rail car, tank truck, or ISO tank containing a heel, as defined in §82.3, of class II controlled substances, must take the following actions:

(1) Indicate on the bill of lading or invoice that the class II controlled substance in the container is a heel.

(2) Report within 30 days of the end of the control period the quantity (in kilograms) brought into the U.S. and certify:

(i) That the residual quantity (in kilograms) in each shipment is no more than 10 percent of the volume of the container;

(ii) That the residual quantity (in kilograms) in each shipment will either:

(A) Remain in the container and be included in a future shipment;

(B) Be recovered and transformed;

(C) Be recovered and destroyed; or

(D) Be recovered for a non-emissive use.

(3) Report on the final disposition of each shipment within 30 days of the end of the control period.

(g) HCFC 141b exemption allowances—Reporting and recordkeeping. (1) Any person allocated HCFC-141b exemption allowances who confers a quantity of the HCFC-141b exemption allowances to a producer or import and places an order for the production or import of HCFC-141b with a verification that the HCFC-141b will only be used for the exempted purpose and not be resold must submit semi-annual reports, due 30 days after the end of the second and fourth respectively, to the Administrator containing the following information:

(i) Total quantity (in kilograms) HCFC-141b received during the 6 month period; and

(ii) The identity of the supplier of HCFC-141b on a shipment-by-shipment basis during the 6 month period.

(2) Any person allocated HCFC-141b exemption allowances must keep records of letters to producers and importers conferring unexpended HCFC-141b exemption allowances for the specified control period in the notice, orders for the production or import of HCFC-141b under those letters and written verifications that the HCFC-141b was produced or imported for the express purpose of meeting HCFC-141b exemption needs in accordance with information submitted under §82.16(h), and that the quantity will not be resold.

[68 FR 2848, Jan. 21, 2003, as amended at 71 FR 41172, July 20, 2006

EPA QA requirements/guidance governing collection: Reporting and record-keeping requirements are published in 40 CFR Part 82, Subpart A, Sections 82.9 through 82.13. These sections of the Stratospheric Ozone Protection Rule specify the required data and accompanying documentation that companies must submit or maintain on-site to demonstrate their compliance with the regulations.

2c. Source Data Reporting:

Form/mechanism for receiving data and entering into EPA system: Data can be submitted on paper form or via EPA's Central Data Exchange. Complete information on reporting options/format can be found at:
http://www.epa.gov/ozone/record/index.html

Timing and frequency of reporting: Quarterly (EPA's regulations specify a quarterly reporting system for U.S. companies) and monthly (for the International Trade Commission).

Quarterly Schedule for US Companies
Quarter 1: January 1 - March 31
Quarter 2: April 1 - June 30
Quarter 3: July 1 - Sept. 30
Quarter 4: October 1 - Dec. 31

3a. Relevant Information Systems:

The Allowance Tracking System (ATS) database is maintained by the Stratospheric Protection Division (SPD). ATS is used to compile and analyze quarterly information from companies on U.S. production, imports, exports, transformations, and allowance trades of ozone-depleting substances (ODS), as well as monthly information on domestic production, imports, and exports from the International Trade Commission.

The Allowance Tracking System contains transformed data.

The Allowance Tracking System meets relevant EPA standards for information system integrity.

3b. Data Quality Procedures:

The ATS is programmed to ensure consistency of the data elements reported by companies. The tracking system flags inconsistent data for review and resolution by the tracking system manager. This information is then cross-checked with compliance data submitted by reporting companies. SPD maintains a user's manual for the ATS that specifies the standard operating procedures for data entry and data analysis.

The data are subject to an annual quality assurance review, coordinated by Office of Air and Radiation (OAR) staff separate from those on the team normally responsible for data collection and maintenance.

Regional inspectors also perform inspections and audits on-site at the producers', importers', and exporters' facilities. These audits verify the accuracy of compliance data submitted to EPA through examination of company records.

The ATS data are subject to a Quality Assurance Plan (Quality Assurance Plan, USEPA Office of Atmospheric Programs, July 2002).

3c. Data Oversight:

Branch Chief, Stratospheric Program Implementation Program, OAP, OAR

3d. Calculation Methodology:

Explanation of Calculations: Data are aggregated across all U.S. companies for each individual ODS to analyze U.S. total consumption and production.

Unit of analysis: Tons of ODP

4a. Oversight and Timing of Final Results Reporting:

Branch Chief, Stratospheric Program Implementation Program, OAP, OAR

4b. Data Limitations/Qualifications:
None, since companies are required by the Clean Air Act to report data.

4c. Third-Party Audits:
The Government Accounting Office (GAO) completed a review of U.S. participation in five international environmental agreements, and analyzed data submissions from the U.S. under the Montreal Protocol on Substances the Deplete the Ozone Layer.   No deficiencies were identified in their January 2003 report. The report may be found at the following website: http://www.gao.gov/new.items/d02960t.pdf

Measure Code: R51 - Percentage of all new single-family homes (SFH) in high radon potential areas built with radon reducing features.

Office of Air and Radiation (OAR)

Goal Number and Title:
1 - Taking Action on Climate Change and Improving Air Quality

Objective Number and Title:
2 - Improve Air Quality

Sub-Objective Number and Title:
4 - Reduce Exposure to Indoor Pollutants

Strategic Target Code and Title:
1 - By 2015, number of future premature lung cancer deaths prevented annually

Managing Office:
OAR

1a. Performance Measure Term Definitions:
New: newly constructed   Single-Family Homes: as defined by NAHB   High Radon Potential Areas (Zone 1): county average indoor radon level predicted to be 4pCi/L or greater   Radon Reducing Features: materials and techniques described in various voluntary consensus standards, primarily ASTM E1465 and ANSI-AARST RRNC 2.0   Background: ·   Historically, about 60% of the new homes built with radon-reducing features in the U.S. are built in Zone 1 areas, the highest risk areas (classified as Zone 1 by EPA). In 2010, an estimated 40% of new homes in Zone 1 were built with radon-reducing features.   ·   Please see EPA's radon website for more information: http://www.epa.gov/radon/

2a. Original Data Source:
National Association of Homebuilders (NAHB)

2b. Source Data Collection:
Calculation Methodology: The National Association of Home Builders (NAHB) Research Center conducts an annual survey of home builders in the United States, most of whom are members of the NAHB, to assess a wide range of builder practices. In January of each year, the survey of building practices for the preceding calendar year is typically mailed out to home builders. The NAHB Research Center voluntarily conducts this survey to maintain an awareness of industry trends in order to improve American housing and to be responsive to the needs of the home building industry. The annual survey gathers information such as types of houses built, lot sizes, foundation designs, types of lumber used, types of doors and windows used, etc. The NAHB Research Center Builder Survey also gathers information on the use of radon-resistant design features in new houses, and

these questions comprise about two percent of the survey questionnaire.

Quality Procedures:
According to the NAHB Research Center, QA/QC procedures have been established, which includes QA/QC by the vendor that is utilized for key entry of survey data. Each survey is manually reviewed, a process that requires several months to complete. The review includes data quality checks to ensure that the respondents understood the survey questions and answered the questions appropriately. NAHB Research Center also applies checks for open-ended questions to verify the appropriateness of the answers. In some cases, where open-ended questions request numerical information, the data are capped between the upper and lower three percent of the values provided in the survey responses.

NAHB Research Center has been conducting its annual builder practices survey for over a decade, and has developed substantial expertise in the survey's design, implementation, and analysis.

Geographical Extent: Zone 1 areas in the United States.

Spatial Detail: http://www.epa.gov/radon/zonemap.html

2c. Source Data Reporting:

Data Submission Instrument: Results are published by the NAHB Research Center in annual reports of radon-resistant home building practices. See http://www.nahbrc.org/ (last accessed 12/15/2009) for more information about NAHB. The most recent reports are "Builder Practices Report: Radon-Resistant Construction Practices in New U.S. Homes 2010." Annual reports with similar titles exist for prior years. NAHB-RC usually delivers the report to EPA in the September-October timeframe annually. Summary annual data for National and Zone 1 are entered into an internal Radon Benefit-Cost spreadsheet.

Data Entry Mechanism: Summary annual data for National and Zone 1 are entered into an internal Radon Benefit-Cost spreadsheet.

Frequency of Data Transmission to EPA NAHB-RC delivers the contracted annual report to EPA annually.

Timing of Data Transmission to EPA: NAHB-RC annual reports are delivered in September-October.

3a. Relevant Information Systems:

Radon Benefit-Cost Excel-based Spreadsheet

System Description: Excel Spreadsheet

Source/Transformed Data: N/A (not a system)

Information System Integrity Standards: N/A (not a system)

3b. Data Quality Procedures:

EPA reviews NAHB's survey methodology. EPA's project officer also quality reviews each year's draft report from the NAHB Research Center. Current report is compared to previous report, and spot checks performed on calculations and arithmetic.	

3c. Data Oversight:

Source Data Reporting Oversight Personnel:
Environmental Protection Specialist in ORIA's Indoor Environments Program

Source Data Reporting Oversight Responsibilities:
Reviews and QA's draft annual NAHB-RC report with comments and corrections to NAHB.

Information Systems Oversight Personnel:

N/A

Information Systems Oversight Responsibilities:

N/A

3d. Calculation Methodology:

Decision Rules for Selecting Data: N/A (Data presented in the NAHB report)

Definitions of Variables: See the NAHB-Research Center (RC) annual report (hard copy only)

Explanation of Calculations: The survey responses are analyzed, with respect to State market areas and Census Divisions in the United States, to assess the percentage and number of homes built each year that incorporate radon-reducing features. The data are also used to assess the percentage and number of homes built with radon-reducing features in high radon potential areas in the United States (high risk areas).

Explanation of Assumptions: See NAHB annual report

Unit of Measure: Percent of new single-family homes

Timeframe of Result: January-December, annually (for calendar year)

Documentation of Methodological Changes: N/A

4a. Oversight and Timing of Final Results Reporting:

Final Reporting Oversight Personnel:
Environmental Protection Specialist in ORIA's Indoor Environments Program

Final Reporting Oversight Responsibilities:

Oversees staff review and QA of NAHB draft and final rerp[eorts; approves public use of the data and as an input to the Radon Excel Benefit-Cost (B-C) Spreadsheet.
Final Reporting Timing: Annual in September-October
4b. Data Limitations/Qualifications:
General Limitations/Qualifications: The NAHB statistical estimates are typically reported with a 95 percent confidence interval. The majority of home builders surveyed are NAHB members, and members construct 80% of the homes built in the United States each year. The NAHB Research Center survey also attempts to capture the activities of builders that are not members of NAHB. Home builders that are not members of NAHB are typically smaller, sporadic builders that in some cases build homes as a secondary profession. To augment the list of NAHB members in the survey sample, NAHB Research Center sends the survey to home builders identified from mailing lists of builder trade publications, such as Professional Builder magazine. There is some uncertainty as to whether the survey adequately characterizes the practices of builders who are not members of NAHB. The effects on the findings are not known. For the most-recently completed survey 2010, NAHB Research Center reported mailing the survey to about 20,000 active United States home-building companies, and received about 1,400 responses, which translates to a response rate of about 7 percent. Although an overall response rate of 7 percent could be considered low, it is the response rate for the entire survey, of which the radon-resistant new construction questions are only a very small portion. Builders responding to the survey would not be doing so principally due to their radon activities. Thus, a low response rate does not necessarily indicate a strong potential for a positive bias under the speculation that builders using radon-resistant construction would be more likely to respond to the survey. NAHB Research Center also makes efforts to reduce the potential for positive bias in the way the radon-related survey questions are presented. Data Lag Length and Explanation: The annual results for any given year are tabulated and delivered to EPA within about 9 months, of the end of the calendar year, i.e., 2010 results were delivered to EPA in October 2011. Methodological Changes: N/A
4c. Third-Party Audits:
N/A

Measure Code: R50 - Percentage of existing homes with an operating radon mitigation system compared to the estimated number of homes at or above EPA's 4pCi/L action level.

Office of Air and Radiation (OAR)

Goal Number and Title:
1 - Taking Action on Climate Change and Improving Air Quality
Objective Number and Title:
2 - Improve Air Quality
Sub-Objective Number and Title:
4 - Reduce Exposure to Indoor Pollutants
Strategic Target Code and Title:
1 - By 2015, number of future premature lung cancer deaths prevented annually
Managing Office:
OAR
1a. Performance Measure Term Definitions:
Cumulative Number: The estimated number of annual net mitigations are summed; from 1986. Existing: New and existing homes Homes: Only individual dwellings are counted to be consistent with the segment of the housing stock delineated in the 1992 EPA Technical Support Document (TSD) for the May 1992 Citizen's Guide to Radon (EPA 400-R-92-011; http://www.epa.gov/nscep/index.html The universe of homes potentially having a radon level of 4pCi/L or more is derived from Census housing data per the TSD criteria. Operating Mitigation System: Defined as a dwelling with a mitigation system that includes an operating radon vent fan. Radon vent fans are presumed to have an average useful life of 10 years. EPA's 4pCi/L* Action Level: Established in 1992 pursuant to publication of the 1992 TSD, EPA Science Advisory Board review, and codified in A Citizen's Guide to Radon: The Guide To Protecting Yourself And Your Family From Radon. EPA and the US Surgeon General: (1) strongly recommend that a home be fixed/mitigated when a radon level of 4pCi/L or more is measured; and (2) occupants consider mitigation when the radon level is between 2-4 pCi/l. EPA's estimate of the 21,100 radon-related lung cancer deaths is based on a long-term exposure to 1.25 pCi/L; the average indoor level in US homes. Background: ·   This performance measure can include existing and new homes. The bulk of the data are applicable to existing homes. Some new home builders are preemptively including radon vent fans in their mitigation systems at the time of construction (primarily) for homes in Zone 1. ·   Please see EPA's radon website for more information: http://www.epa.gov/radon/
2a. Original Data Source:

| | · Manufacturers of radon venting (vent) fans that are used in mitigation systems. |
| | · US Census Bureau, for data on number of homes |

2b. Source Data Collection:

Vent fan manufacturers tabulate and voluntarily provide their annual sales data to EPA. EPA treats the sales data as CBI. All data is rolled up into a single number. That number is adjusted for several assumptions, including a useful life of 10 years, and one fan per dwelling. That adjusted number is then applied to the Radon Benefit-Cost Spreadsheet.

The US Census Bureau is the housing data source from which the number of dwellings that should test for radon is estimated (e.g., 100). That number (100) is adjusted for the Technical Support Document (TSD based assumption that 1 in 15 homes will likely have a radon level of 4 pCi/L or more (e.g., about 7 in every 100 dwellings).

2c. Source Data Reporting:

Data Submission Instrument: Manufacturers voluntarily report data annually to EPA. Manufacturers provide the data once a year, typically in January-February. Data are submitted via an email. After review the data are summed and entered into the Radon Benefit-Cost (B-C) Spreadsheet.

· US Census Bureau publishes the American Housing Survey for the United States; http://www.census.gov/housing/ahs/

Data Entry Mechanism: Information from the manufacturers are entered into ORIA's Radon Benefit/Cost spreadsheet

Frequency of Data Transmission to EPA-Annually

Timing of Data Transmission to EPA- January-February

3a. Relevant Information Systems:

System Description: Information/Data are maintained on/in an internal OAQ/ORIA/IED excel spreadsheet (.xls);

Source/Transformed Data: Yes-excel information/data

Information System Integrity Standards: n/a (a spreadsheet, not a system)

3b. Data Quality Procedures:

EPA receives the manufacturer sales data as provided and has no way of determining whether its quality is other than that attested to by the provider.

3c. Data Oversight:

Source Data Reporting Oversight Personnel:

Environmental Protection Specialist within ORIA's Center for Radon and Air Toxics

Source Data Reporting Oversight Responsibilities:
Contract execution and management; review and comment/correct draft report and submit to NAHB for final draft.

Information Systems Oversight Personnel:
Director, Center for Radon and Air Toxics

Information Systems Oversight Responsibilities:

Assure acceptable quality of final NAHB report

3d. Calculation Methodology:

Decision Rules for Selecting Data: No single data point; several data points drawn from census American Housing Survey (AHS tabulations to construct the Technical Support Document (TSD) equivalent housing population. See attachment A (Analytical Procedures for radon Risk Reduction and Cost-Effectiveness Estimates, March 2010).

Definitions of Variables: See Attachment A.(Analytical Procedures for Radon Risk Reduction and Cost-Effectiveness Estimates (March 2010, unpublished internal EPA document).

Explanation of Calculations: EPA compares the net number of existing homes in a given year that have been mitigated to the total of homes estimated to require mitigation because they equal or exceed the EPA action level of 4pCi/L. The annual homes mitigated number is added to the previous year's cumulative total.

The calculation of the number of homes across the country at or above EPA's 4pCi/L action level is based on methodology in the 1992 technical support document for radon (internal document available upon request) and current census data.

Explanation of Assumptions: When estimating the number of new radon mitigations annually, the data from fan manufacturers is adjusted based on several assumption: (1) that previously-installed radon mitigation systems will require a fan replacement every ten years; (2) only homes at or above the action level are mitigated; (3) only there is one vent fan is used per dwelling; and (4) all vent fans are used for radon.

Unit of Measure: Existing homes mitigated (<4pCi/L) as a percent of those homes that should mitigate (at/equal to 4 pCi/L)

Timeframe of Result: [January-February]

Documentation of Methodological Changes: N/A
Attached Documents:
Attachment A R50 Radon Cost-Effectiveness Analysis.pdf

4a. Oversight and Timing of Final Results Reporting:

Final Reporting Oversight Personnel:

Environmental Protection Specialist in Center for Radon and Air Toxics

Director, Center for Radon and Air Toxics

Final Reporting Oversight Responsibilities:

Review and accept final report.

Enter selected data points into the Radon Benefit-Cost Excel Spreadsheet

Final Reporting Timing:

Final report I delivered in September-October

4b. Data Limitations/Qualifications:

General Limitations/Qualifications: Reporting by radon fan manufacturers is voluntary and may underestimate the number of radon fans sold. Nevertheless, these are the best available data to determine the number of homes mitigated. There are other methods to mitigate radon including: passive mitigation techniques of sealing holes and cracks in floors and foundation walls, installing sealed covers over sump pits, installing one-way drain valves in un-trapped drains, and installing static venting and ground covers in areas like crawl spaces. Because there are no data on the occurrence of these methods, there is again the possibility that the number of radon mitigated homes has been underestimated.

No radon vent fan manufacturer, vent fan motor maker or distributor is required to report to EPA; they provide data/information voluntarily to EPA. There are only four (4) major radon vent fan manufacturers; one of these accounts for an estimated 70% of the market. Radon vent fans are likely to be rarely used for non-radon applications. However, vent fans typically used for non-radon applications are perhaps being installed as substitutes for radon vent fans in some instances, but this is estimated to be less than 1% of the total market. Ascertaining the actual number of radon vent fans used for other applications, and the number of non-radon fans being substituted in radon applications, would be difficult and expensive at this time relative to the benefit of having such data.

Data Lag Length and Explanation: vent fan manufacturers provide sales data to EPA in Jan-Feb timeframe for previous year.

Methodological Changes: N/A

4c. Third-Party Audits:

There are no third party audits for this measure or its inputs.

Measure Code: R37 - Time to approve site changes affecting waste characterization at DOE waste generator sites to ensure safe disposal of transuranic radioactive waste at WIPP.

Office of Air and Radiation (OAR)

Goal Number and Title:
1 - Taking Action on Climate Change and Improving Air Quality

Objective Number and Title:
4 - Reduce Unnecessary Exposure to Radiation

Sub-Objective Number and Title:
1 - Prepare for Radiological Emergencies

Strategic Target Code and Title:
1 - Through 2015, EPA will maintain a level of readiness of radiation program personnel

Managing Office:
OAR

1a. Performance Measure Term Definitions:
Days to Approve: EPA will measure the time between the Department of Energy (DOE) request for approval/notification of change (or the date of the inspection, if applicable) to the date of EPA approval, disapproval or concurrence of the change. Under the requirements of 40 CFR Part 194.8, EPA will perform a baseline inspection of each DOE waste generator site. If all requirements are met, EPA will approve the site's waste characterization program and assign tiers, based on abilities demonstrated during the baseline inspection. The tiering protocol, which applies to waste streams, equipment, and procedures, will require DOE to either notify EPA of changes to the waste characterization program (that can affect the quality of the data required by EPA to ensure the disposal regulations are met) prior to implementation of the change (Tier 1) or to notify EPA of the changes upon implementation (Tier 2). For Tier 1 changes, EPA may request additional information or conduct an inspection prior to issuing a decision. Elapsed time is measured from the EPA evaluation of a complete submission to the date the approval/disapproval is signed.   Site Changes Affecting Waste Characterization: When a DOE site is approved a tiering table is provided by EPA detailing when and how changes to approved systems will be reported and approved/disapproved by EPA.   DOE Waste Generator Sites: Sites where DOE transuranic waste, eligible for WIPP disposal, is generated (e.g. Hanford, Idaho National Lab, etc.)   Compliant Disposal: Disposal of transuranic waste in compliance with 40 CFR 194 using systems approved by EPA.   Transuranic Radioactive Waste: Waste containing more than 100 nanocuries of alpha-emitting transuranic isotopes per gram of waste with half-lives greater than 20 years. TRU elements are heavier than uranium, have several isotopes, and are typically man-made.   Waste Isolation Pilot Plant (WIPP) site: Located in Carlsbad, New Mexico, is a storage site for defense-related

transuranic (TRU) nuclear waste.

Background:

·   This measure provides key information about the time required for EPA to approve the Department of Energy's (DOE's) request to dispose of transuranic waste at the Waste Isolation Pilot Plant (WIPP) site.

Find out more about the Waste Isolation Pilot Plant at http://www.wipp.energy.gov/index.htm The Department of Energy National TRU Waste Management Plan Quarterly Supplement http://www.wipp.energy.gov/shipments.htm contains information on the volumes of waste that are received at the DOE WIPP.

2a. Original Data Source:

Original Data Source: EPA Office of Radiation and Indoor Air

2b. Source Data Collection:

Source Data Collection Methodology and Quality Procedures – Tabulation of records or activities: The example below is excerpted from the excel spreadsheet that is used. The first column lists the activity, the second the complete submission date and the third the signed approval date and the fourth, the number of elapsed days.

The submission date is recorded as soon as the submission is complete. The submission by DOE is considered complete when it meets EPA's criteria for general submissions and the specific requirements necessary for the type of request being submitted for approval.

FY 2011

Activity

Complete Submission Date

Signed Approval Date

Elapsed Days

NRF -INL RH

8/16/2010

11/1/2010

77

INL TRA Sludge RH

8/4/2010

11/1/2010

89

ANL FEW RH

8/28/2010

11/22/2010

86

INL HFEF 4A RH

1/10/2011

3/23/2011

72

HANFORD SHENC
4/1/2011
5/11/2011
40
Bettis RH BL
2/25/2011
5/23/2011
87
SRS RH Sabotage
4/18/2011
6/7/2011
50
WAGS INL Cd
6/14/2011
6/27/2011
13
Sandia BL Pro
7/2/2011
9/6/2011
66

Average
64

2c. Source Data Reporting:

Source Data Reporting – Data Submission Instrument: EPA inspection team's Baseline inspection findings.

Source Data Reporting – Data Entry Mechanism: Data (dates) are entered by the inspection team into the Excel spreadsheet on the WIPP share drive. The dates are determined from correspondence and, as necessary, from the date of additional data submission or final issue resolution.

Source Data Reporting – Frequency of Data Transmission to EPA: The frequency of data generation depends on DOE requests for approval or notification and inspections conducted.

Source Data Reporting – Timing of Data Transmission to EPA: The dates are determined from correspondence and, as necessary, from the date of additional data submission or final issue resolution.

3a. Relevant Information Systems:

Relevant Information Systems – System Description: Internal database stored on the share drive in the WIPP/PART directory and used by the inspection team to calculate measure data. The relevant correspondences are docketed in the EPA Air Docket and are public record. The dates are taken directly from that correspondence. The complete submission date is taken from either e-mail of final document/issue resolution or from the date a data disk is received by EPA with the information requested. No additional QA/QC is needed as the result is an Excel-performed mathematical subtraction of those dates and then an average is generated, reported, entered into the Excel spreadsheet and the file is saved and stored in the WIPP/PART directory on the share drive.

Relevant Information Systems – Source/Transformed Data: Data are drawn from DOE/EPA correspondence and from dates of data submission or issue (waste characterization inspection team's concern and finding) resolution.

Relevant Information Systems – Information System Integrity Standards: Data are backed up regularly by IT staff.

3b. Data Quality Procedures:

Data Quality Procedures: Quality assurance and quality control procedures will follow Agency guidelines and be consistent with EPA Office of Radiation and Indoor Air Quality Management Plan. The relevant correspondences are docketed in the EPA Air Docket and are public record. The dates are taken directly from that correspondence. The complete submission date is taken from either e-mail of final document/issue resolution or from the date a data disk is received by EPA with the information requested. No additional QA/QC is needed as the result is an Excel-performed mathematical subtraction of those dates and then an average is generated and reported.

3c. Data Oversight:

Source Data Reporting of EPA Oversight Personnel: EPA Waste Isolation Pilot Plant (WIPP) Waste Characterization Inspection Team members located in the Office of Radiation and Indoor Air, Radiation Protection Division's Center for Waste Management and Regulation.

Source Data Reporting of EPA Oversight Responsibilities: EPA's WIPP Waste Characterization Inspection Team measures the time between the Department of Energy (DOE) request for approval/notification of change (or the date of the inspection, if applicable) to the date of EPA approval, disapproval or concurrence of the change. Under the requirements of 40 CFR Part 194.8, EPA's Waste Characterization Inspection Team performs a baseline inspection of each DOE waste generator site. If the Inspection Team determines that all requirements are met, EPA approves the site's waste characterization program and assigns tiers, based on abilities demonstrated during the baseline inspection. The tiering protocol, which applies to waste streams, equipment and procedures, requires DOE to either notify EPA of changes to the waste characterization program (that can affect the quality of the data required by EPA to ensure the disposal regulations are met) prior to implementation of the change (Tier 1) or to notify EPA of the changes upon implementation (Tier 2). For Tier 1 changes, EPA may request additional information or conduct an inspection prior to issuing an approval. Elapsed time is measured from the EPA determination of a complete submission to the date EPA signs the approval.

Information Systems of EPA Oversight Personnel: EPA's WIPP Waste Characterization Inspection Team submits the appropriate documentation to the Director of the Center for Waste Management and Regulations for review and concurrence. Once the Center Director formally concurs, the package is delivered to the Deputy Director of the Radiation Protection Division for review, approval and signature. Upon signature, the transmittal letter is dated and the final letter and inspection report are distributed electronically and mailed via the U.S. postal service. EPA then files the complete documentation in the Agency's Air Docket so it is available for public review.

Information Systems of EPA Oversight Responsibilities: Not Applicable.

3d. Calculation Methodology:

Calculation Methodology – Decision Rules for Selecting Data: Activities used are Tier 1 items submitted by the DOE that are completed within the fiscal year of interest.

Calculation Methodology – Definitions of Variables: Not Applicable.

Calculation Methodology – Explanation of Calculations: EPA will measure the time between the DOE request for approval/notification of change (or the date of the inspection, if applicable) to the date of EPA approval, disapproval or concurrence of the change. As stated previously, the dates are determined from correspondence and as necessary from the date of additional data submission or final issue resolution.

Calculation Methodology – Explanation of Assumptions: Not Applicable.

Calculation Methodology – Unit of Measure: Time to approve site changes affecting waste characterization at DOE waste generator sites to ensure safe disposal of transuranic radioactive waste at WIPP measured as percentage reduction from the 2004 baseline of 150 days.

Calculation Methodology – Timeframe of Result: Fiscal Year containing the date of EPA approval, disapproval or concurrence.

Calculation Methodology – Documentation of Methodological Changes: Not Applicable.

4a. Oversight and Timing of Final Results Reporting:

Planning Officer, Office of Radiation and Indoor Air

4b. Data Limitations/Qualifications:

Data Limitations/Qualifications: Not Applicable.

Data Lag Length and Explanation: Not Applicable.
Methodological Changes: Not Applicable.
4c. Third-Party Audits:
Third Party Audits: Not Applicable.

Measure Code: R35 - Level of readiness of radiation program personnel and assets to support federal radiological emergency response and recovery operations.

Office of Air and Radiation (OAR)

Goal Number and Title:	
1 - Taking Action on Climate Change and Improving Air Quality	
Objective Number and Title:	
4 - Reduce Unnecessary Exposure to Radiation	
Sub-Objective Number and Title:	
1 - Prepare for Radiological Emergencies	
Strategic Target Code and Title:	
1 - Through 2015, EPA will maintain a level of readiness of radiation program personnel	
Managing Office:	
OAR	
1a. Performance Measure Term Definitions:	

Level of Readiness: A score indicating the percent (0-100%) of criteria met from a comprehensive list of requirements needed for support of federal radiological emergency response and recovery operations.

Radiation Program: National Air and Radiation Environmental Laboratory, Radiation and Indoor Environments National Laboratory, and Radiation Protection Division of the Office of Radiation and Indoor Air.

Personnel: EPA employees in the three locations listed above who are members of the Radiological Emergency Response Team.

Assets: Equipment and vehicles in the three locations listed above which are utilized as part of Radiological Emergency Response Team activities.

Support: Activities performed by EPA as part of the federal response to a radiological emergency.

Federal Radiological Emergency Response and Recovery Operations: Federal activities addressing the inadvertent release of radioactive material, not including terrorism incidents.

Background:
Radiological Emergency Response Measurement Implementation Plan: Long-Term Outcome Performance Measure, Readiness.

2a. Original Data Source:	
Original Data Source: EPA Office of Radiation and Indoor Air	
2b. Source Data Collection:	

Source Data Collection Methodology and Quality Procedures: EPA developed standardized criteria for readiness levels based on the functional requirements identified in the National Response Framework's (NRF) Nuclear/Radiological Incident Annex and the National Oil and Hazardous Substances Pollution Contingency

Plan (NCP). A baseline analysis for the Radiological Emergency Response Team (RERT) was performed in 2005 for EPA Headquarters and is based on the effectiveness of the RERT during incidents and national exercises.

An evaluation panel consisting of three representatives from the Radiological Emergency Response Team (RERT), one from each Office of Radiation and Indoor Air (ORIA) Laboratory and one from ORIA Headquarters, and ORIA management representatives annually perform a critical evaluation of ORIA's Radiological Emergency Response Program's capabilities versus the standardized criteria, resulting in an overall annual percentage score, as well as component percentage scores. Representatives are not involved in the evaluation of their own location. Members are chosen based on volunteerism and by lottery on an annual basis. The Panel is chaired by the non-RERT management representative.

There are ten elements to the score and each element is comprised of a number of criteria. These criteria are scored from 0-3 points. For the final score, the total received number of points is divided by the total possible number of points to calculate a percentage score over all elements. The criteria may be modified from year to year as operational requirements change for emergency response.

For FY 2014, it is anticipated that the ten elements will be:
1.   Incident Notification, Mobilization and Management
2.   Special Teams Coordination
3.   Professional Development, Training and Exercises
4.   Health and Safety
5.   Public Information and Community Involvement
6.   Field Capabilities
7.   Information and Data Management
8.   Emergency Response and Preparedness Outreach
9.   Law Enforcement Operations and Forensic Evidence Collection
10.   Acquisition Management

A full list of criteria may be obtained from the Office of Emergency Management. (Contact: Bill Finan at finan.bill@epa.gov)

2c. Source Data Reporting:

Source Data Reporting – Data Submission Instrument: The original data are reviewed by the Core National Approach to Response representatives at each of the three locations for completeness and accuracy before reporting to the Washington, D.C. Core National Approach to Response representative. The Washington D.C. Core National Approach to Response representative reviews the combined data for completeness and accuracy before submitting the data to the Office of Emergency Management.

Data Entry Mechanism: The Office of Emergency Management reviews the Office of Radiation and Indoor Air's Radiological Emergency Response Team's data submission and retains the relevant documentation for the Agency.

Frequency of Data Transmission to EPA: Readiness is measured annually. Scoring criteria are made available by 1st Quarter CY. The Office of Radiation and Indoor Air develops responses. The joint meeting of the special teams to alter the scores based on peer assessment and additional documentation is usually held in August. The final scores are available no later than September.

Timing of Data Transmission to EPA: Responses are submitted to the Office of Emergency Management. Using both those responses and the scoring criteria, OEM develops an initial score. The scores for all special teams are shared at a joint meeting, during which the special teams have an opportunity to alter scores (increase or decrease) based on peer assessment and additional supporting documentation. After the meeting, the Office of Emergency Management develops a final score.

3a. Relevant Information Systems:

Relevant Information Systems – System Description: Emergency Management Portal – Field Readiness (https://emp.epa.gov/ Data on personnel training and exercise participation are tracked in this on-line password-protected database maintained by the Office of Emergency Management. Personnel data are entered by the training coordinator or similar position in each of the three locations and the data are quality assured on an annual basis. In addition, personnel have access to review and update their personal information at any time. Personnel training and exercise participation represent a subset of the Core National Approach to Response criteria.

Source/Transformed Data: The original data are reviewed by the Core National Approach to Response representatives at each of the three locations for completeness and accuracy before reporting to the Washington, D.C. Core National Approach to Response representative. The Washington D.C. Core National Approach to Response representative reviews the combined data for completeness and accuracy before submitting the data to the Office of Emergency Management.

Information System Integrity Standards: Office of Emergency Management Information Management Team Lead, Field Readiness Project Officer and Team Personnel. (Contact: Josh Woodyard at woodyard.joshua@epa.gov). ORIA uses the Office of Emergency Management's Emergency Management (EM) Portal which complies with both the Agency's Information System Integrity Standards and Web governance standards and policies.

3b. Data Quality Procedures:

Data Quality Procedures: Results are based on answers provided by subject matter experts in three locations. It is anticipated that the subject matter experts preparing the responses are the best qualified individuals within each location to make a judgment as to the nature of their responses. Data quality is certified by the Laboratory Directors at the Radiation and Indoor Environments National Laboratory and the National Air and Radiation Environmental Laboratory as well as by the Director of the Radiation Protection Division.

The original data are reviewed by the Core National Approach to Response representatives at each of the three locations for completeness and accuracy before reporting to the Washington, D.C. Core National Approach to Response representative. The Washington D.C. Core National Approach to Response representative reviews the combined data for completeness and accuracy before submitting the data to the Office of Emergency Management.

3c. Data Oversight:

Source Data Reporting Oversight Personnel: Core National Approach to Response representatives from Radiation and Indoor Environments National Laboratory, National Air and Radiation Environmental and the Radiation Protection Division.

Source Data Reporting Oversight Responsibilities: Verifying completeness and accuracy of information submitted and transferred to the Washington, D.C. Core National Approach to Response representative.

Information Systems Oversight Personnel: Data are maintained in Microsoft Word and Excel documents by Core National Approach to Response representatives.

Information Systems Oversight Responsibilities: Verifying the completeness and accuracy of the information maintained in those documents.

3d. Calculation Methodology:

Calculation Methodology – Decision Rules for Selecting Data: All data which have been collected by the Core National Approach to Response representatives are used.

Definitions of Variables: There are ten elements to the score.

For FY 2014, it is anticipated that the ten elements will be:
1. Incident Notification, Mobilization and Management
2. Special Teams Coordination
3. Professional Development, Training and Exercises
4. Health and Safety
5. Public Information and Community Involvement
6. Field Capabilities
7. Information and Data Management
8. Emergency Response and Preparedness Outreach
9. Law Enforcement Operations and Forensic Evidence Collection
10. Acquisition Management

Explanation of Calculations: Each element is comprised of a number of criteria, each of which is scored from 0-3 points, where 0 points is "does not meet criteria" and 3 points is "fully meets criteria". For the final score, the total received number of points is divided by the total possible number of points to calculate a percentage score over all elements. The criteria may be modified from year to year as operational requirements change for emergency response. See also page 3.

A full list of criteria may be obtained from the Office of Emergency Management. (Contact: Bill Finan at finan.bill@epa.gov)

Explanation of Assumptions: Not Applicable.

Identification of Unit of Measure: Percent of Radiological Emergency Response Team (RERT) members and assets that meet scenario-based response criteria

Identification of Timeframe of Result: Annual.

Documentation of Methodological Changes: Not Applicable.

4a. Oversight and Timing of Final Results Reporting:

Oversight and Timing of Final Results Reporting – Oversight Personnel: Office of Emergency Management

Roles/Responsibilities of Oversight Personnel: The scores for all special teams are shared at a joint meeting, during which the special teams have an opportunity to alter scores (increase or decrease) based on peer assessment and additional supporting documentation. After the meeting, the Office of Emergency Management develops a final score.

Final Reporting Timing: Annual.

4b. Data Limitations/Qualifications:

Data Limitations/Qualifications: Results are based on answers provided by subject matter experts in three locations. It is anticipated that the subject matter experts preparing the responses are the best qualified individuals within each location to make a judgment as to the nature of their responses.

In the absence of a radiological emergency, this score is considered a good method for assessing emergency response readiness; however, unanticipated factors may affect actual readiness, which are not covered by the score. In the event of a radiological emergency, a comprehensive lessons-learned assessment is conducted and may inform future scoring criteria to account for additional factors that affected readiness.

Data Lag Length and Explanation: Not Applicable.

Methodological Changes: Not Applicable.

4c. Third-Party Audits:

Third Party Audits: Not Applicable.

Measure Code: R17 - Additional health care professionals trained annually on the environmental management of asthma triggers.

Office of Air and Radiation (OAR)

Goal Number and Title:
1 - Taking Action on Climate Change and Improving Air Quality

Objective Number and Title:
2 - Improve Air Quality

Sub-Objective Number and Title:
4 - Reduce Exposure to Indoor Pollutants

Strategic Target Code and Title:
2 - By 2015, reduce exposure to indoor environmental asthma triggers

Managing Office:
OAR

1a. Performance Measure Term Definitions:
Additional: The increment added annually, above the baseline of zero in 2004 (when this measure was adopted).
Health care professionals: Professionally credentialed health care providers delivering care and services to people with asthma (e.g. physicians, physician assistants, nurses, respiratory therapists).
Trained by EPA and partners: Training is defined by health care industry accrediting standards (e.g. CEU, CNE, CME) to be of sufficient quality and duration so as to improve provider knowledge and skills.
Environmental management: One of the 4 components of comprehensive asthma care as defined in the National Guidelines for the Diagnosis and Management of Asthma. Environmental management is the avoidance of asthma triggers either through source control activities (e.g. smoke-free homes and cars), behavior changes (e.g. weekly washing of bedding to reduce dust mite exposure) or prevention practices (e.g. fixing leaks to prevent mold growth).
Asthma triggers: Allergens and irritants that make asthma worse (e.g. secondhand smoke, pet dander, ozone)
Additional background information: www.epa.gov/asthma

2a. Original Data Source:
Data is received from: EPA staff (HQ and Regions) and EPA-funded (cooperative agreements, contracts/procurements) partners (not for profit organizations at the national and local level, universities, community-based organizations).

2b. Source Data Collection:
Data is collected by EPA staff and EPA-funded partners using attendance logs from training sessions.
Data is self-report and is considered to be of sufficient quality.

2c. Source Data Reporting:
Data Submission Instrument: EPA funded partners use a reporting template, or comparable document, to report original data to the EPA project officer (see attachment). Quarterly data reporting by partners is required as a condition of the funding agreement. Data Entry Mechanism: EPA project officers manually enter partner reported data into the OAR/ORIA/IED tracking database (IAQ Impact). Frequency of Data Transmission: Funded partners are required to report quarterly. Data generated as a result of direct training from EPA staff are reported annually. Timing of Data Transmission: Funded partners must submit data 30 days after the end of the quarter. Annually, they are required to submit a report summarizing all accomplishments for the previous year; this report is due 60 days after the end of the project period. The majority of OAR/ORIA/IED and regional partner projects follow the fiscal year calendar.
3a. Relevant Information Systems:
System Description: OAR/ORIA/IED uses an online reporting system (IAQ Impact), built on an Access platform, to log results from EPA and funded partner activities. Templates in the system correspond to program work areas and sorting functions are used to generate reports for specific indicators (e.g. health care professionals trained). Source/Transformed Data: Source data only. Information System Integrity Standards: N/A
3b. Data Quality Procedures:
All data is self report and is assessed to be of sufficient quality. Project officers review data and project reports, conduct meetings with partners to review progress, and conduct formal project reviews as required grants/contracts management.
3c. Data Oversight:
Source Data Reporting Oversight Personnel: OAR/ORIA/IED program office project officers and Regional Air Program project officers. Source Data Reporting Oversight Responsibilities: Check grantee reported data against proposed or target results. Information Systems Oversight Personnel: OAR/ORIA/IED work assignment manager Information Systems Oversight Responsibilities: manage support contract personnel who maintain IAQ Impact; give technical direction for changes to tracking database (e.g. new data fields to accommodate new project outputs/outcomes).

3d. Calculation Methodology:
Decision Rules for Selecting Data: Data are selectedfor simple summation based on coded entries in IAQ Impact (AP2 is the code designation). Definitions of Variables: Not applicable Explanation of Calculations: simple sum Explanation of Assumptions: Not applicable Unit of Measure: persons Timeframe of Result: fiscal year Documentation of Methodological Changes: Not applicable

4a. Oversight and Timing of Final Results Reporting:
Final Reporting Oversight Personnel: Division Director Final Reporting Oversight Responsibilities: Reviews and submits final report to ORIA Program Management Office Final Reporting Timing: standard annual frequency

4b. Data Limitations/Qualifications:
General Limitations/Qualifications: Data limitations are those inherent with self-reporting and are largely thought to be under-reporting of number of professionals trained as a result of attendees failing to sign the attendance log for in person trainings. For online training, data is captured from electronic sign-in that is a requirement for attending the course, so this limitation is averted. Data Lag Length and Explanation: N/A Methodological Changes: N/A

4c. Third-Party Audits:
OMB PART

Office of Air and Radiation (OAR)

Goal Number and Title:
1 - Taking Action on Climate Change and Improving Air Quality
Objective Number and Title:
2 - Improve Air Quality
Sub-Objective Number and Title:
1 - Reduce Criteria Pollutants and Regional Haze
Strategic Target Code and Title:
3 - By 2015, reduce emissions of nitrogen oxides (NOx)
Managing Office:
Office of Transportation and Air Quality
1a. Performance Measure Term Definitions:
Mobile sources: Includes onroad cars/trucks, nonroad engines such as farm/construction, locomotives, commercial marine and aircraft.   Nitrogen oxide: NO2 (nitrogen dioxide) is a combustion product formed from the reaction of nitrogen (in the ambient air) and fuel (gasoline, diesel fuel, - or - for stationary sources - coal) as defined by the EPA National Ambient Air Quality Standard and measurement methods.
2a. Original Data Source:
Estimates for on-road and off-road mobile source emissions are built from inventories fed into the relevant models.   Data for the models are from many sources, including Vehicle Miles Traveled (VMT) estimates by state (Federal Highway Administration), the mix of VMT by type of vehicle (Federal Highway Administration), temperature, gasoline properties, and the designs of Inspection/Maintenance (I/M) programs.
2b. Source Data Collection:
Source Data Collection Methods: Emission tests for engines/vehicles come from EPA, other government agencies (including state/local governments), academic institutions and industry. The data come from actual emission tests measuring HC (HydroCarbon), CO (Carbon Monoxide), NOx (Nitrogen Oxides), and PM (Particulate Matter). It is important to note that total oxides of nitrogen (NO and NO2) are both measured with emission standards applying to the sum of both oxides. Usage survyes for vehicle miles traveled are obtained from DOT surveys and fuel usage for nonroad vehicles/engines are obtained from a variety of sources such as DOE.   Geographical Extent of Source Data: National   Spatial Detail Covered By the Source Data: County
2c. Source Data Reporting:

Form/mechanism for receiving data and entering into EPA system: EPA develops and receives emission data in a g/mile or g/unit Work (or unit fuel consumed) basis.

Timing and frequency of reporting: The inputs to MOVES/MOBILE 6 and NONROAD 2008 and other models are reviewed and updated, sometimes on an annual basis for some parameters. Generally, Vehicle Miles Traveled (VMT), the mix of VMT by type of vehicle (Federal Highway Administration (FHWA)-types), temperature, gasoline properties, and the designs of Inspection/Maintenance (I/M) programs are updated each year.

Emission factors for all mobile sources and activity estimates for non-road sources are revised at the time EPA's Office of Transportation and Air Quality provides new information.

Updates to the inputs to the models means the emissions inventories will change.

3a. Relevant Information Systems:

National Emissions Inventory Database. Obtained by modeling runs using MOBILE/MOVES, NONROAD, and other models.

Please see: http://www.epa.gov/ttnchie1/trends/ for a summary of national emission inventories and how the numbers are obtained in general.

The emission inventory contains source test data as well as usage information compiled from other sources. Also, for consistency from year to year and to provide a baseline over time, the emission inventories are updated for these performance measures only when it is essential to do so. The source data (emissions and usage) are "transformed" into emission inventories.

The models and input undergo peer review receiving scientific input from a variety of sources including academic institutions and public comments.

3b. Data Quality Procedures:

The emissions inventories are reviewed by both internal and external parties, including the states, locals and industries. EPA works with all of these parties in these reviews. Also EPA reviews the inventories comparing them to others derived in earlier years to assure that changes in inputs provide reasonable changes in the inventories themselves.

3c. Data Oversight:

EPA emission inventories for the performance measures are reviewed by various OTAQ Center directors in the Assessment and Standards Division. The Center Directors are responsible for vehicle, engine, fuel, and modeling data used in various EPA programs.

3d. Calculation Methodology:

Explanation of the Calculations:

EPA uses models to estimate mobile source emissions, for both past and future years. The emission inventory estimate is detailed down to the county level and with over 30 line items representing mobile sources.

The MOVES (Motor Vehicle Emission Simulator) model replacing the earlier MOBILE6 vehicle emission factor model is a software tool for predicting gram per mile emissions of hydrocarbons, carbon monoxide, oxides of nitrogen, carbon dioxide, particulate matter, and toxics from cars, trucks, and motorcycles under various conditions. Inputs to the model include fleet composition, activity, temporal information, and control program characteristics. For more information on the MOBILE6 model, please visit http://www.epa.gov/otaq/m6.htm

The NONROAD 2008 emission inventory model replacing an earlier version of NONROAD is a software tool for predicting emissions of hydrocarbons, carbon monoxide, oxides of nitrogen, particulate matter, and sulfur dioxides from small and large off road vehicles, equipment, and engines. Inputs to the model include fleet composition, activity and temporal information. For more information on the NONROAD model, please visit http://www.epa.gov/oms/nonrdmdl.htm

Over the years, improved emission and usage data have led to updated emission inventories more consistent with air quality data.

Additional information:
To keep pace with new analysis needs, new modeling approaches, and new data, EPA is currently working on a new modeling system termed the Multi-scale Motor Vehicles and Equipment Emission System (MOVES). This new system will estimate emissions for on road and off road sources, cover a broad range of pollutants, and allow multiple scale analysis, from fine scale analysis to national inventory estimation. When fully implemented, MOVES will serve as the replacement for MOBILE6 and NONROAD. The new system will not necessarily be a single piece of software, but instead will encompass the necessary tools, algorithms, underlying data and guidance necessary for use in all official analyses associated with regulatory development, compliance with statutory requirements, and national/regional inventory projections. Additional information is available on the Internet at http://www.epa.gov/otaq/ngm.htm

Unit of analysis: tons of emissions, vehicle miles traveled and hours (or fuel) used

4a. Oversight and Timing of Final Results Reporting:
The director for Health Effects, Toxics and Benefits Center, Director of the Air Quality and Modeling Center and the Associate Director of the Assessment and Standards Division are ultimately responsible for the performance measures. These individuals, as well as the other Center Directors, are responsible for assuring that the emission inventory and reduction numbers used in EPA regulatory and other programs are accurate and have obtained extensive academic, public and other review.]
4b. Data Limitations/Qualifications:
The limitations of the inventory estimates for mobile sources come from limitations in the modeled emission factors (based on emission factor testing and models predicting overall fleet emission factors in g/mile) and also in the estimated vehicle miles traveled for each vehicle class (derived from Department of Transportation data)..

For nonroad emissions, the estimates come from a model using equipment populations, emission factors per hour or unit of work, and an estimate of usage. This nonroad emissions model accounts for over 200 types of nonroad equipment. Any limitations in the input data will carry over into limitations in the emission inventory estimates.

Additional information about data integrity for the MOVES/MOBILE6 and NONROAD models is available on the Internet at http://www.epa.gov/otaq/m6.htm and http://www.epa.gov/oms/nonrdmdl.htm respectively.

When the method for estimating emissions changes significantly, older estimates of emissions in years prior to the most recent year are usually revised to avoid a sudden discontinuity in the apparent emissions trend may be revised to be consistent with teh new methodology when possible.

Methods for estimating emission inventories are frequently updated to reflected the most up-to-date inputs and assumptions. Past emission estimates that inform our performance measures frequently do not keep pace with the changing inventories associated with more recent EPA rulemakings. EPA developed the initial numbers for these perfromance measures in 2002, making both current and future year projections for on-road and nonroad. The emission estimates have been updated numerous times since then for rulemaking packages and will be updated for these performance measures.

4c. Third-Party Audits:

All of the inputs for the models, the models themselves and the resultant emission inventories are reviewed as appropriate by academic experts and also by state and local governments which use some of this information for their State Implementation Plans to meet the National Ambient Air Quality Standards.

Office of Air and Radiation (OAR)

Goal Number and Title:
1 - Taking Action on Climate Change and Improving Air Quality

Objective Number and Title:
1 - Address Climate Change

Sub-Objective Number and Title:
1 - Address Climate Change

Strategic Target Code and Title:
0 -

Managing Office:

1a. Performance Measure Term Definitions:
An Annual Greenhouse Gas (GHG) Emission Report is "verified" when it is absent of any substantive errors identified by EPA's electronic verification system or after any substantive error has been addressed with a revised report which corrects the error or an acceptable explanation for why it is not an error.

A "substantive error" is an error that impacts the quantity of GHG emissions reported or otherwise prevents the reported data from being validated or verified. |

2a. Original Data Source:
EPA's Greenhouse Gas Reporting Program (GHGRP) collects annual GHG emission reports from over 8,000 reporters including facilities in nine industry groups (across 41 source categories) that directly emit large quantities of GHGs, as well as suppliers of certain fossil fuels and industrial gases including: Power Plants, Petroleum and Natural Gas Systems, Refineries, Chemicals, Waste, Metals, Minerals, Pulp and Paper.

2b. Source Data Collection:
These reports are collected electronically through EPA's electronic GHG reporting tool ("e-GGRT") which supports an integrated verification program involving EPA subject matter experts (SMEs) that electronically runs the reports against approximately 4,000 verification checks. These verification checks include range checks, algorithm checks, completeness checks, and statistical checks to ensure that the reports are complete, consistent and accurate. The annual reporting deadline is March 31 and EPA's performance measure is to verify 95% of the reports by 150 days after the reporting deadline.

Facilities subject to EPA's GHGRP are required to monitor and report GHG emissions data in accordance with the requirements prescribed under 40 CFR part 98. These include applicability requirements, monitoring and QA/QC requirements, prescribed calculation methodologies and reporting requirements.

Since reporters are conducting the actual sampling and non-direct measurements, EPA relies on the reporter to adhere to the particular standard(s) and calibration procedures required for its industry in 40 CFR part 98. Any technical system assessments and performance audits that are included in the particular standard(s) and |

calibration procedures must be followed as well. EPA requires reporters to include information about the standard(s) used in their GHG report submission.

2c. Source Data Reporting:

Data Submission Instrument: EPA's electronic Greenhouse Gas Reporting Tool (e-GGRT)

EPA's electronic Greenhouse Gas Reporting Tool (e-GGRT) supports approximately 8,000 facilities and suppliers across the US in satisfying their annual requirement to electronically report GHG data to EPA under 40 CFR Part 98 (Mandatory Reporting of Greenhouse Gases Rule) which is implemented under EPA's Clean Air Act authorities. Reporting must be done electronically and e-GGRT provides a user-friendly comprehensive web-based platform for thousands of facilities across 41 subparts (industrial source categories) to conveniently, accurately and efficiently submit detailed GHG data to EPA. e-GGRT includes hundreds of real-time data quality checks that provide data quality feedback to reporters before they submit their data to EPA. The URL for e-GGRT can be found at https://ghgreporting.epa.gov/ Data reporting forms can be found at http://ccdsupport.com/confluence/display/help/Reporting+Form+Instructions

The XML reporting schema can be found at
http://ccdsupport.com/confluence/display/help/XML+Reporting+Instructions

Data Entry Mechanism: Users enter data directly into e-GGRT web forms, or, alternatively, upload standardized XLS data reporting forms or upload XML files that conform to the e-GGRT XML reporting schema.

Frequency of Data Transmission to EPA: Reporters report annually via e-GGRT, re-submittals (corrections) can be provided at any time.

Timing of Data Transmission to EPA: Reporters must sign, certify and submit annual GHG data for the prior year to EPA no later than March 31st.

3a. Relevant Information Systems:

e-GGRT, EPA's electronic Greenhouse Gas Reporting Tool at https://ghgreporting.epa.gov/ is the primary, stand-alone information management system used by EPA to support the collection, verification and publication of GHG data under this measure. It includes front-end reporting and real-time data quality feedback functionality (as described under 2c) as well as back end compliance and verification support. e-GGRT maintains the master data store (Source data). e-GGRT includes two service calls to CDX, the Agency's central data exchange, one for user authentication and the other for CROMERR. These service calls are seamless to the user, i.e. there is no need for the user to log-in to or enter CDX. The entire user experience is handled by e-GGRT. e-GGRT supports EPA's target enterprise architecture, it complies with Agency and federal security regulations; information is validated with XML schemas and business rules prior to the system accepting that information; uses agency standard software such as Java, HTML, Oracle RDBMS, and Tomcat; uses primarily open source development software, is hosted in NCC in their virtual server environment and users only need a Web browser, and appropriate security access rights to interface.

FLIGHT, EPA's Facility Level Information on GreenHouse gases Tool at http://ghgdata.epa.gov/ is the primary, stand-alone publication portal (web page) used to visually display data collected and verified under the program. Transformed data (via ETL from master e-GGRT data store).

ENVIROFACTS, EPA's comprehensive agency-wide repository for environmental data at includes a copy of the GHG dataset. Transformed data (via ETL from master e-GGRT data store).

3b. Data Quality Procedures:

Data quality procedures are documented in the Quality Assurance Project Plan (QAPP) for the GHG Reporting Program dated March 11, 2013 (see attached).

Attached Documents:

GHGRP QAPP_3-11-2013.docx

3c. Data Oversight:

Source Data Reporting Oversight Personnel:

Chief, GHG Reporting Branch, EPA/OAR/OAP/CCD/GGRB: Oversees day to day management and implementation of entire regulatory program under 40 CFR Part 98 to collect and verify GHG data from regulated entities.

Verification Lead, GHG Reporting Branch, EPA/OAR/OAP/CCD/GGRB: Oversees day to day management and implementation of integrated verification program to support the GHG reporting program, including development and execution of electronic data validation and verification checks to assess data quality and subsequent analysis by source category subject matter experts at EPA.

Compliance Lead, GHG Reporting Branch, EPA/OAR/OAP/CCD/GGRB: Oversees day to day management and implementation of compliance and enforcement of GHG reporting program, including coordination with EPA's Office of Compliance Assistance and Enforcement and compliance tracking process.

Subject Matter Experts, GHG Reporting Branch, EPA/OAR/OAP/CCD/GGRB: Multiple personnel covering 41 industrial source categories. Review source category specific data reporting requirements, support software development and testing, review data verification reports, analyze reported data and follow-up with individual facilities, as needed.

Information Systems Oversight Personnel:

Data Management and Integration Lead, GHG Reporting Branch, EPA/OAR/OAP/CCD/GGRB: Coordinates activities that support development, maintenance and interaction of electronic tools that support GHG data collection, verification and publication.

CBI Lead, GHG Reporting Branch, EPA/OAR/OAP/CCD/GGRB: Maintains and supports enforcement of CBI policies and CBI equipment.

Security Lead/ISO, GHG Reporting Branch, EPA/OAR/OAP/CCD/GGRB: Maintains System Security Plan and related reporting/submissions.

3d. Calculation Methodology:

Decision Rules for Selecting Data: Per 40 CFR part 98(

http://www.epa.gov/ghgreporting/reporters/notices/index.html GHGRP collects complete, consistent and accurate GHG emissions data to inform policy and regulatory development. Part 98 provides specific calculation and reporting methods developed through notice-and-comment rulemaking. To ensure that data can be used to support policy and regulatory development, annual GHG emission reports must be verified (i.e. free of substantive errors).

Definitions of Variables: The variables used to calculate the performance measure are total number of "annual GHG reports received" for a "reporting year", and the number of "verified annual GHG reports". Further, a report is determined to be verified when there are no "substantive errors".

Annual GHG report – each facility subject to the GHGRP is required to electronically submit an annual report fulfilling the reporting requirements described under 40 CFR Part 98. There is only one annual GHG report per facility.

Reporting year – the annual reporting cycle extends from January 1 to December 31 and the emissions data collected over this period must be submitted to EPA by March 31 of the following year. For example, Reporting Year 2012 data were collected from January 1, 2012 to December 31, 2012 and reported to EPA on March 31, 2013.

Verified Annual GHG report – an annual GHG report is determined to be verified when it is free of any substantive errors.

Substantive error - an error that impacts the quantity of GHG emissions reported or otherwise prevents the reported data from being validated or verified.

Explanation of Calculations: The percentage of annual GHG reports which are verified is calculated by dividing the number of reports without any substantive errors by the total number of annual GHG reports and multiplying by 100 for a given reporting year.

Explanation of Assumptions: This measurement assumes that once an annual GHG report does not trigger any verification checks and/or any triggered checks are adequately explained by the reporter, the annual report meets the objectives of the GHGRP.

Unit of Measure: Unit of measure is percent of reports that are verified.

Timeframe of Result: This measurement is assessed 150 days after the reporting deadline for a given reporting year.

Documentation of Methodological Changes: N/A

4a. Oversight and Timing of Final Results Reporting:
Branch Chief, GHG Reporting Program

4b. Data Limitations/Qualifications:
General Limitations/Qualifications: Based on current information, GHGRP has high reporting rates but it is difficult for the Program to ensure that 100 percent of the facilities which are subject to the Rule are reporting under certain source categories (e.g., stationary combustion sources). The GHGRP relies upon centralized verification to evaluate whether annual GHG emission reports meet the requirements prescribed under 40 CFR part 98. EPA has implemented a very robust electronic verification system to evaluate these reports to identify reporting errors, inconsistencies, etc. We also have the authority to evaluate the field activities prescribed under 40 CFR part 98. However, given the large number of facilities, we have limitations in our ability to physically confirm implementation of measurement and/or monitoring requirements under Part 98. Data Lag Length and Explanation: Data for a given reporting year is published approximately six months following the reporting deadline. Therefore, there is approximately a 10 month lag between the end of a reporting year and the publication of the respective data. Methodological Changes: None.

4c. Third-Party Audits:
None

Office of Administration and Resource Management (OARM) Record(s)

Measure Code: 098 - Cumulative percentage reduction in energy consumption.

Office of Administration and Resource Management (OARM)

Goal Number and Title:	
0 -	
Objective Number and Title:	
0 -	
Sub-Objective Number and Title:	
0 -	
Strategic Target Code and Title:	
0 -	
Managing Office:	

1a. Performance Measure Term Definitions:

Energy consumption:

Per guidance issued by DOE and CEQ on the implementation of the Energy Policy Act of 2005, Energy Independence Act of 2007, and EO 13514, energy consumption is defined as the electricity, natural gas, steam, high temperature hot water, chilled water, fuel oil, propane, and other energy used in EPA occupied facilities where EPA pays directly for utilities. This group of "reporting facilities" consists of EPA laboratories – either owned by EPA, leased by EPA. or leased by GSA for EPA. This definition of energy consumption matches that used by all federal agencies in implementing the above referenced legislation and EO. Energy consumption reductions are measured using a BTUs/Gross Square Foot/Year metric that is described in the above referenced guidance and used by all federal agencies.

EPA's 34 reporting facilities: The EPA facilities at which the Agency controls building operations, pays utility bills directly to the utility company, and reports annual energy and water consumption data to the U.S. Department of Energy in order to demonstrate compliance with federal energy and water reduction requirements.

FY2003 baseline:

EPA's energy consumption baseline for FY 2003 is 388,190 BTUs/GSF/Year.

Background:

Per statute and EO, EPA must reduce energy use at its "reporting" facilities by 3% annually, for a cumulative reduction of 30% by FY 2015, from a FY 2003 baseline. EPA must reduce its energy use 18% below its FY 2003 baseline by the end of FY 2011, 21% by the end of FY 2012, and 24% by FY 2013. EPA's energy cumulative energy reduction was 18.1% in FY 2011.

2a. Original Data Source:

EPA Contractor

2b. Source Data Collection:

Source Data Collection Methods:

The Agency's contractor requests and collects quarterly energy and water reporting forms, utility invoices, and fuel consumption logs from energy reporters at each of EPA's "reporting" facilities. The reported data are based on metered readings from the laboratory's utility bills for certain utilities (natural gas, electricity, purchased steam, chilled water, high temperature hot water, and potable water) and from on-site consumption logs for other utilities (propane and fuel oil). In instances when data are missing and cannot be retrieved, reported data are based on a proxy or historical average. It is relatively rare for EPA to use proxy data, and even more rare for EPA to use proxy data over a significant period of time. In the relatively few cases where a meter breaks, or an advanced metering system loses data, EPA develops proxy data to substitute for the missing data. For example, if a week's worth of data is missing from a particular meter, an average of the previous week's data and the following week's data is used. These adjustments are similar to those used in the private sector and in most Advanced Metering software systems, which typically flag duplicate data or missing data, and use comparable operating period data to fill in any gaps. Again, the use of proxy data is rare, and would alter EPA's reported energy use by +/- 0.25% at most on an annual basis.

Date/Time Intervals Covered by Source Data:
Quarterly; FY2003 to present

EPA QA Requirements/Guidance Governing Collection:
The contractor is responsible for reviewing and quality assuring/quality checking (QA/QCing) the data. Specifically, the contractor performs an exhaustive review of all invoices and fuel logs to verify that reported consumption and cost data are correct. Once the energy data is reviewed and verified, the contractor will review and verify the GHG equivalents data ensuring they are using the current translation factors.

2c. Source Data Reporting:

Form/mechanism for receiving data and entering into EPA system:

EPA currently relies on a paper based system to collect and report out energy data. A contractor receives hard or PDF copies of all utility bills from reporting locations, assimilates and reports out the data in predetermined quarterly and annual data reports. The standard operating procedures for Energy Reporting include multiple QA/QC practices at each step of the data collection and analysis process.

EPA's contractors use DOE provided conversion factors to convert native fuel units into BTU equivalents. These conversion factors are used by all federal agencies in their mandatory energy reporting. Shortly EPA expects to switch a significant portion of its energy reporting to an advanced metering system (approximately 74% of energy use), but will run the current paper based system for at least a year to ensure quality and continuity of energy data.

Timing and frequency of reporting:

EPA collects and distributes energy data on a quarterly basis. .

3a. Relevant Information Systems:

Energy and Water Database.

The Energy and Water Database is a collection of numerous spreadsheets that track energy consumption and GHG production data supplied by the Agency's contractor.

In addition, beginning on January 31, 2011 and annually thereafter, EPA must enter this data into a Department of Energy Data Portal. This portal gathers energy use data for each federal agency, for the previous fiscal year.

3b. Data Quality Procedures:

EPA's Sustainable Facilities Practices Branch compares reported and verified energy use at each reporting facility against previous years' verified data to see if there are any significant and unexplainable increases or decreases in energy consumption and costs.

3c. Data Oversight:

The Chief, Sustainable Facilities Practices Branch, is responsible for overseeing the energy and water data collection system. This position manages EPA's energy conservation program, including forecasting, project development, and data reporting.

Source Data Reporting Oversight Personnel:

Detailed Standard Operating Procedures have been developed, that includes specific requirements for quality control of energy data collection and reporting, covering areas such as data verification, data entry, and other steps in the energy data reporting process.

Information Systems Oversight Personnel:

While EPA is still developing experience with advanced metering systems, it has procedures in place to insure data accuracy. These include running manual data collection and advanced metering data collection in parallel, typically for at least one year, to confirm accuracy of advanced metered data. We also compare current period information with historic information to identify any variances.

3d. Calculation Methodology:

Timeframe:

Cumulative from FY2003 to end of most recent fiscal year

Generally, any change in energy data reporting procedures involves running the previous method in parallel with the new methof for at least a year, prior to standardizing a new methodology. For example, when our Research Triangle Park, North Carolina laboratory installed an advanced metering system, we ran the old and the new data streams for two years in ensure accuracy/continuity of data.

See attached Standard Operating Procedures.

Attached Documents:
EPA Energy Database SOP 1st Q FY 2012.pdf

4a. Oversight and Timing of Final Results Reporting:
The Chief, Sustainable Facilities Practices Branch, is responsible for overseeing the energy and water data collection system. This position manages EPA's energy conservation program, including forecasting, project development, and data reporting. EPA reports energy data internally to facility managers and staff involved in energy management, and annually to DOE and CEQ.

4b. Data Limitations/Qualifications:
EPA does not currently have a formal meter verification program to ensure that an on-site utility meter reading corresponds to the charges included in the utility bill. However, as EPA implements the advance metering requirements of the Energy Policy Act of 2005 and the Energy Independence and Security Act of 2007, which is currently underway, EPA will move to annual calibration of advanced meters.

4c. Third-Party Audits:
EPA reports energy data internally to facility managers and staff involved in energy management, and annually to DOE and CEQ.

Measure Code: 009 - Increase in number and percentage of certified acquisition staff (1102)

Office of Administration and Resource Management (OARM)

Goal Number and Title:
0 -

Objective Number and Title:
0 -

Sub-Objective Number and Title:
0 -

Strategic Target Code and Title:
0 -

Managing Office:

1a. Performance Measure Term Definitions:

Certified acquisition staff (1102): The GS-1102 series includes positions that manage, supervise, perform, or develop policies and procedures for professional work involving the procurement of supplies, services, construction, or research and development using formal advertising or negotiation procedures; the evaluation of contract price proposals; and the administration or termination and close out of contracts. The work requires knowledge of the legislation, regulations, and methods used in contracting; and knowledge of business and industry practices, sources of supply, cost factors, and requirements characteristics. The purpose of the Federal Acquisition Certification in Contracting (FAC-C) program is to establish core requirements for education, training, and experience for contracting professionals in civilian agencies. The federal certification in contracting is not mandatory for all GS-1102s; however, members of the workforce issued new Contracting Officer (CO) warrants on or after January 1, 2007, regardless of GS series, must be certified at an appropriate level to support their warrant obligations, pursuant to agency policy.
 	
Background:
   It is essential that the Federal Government have the capacity to carry out robust and thorough management and oversight of its contracts in order to achieve programmatic goals, avoid significant overcharges, and curb wasteful spending. A GAO study last year of 95 major defense acquisitions projects found cost overruns of 26 percent, totaling $295 billion over the life of the projects. Improved contract oversight could reduce such sums significantly.
   Executive Agencies were requested to propose plans to increase the Acquisition Workforce by 5%. OMB provided tools to the Agencies to determine what the appropriate size would be for the acquisition workforce which is how EPA determined that we need 351 1102s by FY2014. We proposed adding new contracting personnel annually, in even increments, through 2014 in order to reach this goal. Since EPA is always working on certifying our contracting personnel, the target certification levels for FY2012 include certifying the personnel that EPA is bringing onboard to satisfy the increase in the acquisition workforce and certifying those already at EPA. Since EPA's proposed plan included bringing on mid- and senior-level 1102s, it is expected that many will already be certified.

Certification and warranting procedures are initiated by the individual seeking the certification/warrant. There may be eligible individuals already in the acquisition workforce who have not yet applied for certification that EPA is unable to track.

For more information, please see:

Presidential Memorandum for the Heads of Executive Departments and Agencies – Subject: Government Contracting, http://www.whitehouse.gov/the_press_office/Memorandum-for-the-Heads-of-Executive-Departments-and-Agencies-Subject-Government/ March 4, 2009

October 27, 2009 OMB Memorandum for Chief Acquisition Officers, Senior Procurement Executives, Chief Financial Officers, Chief Human Capital Officers – Subject: Acquisition Workforce Development Strategic Plan for Civilian Agencies – FY 2010 – 2014.
http://www.whitehouse.gov/sites/default/files/omb/assets/procurement_workforce/AWF_Plan_10272009.pdf
The link is correct as it applies to the Acquisition Workforce Strategic Plan for Civilian Agencies-FY 2010- 2014 relative to increasing the by 5% as stated in the Background summary for EPA.

2a. Original Data Source:

The Agency Acquisition Career Manager (ACM) reviews and approves the final completed package for an applicant's certification. The EPA has a Certification and Warrant Database that is used as the tool for approval and tracking the number of FAC-C and warrants issued in the Agency. This data is reported as the total assigned number of EPA 1102s assigned and the percentage of the total 1102 staff the certified. The baseline is 324 assigned 1102s in FY 09 with 70% of the total 1102s assigned in FY 09 certified.

2b. Source Data Collection:

Source Data Collection Methods:
Before an individual is certified, there are three levels of review and approval of documentation proving certification eligibility. An initial review is performed on every individual's documentation for certification by an EPA Policy Analyst that specializes in FAC-C certification eligibility. The Analyst aids the applicant in preparing a complete package to be reviewed for approval. Once the package is completed, it is provided to the Policy Analyst's Team Leader for review and approval. Once it is determined that the package is ready for final review by the Agency Acquisition Career Manager (ACM) the final completed package is sent forward for review and approval. Once approved, FAC-C level I, II, or III is granted based on the information provided and applied for. The FAC-C certification allows for a warrant to be applied for and issued.

2c. Source Data Reporting:

Form/mechanism for receiving data and entering into EPA system:
The data in the "Federal Acquisition Certification, Warrants, and BPAs" database is reviewed and inputted by EPA Procurement Analysts who are trained to verify documents submitted by employees for Federal Acquisition Certification in Contracting (FAC-C) certification and approval. The individual uploads his or her documents for review and approval into the email the FAC-C mailbox where the EPA Procurement Analyst can review the uploaded documentation to support the education, experience and training requirements for FAC-C certification. Once this review is completed the Procurement Analyst releases the file to the supervisor of

record for approval/disapproval. After the supervisor's approval/disapproval, the system notifies the ACM that the file is ready for review and approval/disapproval. After the ACM approves the application, the FAC-C certificate is then ready for printing and signature by the ACM.

Timing and frequency of reporting:
Once the individual uploads all the documents in their application request for certification, there are system notifications generated that flow in the review and approval to the Procurement Analyst, Supervisor, and ACM. After the FAC-C Level I, II, or III certificate is signed by the ACM, it is scanned and emailed to the applicant in advance of receiving the original in the mail. The 1102 certification data is reported annually consistent with the OMB, OFPP reporting guidance for the Annual Acquisition Human Plan (AHCP).

3a. Relevant Information Systems:

The information for tracking the certification targets is currently maintained in the EPA's "Federal Acquisition Certification, Warrants, and BPAs" database.

The EPA's "Federal Acquisition Certification, Warrants, and BPAs" database Warrants/Certifications is a Lotus Notes Database which contains scanned copies of EPA Warrants. For reporting purposes, information is pulled manually from the scanned Warrant and placed on each record. This information includes Warrant Number, Level, Type, Authority (name and title), Issue Date, Limitation, Start Date, AAShip and Division. Access is closely kept; each record can only be accessed by the FAC/C and warrant holder, the supervisor, and such administrative officers as are listed in the configuration. Contents are reviewed and updated twice yearly by a designated PTOD POC.

As Warrants are added or cancelled, a group of specialists in OCFO and ITSC are notified so as to keep records up to date in other systems. Updates to other systems are manual. The source data exists on the paper documents. There is no transformation i.e., aggregated, modeled, normalized, etc.).
EXAMPLES of system integrity standards include the System Life Cycle Management Policy and the IT security policy. This is a stand-alone reporting system built on the EPA approved Lotus Notes platform. It is in the Operations and Maintenance portion of the System Life Cycle Management. It rests on secured, internal EPA server and does not replicate. Proper access is applied to each document. All reporting is done in the Notes Client in canned reporting views. There is no web access.

3b. Data Quality Procedures:

This is not public data viewable outside of EPA information system. The data in the "Federal Acquisition Certification, Warrants, and BPAs" database is reviewed and inputted by EPA Procurement Analysts who are trained to verify documents submitted by employees for Federal Acquisition Certification in Contracting (FAC-C) certification and approval. Once this review is completed the Procurement Analyst releases the file to the supervisor of record for approval/disapproval. After the supervisor's approval/disapproval, the system notifies the ACM that the file is ready for review and approval/disapproval. After the ACM approves the application, the FAC-C certificate is then ready for printing and signature by the ACM.

3c. Data Oversight:

Source Data Reporting Oversight Personnel: The Agency Senior Procurement Executive (SPE) oversees the final reporting of 1102 certification data consistent with the OMB, OFPP reporting guidance in the Annual

Acquisition Human Plan (AHCP). The Agency Acquisition Career Manager (ACM) is responsible for data research, data collection, data validation, and preparation of the Annual AHCP.

Information system Oversight Personnel: The Senior Procurement Executive (SPE) of the Environmental Protection Agency (EPA) is responsible for establishing an effective acquisition management system which ensures that quality goods and services are obtained at reasonable prices, in a timely fashion, and in accordance with the statutory and regulatory requirements and the programmatic needs of the agency. The Agency Senior Procurement Executive (SPE) oversees the final reporting of 1102 certification data consistent with the OMB, OFPP reporting guidance in the Annual Acquisition Human Plan (AHCP). As warrants are added or cancelled in the EPA "Federal Acquisition Certification, Warrants, and BPAs" database, a group of specialists in OCFO and ITSC are notified so as to keep records up to date in other systems. As warrants are added or cancelled, a group of specialists in OCFO and ITSC are notified so as to keep records up to date in other systems.

3d. Calculation Methodology:

This data is reported as the total assigned number of EPA 1102s assigned and the percentage of the total 1102 staff the certified. The baseline is 324 assigned 1102s in FY 09 with 70% of the total 1102s assigned in FY 09 certified. The projected target for 2012 for total assigned 1102s is 335 with a projected 80% of the total assigned staff certified. EPA is continually working on certifying our 1102 acquisition workforce; however, the estimates proposed targets rely upon receiving the additional FTEs for the acquisition workforce.

4a. Oversight and Timing of Final Results Reporting:

The Agency Senior Procurement Executive (SPE) oversees the final reporting of 1102 certification data consistent with the OMB, OFPP reporting guidance in the Annual Acquisition Human Plan (AHCP).

4b. Data Limitations/Qualifications:

An error estimate has not been calculated for this measure. The EPA has a Certification and Warrant Database that is used as the tool for approval and tracking the number of FAC-C and warrants issued in the Agency. The database is a stand-alone reporting system built on the EPA approved Lotus Notes platform. It is in the Operations and Maintenance portion of the System Life Cycle Management. It rests on secured, internal EPA server and does not replicate. Proper access is applied to each document. All reporting is done in the Notes Client in canned reporting views. There is no web access. The source data exist on paper documents. There is no transformation of data (i.e., aggregated, modeled, normalized, etc.).

4c. Third-Party Audits:

There are no independent third party audits of the data flow for this performance measure at this time. However, future audits could be conducted by relevant OIG, GAO, and OMB.

As an internal management control tool, the Senior Procurement Executive (SPE) has established the Balanced Scorecard Performance Measurement and Performance Management Program (Balanced Scorecard- BSC). The purpose of the BSC program establishes an Acquisition System Performance Management Plan framework under which the Office of Acquisition Management (OAM) may ensure that business systems adhere to EPA's mission and vision, and strategy statements follow best business management practices, and comply with applicable statutes, regulations, and contract terms and conditions. Through the utilization of the Balance Scorecard framework, OAM will be able to identify opportunities to strengthen the EPA's Acquisition Workforce Strategic Human Capital Plan, thus allowing EPA to purse all available authorities and strategies to ensure that the Agency appropriate resources and the best qualified staff to provide mission support. The BSC program operates with performance measures, self-assessment, and peer review/oversight components.

Office of Administration and Resource Management (OARM)

Goal Number and Title:
0 -
Objective Number and Title:
0 -
Sub-Objective Number and Title:
0 -
Strategic Target Code and Title:
0 -
Managing Office:

1a. Performance Measure Term Definitions:

GS employees: The General Schedule (GS) classification and pay system covers the majority of civilian white-collar Federal employees. GS classification standards, qualifications, pay structure, and related human resources policies (e.g., general staffing and pay administration policies) are administered by the U.S. Office of Personnel Management (OPM) on a Government-wide basis. Each agency classifies its GS positions and appoints and pays its GS employees filling those positions following statutory and OPM guidelines. The General Schedule has 15 grades--GS-1 (lowest) to GS-15 (highest).

DEU: This measure will track the hiring timeliness for non-federal applicants using the delegated examining recruitment process. Delegated examining authority is an authority OPM grants to agencies to fill competitive civil service jobs with applicants applying from outside the Federal workforce, Federal employees who do not have competitive service status, or Federal employees with competitive service status. Appointments made by agencies through delegated examining authority are subject to civil service laws and regulations. This is to ensure fair and open competition, recruitment from all segments of society, and selection on the basis of the applicants' competencies or knowledge, skills, and abilities (see 5 U.S.C. § 2301).

Hired within 80 calendar days:
This is the measure used to track the time to hire for all Job Opportunity Announcements (JOAs) posted on USAJobs from the time the announcement is drafted until the time of entry on duty (EOD) .

Background:
   OPM's original End-to-End 80-day hiring initiative focused on the Agency's entire hiring process from the time a hiring request is initiated until the employee comes on board; the 80-day hiring initiative focused on those non-federal employees hired through the delegated examining recruitment process.
   OPM's 80-day hiring model is designed to assess the time to hire federal employees where a job opportunity announcement was posted on USAJOBs.

The President's May 2010 "Hiring Reform Initiative" memo seeks agencies to improve the timeliness of "all" hiring actions and in particular hiring actions for Mission Critical Occupations and commonly-filled positions. Agency specific reporting requirements for time to hire statistics are uncertain and not yet finalized (please see http://www.whitehouse.gov/the-press-office/presidential-memorandum-improving-federal-recruitment-and-hiring-process

For more information, please see http://www.opm.gov/publications/EndToEnd-HiringInitiative.pdf

2a. Original Data Source:
The original data source is EPA employees who request, prepare, and process SF-52s, Requests for Personnel Actions, and other documents, (e.g., staffing requisition, position description, job analysis, etc.) associated with processing hiring actions.

2b. Source Data Collection:
The source data is collected from the SF-52, Request for Personnel Action, and other documents associated (e.g., staffing requisition, position description, job analysis, etc.) with processing hiring actions, as well as steps taken by staff in processing these actions. Staff in the three Human Resources Shared Service Centers use dates on the SF-52s to enter dates in the Human Resources Activities and Communication Tracking System (HRACTS). They also record information, such as vacancy announcement numbers and comments in HRACTS. Data in HRACTS is reviewed quarterly by the SSC staff to ensure completeness and accuracy. Customers serve as an additional review layer as they have access to HRACTS and can raise any inconsistencies in data entered.

2c. Source Data Reporting:
Form/mechanism for receiving data and entering into EPA system: The servicing human resources personnel at EPA's 3 Shared Service Centers enter data into the system. Data is typically transmitted through scanning and emailing to a designated email box from the hiring decision-makers to the SSC staff. Once received, the servicing human resources personnel at EPA's 3 Shared Service Centers enter data into the system. Timing and frequency of reporting: The data is reported quarterly to the Office of Personnel Management. In addition, Agency-wide, Office-level, and SSC reports can be prepared on an annual, quarterly, or selected time period basis.

3a. Relevant Information Systems:
Office of Human Resources (OHR) HRACTS. Office of Human Resources (OHR) Human Resources Activity Communication Tracking System (HRACTS). EPA's Human Resources Activity and Communication Tracking System (HRACTS) is an in-house, lotus-notes based system designed to track and monitor HR workload including recruitment actions at the Agency's Shared Service Centers. HRACTS also tracks other HR workload activity including awards, reassignment, etc.; tracks EPA's status towards achieving OPM's original 80-day hiring goal for delegated examining recruitment actions and provides status reports to customers. HRACTS has multiple date fields for inputting the date for each step in the hiring process. HRACTS can track the time throughout EPA's hiring process from the time a

hiring request is initiated until the employee comes on board. Upon HR office consolidation to the Shared Service Center in FY09, HRACTS was refined to be useful in tracking Agency-wide hiring timeliness, standards for data quality were developed; and types of hiring methods used (e.g. MP, DEU, etc) were incorporated.

HRACTS is continually undergoing changes and modifications to meet the constant clarification and unique needs of the 80-day end-to-end hiring model. HRACTS has been revised to meet the diverse demands for easy access by Agency-wide managers to track the status of hiring actions. HRACTS reports are being revised to provide organizations with in-depth information on the status of their pending recruitment actions in a secure and controlled environment. The system was refined to notify applicants of the status of their vacancy application throughout the hiring process and also provide managers with a link to survey their perspective of the overall hiring process. Revisions also include better reporting templates to track trends and anomalies along the hiring process timeline.

Agency-wide, Office-level, and SSC reports can be prepared on an annual, quarterly, or selected time period basis. Manager access was made available to better enable tracking of the status of their individual recruitment actions.

While HRACTS can track by the type of recruitment action (DEU, MP, etc), HRACTS is currently not capable of tracking by occupational series (e.g. Mission Critical Occupations and commonly-filled positions).

The system meets the quality control standards of lotus notes.

Additional information:
Further system enhancements may be needed to track hiring timeliness for MCOs and commonly-filled positions to meet the President's Hiring Reform Initiatives.

3b. Data Quality Procedures:

SSC / OHR staff review and analyze the reports to determine trends and assess workload. SSC staff review and validate the data, identify anomalies or data-entry errors, make corrections, and provide the updated information so that the system's reports can be current and accurate. Agency managers can be provided with system access to further enhance data integrity. Questions about the data or resolution of data issues are frequently resolved through discussion and consultation with the SSC and OHR.

3c. Data Oversight:

The Lotus Notes Manager of the Information Resources Management Division is responsible for overseeing the source data reporting and making changes/modifications to the system to further improve tracking and reporting; run reports; train authorized staff on the use of the system, and makes enhancements to the system to meet time to hire goals.

3d. Calculation Methodology:

Data is entered to track all hires where a JOA was posted on USAJOBs. The system tracks each step of the hiring process. The steps included in the metrics are: SSC drafts/posts JOA; JOA open period; SSC prepares certificates; customer has certificates (interview/selection process; SSC makes tentative offer; conduct background check; make formal job offer; selectee enters on duty. We were instructed to track the Senior Executive Service (SES) hiring process as well, although these are two very different hiring processes.

4a. Oversight and Timing of Final Results Reporting:

The Reporting Oversight Personnel is the HR Director. Responsibilities include monitoring progress against milestones and measures; work with OPM and HR community to achieve timelines and targets for correcting agency hiring by reducing substantially the time to hire for Mission Critical Occupations (MCOs) and commonly filled positions; measuring/improving the quality and speed of the hiring process, and analyzing the causes of agency hiring problems and establishing timelines/targets for reducing them. Time to hire information is reported on a quarterly basis.

4b. Data Limitations/Qualifications:

HRACTS is not integrated with the Agency's People Plus System, the Agency's official personnel system, therefore, discrepancies may arise such as the total number of hires. While HRACTS can track by the type of recruitment action (DEU, MP, etc.), HRACTS is currently not capable of tracking by occupational series (e.g., Mission Critical Occupations and commonly-filled positions.)

4c. Third-Party Audits:

EPA OIG released a report on OARM's revised hiring process, including timing and technological capability, in 2010. Please see http://www.epa.gov/oig/reports/2010/20100809-10-P-0177.pdf

OPM conducted a review of EPA's hiring process. Please see
http://www.opm.gov/hiringtoolkit/docs/EPAcasestudy.pdf

Measure Code: 010 - Cumulative percentage reduction in GreenHouse Gas (GHG) Scopes 1 & 2 emissions.

Office of Administration and Resource Management (OARM)

Goal Number and Title:
0 -
Objective Number and Title:
0 -
Sub-Objective Number and Title:
0 -
Strategic Target Code and Title:
0 -
Managing Office:
1a. Performance Measure Term Definitions:
GreenHouse Gas (GHG) Scope 1 emissions: Scope 1 GHG emissions are emissions associated with fossil fuel burned at EPA facilities or in EPA vehicles and equipment. Sources of Scope 1 GHG emissions include fuel oil and natural gas burned in boilers, gasoline used in vehicles, and diesel fuel used in emergency generators.

GreenHouse Gas (GHG) Scope 2 emissions: Scope 2 GHG emissions are emissions associated with indirect sources of energy such as electricity, chilled water, or purchased steam. For example, the GHG emissions from the coal and natural gas used to generate the electricity supplied to EPA facilities are considered EPA Scope 2 GHG emissions.

Note: This measure reports cumulative percentage reduction in Scope 1 and 2 emissions aggregately.

EPA's 34 reporting facilities: The EPA facilities at which the Agency controls building operations, pays utility bills directly to the utility company, and reports annual energy and water consumption data to the U.S. Department of Energy in order to demonstrate compliance with federal energy and water reduction requirements.

1)   Research Triangle Park, NC New Main
2)   Research Triangle Park, NC RTF
3)   Research Triangle Park, NC National Computer Center
4)   Research Triangle Park, NC Incinerator
5)   Research Triangle Park, NC Child Care Center
6)   Research Triangle Park, NC Page Road
7)   Chapel Hill, NC
8)   Cincinnati – AWBERC, OH
9)   Cincinnati- T and E, OH
10)   Cincinnati- Center Hill, OH
11)   Cincinnati – Child Care |

12) Cincinnati – PUBS, OH

13) Ann Arbor, MI

14) Fort Meade, MD

15) Edison, NJ

16) Edison – REAC, NJ

17) Duluth, MN

18) Las Vegas, NV

19) Narragansett, RI

20) Richmond, CA

21) Corvallis-Main, OR

22) Corvallis-WRS, OR

23) Houston, TX

24) Athens-ORD, GA

25) Athens SESD, GA

26) Manchester, WA

27) Kansas City STC, KS

28) Golden, CO

29) Chelmsford, MA

30) Gulf Breeze, FL

31) Newport, OR

32) Ada, OK

33) Montgomery, AL

34) Grosse Ile, MI

FY 2008 baseline: 140,911 metric tons of carbon dioxide equivalent (MTCO2e). A breakdown of this baseline is available at http://www.epa.gov/oaintrnt/documents/epa_ghg_targets_letter_omb.pdf

Background: This measure tracks EPA's performance in meeting Executive Order 13514 (Federal Leadership in Environmental, Energy, and Economic Performance) and demonstrating leadership in GHG emissions reductions. For more information on Executive Order 13514, please see http://www.epa.gov/oaintrnt/practices/eo13514.htm More information on EPA's GHG reduction goals and strategies is available at http://www.epa.gov/oaintrnt/ghg/strategies.htm and EPA's letter informing OMB of the Agency's Scope 1 and 2 GHG emissions reduction goal is available at http://www.epa.gov/oaintrnt/documents/epa_ghg_targets_letter_omb.pdf An OIG evaluation of EPA's progress in meeting its GHG reduction goals is available at http://www.epa.gov/oig/reports/2011/20110412-11-P-0209.pdf

2a. Original Data Source:
EPA Contractor

2b. Source Data Collection:
Source Data Collection Methods: Scope 1 emissions. See section on Energy Consumption Goal for detail on Enegy and Water Data collection.

For other foundation information needed for GHG emissions calculations, EPA relies primarily on federal wide data systems to collect other information necessary to collect foundation data for GHG Scope 1 and 2 emissions. These data systems are used by all federal agencies, with some minor exceptions. For example, EPA utilizes GSA's FAS system to gather fleet fuel use; however EPA keeps a separate parallel system to ensure data quality.

Scope 2 emissions. See section on Energy Consumption Goal for detail on Enegy and Water Data collection.

EPA uses the DOE data portal to convert foundation information into GHG emissions equivalents.

Date/Time Intervals Covered by Source Data:

Quarterly; FY2008 to present

While EPA collects energy and water use data quarterly, use of the DOE Data Portal to calculate GHG Scope 1 and 2 emissions is done once each Fiscal Year.

EPA QA Requirements/Guidance Governing Collection:

The contractor is responsible for reviewing and quality assuring/quality checking (QA/QCing) the data. Specifically, the contractor performs an exhaustive review of all invoices and fuel logs to verify that reported consumption and cost data are correct. Once the energy data is reviewed and verified, the contractor will review and verify the GHG equivalents data ensuring they are using the current translation factors.

2c. Source Data Reporting:

Form/mechanism for receiving data and entering into EPA system:

EPA has abandoned its earlier system of GHG emissions calculations and relies primarily on the DOE Data Portal to calculate its GHG emissions. EPA merely reports out the DOE generated data as it's performance metrics.

Scope 1 emissions. See section on Energy Consumption Goal for detail on Enegy and Water Data collection

Scope 2 emissions. See section on Energy Consumption Goal for detail on Enegy and Water Data collection.

For other foundation information needed for GHG emissions calculations, EPA relies primarily on federal wide data systems to collect other information necessary to collect foundation data for GHG Scope 1 and 2 emissions. These data systems are used by all federal agencies, with some minor exceptions. For example, EPAUtilizes GSA's FAS system to gather fleet fuel use; however EPA keeps a separate parallel system to ensure data quality.

Timing and frequency of reporting:

The contractor provides GHG production information to the Agency quarterly and annually.

3a. Relevant Information Systems:

Energy and Water Database.

The Energy and Water Database is a collection of numerous spreadsheets that track energy consumption and GHG production data supplied by the Agency's contractor.

Beginning on January 31, 2011 and annually thereafter, EPA contractors enter basic energy use and green power purchase information into a new Department of Energy Data Portal. This portal takes the energy use data and green power purchase information for each federal agency, for the previous fiscal year, and calculates Scope 1 and 2 GHG emissions.

3b. Data Quality Procedures:

EPA's Sustainable Facilities Practices Branch compares reported and verified energy use at each reporting facility against previous years' verified data to see if there are any significant and unexplainable increases or decreases in energy consumption and costs.

3c. Data Oversight:

The Chief, Sustainable Facilities Practices Branch, is responsible for overseeing the data entry into the DOE Data Portal. This position manages EPA's energy conservation program, including forecasting, project development, data reporting, and EPA's GHG inventory.

Source Data Reporting Oversight Personnel:

Detailed Standard Operating Procedures have been developed, that includes specific requirements for quality control of energy data collection and reporting, covering areas such as data verification, data entry, and other steps in the energy data reporting process
Information Systems Oversight Personnel:
While EPA is still developing experience with advanced metering systems, it has procedures in place to insure data accuracy. These include running manual data collection and advanced metering data collection in parallel, typically for at least one year, to confirm accuracy of advanced metered data. We also compare current period information with historic information to identify any variances.

Agency feedback to DOE serves as a QA/QC mechanism for formula and conversion factor changes in the DOE Data Portal system..

3d. Calculation Methodology:

Timeframe: Cumulative from FY2008 to end of most recent fiscal year

The Department of Energy, EPA, and GSA in cooperation with CEQ and OMB developed Greenhouse Gas Accounting Guidance for federal government GHG reporting in 2010. DOE developed a data portal for federal GHG reporting in the same year. This Data Portal receives foundation data (i.e. energy use) and converts the data into GHG emissions for each federal agency. In January 2011, EPA entered the various energy, water, transportation, travel, and commuting data for FY 2008 and FY 2010 into the DOE Data Portal. While some calculations or conversion factors change periodically in the Data Portal, each change is vetted by federal government working groups, DOE, CEQ and OMB. EPA is currently in the process of uploading FY 2011 foundation data into the DOE Data Portal, and will complete this by no later than January 31, 2012.

4a. Oversight and Timing of Final Results Reporting:
The Chief, Sustainable Facilities Practices Branch, is responsible for overseeing the data entry into the DOE Data Portal. This position manages EPA's energy conservation program, including forecasting, project development, data reporting, and EPA's GHG inventory.
4b. Data Limitations/Qualifications:
EPA does not currently have a formal meter verification program to ensure that an on-site utility meter reading corresponds to the charges included in the utility bill. However, as EPA implements the advance metering requirements of the Energy Policy Act of 2005 and the Energy Independence and Security Act of 2007, which is currently underway, EPA will move to annual calibration of advanced meters.
4c. Third-Party Audits:
Currently, EPA relies on DOE to maintain the appropriate conversion formulas to calculate GHG emissions.

Office of Administration and Resource Management (OARM)

Goal Number and Title:
0 -
Objective Number and Title:
0 -
Sub-Objective Number and Title:
0 -
Strategic Target Code and Title:
0 -
Managing Office:

1a. Performance Measure Term Definitions:

GS employees: The General Schedule (GS) classification and pay system covers the majority of civilian white-collar Federal employees. GS classification standards, qualifications, pay structure, and related human resources policies (e.g., general staffing and pay administration policies) are administered by the U.S. Office of Personnel Management (OPM) on a Government-wide basis. Each agency classifies its GS positions and appoints and pays its GS employees filling those positions following statutory and OPM guidelines. The General Schedule has 15 grades--GS-1 (lowest) to GS-15 (highest).

Other than DEU:
This measure will track the hiring timeliness for all hires not using the delegated examining recruitment process. Delegated examining authority is an authority OPM grants to agencies to fill competitive civil service jobs with applicants applying from outside the Federal workforce, Federal employees who do not have competitive service status, or Federal employees with competitive service status. Appointments made by agencies through delegated examining authority are subject to civil service laws and regulations. This is to ensure fair and open competition, recruitment from all segments of society, and selection on the basis of the applicants' competencies or knowledge, skills, and abilities (see 5 U.S.C. § 2301).

Hired within 80 calendar days:
This is the measure used to track the time to hire for all Job Opportunity Announcements (JOAs) posted on USAJobs from the time the announcement is drafted until the time of entry on duty (EOD) .

Background:
    OPM's original End-to-End 80-day hiring initiative focused on the Agency's entire hiring process from the time a hiring request is initiated until the employee comes on board; the 80-day hiring initiative focused on those non-federal employees hired through the delegated examining recruitment process.
    OPM's 80-day hiring model is designed to assess the time to hire federal employees where a job opportunity announcement was posted on USAJOBs.

The President's May 2010 "Hiring Reform Initiative" memo seeks agencies to improve the timeliness of "all" hiring actions and in particular hiring actions for Mission Critical Occupations and commonly-filled positions. Agency specific reporting requirements for time to hire statistics are uncertain and not yet finalized (please see http://www.whitehouse.gov/the-press-office/presidential-memorandum-improving-federal-recruitment-and-hiring-process

For more information, please see http://www.opm.gov/publications/EndToEnd-HiringInitiative.pdf

2a. Original Data Source:

The original data source is EPA employees who request, prepare, and process SF-52s, Requests for Personnel Actions, and other documents, (e.g., staffing requisition, position description, job analysis, etc.) associated with processing hiring actions.

2b. Source Data Collection:

The source data is collected from the SF-52, Request for Personnel Action, and other documents associated (e.g., staffing requisition, position description, job analysis, etc.) with processing hiring actions, as well as steps taken by staff in processing these actions. Staff in the three Human Resources Shared Service Centers use dates on the SF-52s to enter dates in the Human Resources Activities and Communication Tracking System (HRACTS). They also record information, such as vacancy announcement numbers and comments in HRACTS. Data in HRACTS is reviewed quarterly by the SSC staff to ensure completeness and accuracy. Customers serve as an additional review layer as they have access to HRACTS and can raise any inconsistencies in data entered.

2c. Source Data Reporting:

Form/mechanism for receiving data and entering into EPA system:
The servicing human resources personnel at EPA's 3 Shared Service Centers enter data into the system. Data is typically transmitted through scanning and emailing to a designated email box from the hiring decision-makers to the SSC staff. Once received, the servicing human resources personnel at EPA's 3 Shared Service Centers enter data into the system.

Timing and frequency of reporting:
The data is reported quarterly to the Office of Personnel Management. In addition, Agency-wide, Office-level, and SSC reports can be prepared on an annual, quarterly, or selected time period basis.

3a. Relevant Information Systems:

Office of Human Resources (OHR) HRACTS.
Office of Human Resources (OHR) Human Resources Activity Communication Tracking System (HRACTS).

EPA's Human Resources Activity and Communication Tracking System (HRACTS) is an in-house, lotus-notes based system designed to track and monitor HR workload including recruitment actions at the Agency's Shared Service Centers. HRACTS also tracks other HR workload activity including awards, reassignment, etc.; tracks EPA's status towards achieving OPM's original 80-day hiring goal for delegated examining recruitment actions and provides status reports to customers. HRACTS has multiple date fields for inputting the date for each step in the hiring process. HRACTS can track the time throughout EPA's hiring process from the time a hiring request is initiated until the employee comes on board. Upon HR office consolidation to the Shared

Service Center in FY09, HRACTS was refined to be useful in tracking Agency-wide hiring timeliness, standards for data quality were developed; and types of hiring methods used (e.g. MP, DEU, etc) were incorporated.

HRACTS is continually undergoing changes and modifications to meet the constant clarification and unique needs of the 80-day end-to-end hiring model. HRACTS has been revised to meet the diverse demands for easy access by Agency-wide managers to track the status of hiring actions. HRACTS reports are being revised to provide organizations with in-depth information on the status of their pending recruitment actions in a secure and controlled environment. The system was refined to notify applicants of the status of their vacancy application throughout the hiring process and also provide managers with a link to survey their perspective of the overall hiring process. Revisions also include better reporting templates to track trends and anomalies along the hiring process timeline.

Agency-wide, Office-level, and SSC reports can be prepared on an annual, quarterly, or selected time period basis. Manager access was made available to better enable tracking of the status of their individual recruitment actions.

While HRACTS can track by the type of recruitment action (DEU, MP, etc), HRACTS is currently not capable of tracking by occupational series (e.g. Mission Critical Occupations and commonly-filled positions).

The system meets the quality control standards of Lotus Notes.

Additional information:
Further system enhancements may be needed to track hiring timeliness for MCOs and commonly-filled positions to meet the President's Hiring Reform Initiatives.

3b. Data Quality Procedures:
SSC / OHR staff review and analyze the reports to determine trends and assess workload. SSC staff review and validate the data, identify anomalies or data-entry errors, make corrections, and provide the updated information so that the system's reports can be current and accurate. Agency managers can be provided with system access to further enhance data integrity. Questions about the data or resolution of data issues are frequently resolved through discussion and consultation with the SSC and OHR.

3c. Data Oversight:
The Lotus Notes Manager of the Information Resources Management Division is responsible for overseeing the source data reporting and making changes/modifications to the system to further improve tracking and reporting; run reports; train authorized staff on the use of the system, and makes enhancements to the system to meet time to hire goals.

3d. Calculation Methodology:
Data is entered to track all hires where a JOA was posted on USAJOBs. The system tracks each step of the hiring process. The steps included in the metrics are: SSC drafts/posts JOA; JOA open period; SSC prepares certificates; customer has certificates (interview/selection process; SSC makes tentative offer; conduct background check; make formal job offer; selectee enters on duty. We were instructed to track the Senior

Executive Service (SES) hiring process as well, although these are two very different hiring processes.

4a. Oversight and Timing of Final Results Reporting:

The Reporting Oversight Personnel is the HR Director. Responsibilities include monitoring progress against milestones and measures; work with OPM and HR community to achieve timelines and targets for correcting agency hiring by reducing substantially the time to hire for Mission Critical Occupations (MCOs) and commonly filled positions; measuring/improving the quality and speed of the hiring process, and analyzing the causes of agency hiring problems and establishing timelines/targets for reducing them. Time to hire information is reported on a quarterly basis.

4b. Data Limitations/Qualifications:

HRACTS is not integrated with the Agency's People Plus System, the Agency's official personnel system, therefore, discrepancies may arise such as the total number of hires. While HRACTS can track by the type of recruitment action (DEU, MP, etc.), HRACTS is currently not capable of tracking by occupational series (e.g., Mission Critical Occupations and commonly-filled positions.)

4c. Third-Party Audits:

EPA OIG released a report on OARM's revised hiring process, including timing and technological capability, in 2010. Please see http://www.epa.gov/oig/reports/2010/20100809-10-P-0177.pdf

OPM conducted a review of EPA's hiring process. Please see
http://www.opm.gov/hiringtoolkit/docs/EPAcasestudy.pdf

Office of Enforcement and Compliance Assurance (OECA) Record(s)

Measure Code: 400 - Millions of pounds of air pollutants reduced, treated, or eliminated through concluded enforcement actions.

Office of Enforcement and Compliance Assurance (OECA)

Goal Number and Title:
5 - Enforcing Environmental Laws
Objective Number and Title:
1 - Enforce Environmental Laws
Sub-Objective Number and Title:
2 - Support Taking Action on Climate Change and Improving Air Quality
Strategic Target Code and Title:
1 - By 2015, reduce, treat, or eliminate 2,400 million estimated cumulative pounds of air pollutants
Managing Office:
Office of Compliance
1a. Performance Measure Term Definitions:
Air pollutants:

Air pollutants:

The Clean Air Act lists the pollutants and sources of pollutants that are to be regulated by EPA. Pollutants include hazardous air pollutants, criteria pollutants, and chemicals that destroy stratospheric ozone. Sources of pollutants include stationary sources (e.g., chemical plants, gas stations, and power plants) and mobile sources (e.g., cars, trucks, and planes).

For more information, see: http://www.epa.gov/air/airpollutants.html

Reduced, Treated or Eliminated: Reduced, treated, or eliminated is the quantity of pollutant(s) that will no longer be released to the environment as a result of a non-complying facility returning to its allowable permit limits through the successful completion of an enforcement settlement. Facilities may further reduce, treat or eliminate pollutants by carrying out voluntary Supplemental Environmental Projects.

Concluded enforcement actions: For purposes of this measure, there are two categories of concluded enforcement actions counted.

The first are administrative enforcement actions which are undertake by EPA through authority granted to it under various federal environmental statutes, such as CERCLA, RCRA, CAA, CWA, TSCA, and others. Administrative enforcement actions can take several forms, including EPA issuing an administrative order requiring a facility to implement specific corrective measures to filing an administrative complaint commencing a formal administrative adjudication. An administrative action is concluded when a written agreement between the defendant/respondent and EPA resolving the complaint is documented, signed by the Regional Administrator or designee, and is filed with the regional hearing clerk.

The second type of enforcement action is known as a civil judicial action which is a formal lawsuit, filed in court, against a person who has either failed to comply with a statutory or regulatory requirement or an administrative order. Civil judicial actions attorneys from the U.S. Department of Justice prosecute civil cases for EPA. A concluded action occurs when a consent decree is signed by all parties to the action and filed in the appropriate court and signed by a judge or a written ruling or decision is made by a judge after a full trial.

2a. Original Data Source:

EPA Regional Enforcement Organizations

EPA Regional Program Organizations

EPA Headquarters Enforcement Organizations

Facility Personnel and Facility Contractors

DOJ

2b. Source Data Collection:

EPA calculates the estimated pollutant reductions after case settlement or during discussions with the facility personnel over specific plans for compliance. The final enforcement documents often spell out the terms and methodologies the facility must follow to mitigate and prevent the future release of pollutants. These documents serve as the starting point for EPA's calculations.

Example of consent decree document containing pollutant mitigation instructions to the facility:

http://www.epa.gov/compliance/resources/cases/civil/caa/essroc.html

2c. Source Data Reporting:

When a formal administrative or judicial enforcement case is "concluded" enforcement staff enters information into ICIS to document the environmental benefits achieved by the concluded enforcement case. Original source documents may include facility permits, legal documents such as consent decrees and administrative orders, inspection reports, case engineer reports and facility reports. For civil judicial cases, the information is reported when a consent decree or court order, or judgment is entered (not lodged). For administrative cases, information is reported when an administrative order or final agreement is signed.

Environmental benefits should be reported in the year the case is settled, regardless of when the benefits will occur. Reductions are calculated after the judicial consent decree is lodged or entered, or when the administrative compliance order is signed by the region designee and filed with the regional hearing clerk.

Attached Documents:

FY2012 CCDS.docx

3a. Relevant Information Systems:

The ICIS FE&C data system meets Office of Environmental Information (OEI) Lifecycle Management Guidance, which includes data validation processes, internal screen audit checks and verification, system and user documents, data quality audit reports, third party testing reports, and detailed report specifications data calculation methodology. Reference: Quality Assurance and Quality Control procedures: Data Quality: Life Cycle Management Policy, (EPA CIO2121, April 7, 2006)

The Integrated Compliance Information System (ICIS) is a three phase multi-year modernization project that improves the ability of EPA and the states to ensure compliance with the nation's environmental laws with the collection of comprehensive enforcement and compliance information. Phase I, implemented in FY02, replaced several legacy systems, and created an integrated system to support federal enforcement and compliance tracking, targeting and reporting, including GPRA reporting. Phase II, also called Permit Compliance System (PCS) Modernization, expands ICIS to include the National Pollutant Discharge Elimination System (NPDES) program and enables improved management of the complete program (e.g., stormwater) as well as replacing the legacy PCS. PCS is currently identified as an Agency Federal Managers' Financial Integrity Act (FMFIA) weakness, and the modernization of the system is critical to address the weakness. Phase II was first implemented in FY06 for 21 states and 11 tribes/territories that use ICIS to directly manage their NPDES programs. In FY08, seven more states moved to ICIS from the legacy PCS and began electronically flowing their Discharge Monitoring Report (DMR) data from their states systems via the Exchange Network and CDX to ICIS. In FY09, Phase II continued with implementation of the National Installation of NetDMR allowing NPDES permittees to electronically submit DMR data from permitted facility systems via the Exchange Network to ICIS and migrated three additional states. In FY11 OECA implemented Full-Batch Release 1 of Phase II allowing Batch Flows of permits and facility data from states. FY12 will include Full-Batch Release 2 enabling batch flow will allow Batch Flows of inspection data from states. Inspection information and was implemented early in FY12. The final part of Phase II which will add the remaining NPDES Batch Flows and migrate and all remaining states is projected to be completed in FY13. Phase III will modernize the Air Facility System (AFS) into ICIS. AFS is used by EPA and States to track Clean Air Act enforcement and compliance activities. Integration of AFS into ICIS will modernize and replace a legacy system that does not meet current business needs. Implementation of this phase is projected for FY14.

ICIS contains both source data and transformed data.

OECA's Data System Quality Assurance Plan

Attached Documents:

Data System Quality Assurance Plan (ICIS).doc

3b. Data Quality Procedures:

Annual Data Certification Process - OECA has instituted a semi-annual data certification process for the collection and reporting of enforcement and compliance information. The certification process was set up to ensure all reporting entities are aware of the reporting deadlines, receive the most up-to-date reporting instructions for select measures, follow best data management practices to assure reporting accuracy, and have access to the recent methodologies for calculating pounds of pollutants reduced. The air pounds of pollutants reduced measure is covered by the annual data certification process.

As part of the annual data certification process, regions are provided a checklist to assist them in their data quality procedures.

Quality Management Plan - September 2011

Attached Documents:

FY11 Data Quality Check List.pdf

Data System Quality Assurance Plan (ICIS).doc

OC QMP Concurrence Signatures.pdf

OC QMP 2011 Final.docx

3c. Data Oversight:

Source Data Reporting Oversight

HQ - Director, Enforcement Targeting and Data Division

Region 1 - Division Director, Office of Environmental Stewardship

Region 2 - Director, Office of Enforcement and Compliance Assistance

Region 3 - Director, Office of Enforcement, Compliance and Environmental Justice

Region 4 - Regional Counsel and Director, Office of Environmental Accountability

Region 5 - Director, Office of Enforcement and Compliance Assurance

Region 6 - Compliance Assurance and Enforcement Division Director

Region 7 - Enforcement Coordinator

Region 8 - Director, Policy, Information Management and Environmental Justice

Region 9 - Enforcement Coordinator

Region 10 - Director, Office of Compliance and Enforcement

Information Systems Oversight Personnel

HQ - ICIS System Administrator

Region 1 - ICIS Steward and Data Systems Administrator

Region 2 - ICIS System Administrator

Region 3 - ICIS Data Steward and System Administrator

Region 4 - ICIS System Administrator, Regional Compliance and Enforcement Data Steward

Region 5 - ICIS Data Steward and Systems Administrator

Region 6 - ICIS Data Steward

Region 7 - ICIS Data Steward and Systems Administrator

Region 8 - ICIS System Administrator

Region 9 - ICIS System Administrator

Region 10 - ICIS System Administrator and Data Steward

3d. Calculation Methodology:

The Case Conclusion Data Sheet (CCDS) is a manual data collection tool HQ implemented in FY 1996, updated in FY 2012, to collect information on concluded federal enforcement cases including the case name and identification number, injunctive relief, environmental benefits (including environmental benefits from Supplemental Environmental Projects [SEPs]), and assessed penalties. The CCDS data are entered into the Integrated Information and Compliance System (ICIS). OECA uses data obtained from the CCDS via ICIS to assess the environmental outcomes of its enforcement program.

The CCDS guidance provides detailed calculation methodologies for estimating the environmental benefits on a variety of environmental statutes including air, water, waste, toxics and pesticides. Additionally, the CCDS provides specific instruction on how to enter the environmental benefits information into ICIS.

To view the the CCDS guidance in its entirety go to:

Attached Documents:

CCDS.xps

4a. Oversight and Timing of Final Results Reporting:

Oversight of Final Reporting:

The Deputy Regional Administrators, the Office of Civil Enforcement Director, and the Monitoring, Assistance and Program Division Director all must sign the attached certification form.

Timing of Results Reporting: Semiannually

Attached Documents:

Data Certification Form.pdf

4b. Data Limitations/Qualifications:

Pollutant reductions or eliminations reported in ICIS project an estimate of pollutants to be reduced or eliminated if the defendant carries out the requirements of the settlement. The estimates use information available at the time a case settles or an order is issued. In some instances, EPA develops and enters this information on pollutant reduction estimates after the settlement or during continued discussions over specific plans for compliance. Due to the time required for EPA to negotiate a settlement agreement with a defendant, there may be a delay in completing the CCDS. Additionally, because of unknowns at the time of settlement, different levels of technical proficiency, or the nature of a case, OECA's expectation is that the overall amount of pollutants reduced or eliminated is prudently underestimated based on CCDS information. EPA also bases the pollutant estimates on the expectation that the defendant/respondent implements the negotiated settlement agreement.

4c. Third-Party Audits:

Inspector General Report on Pounds of Pollutants Reduced Estimates:

Attached Documents:

Projected Lbs of Pollutants Reduced.pdf

Measure Code: 402 - Millions of pounds of water pollutants reduced, treated, or eliminated through concluded enforcement actions.

Office of Enforcement and Compliance Assurance (OECA)

Goal Number and Title:
5 - Enforcing Environmental Laws
Objective Number and Title:
1 - Enforce Environmental Laws
Sub-Objective Number and Title:
3 - Support Protecting America's Waters
Strategic Target Code and Title:
1 - By 2015, reduce, treat, or eliminate 1,600 million estimated cumulative pounds of water pollutants
Managing Office:
Office of Compliance
1a. Performance Measure Term Definitions:
Water pollutants:

EPA divides water pollution sources into two categories: point and non-point. Point sources of water pollution are stationary locations such as sewage treatment plants, factories and ships. Non-point sources are more diffuse and include agricultural runoff, mining activities and paved roads. Under the Clean Water Act, the National Pollutant Discharge Elimination System (NPDES) permit program controls water pollution by regulating point sources that discharge pollutants into waters of the United States. EPA works with state and local authorities to monitor pollution levels in the nations water and provide status and trend information on a representative variety of ecosystems.

The Clean Water Act (CWA) establishes the basic structure for regulating discharges of pollutants into the waters of the United States and regulating quality standards for surface waters. The basis of the CWA was enacted in 1948 and was called the Federal Water Pollution Control Act, but the Act was significantly reorganized and expanded in 1972. "Clean Water Act" became the Act's common name with amendments in 1977.

Under the CWA, EPA has implemented pollution control programs such as setting wastewater standards for industry. We have also set water quality standards for all contaminants in surface waters.

The CWA made it unlawful to discharge any pollutant from a point source into navigable waters, unless a permit was obtained. EPA's National Pollutant Discharge Elimination System (NPDES) permit program controls discharges. Point sources are discrete conveyances such as pipes or man-made ditches. Individual homes that are connected to a municipal system, use a septic system, or do not have a surface discharge do not need an NPDES permit; however, industrial, municipal, and other facilities must obtain permits if their discharges go directly to surface waters.

Nonpoint source (NPS) pollution, or polluted runoff, is the major source and cause of water quality impairment for waters on the state water quality limited segment lists required under CWA 303(d). Polluted runoff occurs when rain, snowmelt, irrigation water, and other water sources move across and through land, picking up pollutants and carrying them into lakes, rivers, wetlands, coastal waters and underground sources of drinking water. Taking a watershed approach to environmental issues provides an excellent opportunity for communities and agencies to work together to achieve water quality improvements.

Reduced, Treated or Eliminated: Reduced, treated, or eliminated is the quantity of pollutant(s) that will no longer be released to the environment as a result of a non-complying facility returning to its allowable permit limits through the successful completion of an enforcement settlement. Facilities may further reduce, treat or eliminate pollutants by carrying out voluntary Supplemental Environmental Projects.

Concluded enforcement actions: For purposes of this measure, there are two categories of concluded enforcement actions counted.

The first are administrative enforcement actions which are undertake by EPA through authority granted to it under various federal environmental statutes, such as CERCLA, RCRA, CAA, CWA, TSCA, and others. Administrative enforcement actions can take several forms, including EPA issuing an administrative order requiring a facility to implement specific corrective measures to filing an administrative complaint commencing a formal administrative adjudication. An administrative action is concluded when a written agreement between the defendant/respondent and EPA resolving the complaint is documented, is signed by the Regional Administrator or designee, and is filed with the regional hearing clerk.

The second type of enforcement action is known as a civil judicial action which is a formal lawsuit, filed in court, against a person who has either failed to comply with a statutory or regulatory requirement or an administrative order. Civil judicial actions attorneys from the U.S. Department of Justice prosecute civil cases for EPA. A concluded action occurs when a consent decree is signed by all parties to the action and filed in the appropriate court and signed by a judge or a written ruling or decision is made by a judge after a full trial.

2a. Original Data Source:
EPA Regional Enforcement Organizations EPA Regional Program Organizations EPA Headquarters Enforcement Organizations Facility Personnel and Facility Contractors DOJ
2b. Source Data Collection:
EPA calculates the estimated pollutant reductions after case settlement or during discussions with the facility personnel over specific plans for compliance. The final enforcement documents often spell out the terms and methodologies the facility must follow to mitigate and prevent the future release of pollutants. These documents serve as the starting point for EPA's calculations. Example of consent decree document containing pollutant mitigation instructions to the facility: http://www.epa.gov/compliance/resources/cases/civil/caa/essroc.html

2c. Source Data Reporting:

When a formal administrative or judicial enforcement case is "concluded" enforcement staff enters information into ICIS to document the environmental benefits achieved by the concluded enforcement case. Original source documents may include facility permits, legal documents such as consent decrees and administrative orders, inspection reports, case engineer reports and facility reports. For civil judicial cases, the information is reported when a consent decree or court order, or judgment is entered (not lodged). For administrative cases, information is reported when an administrative order or final agreement is signed.

Environmental benefits should be reported in the year the case is settled, regardless of when the benefits will occur. Reductions are calculated after the judicial consent decree is lodged or entered, or when the administrative compliance order is signed by the region designee and filed with the regional hearing clerk.

Attached Documents:

FY2012 CCDS.docx

3a. Relevant Information Systems:

The ICIS FE&C data system meets Office of Environmental Information (OEI) Lifecycle Management Guidance, which includes data validation processes, internal screen audit checks and verification, system and user documents, data quality audit reports, third party testing reports, and detailed report specifications data calculation methodology. Reference: Quality Assurance and Quality Control procedures: Data Quality: Life Cycle Management Policy, (EPA CIO2121, April 7, 2006)

The Integrated Compliance Information System (ICIS) is a three phase multi-year modernization project that improves the ability of EPA and the states to ensure compliance with the nation's environmental laws with the collection of comprehensive enforcement and compliance information. Phase I, implemented in FY02, replaced several legacy systems, and created an integrated system to support federal enforcement and compliance tracking, targeting and reporting, including GPRA reporting. Phase II, also called Permit Compliance System (PCS) Modernization, expands ICIS to include the National Pollutant Discharge Elimination System (NPDES) program and enables improved management of the complete program (e.g., stormwater) as well as replacing the legacy PCS. PCS is currently identified as an Agency Federal Managers' Financial Integrity Act (FMFIA) weakness, and the modernization of the system is critical to address the weakness. Phase II was first implemented in FY06 for 21 states and 11 tribes/territories that use ICIS to directly manage their NPDES programs. In FY08, seven more states moved to ICIS from the legacy PCS and began electronically flowing their Discharge Monitoring Report (DMR) data from their states systems via the Exchange Network and CDX to ICIS. In FY09, Phase II continued with implementation of the National Installation of NetDMR allowing NPDES permittees to electronically submit DMR data from permitted facility systems via the Exchange Network to ICIS and migrated three additional states. In FY11 OECA implemented Full-Batch Release 1 of Phase II allowing

Batch Flows of permits and facility data from states. FY12 will include Full-Batch Release 2 enabling batch flow will allow Batch Flows of inspection data from states. Inspection information and was implemented early in FY12. The final part of Phase II which will add the remaining NPDES Batch Flows and migrate and all remaining states is projected to be completed in FY13. Phase III will modernize the Air Facility System (AFS) into ICIS. AFS is used by EPA and States to track Clean Air Act enforcement and compliance activities. Integration of AFS into ICIS will modernize and replace a legacy system that does not meet current business needs. Implementation of this phase is projected for FY14.

ICIS contains both source data and transformed data.

OECA's Data System Quality Assurance Plan

Attached Documents:
Data System Quality Assurance Plan (ICIS).doc

3b. Data Quality Procedures:

Annual Data Certification Process - OECA has instituted a semi-annual data certification process for the collection and reporting of enforcement and compliance information. The certification process was set up to ensure all reporting entities are aware of the reporting deadlines, receive the most up-to-date reporting instructions for select measures, follow best data management practices to assure reporting accuracy, and have access to the recent methodologies for calculating pounds of pollutants reduced. The air pounds of pollutants reduced measure is covered by the annual data certification process.

As part of the annual data certification process, regions are provided a checklist to assist them in their data quality procedures.

OECA's Quality Management Plan - September 2011

Attached Documents:
FY11 Data Quality Check List.pdf
OC QMP Concurrence Signatures.pdf
OC QMP 2011 Final.docx

3c. Data Oversight:

Source Data Reporting Oversight
HQ - Director, Enforcement Targeting and Data Division
Region 1 - Division Director, Office of Environmental Stewardship
Region 2 - Director, Office of Enforcement and Compliance Assistance

Region 3 - Director, Office of Enforcement, Compliance and Environmental Justice

Region 4 - Regional Counsel and Director, Office of Environmental Accountability

Region 5 - Director, Office of Enforcement and Compliance Assurance

Region 6 - Compliance Assurance and Enforcement Division Director

Region 7 - Enforcement Coordinator

Region 8 - Assistant Regional Administrator for Enforcement, Compliance and Environmental Justice

Region 9 - Enforcement Coordinator

Region 10 - Director, Office of Compliance and Enforcement

Information Systems Oversight Personnel

HQ - ICIS System Administrator

Region 1 - ICIS Steward and Data Systems Administrator

Region 2 - ICIS System Administrator

Region 3 - ICIS Data Steward and System Administrator

Region 4 - ICIS System Administrator, Regional Compliance and Enforcement Data Steward

Region 5 - ICIS Data Steward and Systems Administrator

Region 6 - ICIS Data Steward

Region 7 - ICIS Data Steward and Systems Administrtor

Region 8 - ICIS System Administrator

Region 9 - ICIS System Administrator

Region 10 - ICIS System Administrator and Data Steward

3d. Calculation Methodology:

The Case Conclusion Data Sheet (CCDS) is a manual data collection tool HQ implemented in FY 1996, updated in FY 2012, to collect information on concluded federal enforcement cases including the case name and identification number, injunctive relief, environmental benefits (including environmental benefits from Supplemental Environmental Projects [SEPs]), and assessed penalties. The CCDS data are entered into the Integrated Information and Compliance System (ICIS). OECA uses data obtained from the CCDS via ICIS to assess the environmental outcomes of its enforcement program.

The CCDS guidance provides detailed calculation methodologies for estimating the environmental benefits on a variety of environmental statutes including air, water, waste, toxics and pesticides. Additionally, the CCDS provides specific instruction on how to enter the environmental benefits information into ICIS.

To view the the CCDS guidance in its entirety go to:

Attached Documents:

CCDS.xps

4a. Oversight and Timing of Final Results Reporting:

Oversight of Final Reporting:

The Deputy Regional Administrators, the Office of Civil Enforcement Director, and the Monitoring, Assistance and Program Division Director all must sign the attached certification form.

Timing of Results Reporting: Semiannually

Attached Documents:

Data Certification Form.pdf

4b. Data Limitations/Qualifications:

Pollutant reductions or eliminations reported in ICIS project an estimate of pollutants to be reduced or eliminated if the defendant carries out the requirements of the settlement. (Information on expected outcomes of state enforcement is not available.) The estimates use information available at the time a case settles or an order is issued. In some instances, EPA develops and enters this information on pollutant reduction estimates after the settlement or during continued discussions over specific plans for compliance. Due to the time required for EPA to negotiate a settlement agreement with a defendant, there may be a delay in completing the CCDS. Additionally, because of unknowns at the time of settlement, different levels of technical proficiency, or the nature of a case, OECA's expectation is that the overall amount of pollutants reduced or eliminated is prudently underestimated based on CCDS information. EPA also bases the pollutant estimates on the expectation that the defendant/respondent implements the negotiated settlement agreement.

4c. Third-Party Audits:

Inspector General Report on Pounds of Pollutants Reduced:

Attached Documents:

Projected Lbs of Pollutants Reduced.pdf

Measure Code: 404 - Millions of pounds of toxic and pesticide pollutants reduced, treated, or eliminated through concluded enforcement actions.

Office of Enforcement and Compliance Assurance (OECA)

Goal Number and Title:
5 - Enforcing Environmental Laws
Objective Number and Title:
1 - Enforce Environmental Laws
Sub-Objective Number and Title:
5 - Support Ensuring the Safety of Chemicals and Preventing Pollution
Strategic Target Code and Title:
1 - By 2015, reduce, treat, or eliminate 19 million estimated cumulative pounds of toxic and pesticide
Managing Office:
Office of Compliance
1a. Performance Measure Term Definitions:
Toxic and pesticide pollutants:

The Toxic Substances Control Act of 1976 provides EPA with authority to require reporting, record-keeping and testing requirements; and restrictions relating to chemical substances and/or mixtures; and the production, importation, use, and disposal of specific chemicals, including lead-based paint, polychlorinated biphenyls (PCBs), and asbestos. Lead-based paint is particularly dangerous to children: exposure may cause reduced intelligence, learning disabilities, behavior problems and slowed physical development. Because LBP is found in pre-1978 buildings, it is more common in communities predominated by older housing, which usually are low-income, minority and EJ communities. Asbestos in schools, if not properly managed, can expose children, teachers and other school staff to harm that may not manifest for years. PCBs bioaccumulate and thus cause a variety of adverse health effects. Asbestos and PCBs are also generally found in older buildings. Additionally, PCBs are generally found in older transformers, capacitors and some hydraulic equipment and more recently in recycled and used oil. Inappropriate abatement and disposal of asbestos and PCBs can be dangerous. For more information on the Toxics program go to:
http://www.epa.gov/compliance/civil/tsca/tscaenfstatreq.html

The Federal Insecticide, Fungicide and Rodenticide Act (FIFRA) provides EPA the authority to regulate pesticides to prevent unreasonable adverse affects on the environment. The term "unreasonable adverse effects on the environment" means: "(1) any unreasonable risk to man or the environment, taking into account the economic, social, and environmental costs and benefits of the use of any pesticide, or (2) a human dietary risk from residues that result from a use of a pesticide in or on any food inconsistent with the standard under section 408 of the Federal Food, Drug, and Cosmetic Act." The term pesticide includes many kinds of ingredients in products, such as insect repellants, weed killers, disinfectants, and swimming pool chemicals which are designed to prevent, destroy, repel or reduce pests of any sort. Pesticides are found in nearly every home, business, farm, school, hospital and park in the United States. EPA must evaluate pesticides thoroughly before they can be marketed and used in the United States to ensure that they will meet federal safety standards to protect human health and the environment. Pesticides that meet the requirements are granted a license or "registration" which permits their distribution, sale, and use according to specific use directions and requirements identified on the label. For more information on the pesticide program go to: http://www.epa.gov/compliance/civil/fifra/fifraenfstatreq.html

Reduced, Treated or Eliminated: Reduced, treated, or eliminated is the quantity of pollutant(s) that will no longer be released to the environment as a result of a non-complying facility returning to its allowable permit limits through the successful completion of an enforcement settlement. Facilities may further reduce, treat or eliminate pollutants by carrying out voluntary Supplemental Environmental Projects.

Concluded enforcement actions: For purposes of this measure, there are two categories of concluded enforcement actions counted.

The first are administrative enforcement actions which are undertake by EPA through authority granted to it under various federal environmental statutes, such as CERCLA, RCRA, CAA, CWA, TSCA, and others. Administrative enforcement actions can take several forms, including EPA issuing an administrative order requiring a facility to implement specific corrective measures to filing an administrative complaint commencing a formal administrative adjudication. An administrative action is concluded when a written agreement between the defendant/respondent and EPA resolving the complaint is documented, signed by the Regional Administrator or designee, and is filed with the regional hearing clerk.

The second type of enforcement action is known as a civil judicial action which is a formal lawsuit, filed in court, against a person who has either failed to comply with a statutory or regulatory requirement or an administrative order. Civil judicial actions attorneys from the U.S. Department of Justice prosecute civil cases for EPA. A concluded action occurs when a consent decree is signed by all parties to the action and filed in the appropriate court and signed by a judge or a written ruling or decision is made by a judge after a full trial.

2a. Original Data Source:
EPA Regional Enforcement Organizations EPA Regional Program Organizations EPA Headquarters Enforcement Organizations Facility Personnel and Facility Contractors DOJ
2b. Source Data Collection:
EPA calculates the estimated pollutant reductions after case settlement or during discussions with the facility personnel over specific plans for compliance. The final enforcement documents often spell out the terms and methodologies the facility must follow to mitigate and prevent the future release of pollutants. These documents serve as the starting point for EPA's calculations. Example of consent decree document containing pollutant mitigation instructions to the facility: http://www.epa.gov/compliance/resources/cases/civil/caa/essroc.html
2c. Source Data Reporting:

When a formal administrative or judicial enforcement case is "concluded" enforcement staff enters information into ICIS to document the environmental benefits achieved by the concluded enforcement case. Original source documents may include facility permits, legal documents such as consent decrees and administrative orders, inspection reports, case engineer reports and facility reports. For civil judicial cases, the information is reported when a consent decree or court order, or judgment is entered (not lodged). For administrative cases, information is reported when an administrative order or final agreement is signed.

Environmental benefits should be reported in the year the case is settled, regardless of when the benefits will occur. Reductions are calculated after the judicial consent decree is lodged or entered, or when the administrative compliance order is signed by the region designee and filed with the regional hearing clerk.

Attached Documents:
FY2012 CCDS.docx

3a. Relevant Information Systems:

The ICIS FE&C data system meets Office of Environmental Information (OEI) Lifecycle Management Guidance, which includes data validation processes, internal screen audit checks and verification, system and user documents, data quality audit reports, third party testing reports, and detailed report specifications data calculation methodology. Reference: Quality Assurance and Quality Control procedures: Data Quality: Life Cycle Management Policy, (EPA CIO2121, April 7, 2006)

The Integrated Compliance Information System (ICIS) is a three phase multi-year modernization project that improves the ability of EPA and the states to ensure compliance with the nation's environmental laws with the collection of comprehensive enforcement and compliance information. Phase I, implemented in FY02, replaced several legacy systems, and created an integrated system to support federal enforcement and compliance tracking, targeting and reporting, including GPRA reporting. Phase II, also called Permit Compliance System (PCS) Modernization, expands ICIS to include the National Pollutant Discharge Elimination System (NPDES) program and enables improved management of the complete program (e.g., stormwater) as well as replacing the legacy PCS. PCS is currently identified as an Agency Federal Managers' Financial Integrity Act (FMFIA) weakness, and the modernization of the system is critical to address the weakness. Phase II was first implemented in FY06 for 21 states and 11 tribes/territories that use ICIS to directly manage their NPDES programs. In FY08, seven more states moved to ICIS from the legacy PCS and began electronically flowing their Discharge Monitoring Report (DMR) data from their states systems via the Exchange Network and CDX to ICIS. In FY09, Phase II continued with implementation of the National Installation of NetDMR allowing NPDES permittees to electronically submit DMR data from permitted facility systems via the Exchange Network to ICIS and migrated three additional states. In FY11 OECA implemented Full-Batch Release 1 of Phase II allowing Batch Flows of permits and facility data from states. FY12 will include Full-Batch Release 2 enabling batch flow will allow Batch Flows of inspection data from states. Inspection information and was implemented early in FY12. The final part of Phase II which will add the remaining NPDES Batch Flows and migrate and all remaining states is projected to be completed in FY13. Phase III will modernize the Air Facility System (AFS) into ICIS. AFS is used by EPA and States to track Clean Air Act enforcement and compliance activities. Integration of AFS into ICIS will modernize and replace a legacy system that does not meet current business needs. Implementation of this phase is projected for FY14.

ICIS contains both source data and transformed data.

OECA's Data System Quality Assurance Plan

Attached Documents:
Data System Quality Assurance Plan (ICIS).doc

3b. Data Quality Procedures:

Annual Data Certification Process - OECA has instituted a semi-annual data certification process for the collection and reporting of enforcement and compliance information. The certification process was set up to ensure all reporting entities are aware of the reporting deadlines, receive the most up-to-date reporting instructions for select measures, follow best data management practices to assure reporting accuracy, and have access to the recent methodologies for calculating pounds of pollutants reduced. The toxics and pesticides pounds of pollutants reduced measure is covered by the annual data certification process.

As part of the annual data certification process, regions are provided a checklist to assist them in their data quality procedures.

OECA's QMP - September 2011

Attached Documents:
FY11 Data Quality Check List.pdf
OC QMP Concurrence Signatures.pdf
OC QMP 2011 Final.docx

3c. Data Oversight:

Source Data Reporting Oversight:
HQ - Director, Enforcement Targeting and Data Division
Region 1 - Division Director, Office of Environmental Stewardship
Region 2 - Director, Office of Enforcement and Compliance Assistance
Region 3 - Director, Office of Enforcement, Compliance and Environmental Justice
Region 4 - Regional Counsel and Director, Office of Environmental Accountability
Region 5 - Director, Office of Enforcement and Compliance Assurance
Region 6 - Compliance Assurance and Enforcement Division Director

Region 7 - Enforcement Coordinator

Region 8 - Director, Policy, Information Management and Environmental Justice

Region 9 - Enforcement Coordinator

Region 10 - Director, Office of Compliance and Enforcement

Information Systems Oversight Personnel

HQ - ICIS System Administrator

Region 1 - ICIS Steward and Data Systems Administrator

Region 2 - ICIS System Administrator

Region 3 - ICIS Data Steward and System Administrator

Region 4 - ICIS System Administrator, Regional Compliance and Enforcement Data Steward

Region 5 - ICIS Data Steward and Systems Administrator

Region 6 - ICIS Data Steward

Region 7 - ICIS Data Steward and Systems Administrtor

Region 8 - ICIS System Administrator

Region 9 - ICIS System Administrator

Region 10 - ICIS System Administrator and Data Steward

3d. Calculation Methodology:

The Case Conclusion Data Sheet (CCDS) is a manual data collection tool HQ implemented in FY 1996, updated in FY 2012, to collect information on concluded federal enforcement cases including the case name and identification number, injunctive relief, environmental benefits (including environmental benefits from Supplemental Environmental Projects [SEPs]), and assessed penalties. The CCDS data are entered into the Integrated Information and Compliance System (ICIS). OECA uses data obtained from the CCDS via ICIS to assess the environmental outcomes of its enforcement program.

The CCDS guidance provides detailed calculation methodologies for estimating the environmental benefits on a variety of environmental statutes including air, water, waste, toxics and pesticides. Additionally, the CCDS provides specific instruction on how to enter the environmental benefits information into ICIS.

To view the the CCDS guidance in its entirety go to:

Attached Documents:

CCDS.xps

4a. Oversight and Timing of Final Results Reporting:

Oversight of Final Reporting: The Deputy Regional Administrators, the Office of Civil Enforcement Director, and the Monitoring, Assistance and Program Division Director all must sign the attached certification form.

Timing of Results Reporting: Semiannually

Attached Documents:

Data Certification Form.pdf

4b. Data Limitations/Qualifications:

Pollutant reductions or eliminations reported in ICIS project an estimate of pollutants to be reduced or eliminated if the defendant carries out the requirements of the settlement. (Information on expected outcomes of state enforcement is not available.) The estimates use information available at the time a case settles or an order is issued. In some instances, EPA develops and enters this information on pollutant reduction estimates after the settlement or during continued discussions over specific plans for compliance. Due to the time required for EPA to negotiate a settlement agreement with a defendant, there may be a delay in completing the CCDS. Additionally, because of unknowns at the time of settlement, different levels of technical proficiency, or the nature of a case, OECA's expectation is that the overall amount of pollutants reduced or eliminated is prudently underestimated based on CCDS information. EPA also bases the pollutant estimates on the expectation that the defendant/respondent implements the negotiated settlement agreement.

4c. Third-Party Audits:

Inspector General Report on Pounds of Pollution Reduced Estimates:

Attached Documents:

Projected Lbs of Pollutants Reduced.pdf

Measure Code: 405 - Millions of pounds of hazardous waste reduced, treated, or eliminated through concluded enforcement actions.

Office of Enforcement and Compliance Assurance (OECA)

Goal Number and Title:
5 - Enforcing Environmental Laws

Objective Number and Title:
1 - Enforce Environmental Laws

Sub-Objective Number and Title:
4 - Support Cleaning Up Communities and Advancing Sustainable Development

Strategic Target Code and Title:
1 - By 2015, reduce, treat, or eliminate 32,000 million estimated pounds of hazardous waste

Managing Office:
Office of Compliance

1a. Performance Measure Term Definitions:
Hazardous waste: Hazardous waste is defined as liquid, solid, contained gas, or sludge wastes that contain properties that are dangerous or potentially harmful to human health or the environment. Hazardous wastes are generally regulated by the Resource Conservation and Recovery Act (RCRA) and cleaned up under the RCRA Corrective Action Program or CERCLA (Comprehensive Environmental Response, Compensation, and Liability Act; also known as Superfund). RCRA is comprised of three major programs: Subtitle C (the hazardous waste management program), Subtitle D (the solid waste program), and Subtitle I (the UST program). Under Subtitle C, EPA has developed a comprehensive program to ensure that all hazardous waste is safely managed from the time it is generated to its final disposition at a Treatment, Storage, or Disposal (TSD) facility. The objective of the "cradle-to-grave" management system is to ensure that hazardous waste is handled in a manner that protects human health and the environment. To this end, there are Subtitle C regulations for the generation, transportation, and treatment, storage, or disposal of hazardous wastes.
Through the RCRA Corrective Action Program, EPA requires the investigation and cleanup, or in-situ or ex-situ treatment of hazardous releases at RCRA facilities. The corrective action program is structured around elements common to most cleanups under other EPA programs: an initial site assessment, characterization of the contamination, and the evaluation and implementation of cleanup alternatives, both immediate and long-term. Components of a cleanup action can impact all media types, including releases to the air, surface or groundwater, and cleanup of contaminated soil.
For more information on the different types of hazardous waste go to:
http://www.epa.gov/wastes/hazard/wastetypes/index.htm
Reduced, Treated or Eliminated: Reduced, treated, or eliminated is the quantity of pollutant(s) that will no longer be released to the environment as a result of a non-complying facility returning to its allowable permit limits through the successful completion of an enforcement settlement. Facilities may further reduce, treat or eliminate pollutants by carrying out voluntary Supplemental Environmental Projects.

Concluded enforcement actions: For purposes of this measure, there are two categories of concluded enforcement actions counted.

The first are administrative enforcement actions which are undertake by EPA through authority granted to it under various federal environmental statutes, such as CERCLA, RCRA, CAA, CWA, TSCA, and others. Administrative enforcement actions can take several forms, including EPA issuing an administrative order requiring a facility to implement specific corrective measures to filing an administrative complaint commencing a formal administrative adjudication. An administrative action is concluded when a written agreement between the defendant/respondent and EPA resolving the complaint is documented, signed by the Regional Administrator or designee, and is filed with the regional hearing clerk.

The second type of enforcement action is known as a civil judicial action which is a formal lawsuit, filed in court, against a person who has either failed to comply with a statutory or regulatory requirement or an administrative order. Civil judicial actions attorneys from the U.S. Department of Justice prosecute civil cases for EPA. A concluded action occurs when a consent decree is signed by all parties to the action and filed in the appropriate court and signed by a judge or a written ruling or decision is made by a judge after a full trial.

2a. Original Data Source:

 EPA Regional Enforcement Organizations
EPA Regional Program Organizations
EPA Headquarters Enforcement Organizations
Facility Personnel and Facility Contractors
DOJ

2b. Source Data Collection:

EPA calculates the estimated pollutant reductions after case settlement or during discussions with the facility personnel over specific plans for compliance. The final enforcement documents often spell out the terms and methodologies the facility must follow to mitigate and prevent the future release of pollutants.   These documents serve as the starting point for EPA's calculations.

Example of consent decree document containing pollutant mitigation instructions to the facility:
http://www.epa.gov/compliance/resources/cases/civil/caa/essroc.html

2c. Source Data Reporting:

When a formal administrative or judicial enforcement case is "concluded" enforcement staff enters information into ICIS to document the environmental benefits achieved by the concluded enforcement case. Original source documents may include facility permits, legal documents such as consent decrees and administrative orders, inspection reports, case engineer reports and facility reports. For civil judicial cases, the information is reported when a consent decree or court order, or judgment is entered (not lodged). For administrative cases, information is reported when an administrative order or final agreement is signed.

Environmental benefits should be reported in the year the case is settled, regardless of when the benefits will occur. Reductions are calculated after the judicial consent decree is lodged or entered, or when the administrative compliance order is signed by the region designee and filed with the regional hearing clerk.

Attached Documents:
FY2012 CCDS.docx

3a. Relevant Information Systems:

The ICIS FE&C data system meets Office of Environmental Information (OEI) Lifecycle Management Guidance, which includes data validation processes, internal screen audit checks and verification, system and user documents, data quality audit reports, third party testing reports, and detailed report specifications data calculation methodology. Reference: Quality Assurance and Quality Control procedures: Data Quality: Life Cycle Management Policy, (EPA CIO2121, April 7, 2006)

The Integrated Compliance Information System (ICIS) is a three phase multi-year modernization project that improves the ability of EPA and the states to ensure compliance with the nation's environmental laws with the collection of comprehensive enforcement and compliance information. Phase I, implemented in FY02, replaced several legacy systems, and created an integrated system to support federal enforcement and compliance tracking, targeting and reporting, including GPRA reporting. Phase II, also called Permit Compliance System (PCS) Modernization, expands ICIS to include the National Pollutant Discharge Elimination System (NPDES) program and enables improved management of the complete program (e.g., stormwater) as well as replacing the legacy PCS. PCS is currently identified as an Agency Federal Managers' Financial Integrity Act (FMFIA) weakness, and the modernization of the system is critical to address the weakness. Phase II was first implemented in FY06 for 21 states and 11 tribes/territories that use ICIS to directly manage their NPDES programs. In FY08, seven more states moved to ICIS from the legacy PCS and began electronically flowing their Discharge Monitoring Report (DMR) data from their states systems via the Exchange Network and CDX to ICIS. In FY09, Phase II continued with implementation of the National Installation of NetDMR allowing NPDES permittees to electronically submit DMR data from permitted facility systems via the Exchange Network to ICIS and migrated three additional states. In FY11 OECA implemented Full-Batch Release 1 of Phase II allowing Batch Flows of permits and facility data from states. FY12 will include Full-Batch Release 2 enabling batch flow will allow Batch Flows of inspection data from states. Inspection information and was implemented early in FY12. The final part of Phase II which will add the remaining NPDES Batch Flows and migrate and all remaining states is projected to be completed in FY13. Phase III will modernize the Air Facility System (AFS) into ICIS. AFS is used by EPA and States to track Clean Air Act enforcement and compliance activities. Integration of AFS into ICIS will modernize and replace a legacy system that does not meet current business needs. Implementation of this phase is projected for FY14.

ICIS contains both source data and transformed data.

OECA's Data System Quality Assurance Plan

Attached Documents:
Data System Quality Assurance Plan (ICIS).doc

3b. Data Quality Procedures:

Annual Data Certification Process - OECA has instituted a semi-annual data certification process for the collection and reporting of enforcement and compliance information. The certification process was set up to ensure all reporting entities are aware of the reporting deadlines, receive the most up-to-date reporting instructions for select measures, follow best data management practices to assure reporting accuracy, and have access to the recent methodologies for calculating pounds of pollutants reduced. The hazardous waste pounds of pollutants reduced measure is covered by the annual data certification process.

As part of the annual data certification process, regions are provided a checklist to assist them in their data quality procedures.

OECA's Quality Management Plan - September 2011

Attached Documents:
FY11 Data Quality Check List.pdf
OC QMP Concurrence Signatures.pdf
OC QMP 2011 Final.docx

3c. Data Oversight:

Source Data Reporting Oversight
HQ - Director, Enforcement Targeting and Data Division
Region 1 - Division Director, Office of Environmental Stewardship
Region 2 - Director, Office of Enforcement and Compliance Assistance
Region 3 - Director, Office of Enforcement, Compliance and Environmental Justice
Region 4 - Regional Counsel and Director, Office of Environmental Accountability
Region 5 - Director, Office of Enforcement and Compliance Assurance
Region 6 - Compliance Assurance and Enforcement Division Director
Region 7 - Enforcement Coordinator
Region 8 - Assistant Regional Administrator for Enforcement, Compliance and Environmental Justice
Region 9 - Enforcement Coordinator
Region 10 - Director, Office of Compliance and Enforcement

Information Systems Oversight Personnel
HQ - ICIS System Administrator
Region 1 - ICIS Steward and Data Systems Administrator
Region 2 - ICIS System Administrator
Region 3 - ICIS Data Steward and System Administrator
Region 4 - ICIS System Administrator, Regional Compliance and Enforcement Data Steward
Region 5 - ICIS Data Steward and Systems Administrator
Region 6 - ICIS Data Steward
Region 7 - ICIS Data Steward and Systems Administrtor
Region 8 - ICIS System Administrator
Region 9 - ICIS System Administrator
Region 10 - ICIS System Administrator and Data Steward

3d. Calculation Methodology:

The Case Conclusion Data Sheet (CCDS) is a manual data collection tool HQ implemented in FY 1996, updated in FY 2012, to collect information on concluded federal enforcement cases including the case name and identification number, injunctive relief, environmental benefits (including environmental benefits from Supplemental Environmental Projects [SEPs]), and assessed penalties. The CCDS data are entered into the Integrated Information and Compliance System (ICIS). OECA uses data obtained from the CCDS via ICIS to assess the environmental outcomes of its enforcement program.

The CCDS guidance provides detailed calculation methodologies for estimating the environmental benefits on a variety of environmental statutes including air, water, waste, toxics and pesticides. Additionally, the CCDS provides specific instruction on how to enter the environmental benefits information into ICIS.

To view the the CCDS guidance in its entirety go to:

Attached Documents:

CCDS.xps

4a. Oversight and Timing of Final Results Reporting:

Oversight of Final Reporting: The Deputy Regional Administrators, the Office of Civil Enforcement Director, and the Monitoring, Assistance and Program Division Director all must sign the attached certification form.

Timing of Results Reporting: Semiannually

Attached Documents:

Data Certification Form.pdf

4b. Data Limitations/Qualifications:

Pollutant reductions or eliminations reported in ICIS project an estimate of pollutants to be reduced or eliminated if the defendant carries out the requirements of the settlement. (Information on expected outcomes of state enforcement is not available.) The estimates use information available at the time a case settles or an order is issued. In some instances, EPA develops and enters this information on pollutant reduction estimates after the settlement or during continued discussions over specific plans for compliance. Due to the time required for EPA to negotiate a settlement agreement with a defendant, there may be a delay in completing the CCDS. Additionally, because of unknowns at the time of settlement, different levels of technical proficiency, or the nature of a case, OECA's expectation is that the overall amount of pollutants reduced or eliminated is prudently underestimated based on CCDS information. EPA also bases the pollutant estimates on the expectation that the defendant/respondent implements the negotiated settlement agreement.

4c. Third-Party Audits:

Inspector General Report on Pounds of Pollutants Reduced estimates:

Attached Documents:

Projected Lbs of Pollutants Reduced.pdf

Measure Code: 409 - Number of federal inspections and evaluations.

Office of Enforcement and Compliance Assurance (OECA)

Goal Number and Title:
5 - Enforcing Environmental Laws

Objective Number and Title:
1 - Enforce Environmental Laws

Sub-Objective Number and Title:
1 - Maintain Enforcement Presence

Strategic Target Code and Title:
1 - By 2015, conduct 105,000 federal inspections and evaluations

Managing Office:

1a. Performance Measure Term Definitions:
Conduct: Performance of activities involving observation of facility operations and collection of data for the purpose of determining compliance status. Federal: Activities authorized by, and conducted on behalf of, EPA. Inspections: On-site activities conducted for the purpose of establishing the compliance status of facilities or sites with applicable laws, standards, regulations, permits, and/or of supporting appropriate enforcement action (administrative, civil judicial or criminal) including: (1) Observation of pollution abatement equipment, facility operations, maintenance practices, self monitoring practices, records, and laboratory equipment; (2) Collection of evidence, including but not limited to emission monitoring measurements, other analytical field procedures such as sampling, the associated quality assurance procedures, and in-depth engineering evaluations. Evaluations: Clean Air Act Evaluations can be either a Full Compliance Evaluation (FCE) or a Partial Compliance Evaluation (PCE). A Full Compliance Evaluation (FCE) is a comprehensive evaluation of the compliance status of the facility. An FCE includes: 1) a review of all required reports and the underlying records; 2) an assessment of air pollution control devices and operating conditions; 3) observing visible emissions; 4) a review of facility records and operating logs; 5) an assessment of process parameters, such as feed rates, raw material compositions, and process rates; and 6) a stack test if there is no other way to determine compliance with the emission limits. A Partial Compliance Evaluation (PCE) is a documented compliance assessment focusing on a subset of regulated pollutants, regulatory requirements, or emission units at a given facility.

2a. Original Data Source:
EPA Regional Enforcement Organizations EPA Regional Program Organizations

| EPA Headquarters Enforcement Organizations |
| Facility Personnel and Facility Contractors |
| EPA designated State or Tribal Government Personnel |

2b. Source Data Collection:

Collection Methodology: The source data for this measure is found in the inspector's report and documented on the Inspection Conclusion Data Sheet (ICDS). The ICDS reporting instructions may be found at:

http://intranet.epa.gov/oeca/oc/resources/etdd/reporting/fy2012/reportingplanfy12neiguide.pdf

Quality Procedures: As part of the annual data certification process, regions are provided a checklist to assist them in their data quality procedures. The checklist is found at:

http://intranet.epa.gov/oeca/oc/resources/etdd/reporting/fy2012/reportingplanfy12-attachment7-fy2012bestpractices.pdf

Geographical Extent: Site or facility specific data are reported by the EPA Regional offices and aggregated at the national level.

Spatial Detail: Site specific data identified by street address, city, state and zip code are reported by EPA Regional offices and aggregated at the national level.

2c. Source Data Reporting:

Data Submission and Data Entry: The Inspection Conclusion Data Sheet (ICDS) is used to record key activities and outcomes at facilities during on-site inspections and evaluations. Inspectors use the ICDS form while performing inspections or investigation to collect information on on-site complying actions taken by facilities, deficiencies observed, and compliance assistance provided. The information from the completed ICDS form is entered into ICIS or reported manually.

Frequency and Timing of Data Transmission: Enforcement staff report data on a semiannual fiscal year basis. Data must be submitted no later than April 15th (mid-year) and October 15th (end-of-year) for activity conducted in the preceding fiscal year.

3a. Relevant Information Systems:

The Integrated Compliance Information System Federal Enforcement & Compliance (ICIS FE&C) database tracks a portion of EPA's federal inspections. PCS and ICIS-NPDES track federal CWA/NPDES inspections. The Airs Facility Subsystem (AFS) tracks CAA stationary courses inspections and evaluations. The RCRAInfo database tracks RCRA Subtitle C inspections. UIC inspections are reported to the OW UIC database and can be reported manually to ICIS-FE&C by EPA regional staff.

ICIS. The Integrated Compliance Information System (ICIS) is a three phase multi-year modernization project that improves the ability of EPA and the states to ensure compliance with the nation's environmental laws with the collection of comprehensive enforcement and compliance information. Phase I, implemented in FY02, replaced several legacy systems, and created an integrated system to support federal enforcement and compliance tracking, targeting and reporting, including GPRA reporting. Phase II, also called Permit Compliance System (PCS) Modernization, expands ICIS to include the National Pollutant Discharge Elimination System (NPDES) program and enables improved management of the complete program (e.g., stormwater) as well as replacing the legacy PCS. PCS is currently identified as an Agency Federal Managers' Financial Integrity Act (FMFIA) weakness, and the modernization of the system is critical to address the weakness. Phase II was first implemented in FY06 for 21 states and 11 tribes/territories that use ICIS to directly manage their NPDES programs. In FY08, seven more states moved to ICIS from the legacy PCS and began electronically flowing their Discharge Monitoring Report (DMR) data from their states systems via the Exchange Network and CDX to ICIS. In FY09, Phase II continued with implementation of the National Installation of NetDMR allowing NPDES permittees to electronically submit DMR data from permitted facility systems via the Exchange Network to ICIS and migrated three additional states. In FY11 OECA implemented Full-Batch Release 1 of Phase II allowing Batch Flows of permits and facility data from states. FY12 will include Full-Batch Release 2 enabling batch flow will allow Batch Flows of inspection data from states. Inspection information and was implemented early in FY12. The final part of Phase II which will add the remaining NPDES Batch Flows and migrate and all remaining states is projected to be completed in FY13. Phase III will modernize the Air Facility System (AFS) into ICIS. AFS is used by EPA and States to track Clean Air Act enforcement and compliance activities. Integration of AFS into ICIS will modernize and replace a legacy system that does not meet current business needs. Implementation of this phase is projected for FY14.

ICIS contains both source data and transformed data.

The ICIS FE&C data system meets Office of Environmental Information (OEI) Lifecycle Management Guidance, which includes data validation processes, internal screen audit checks and verification, system and user documents, data quality audit reports, third party testing reports, and detailed report specifications data calculation methodology. Reference: Quality Assurance and Quality Control procedures: Data Quality: Life Cycle Management Policy, (EPA CIO2121, April 7, 2006)

To support the Government Performance and Results Act (GPRA), the Agency's information quality guidelines, and other significant enforcement and compliance policies on performance measurement, OECA instituted a semiannual certification of the overall accuracy of ICIS information.

ICIS contains both source data and transformed data.

Attached Documents:
Data System Quality Assurance Plan (ICIS).doc

3b. Data Quality Procedures:

Annual Data Certification Process - OECA has instituted a semi-annual data certification process for the collection and reporting of enforcement and compliance information. The certification process was set up to ensure all reporting entities are aware of the reporting deadlines, receive the most up-to-date reporting instructions for select measures, follow best data management practices to assure reporting accuracy, and have access to the recent methodologies for calculating pounds of pollutants reduced. The air pounds of pollutants reduced measure is covered by the annual data certification process. Each office within the Office of Enforcement and Compliance Assurance (OECA) prepares Quality Management Plans (QMPs) every five years. As part of the annual data certification process, regions are provided a checklist to assist them in their data quality procedures (attached).

Attached Documents:

OC QMP Concurrence Signatures.pdf

FY12 Data Quality Check List.pdf

OC QMP 2011 Final.docx

3c. Data Oversight:

Source Data Reporting Oversight

HQ - Director, Enforcement Targeting and Data Division

Region 1 - Division Director, Office of Environmental Stewardship

Region 2 - Director, Office of Enforcement and Compliance Assistance

Region 3 - Director, Office of Enforcement, Compliance and Environmental Justice

Region 4 - Regional Counsel and Director, Office of Environmental Accountability

Region 5 - Director, Office of Enforcement and Compliance Assurance

Region 6 - Compliance Assurance and Enforcement Division Director

Region 7 - Enforcement Coordinator

Region 8 - Director, Policy, Information Management and Environmental Justice

Region 9 - Enforcement Coordinator

Region 10 - Director, Office of Compliance and Enforcement

Information Systems Oversight Personnel

HQ - ICIS System Administrator

 Region 1 - ICIS Steward and Data Systems Administrator

Region 2 - ICIS System Administrator

Region 3 - ICIS Data Steward and System Administrator

Region 4 - ICIS System Administrator, Regional Compliance and Enforcement Data Steward

Region 5 - ICIS Data Steward and Systems Administrator

Region 6 - ICIS Data Steward

Region 7 - ICIS Data Steward and Systems Administrator

Region 8 - ICIS System Administrator

Region 9 - ICIS System Administrator

Region 10 - ICIS System Administrator and Data Steward

3d. Calculation Methodology:

Decision Rules for Selecting Data: The following federal program area inspection and evaluation types are counted:

CAA: CAA Stationary Source FCEs, CAA Stationary Source PCEs, CAA CFC only PCEs, CAA Mobile Sources, CAA CFC only FCEs, CAA 112r, Asbestos D & R, Wood Heater Evaluation

CWA: NPDES Minors, NPDES Majors, Pretreatment IUs, Pretreatment POTWs, CWA-311(FRP), CWA-311(SPCC), CWA-404 (Wetlands)

EPRCA: EPCRA 313 Data Quality, EPCRA 313 Non Data Quality (Non-Reporters), EPCRA non-313

FIFRA: FIFRA, FIFRA GLP

MPRSA: MPRSA

RCRA: RCRA-HW, RCRA-UST, RCRA 4005 Subtitle D

SDWA: SDWA-PWSS, SDWA-UIC

TCSA: TSCA GLP, TSCA Core, TSCA PCBs, TSCA Asbestos/ AHERA, TSCA Lead-based Paint

Definitions of Variables: "Not applicable."

Explanation of Calculations: "Not applicable."

Explanation of Assumptions: "Not applicable."

Unit of Measure: Numerical

Timeframe of Result: Annually by fiscal year.

Documentation of Methodological Changes: "Not applicable."

4a. Oversight and Timing of Final Results Reporting:

Final Reporting Oversight Personnel:

Oversight of Final Reporting:
The Deputy Regional Administrators, the Office of Civil Enforcement Director, and the Monitoring, Assistance and Program Division Director all must sign the attached certification form.

Frequency and Timing of Results Reporting: Semiannually.

Attached Documents:

OC QMP Concurrence Signatures.pdf
4b. Data Limitations/Qualifications:
GenerGeneral Limitations/Qualifications: No error estimate is available. Data Lag Length and Explanation: From the October 31 end-of-year date, approximately 10-12 weeks to allow for QA/QC of the data. Methodological Changes: "Not Applicable."
4c. Third-Party Audits:
Not Applicable

Measure Code: 421 - Percentage of conviction rate for criminal defendants.

Office of Enforcement and Compliance Assurance (OECA)

Goal Number and Title:	
5 - Enforcing Environmental Laws	
Objective Number and Title:	
1 - Enforce Environmental Laws	
Sub-Objective Number and Title:	
1 - Maintain Enforcement Presence	
Strategic Target Code and Title:	
7 - By 2015, maintain an 85 percent conviction rate for criminal defendants	
Managing Office:	
Office of Criminal Enforcement;Forensics and Training	
1a. Performance Measure Term Definitions:	

Criminal Cases: A criminal case exists when EPA's criminal enforcement program, specifically special agents in the Criminal Investigation Division (CID), investigate allegations of criminal violations of environmental law. The EPA active ("open") criminal case docket consists of cases in all stages of the legal process – from initial investigations to charged cases to convicted cases that are awaiting sentencing or are on appeal.

A criminal case with charges filed is one in which, based upon an investigation by the EPA criminal enforcement program, the U.S. Department of Justice formally files charges against one or more defendants (either a person, company or both) alleging a criminal violation of one or more of the environmental statutes and/or associated violations of the U.S. Criminal Code in U.S. District Court.

Conviction: A defendant (either a person or company) who has been previously charged with committing one or more environmental crimes is found legally "guilty" of at least one of those crimes. Legal guilt (conviction) occurs either when the defendant pleads guilty or is convicted following a trial.

For more information about EPA's Criminal Enforcement Program, visit http://www.epa.gov/compliance/criminal/

2a. Original Data Source:

As part of the investigative process, the Criminal Investigation Division (CID) special agent assigned completes an Investigation Activity Report (IAR). The IAR is the primary means used to document all investigative activity, operational activities, judicial activities, or responses to investigative tasking or leads. Investigative activities include interviews, surveillance, electronic monitoring, arrests, searches, evidence handling and disposition, and document reviews. Operational activities include undercover reports, and consensual monitoring. Judicial activities include indictments, criminal informations, criminal complaints, guilty pleas, trials, convictions, and sentencing hearings and results. Investigative tasking relates to collateral requests from CID headquarters and other offices, as well as memorializing activity conducted in furtherance of lead inquiries. All relevant data is entered into the Criminal Case Reporting System (CCRS, cf section 3a), which tracks a criminal investigation from the time it is first opened through all stages of the legal process to a conclusion (e.g., when the case is indicted, when a defendant is convicted, sentenced or acquitted).CCRS is used to create

the IAR
The data used to compile the measure is based upon the legal documents filed in the U.S. District Court where the defendant is prosecuted. Charges can be dismissed after exculpatory evidence in their favor was entered into the record or the legal process can results in either a conviction or an acquittal. A conviction is also reaffirmed at the subsequent sentencing of a convicted defendant, when the judge imposes the sentence through a legal document known as the Judgment and Commitment Notice. (J&C).

2b. Source Data Collection:

Source Data Collection Methods:
The measure is based upon enforcement and legal documents which memorialize the status of a criminal prosecution. As noted above, the data for the measure are formally compiled through the IARs and DOJ legal documents entered into CCRS. In addition, all public legal documents relating to a charged case, including the conviction, are also entered into and are publicly available through Public Access to Court Electronic Records (PACER), an electronic public access service that allows users to obtain case and docket information from federal appellate, district and bankruptcy courts

(http://www.pacer.gov/
Date/Time Intervals Covered by Source Data:
Ongoing.

Geographical Extent of Source Data:
National.

2c. Source Data Reporting:

The status of the case is updated as the legal process proceeds. The case agents update and enter into CCRS or submit to their superior IARs which highlight changes in the case and all subsequent stages of the criminal enforcement process (e.g., a case is dismissed or the defendants are either acquitted or convicted and sentenced)
Timing and frequency of reporting: The status of the case is updated as the legal process proceeds

3a. Relevant Information Systems:

The Criminal Case Reporting System (CCRS) stores criminal enforcement data in an enforcement sensitive database which contains historical data on all criminal enforcement prosecutions as well as information about the pollutants involved and the impact on the public and the environment. CCRS maintains information pertaining to individuals and companies associated with the Criminal Investigation Division's criminal leads and cases, as well as other information related to the conduct of criminal investigations.
 The data is used to document the progress and results of criminal investigations. The data used for all criminal enforcement performance measures are in the CCRS database.
The status of the case is updated on CCRS as the legal process proceeds. All legal documents relating to a prosecution are entered into the system.

3b. Data Quality Procedures:

The Criminal Investigations Division (CID) has a process for document control and records management and has Quality Management Plans in place. The information on defendant dismissals, convictions or acquittals that is entered into CCRS goes through several layers of review. Initial verification of the quality and accuracy of case information is the responsibility of the Special Agent-in-Charge (SAC) of the office that is managing the case. HQ responsibility for QA/QC is conducted by the System Administrator of CCRS.

3c. Data Oversight:

Initial oversight, review and quality assurance at the field level is the responsibility of the Special Agent-in-Charge (SAC) and Assistant Special Agent-in-Charge (ASAC) of the criminal enforcement office managing the case. That information is further reviewed by OCEFT HQ through semi-annual case management reviews conducted by the Assistant Director of Investigations, CID, and quarterly reports by the System Administrator of CCRS. The System Administrator, who creates all statistical and management reports based on information in CCRS, conducts regular oversight of the data entered by the criminal enforcement field offices to ensure that all data entered into CCRS is complete and accurate.

3d. Calculation Methodology:

The methodology for the criminal enforcement measure "Conviction rate for criminal defendants" employed a five year analysis (FY2006-2010) to develop the baseline and targets. The decision rules reflect the legal status of the defendants. The data files relevant to this analysis include defendant names and type (individual or company), date of charges filed and the results (convicted, acquitted, or charges dismissed) of the prosecution regarding each of the charges on which the defendant was found guilty or not guilty (either or both environmental law or general U.S. Criminal Code). A defendant is defined as having been "convicted" if he is guilty of at least one of the criminal counts of which he has been charged.

There are no "assumptions" or "quantifiers" used in calculating the measure. The measure is based upon the legal status of cases, i.e., whether the defendant has been convicted, acquitted or had the charges dismissed after exculpatory evidence in their favor was entered into the record. The measure is calculated by dividing the total number of defendants who have been convicted during the current Fiscal Year (numerator) by the total number of defendants with a legal result of their case in the current Fiscal Year (denominator). The "legal result" denominator includes all defendants whose charges were dismissed, who were acquitted or had their charges overturned on appeal following conviction.
Semiannual reporting.
Unit of analysis: Percent.

4a. Oversight and Timing of Final Results Reporting:

Oversight of Final Reporting: The System Administrator of the OCEFT CCRS has the responsibility for compiling and verifying the accuracy of the report on the percentage of convicted defendants. Once compiled, data goes through a second level of verification through the Assistant Director of Investigations, CID. While data is verified on an on-going basis, final verification is conducted at the end of the year.

Timing of Results Reporting:
Semiannually.

4b. Data Limitations/Qualifications:

The only data limitations that result (although infrequently) occur when a defendant who has been initially convicted of one or more environmental crimes has all of his charges overturned by the U.S. Appellate Court on appeal in a subsequent fiscal year than the one in which the measure is being reported. The conviction rate for charged defendants has historically been in the 90% range, and is not materially affected by post-conviction appeals, so the low incidence of defendants having their convictions eventually overturned does not limit the suitability of the performance measure.

4c. Third-Party Audits:

N/A

Office of Enforcement and Compliance Assurance (OECA)

Goal Number and Title:
5 - Enforcing Environmental Laws
Objective Number and Title:
1 - Enforce Environmental Laws
Sub-Objective Number and Title:
1 - Maintain Enforcement Presence
Strategic Target Code and Title:
6 - By 2015, increase the number of criminal cases with charges filed
Managing Office:
Office of Criminal Enforcement;Forensics and Training
1a. Performance Measure Term Definitions:
Criminal Cases: A criminal case exists when EPA's criminal enforcement program, specifically special agents in the Criminal Investigation Division (CID), investigate allegations of criminal violations of environmental law. The EPA active ("open") criminal case docket consists of cases in all stages of the legal process – from initial investigations to charged cases to convicted cases that are awaiting sentencing or are on appeal. Charges Filed: A criminal case with charges filed is one in which, based upon an investigation by the EPA criminal enforcement program, the U.S. Department of Justice formally files charges against one or more defendants (either a person, company or both) alleging a criminal violation of one or more of the environmental statutes and/or associated violations of the U.S. Criminal Code in U.S. District Court. For more information about EPA's Criminal Enforcement Program, visit http://www.epa.gov/compliance/criminal/
2a. Original Data Source:
As part of the investigative process, the Criminal Investigation Division (CID) special agent assigned to the case completes an Investigation Activity Report (IAR). The IAR is the primary means used to document all investigative activity, operational activities, judicial activities, or responses to investigative tasking or leads. Investigative activities include interviews, surveillance, electronic monitoring, arrests, searches, evidence handling and disposition, and document reviews. Operational activities include undercover reports, and consensual monitoring. Judicial activities include indictments, criminal informations, criminal complaints, guilty pleas, trials, convictions, and sentencing hearings and results. Investigative tasking relates to collateral requests from CID headquarters and other offices, as well as memorializing activity conducted in furtherance of lead inquiries. All relevant data is entered into the Criminal Case Reporting System (CCRS, cf section 3a), which tracks a criminal investigation from the time it is first opened through all stages of the legal process to a conclusion (e.g., when the case is indicted, when a defendant is found guilty, sentenced or acquitted.) CCRS is used to create the IAR. Once the defendants are charged, the data used to compile the measure is based upon the legal documents

outlining the criminal charges (which can either take the form of a criminal information or criminal indictment) that is filed by either the Office of the U.S. Attorney or the Environmental Crimes Section at DOJ HQ and filed in the U.S. District Court in which the alleged criminal violations occurred. The charges are part of the case file.

2b. Source Data Collection:

Source Data Collection Methods: The measure is based upon enforcement and legal documents which memorialize the status of a criminal prosecution. As noted above, the data for the measure are formally compiled through the IARs and DOJ legal documents entered into CCRS. In addition, all public legal documents relating to a charged case (e.g., the indictment or criminal information), including the names of all defendants, is also entered into and are publicly available through Public Access to Court Electronic Records (PACER), an electronic public access service that allows users to obtain case and docket information from federal appellate, district and bankruptcy courts (http://www.pacer.gov/

Date/time Intervals Covered by Source Data:
Ongoing.

EPA QA Requirements/Guidance Governing Collection:
All criminal enforcement special agents receive training on the accurate completion of IAR reports and the entry of criminal case data into the CCRS.

Geographical Extent of Source Data:
National.

2c. Source Data Reporting:

After DOJ formally charges the defendants, the information is entered into CCRS (e.g., all the violations alleged, all of the defendants charged, as well as forensic information about the pollutants involved and the impact on the public and the environment.) The status of the case is updated as the legal process proceeds. The case agents update and enter into CCRS or submit to their superior IARs which highlight changes in the case and all subsequent stages of the criminal enforcement process (e.g., a case is dismissed or the defendants are either acquitted or convicted and sentenced.)
Timing and frequency of reporting: The status of the case is updated as the legal process proceeds

3a. Relevant Information Systems:

The Criminal Case Reporting System (CCRS) stores criminal enforcement information and data in an enforcement sensitive database which contains historical data on all criminal enforcement prosecutions as well as information about the pollutants involved and the impact on the public and the environment. CCRS maintains information pertaining to individuals and companies associated with the Criminal Investigation Division's criminal leads and cases, as well as other information related to the conduct of criminal investigations.
. The data is used to document the progress and results of criminal investigations. The data used for all criminal enforcement performance measures are in the CCRS database.

The status of the case is updated on CCRS as the legal process proceeds. All legal documents relating to a prosecution are entered into the system

3b. Data Quality Procedures:

The Criminal Investigations Division (CID) has a process for document control and records and has Quality Management Plans in place. The information on charged cases that is entered into CCRS goes through several layers of review. Initial verification of the quality and accuracy of case information is the responsibility of the Special Agent-in-Charge (SAC) of the office that is managing the case. HQ responsibility for QA/QC is conducted by the System Administrator of CCRS

3c. Data Oversight:

Initial oversight at the field level is the responsibility of the Assistant Special Agent-in-Charge (ASAC) and Special Agent-in-Charge (SAC) of the criminal enforcement office managing the case. That information is further reviewed by OCEFT HQ through semi-annual case management reviews conducted by the Assistant Director of Investigations, CID, and quarterly reports by the System Administrator of CCRS. The System Administrator, who creates all statistical and management reports based on information in CCRS, conducts regular oversight of the data entered by the criminal enforcement field offices to ensure that all data entered into CCRS is complete and accurate.

3d. Calculation Methodology:

The methodology for the criminal enforcement measure "Percent of criminal cases with charges filed" employed a five year analysis (FY2006-2010) to develop the baseline and targets. The decision rules reflect the legal status of the defendants charged. The data files relevant to this analysis include defendant names and type (individual or company), date of charges filed and the actual statutes (either or both environmental or U.S. Criminal Code) listed in the criminal indictment or criminal information.

There are no "assumptions" or "quantifiers" used in calculating the measure. The measure is based upon the legal status of cases, i.e., whether the case has been closed without prosecution or is being prosecuted. The measure is calculated by dividing the number of cases that have been charged (i.e., with an indictment or criminal information) during the current Fiscal Year (numerator) by the total number of criminal cases that were closed during the current Fiscal Year (denominator).

Time frame: Semiannual reporting.

Unit of analysis: Percent.

4a. Oversight and Timing of Final Results Reporting:

The System Administrator of the OCEFT CCRS has the responsibility for compiling and verifying the accuracy of the report on charged defendants. Once compiled, data goes through a second level of verification through the Assistant Director of Investigations, CID. While data is verified on an on-going basis, final verification is conducted at the end of the fiscal year.

Timing of Results Reporting:
Semiannually.

4b. Data Limitations/Qualifications:
N/A since the measure is based upon the legal status of charged cases

4c. Third-Party Audits:
N/A

Measure Code: 419 - Percentage of criminal cases with individual defendants.

Office of Enforcement and Compliance Assurance (OECA)

Goal Number and Title:
5 - Enforcing Environmental Laws
Objective Number and Title:
1 - Enforce Environmental Laws
Sub-Objective Number and Title:
1 - Maintain Enforcement Presence
Strategic Target Code and Title:
7 - By 2015, maintain an 85 percent conviction rate for criminal defendants
Managing Office:
Office of Criminal Enforcement;Forensics and Training
1a. Performance Measure Term Definitions:
Criminal Cases: A criminal case exists when EPA's criminal enforcement program, specifically special agents in the Criminal Investigation Division (CID), investigate allegations of criminal violations of environmental law. The EPA active ("open") criminal case docket consists of cases in all stages of the legal process – from initial investigations to charged cases to convicted cases that are awaiting sentencing or are on appeal. A criminal case with charges filed is one in which, based upon an investigation by the EPA criminal enforcement program, the U.S. Department of Justice formally files charges against one or more defendants (either a person, company or both) alleging a criminal violation of one or more of the environmental statutes and/or associated violations of the U.S. Criminal Code in U.S. District Court. Individual Defendants: An individual defendant is a person, as opposed to a company. Criminal enforcement can be employed against persons and companies. Individuals, unlike companies, can be sentenced to prison, as well as paying a monetary fine, for breaking the criminal law. It is the possibility of incarceration that most distinguishes criminal law from civil law and, therefore, enables criminal law to provide the most deterrence. For more information about EPA's Criminal Enforcement Program, visit http://www.epa.gov/compliance/criminal/
2a. Original Data Source:
As part of the investigative process, the Criminal Investigation Division (CID) special agent assigned creates an Investigative Activity Report (IAR). The IAR is the primary means used to document all investigative activity, operational activities, judicial activities, or responses to investigative tasking or leads. Investigative activities include interviews, surveillance, electronic monitoring, arrests, searches, evidence handling and disposition, and document reviews. Operational activities include undercover reports, and consensual monitoring. Judicial activities include indictments, criminal informations, criminal complaints, guilty pleas, trials, convictions, and sentencing hearings and results. Investigative tasking relates to collateral requests from CID headquarters and other offices, as well as memorializing activity conducted in furtherance of lead inquiries. All relevant data is entered into the Criminal Case Reporting System (CCRS, cf section 3a), which tracks a

criminal investigation from the time it is first opened through all stages of the legal process to a conclusion (e.g., when the case is indicted, when a defendant is found guilty, sentenced or acquitted.) CCRS is used to create the IAR.

Once the defendants are charged, the data used to compile the measure is based upon the legal documents outlining the criminal charges (which can either take the form of a criminal information or criminal indictment) that is filed by either the Office of the U.S. Attorney or the Environmental Crimes Section at DOJ HQ and filed in the U.S. District Court in which the alleged criminal violations occurred. The charges are part of the case file.

2b. Source Data Collection:

Source Data Collection Methods: The measure is based upon enforcement and legal documents which memorialize the status of a criminal prosecution. As noted above, the data for the measure are formally compiled through the IARs and DOJ legal documents entered into CCRS. In addition, all public legal documents relating to a charged case (e.g., the indictment or criminal information), including the names of all defendants, is also entered into and are publicly available through Public Access to Court Electronic Records (PACER), an electronic public access service that allows users to obtain case and docket information from federal appellate, district and bankruptcy courts http://www.pacer.gov/

Date/Time Intervals Covered by Source Data:
Ongoing.

EPA QA Requirements/Guidance Governing collection:
All criminal enforcement special agents receive training on the accurate completion of IAR reports and the entry of criminal case data into the CCRS.

Geographical Extent of Source Data:
National.

2c. Source Data Reporting:

After DOJ formally charges the defendants, the information is entered into CCRS (e.g., all the violations alleged, all of the defendants charged, as well as forensic information about the pollutants involved and the impact on the public and the environment.) The status of the case is updated as the legal process proceeds. The case agents update and enter into CCRS or submit to their superior IARs which highlight changes in the case and all subsequent stages of the criminal enforcement process (e.g., a case is dismissed or the defendants are either acquitted or convicted and sentenced.)
Timing and frequency of reporting: The status of the case is updated as the legal process proceeds.

3a. Relevant Information Systems:

The Criminal Case Reporting System (CCRS) stores criminal enforcement information and data in an enforcement sensitive database which contains historical data on all criminal enforcement prosecutions as well as information about the pollutants involved and the impact on the public and the environment. CCRS maintains information pertaining to individuals and companies associated with the Criminal Investigation Division's criminal leads and cases, as well as other information related to the conduct of criminal investigations.

The data is used to document the progress and results of criminal investigations. The data used for all criminal enforcement performance measures are in the CCRS database.

The status of the case is updated on CCRS as the legal process proceeds. All legal documents relating to a prosecution are entered into the system

3b. Data Quality Procedures:

The Criminal Investigations Division (CID) has a process for document control and records and has Quality Management Plans in place. The information on charged cases that is entered into CCRS goes through several layers of review. Initial verification of the quality and accuracy of case information is the responsibility of the Special Agent-in-Charge (SAC) of the office that is managing the case. HQ responsibility for QA/QC is conducted by the System Administrator of CCRS.

3c. Data Oversight:

Initial oversight at the field level is the responsibility of the Assistant Special Agent-in-Charge (ASAC) and Special Agent-in-Charge (SAC) of the criminal enforcement office managing the case. That information is further reviewed by OCEFT HQ through semi-annual case management reviews conducted by the Assistant Director of Investigations, CID and quarterly reports by the System Administrator of CCRS. The System Administrator, who creates all statistical and management reports based on information in CCRS, conducts regular oversight of the data entered by the criminal enforcement field offices to ensure that all data entered into CCRS is complete and accurate.

3d. Calculation Methodology:

The methodology for the criminal enforcement measure "Percent of criminal cases with individual defendants" employed a three year analysis (FY2008-2010) to develop the baseline and targets.

The decision rules reflect the legal status of the individuals who are named as charged defendants. The data files relevant to this analysis include defendant names and type (individual or company), date of charges filed and the actual statutes (either or both environmental or U.S. Criminal Code) listed in the criminal indictment or criminal information.

There are no assumptions or "quantifiers" used in calculating the measure. The measure is based upon the legal status of cases, i.e., whether the case has at least one individual person charged as a defendant that is being prosecuted. The measure is calculated by dividing the number of charged cases that have at least one individual defendant during the current Fiscal Year (numerator) by the total number of charged criminal cases during the current Fiscal Year (denominator).

Timeframe: Fiscal Year (October – September) Semiannual reporting.

Unit of analysis: Percent.

4a. Oversight and Timing of Final Results Reporting:

The System Administrator of the CCRS has the responsibility for compiling and verifying the accuracy of the report on charged defendants. Once compiled, data goes through a second level of verification through the Assistant Director of Investigations, CID. While data is verified on an on-going basis, final verification is

conducted at the end of the fiscal year.
Timing of Results Reporting:
Semiannually.
4b. Data Limitations/Qualifications:
N/A, since the measure is based on the legal status of prosecuted individual defendants
4c. Third-Party Audits:
N/A

Measure Code: 418 - Percentage of criminal cases having the most significant health, environmental, and deterrence impacts.

Office of Enforcement and Compliance Assurance (OECA)

Goal Number and Title:
5 - Enforcing Environmental Laws
Objective Number and Title:
1 - Enforce Environmental Laws
Sub-Objective Number and Title:
1 - Maintain Enforcement Presence
Strategic Target Code and Title:
5 - Each year through 2015, support cleanups and save federal dollars for sites
Managing Office:
Office of Criminal Enforcement
1a. Performance Measure Term Definitions:
Criminal Case Docket: A criminal case exists when EPA's criminal enforcement program, specifically special agents in the Criminal Investigation Division (CID), investigate allegations of criminal violations of environmental law. The EPA active ("open") criminal case docket consists of cases in all stages of the legal process – from initial investigations to charged cases to convicted cases that are awaiting sentencing or are on appeal.   Most Significant Health, Environmental, and Deterrence Impacts: The most significant cases are defined by the categories of health effects (e.g., death, serious injury, or exposure, etc.), pollutant release and discharge characteristics (e.g., documented exposure, need for remediation, etc.) and defendant profiles (e.g., size of business, compliance history, etc.) The cases with the most significant health, environmental and deterrent impacts fall into Tier 1 and Tier 2 of four possible categories of tiers (as calculated by the tiering methodology (cf section 3d). The tier designation is used throughout the investigative process including case selection and prosecution.   For more information about EPA's Criminal Enforcement Program, visit http://www.epa.gov/compliance/criminal/
2a. Original Data Source:
All data used to calculate and classify the "most significant cases" result from evidence collected during the investigative process. The Criminal Investigation Division (CID) special agent assigned to the case creates an Investigative Activity Report (IAR, cf 419,420, 421). The IAR is the primary means used to document all investigative activity operational activities, judicial activities, or responses to investigative tasking or leads. Investigative activities include interviews, surveillance, electronic monitoring, arrests, searches, evidence handling and disposition, and document reviews. Operational activities include undercover reports, and consensual monitoring. Judicial activities include legal documents such as indictments, criminal informations, criminal complaints, guilty pleas, trials, convictions, and sentencing hearings and results. Investigative tasking relates to collateral requests from CID headquarters and other offices, as well as memorializing activity conducted in furtherance of lead inquiries.

2b. Source Data Collection:

Source Data Collection Methods:

Tabulation of records or activities. Information used for the case tiering methodology (cf section 3d) comes from the evidence collected during the course of the investigation. Forensic evidence gathering (e.g., environmental sampling and analysis) is conducted by the National Enforcement Investigations Center (NEIC) or other EPA laboratories or programs in conformity with their established protocols.

The data for case tiering is compiled through the IARs and legal documents which are collected and entered into the Criminal Case Reporting System (CCRS, cf section 3a). OCEFT collects data on a variety of case attributes to describe the range, complexity, and quality of the national docket. Data for selected attributes are being used to categorize the cases into four tiers based on the severity of the crime associated with the alleged violation.

Date/Time Intervals Covered by Source Data:

Ongoing.

EPA QA requirements/guidance governing collection:

All criminal enforcement special agents receive training on the accurate completion of IAR reports and the entry of criminal case data into the CCRS.

Geographical Extent of Source Data:

National.

2c. Source Data Reporting:

Form/mechanism for receiving data and entering into EPA system:

After a criminal case is opened, all major data and information is entered into CCRS and is tracked through all subsequent stages of the criminal enforcement process. All case information and data that will be used for the case tiering methodology is entered into CCRS, including information about the pollutants involved and the impact on the public and the environment that result from forensic sampling and analysis undertaken as a routine part of the investigation of the alleged violations.

Timing and frequency of reporting: The status of the case is updated as the legal process proceeds.

3a. Relevant Information Systems:

CCRS stores criminal enforcement data in an enforcement sensitive database which contains historical data on all criminal enforcement prosecutions as well as information about the pollutants involved and the impact on the public and the environment. CCRS contains a drop down menu for entering all data used to assign a case to a specific tier. When all required fields are populated, the system automatically determines the tier for the case. Designating a tier is mandatory for all open criminal cases.

CCRS is an internal EPA database; All public legal documents relating to prosecuted criminal cases (e.g., the indictments, guilty pleas, trial verdicts and judge's sentencing decisions) are publicly available through Public Access to Court Electronic Records (PACER), an electronic public access service that allows users to obtain case and docket information from federal appellate, district and bankruptcy courts (http://www.pacer.gov/

3b. Data Quality Procedures:

Environmental and forensic data used to conduct case tiering is supplied from EPA's National Enforcement Investigations Center (NEIC), national databases, and other EPA programs. This data has been QA/QCd following the protocols established by those programs. It should be noted that the data will often serve as evidence in criminal judicial enforcement proceedings, so the quality and sufficiency of the data is carefully reviewed.

3c. Data Oversight:

Initial oversight at the field level is the responsibility of the Special Agent-in-Charge and Assistant Special Agent-in-Charge of the criminal office managing the case. That information is further reviewed by OCEFT HQ through semi-annual case management reviews conducted by the Assistant Director for Investigations, CID.

3d. Calculation Methodology:

The methodology for the measure "percent of criminal cases with the most significant health, environmental and deterrence impact" used the FY 2010 criminal enforcement docket to develop the baseline and targets for FY 2011-15.. The cases are analyzed and scored on a variety of case attributes describing the range, complexity and quality of the criminal enforcement docket. Cases are then entered into one of four categories ("tiers") depending upon factors such as the human health (e.g., death, serious injury) and environmental impacts, the nature of the pollutant and its release into the environment, and violator characteristics (e.g., repeat violator, size and location(s) of the regulated entity)

Many of the data elements used in the tier method are directly linked to the Federal Sentencing Guidelines:
http://www.ussc.gov/guidelines/2010_guidelines/index.cfm

See the two attachments for graphic representations of the criminal case tier methodology and the explanations of the categories. They indicate the process used to assign a case to one of the four tiers.

Tiering is based upon these decision rules:

Tier 1 (1st or highest): any case involving death or actual serious injury; otherwise a case that possesses specified attributes in at least three of four established categories.

Tier 2 (second): two categories out of four

Tier 3 (third): one category out of four

Tier 4 (fourth): no category

Tier 1 and Tier 2 cases added together and divided by the total number of open cases in the criminal case docket is how the "most significant cases" cases measure, that also serves as the Key Performance indicator for the criminal enforcement program, is calculated. The measure only reflects the percentage of cases in the upper two tiers.

Time frame: Updated throughout the fiscal year as the case docket changes. Fiscal Year (October – September) Semiannual reporting.

Unit of analysis: Percent.
Attached Documents:
tiering.pptx
tieirngmethodology2012.ppt

4a. Oversight and Timing of Final Results Reporting:
Oversight of Final Reporting:: Once initial case tiering has been conducted by the case agent, initial oversight, review and quality assurance at the field level is the responsibility of the Special Agent-in-Charge and Assistant Special Agent–in-Charge of the criminal enforcement office managing the case. It receives a second round of review in HQ by CID's Assistant Director for Investigations, who also conducts a semi-annual review of all cases in the criminal case docket . The review includes discussions of any new evidence or information that would potentially affect or change the tier in which a case had been assigned. Any decision to categorize a case as being a Tier 4 (lowest level) case must be approved by both the SAC and the Assistant Director for Investigations. Data is verified on an on-going basis. Timing of Results Reporting: Semiannually.
4b. Data Limitations/Qualifications:
A case's tier classification may change as cases are investigated and additional information uncovered. Potential data limitations include inaccurate environmental sampling or mistakes in evidence gathering that can result in improper classification or "tiering" of an individual case. Determining data for some characteristics used in tiering may be based upon ranges or estimates (e.g., the extent of documented human population exposure to a toxic pollutant may be based upon a consensus or "best estimate" of the geographic area surrounding the release rather than a detailed examination of all people potentially exposed).
4c. Third-Party Audits:
N/A

Measure Code: 412 - Percentage of open consent decrees reviewed for overall compliance status.

Office of Enforcement and Compliance Assurance (OECA)

Goal Number and Title:
5 - Enforcing Environmental Laws

Objective Number and Title:
1 - Enforce Environmental Laws

Sub-Objective Number and Title:
1 - Maintain Enforcement Presence

Strategic Target Code and Title:
4 - By 2015, review the overall compliance status of 100 percent of the open consent decrees

Managing Office:

1a. Performance Measure Term Definitions:

Review Overall Compliance Status: For each open judicial non-Superfund, non-bankruptcy judicial consent decree, the EPA Regions will track up to four milestones, depending on the content of the consent decrees and the length of their compliance schedules. Three of the milestones address specific, one-time events to be tracked in ICIS as Compliance Schedule Events: 1) Pay Required Penalty Amount in Full; 2) Complete Required Supplemental Environmental Project (SEP); 3) Achieve Final Compliance With All Obligations Under This Order. The fourth milestone addresses overall consent decree compliance status.

Open Consent Decrees: The consent decree tracking measures apply to open, non-Superfund, non-bankruptcy, judicial consent decrees, coded in ICIS with the Enforcement Action Types "Civil Judicial Actions," "Pre-Referral Negotiations," and "Collection Actions," entered by the courts in FY 2007 and later. Consent decrees that have been open for fewer than three years are excluded from the 4th measure, since Regions are expected to review the overall compliance status of such consent decrees beginning no later than on the first 3-year anniversary of their entry dates and repeat the reviews at least once every three years from the dates of the most recent reviews until the consent decrees are closed. ICIS tracking for review of open Consent Decrees every three years continues until the CD is closed.

The data collected and reported on consists of three specific, one-time event critical milestones, and one overall consent decree critical milestone. The milestones are:

·   number and percentage of open consent decrees in ICIS with a completed Federal Penalty Required field, but no corresponding Schedule Event or, the Schedule Date has passed with no Actual Date entered;

·   number and percentage of open consent decrees in ICIS with a SEP entered in the SEP screen, but no corresponding Schedule Event or, the Schedule Date has passed with no Actual Date entered;

·   number and percentage of all open consent decrees in ICIS without an Achieve Final Compliance With All Obligations Under This Order or, the corresponding Schedule Date has passed with no Actual Date entered.

·   number and percentage of open consent decrees, more than 3 years old for which a timely overall consent decree compliance status review has not been conducted.

The following new ICIS consent decree milestone reports have been created and are available in the Public folder under Federal Enforcement and Compliance (FE&C) – National Standards Reports:

- Consent Decree Tracking Measures 1 – 3 (Penalty, SEP, Final Compliance)
- Consent Decree Tracking Measure 4 (Overall Compliance Review)

These reports provide both cross-tab results for the new measures and detailed listings of the cases that underlie the cross-tab numbers. Both reports (as revised) will be used at end-of-year processing for populating the related table in the Certification Workbook.

Background: For more information on EPA's tracking of consent decrees, see Consent Decree Tracking Guidance (attached).

2a. Original Data Source:

EPA HQ and Regional enforcement staff

2b. Source Data Collection:

Collection Methodology: Data on Consent Decree tracking is collected by the regions (and for some cases HQ) from the Consent Decrees. These are the date of entry of the CD, the date a penalty is due, the date a SEP is to be completed, and the date all activities required under the CD are to be completed. This information is entered into ICIS by regional and HQ enforcement personnel. The same is true, later, for entry to ICIS of the actual dates when CD compliance was actually tracked, when the payment of the penalty actually occurred (or did not occur), when the SEP was completed, and when the activities required under the CD had all been completed.

Quality Procedures: Quality procedures are implemented by each EPA regional office and thru the Office of Compliance data certification process that requires review of enforcement data and certification by the Regional DRAs that the enforcement data has been reviewed and determined to be accurate.

2c. Source Data Reporting:

Data Submission: The EPA regions and OECA offices enter the CD tracking data (see above) to ICIS at the time all of the other enforcement case conclusion data is entered into ICIS. As described above, the data comes from the Consent Decree.

Frequency and Timing of Data Transmission: As described above, the data comes from the Consent Decree and is entered within 10 business days of conclusion of the case. The data is available in ICIS as it is entered into the system.

3a. Relevant Information Systems:

ICIS FE&C. Data for this measure is housed in the Integrated Compliance Information System (ICIS), in the FE&C subsystem. The Integrated Compliance Information System Federal Enforcement & Compliance (ICIS FE&C) database tracks EPA judicial and administrative civil enforcement actions. For more information, see: http://www.epa.gov/compliance/data/systems/icis/index.html The ICIS FE&C data system meets Office of Environmental Information (OEI) Lifecycle Management Guidance, which includes data validation processes, internal screen audit checks and verification, system and user documents, data quality audit reports, third party testing reports, and detailed report specifications data calculation methodology. Reference: Quality

Assurance and Quality Control procedures: Data Quality: Life Cycle Management Policy, (EPA CIO2121, April 7, 2006)

To support the Government Performance and Results Act (GPRA), the Agency's information quality guidelines, and other significant enforcement and compliance policies on performance measurement, OECA instituted a semiannual certification of the overall accuracy of ICIS information.

ICIS contains both source data and transformed data.

Attached Documents:
Data System Quality Assurance Plan (ICIS).doc

3b. Data Quality Procedures:

Annual Data Certification Process - OECA has instituted a semi-annual data certification process for the collection and reporting of enforcement and compliance information. The certification process was set up to ensure all reporting entities are aware of the reporting deadlines, receive the most up-to-date reporting instructions for select measures, follow best data management practices to assure reporting accuracy, and have access to the recent methodologies for calculating pounds of pollutants reduced. The Consent Decree Tracking measures are covered by this data certification process.

Each office within the Office of Enforcement and Compliance Assurance (OECA) prepares Quality Management Plans (QMPs) every five years. As part of the annual data certification process, regions are provided a checklist to assist them in their data quality procedures (attached).

Quality Management Plan - September 2011

Attached Documents:
FY12 Data Quality Check List.pdf
OC QMP Concurrence Signatures.pdf
OC QMP 2011 Final.docx

3c. Data Oversight:

Source Data Reporting Oversight

HQ - Director, Enforcement Targeting and Data Division
Region 1 - Division Director, Office of Environmental Stewardship
Region 2 - Director, Office of Enforcement and Compliance Assistance
Region 3 - Director, Office of Enforcement, Compliance and Environmental Justice
Region 4 - Regional Counsel and Director, Office of Environmental Accountability
Region 5 - Director, Office of Enforcement and Compliance Assurance
Region 6 - Compliance Assurance and Enforcement Division Director
Region 7 - Enforcement Coordinator

Region 8 - Director, Policy, Information Management and Environmental Justice

Region 9 - Enforcement Coordinator

Region 10 - Director, Office of Compliance and Enforcement

Information Systems Oversight Personnel

HQ - ICIS System Administrator

 Region 1 - ICIS Steward and Data Systems Administrator

Region 2 - ICIS System Administrator

Region 3 - ICIS Data Steward and System Administrator

Region 4 - ICIS System Administrator, Regional Compliance and Enforcement Data Steward

Region 5 - ICIS Data Steward and Systems Administrator

Region 6 - ICIS Data Steward

Region 7 - ICIS Data Steward and Systems Administrator

Region 8 - ICIS System Administrator

Region 9 - ICIS System Administrator

Region 10 - ICIS System Administrator and Data Steward

3d. Calculation Methodology:

DeciDecision Rules for Selecting Data: Below is select logic for extracting Consent Decree Tracking data from ICIS

Measure 1:

  # of CDs in denominator that either (1) lack a "Pay Required Civil Penalty Amount in Full" milestone - or - (2) have the milestone but the "Schedule Date" has passed with no "Actual Date" entered_____

          # of open CDs that have a penalty entered in the Federal Penalty required" filed in the Penalty Screen

Measure 2:

  # of CDs in denominator that either (1) lack a "Complete Required SEP" milestone - or – (2) have the milestone but the "Schedule Date" has passed with no "Actual Date" entered_____

              # of open CDs that have a SEP entered in the "SEP" screen

Measure 3:

  # of CDs in denominator that either (1) lack an "Achieve Final Compliance..." milestone - or – (2) have the milestone but the "Schedule Date" has passed with no "Actual Date" entered_____

# of all open CDs

Measure 4:

# of CDs in denominator, for which (1) 12 or fewer Quarters have transpired from the consent decree's date of entry – or –
(2) an appropriate Final Order status designations for the most recently applicable FY and Quarter has been entered into ICIS and designated in the ICIS Final Order Basic Information screen as: (1) Overall CD Compliance Status Reviewed/Defendant in General Compliance with all critical milestones – or – (2) Overall CD Compliance Status Reviewed/Defendant in Not General Compliance with all critical milestones/appropriate response Initiated – or – (3) Overall CD Compliance Status Cannot be Determined/Appropriate Response Initiated_____

# of all open CDs in 13th Quarter or beyond

Definitions of Variables: See Consent Decree Tracking guidance (attached).

Explanation of Calculations: Each of the four CD tracking measures are calculated based on this equation: Number of open non-Superfund, non-bankruptcy consent decrees NOT tracked divided by the full number of open non-Superfund, non-bankruptcy consent decrees times 100.

Explanation of Assumptions: N/A
Unit of Measure: Percent

Timeframe of Result: Snapshot in time of open consent decree tracking status at mid-year and at end-of-year.

Documentation of Methodological Changes: N/A

4a. Oversight and Timing of Final Results Reporting:
Final Reporting Oversight Personnel: Oversight of Final Reporting: The Deputy Regional Administrators, the Office of Civil Enforcement Director, and the Monitoring, Assistance and Program Division Director all must sign the attached certification form at MY and EOY certifying that their enforcement data is accurate, reliable and complete.   Frequency and Timing of Results Reporting: Semiannually, at mid-year and end-of-year.. Attached Documents: OC QMP Concurrence Signatures.pdf

4b. Data Limitations/Qualifications:
General Limitations/Qualifications: No error estimate is available. Data Lag Length and Explanation: From the October 31 end-of-year date, approximately 10-12 weeks to allow for QA/QC of the data. Methodological Changes: "Not Applicable."
4c. Third-Party Audits:
Beginning in FY2010, the Integrated Compliance Information System (ICIS) produced data to report on the four ICIS consent decree tracking measures. This data supports OECA's response in fulfilling its commitment to the Office of Inspector General (OIG) Audit Report No. 2001-P-00006, Compliance With Enforcement Instruments. (www.epa.gov/oig/reports/2001/enforce.pdf

Measure Code: 411 - Number of civil judicial and administrative enforcement cases concluded.

Office of Enforcement and Compliance Assurance (OECA)

Goal Number and Title:
5 - Enforcing Environmental Laws

Objective Number and Title:
1 - Enforce Environmental Laws

Sub-Objective Number and Title:
1 - Maintain Enforcement Presence

Strategic Target Code and Title:
3 - By 2015, conclude 19,000 civil judicial and administrative enforcement cases

Managing Office:
Office of Compliance

1a. Performance Measure Term Definitions:
Civil Judicial Enforcement Cases: a civil judicial enforcement case is a formal lawsuit, filed in court, against a person who has either failed to comply with a statutory or regulatory requirement, administrative order, or against a person who has contributed to a release. Civil judicial actions are often employed in situations that present repeated or significant violations or where there are serious environmental concerns. Attorneys from the U.S. Department of Justice prosecute civil judicial enforcement cases for the Agency.

Civil Administrative Enforcement Cases: A civil administrative enforcement case is an enforcement action taken by EPA under its own authority. Administrative enforcement cases can take several forms, including EPA issuing an administrative order requiring a facility to implement specific corrective measures to filing an administrative complaint commencing a form administrative adjudication. Administrative actions tend to be resolved quickly and can ofte be quite effective in bringing the facility into compliance with the regulations or in remedying a potential threat to human health of the environment.

Concluded: For purposes of this measure, there are two types of concluded enforcement actions counted.

The first are administrative enforcement actions which are undertake by EPA through authority granted to it under various federal environmental statutes, such as CERCLA, RCRA, CAA, CWA, TSCA, and others. An administrative action is concluded when a written agreement between the defendant/respondent and EPA resolving the complaint is documented in a Consent Agreement/Final Order (CA/FOs), is signed by the Regional Administrator or designee, and is filed with the regional hearing clerk.

The second type of enforcement action is known as a civil judicial action. Civil judicial actions attorneys from the U.S. Department of Justice prosecute civil cases for EPA. A concluded action occurs when a consent decree is signed by all parties to the action and filed in the appropriate court and signed by a judge or a written ruling or decision is made by a judge after a full trial.

2a. Original Data Source:

EPA attorneys

EPA regional hearing clerks

DOJ attorneys

Federal and state courts

2b. Source Data Collection:

The source data for this measure is found on completed enforcement documents. For example, the attached final consent agreement and final order (CAFO) contains the final date stamp affixed by the regional hearing clerk. An enforcement record is created in ICIS with the CAFO's final date indicating the case has been concluded.

Example of a concluded enforcement case document:

Attached Documents:

CAFO.pdf

2c. Source Data Reporting:

Administrative Penalty Orders

Administrative Penalty Orders on Consent

Consent Decrees

Notice of Determination

Unilateral Administrative Orders

3a. Relevant Information Systems:

The ICIS FE&C data system meets Office of Environmental Information (OEI) Lifecycle Management Guidance, which includes data validation processes, internal screen audit checks and verification, system and user documents, data quality audit reports, third party testing reports, and detailed report specifications data calculation methodology. Reference: Quality Assurance and Quality Control procedures: Data Quality: Life Cycle Management Policy, (EPA CIO2121, April 7, 2006)

The Integrated Compliance Information System (ICIS) is a three phase multi-year modernization project that improves the ability of EPA and the states to ensure compliance with the nation's environmental laws with the collection of comprehensive enforcement and compliance information. Phase I, implemented in FY02, replaced several legacy systems, and created an integrated system to support federal enforcement and compliance tracking, targeting and reporting, including GPRA reporting. Phase II, also called Permit Compliance System (PCS) Modernization, expands ICIS to include the National Pollutant Discharge Elimination System (NPDES) program and enables improved management of the complete program (e.g., stormwater) as well as replacing the legacy PCS. PCS is currently identified as an Agency Federal Managers' Financial Integrity Act (FMFIA) weakness, and the modernization of the system is critical to address the weakness. Phase II was first implemented in FY06 for 21 states and 11 tribes/territories that use ICIS to directly manage their NPDES programs. In FY08, seven more states moved to ICIS from the legacy PCS and began electronically flowing their Discharge Monitoring Report (DMR) data from their states systems via the Exchange Network and CDX to ICIS.

In FY09, Phase II continued with implementation of the National Installation of NetDMR allowing NPDES permittees to electronically submit DMR data from permitted facility systems via the Exchange Network to ICIS and migrated three additional states. In FY11 OECA implemented Full-Batch Release 1 of Phase II allowing Batch Flows of permits and facility data from states. FY12 will include Full-Batch Release 2 enabling batch flow will allow Batch Flows of inspection data from states. Inspection information and was implemented early in FY12. The final part of Phase II which will add the remaining NPDES Batch Flows and migrate and all remaining states is projected to be completed in FY13. Phase III will modernize the Air Facility System (AFS) into ICIS. AFS is used by EPA and States to track Clean Air Act enforcement and compliance activities. Integration of AFS into ICIS will modernize and replace a legacy system that does not meet current business needs. Implementation of this phase is projected for FY14.

ICIS contains both source data and transformed data.

Data System Quality Assurance Plan

Attached Documents:
Data System Quality Assurance Plan (ICIS).doc

3b. Data Quality Procedures:

Annual Data Certification Process - OECA has instituted a semi-annual data certification process for the collection and reporting of enforcement and compliance information. The certification process was set up to ensure all reporting entities are aware of the reporting deadlines, receive the most up-to-date reporting instructions for select measures, follow best data management practices to assure reporting accuracy, and have access to the recent methodologies for calculating pounds of pollutants reduced. The cases concluded measure is covered by the annual data certification process.

As part of the annual data certification process, regions are provided a checklist to assist them in their data quality procedures.

OECA's Quality Management Plan - September 2011

Attached Documents:
FY11 Data Quality Check List.pdf
OC QMP Concurrence Signatures.pdf
OC QMP 2011 Final.docx

3c. Data Oversight:

Source Data Reporting Oversight:

HQ - Director, Enforcement Targeting and Data Division

Region 1 - Division Director, Office of Environmental Stewardship

Region 2 - Director, Office of Enforcement and Compliance Assistance

Region 3 - Director, Office of Enforcement, Compliance and Environmental Justice

Region 4 - Regional Counsel and Director, Office of Environmental Accountability

Region 5 - Director, Office of Enforcement and Compliance Assurance

Region 6 - Compliance Assurance and Enforcement Division Director

Region 7 - Enforcement Coordinator

Region 8 - Director, Policy, Information Management and Environmental Justice

Region 9 - Enforcement Coordinator

Region 10 - Director, Office of Compliance and Enforcement

Information Systems Oversight Personnel

HQ - ICIS System Administrator

 Region 1 - ICIS Steward and Data Systems Administrator

Region 2 - ICIS System Administrator

Region 3 - ICIS Data Steward and System Administrator

Region 4 - ICIS System Administrator, Regional Compliance and Enforcement Data Steward

Region 5 - ICIS Data Steward and Systems Administrator

Region 6 - ICIS Data Steward

Region 7 - ICIS Data Steward and Systems Administrtor

Region 8 - ICIS System Administrator

Region 9 - ICIS System Administrator

Region 10 - ICIS System Administrator and Data Steward

3d. Calculation Methodology:

A civil or judicial case is counted as concluded when one instance of the following occurs:

An administrative action is concluded when a written agreement between the defendant/respondent and EPA resolving the complaint is documented in a Consent Agreement/Final Order (CA/FOs), is signed by the Regional Administrator or designee, and is filed with the regional hearing clerk.

A civil judicial action is concluded when a consent decree is signed by all parties to the action and filed in the appropriate court and signed by a judge or a written ruling or decision is made by a judge after a full trial.

4a. Oversight and Timing of Final Results Reporting:

The Deputy Regional Administrators, the Office of Civil Enforcement Director, and the Monitoring, Assistance and Program Division Director all must sign the attached certification form.

Attached Documents:
Data Certification Form.pdf

4b. Data Limitations/Qualifications:
The potential always exists that there are facilities, not yet identified as part of the regulated universe, subject to an EPA enforcement action.

4c. Third-Party Audits:
None to-date.

Measure Code: 410 - Number of civil judicial and administrative enforcement cases initiated.

Office of Enforcement and Compliance Assurance (OECA)

Goal Number and Title:
5 - Enforcing Environmental Laws

Objective Number and Title:
1 - Enforce Environmental Laws

Sub-Objective Number and Title:
1 - Maintain Enforcement Presence

Strategic Target Code and Title:
2 - By 2015, initiate 19,500 civil judicial and administrative enforcement cases

Managing Office:
Office of Compliance

1a. Performance Measure Term Definitions:
Civil Judicial Enforcement Cases: a civil judicial enforcement case is a formal lawsuit, filed in court, against a person who has either failed to comply with a statutory or regulatory requirement, administrative order, or against a person who has contributed to a release. Civil judicial actions are often employed in situations that present repeated or significant violations or where there are serious environmental concerns. Attorneys from the U.S. Department of Justice prosecute civil judicial enforcement cases for the Agency.   Civil Administrative Enforcement Cases: A civil administrative enforcement case is an enforcement action taken by EPA under its own authority. Administrative enforcement cases can take several forms, including EPA issuing an administrative order requiring a facility to implement specific corrective measures to filing an administrative complaint commencing a form administrative adjudication. Administrative actions tend to be resolved quickly and can often be quite effective in bringing the facility into compliance with the regulations or in remedying a potential threat to human health of the environment.   Initiated: A civil judicial enforcement case is considered initiated when it has been referred to DOJ. A referral is a formal written request to another agency or unit of government to proceed with judicial enforcement relating to the violation(s) in question.   Civil administrative enforcement cases are considered initiated when an administrative order or an administrative penalty order on consent has been issued by a Regional Administrator or designee.

2a. Original Data Source:
EPA attorneys   EPA regional hearing clerks   DOJ attorneys   Federal and state courts

2b. Source Data Collection:
The source data for this measure is found on initiated enforcement documents. For example, the attached initiated administrative order was issued by the Region 4 Assistant Administrator. An enforcement record is created in ICIS with the regional administrator's signature date which indicates the case has been initiated. Example of an initiated case document: Attached Documents: Admin Order.pdf

2c. Source Data Reporting:
Referral Letters Administrative Penalty Orders Administrative Compliance Orders Unilateral Administrative Orders

3a. Relevant Information Systems:
The ICIS FE&C data system meets Office of Environmental Information (OEI) Lifecycle Management Guidance, which includes data validation processes, internal screen audit checks and verification, system and user documents, data quality audit reports, third party testing reports, and detailed report specifications data calculation methodology. Reference: Quality Assurance and Quality Control procedures: Data Quality: Life Cycle Management Policy, (EPA CIO2121, April 7, 2006) The Integrated Compliance Information System (ICIS) is a three phase multi-year modernization project that improves the ability of EPA and the states to ensure compliance with the nation's environmental laws with the collection of comprehensive enforcement and compliance information. Phase I, implemented in FY02, replaced several legacy systems, and created an integrated system to support federal enforcement and compliance tracking, targeting and reporting, including GPRA reporting. Phase II, also called Permit Compliance System (PCS) Modernization, expands ICIS to include the National Pollutant Discharge Elimination System (NPDES) program and enables improved management of the complete program (e.g., stormwater) as well as replacing the legacy PCS. PCS is currently identified as an Agency Federal Managers' Financial Integrity Act (FMFIA) weakness, and the modernization of the system is critical to address the weakness. Phase II was first implemented in FY06 for 21 states and 11 tribes/territories that use ICIS to directly manage their NPDES programs. In FY08, seven more states moved to ICIS from the legacy PCS and began electronically flowing their Discharge Monitoring Report (DMR) data from their states systems via the Exchange Network and CDX to ICIS. In FY09, Phase II continued with implementation of the National Installation of NetDMR allowing NPDES permittees to electronically submit DMR data from permitted facility systems via the Exchange Network to ICIS and migrated three additional states. In FY11 OECA implemented Full-Batch Release 1 of Phase II allowing Batch Flows of permits and facility data from states. FY12 will include Full-Batch Release 2 enabling batch flow will allow Batch Flows of inspection data from states. Inspection information and was implemented early in

FY12. The final part of Phase II which will add the remaining NPDES Batch Flows and migrate and all remaining states is projected to be completed in FY13. Phase III will modernize the Air Facility System (AFS) into ICIS. AFS is used by EPA and States to track Clean Air Act enforcement and compliance activities. Integration of AFS into ICIS will modernize and replace a legacy system that does not meet current business needs. Implementation of this phase is projected for FY14.

ICIS contains both source data and transformed data.

OECA's Data System Quality Assurance Plan

Attached Documents:
Data System Quality Assurance Plan (ICIS).doc

3b. Data Quality Procedures:

Annual Data Certification Process - OECA has instituted a semi-annual data certification process for the collection and reporting of enforcement and compliance information. The certification process was set up to ensure all reporting entities are aware of the reporting deadlines, receive the most up-to-date reporting instructions for select measures, follow best data management practices to assure reporting accuracy, and have access to the recent methodologies for calculating pounds of pollutants reduced. The cases initiated measure is covered by the annual data certification process.

As part of the annual data certification process, regions are provided a checklist to assist them in their data quality procedures.

OECA's Quality Management Plan - September 2011

Attached Documents:
FY11 Data Quality Check List.pdf
OC QMP Concurrence Signatures.pdf
OC QMP 2011 Final.docx

3c. Data Oversight:

Source Data Reporting Oversight
HQ - Director, Enforcement Targeting and Data Division
Region 1 - Division Director, Office of Environmental Stewardship
Region 2 - Director, Office of Enforcement and Compliance Assistance
Region 3 - Director, Office of Enforcement, Compliance and Environmental Justice
Region 4 - Regional Counsel and Director, Office of Environmental Accountability
Region 5 - Director, Office of Enforcement and Compliance Assurance

Region 6 - Compliance Assurance and Enforcement Division Director

Region 7 - Enforcement Coordinator

Region 8 - Director, Policy, Information Management and Environmental Justice

Region 9 - Enforcement Coordinator

Region 10 - Director, Office of Compliance and Enforcement

Information Systems Oversight Personnel

HQ - ICIS System Administrator

 Region 1 - ICIS Steward and Data Systems Administrator

Region 2 - ICIS System Administrator

Region 3 - ICIS Data Steward and System Administrator

Region 4 - ICIS System Administrator, Regional Compliance and Enforcement Data Steward

Region 5 - ICIS Data Steward and Systems Administrator

Region 6 - ICIS Data Steward

Region 7 - ICIS Data Steward and Systems Administrtor

Region 8 - ICIS System Administrator

Region 9 - ICIS System Administrator

Region 10 - ICIS System Administrator and Data Steward

3d. Calculation Methodology:

A civil or judicial case is counted as initiated when one instance of the following occurs:

Civil judicial enforcement cases are considered initiated when a referral has been made to DOJ.

Civil administrative enforcement cases are considered initiated when an administrative order or an administrative penalty order on consent has been issued by a Regional Administrator or designee.

4a. Oversight and Timing of Final Results Reporting:

The Deputy Regional Administrators, the Office of Civil Enforcement Director, and the Monitoring, Assistance and Program Division Director all must sign the attached certification form.

Attached Documents:

Data Certification Form.pdf

4b. Data Limitations/Qualifications:

The potential always exists that there are facilities, not yet identified as part of the regulated universe, subject to an EPA enforcement action.

4c. Third-Party Audits:

None to-date.

Office of Environmental Information (OEI) Record(s)

Measure Code: 052 - Number of major EPA environmental systems that use the CDX electronic requirements enabling faster receipt, processing, and quality checking of data.

Office of Environmental Information (OEI)

Goal Number and Title:
0 -
Objective Number and Title:
0 -
Sub-Objective Number and Title:
0 -
Strategic Target Code and Title:
0 -
Managing Office:
Office of Information Collection
1a. Performance Measure Term Definitions:
Major EPA Environmental Systems: Major environmental systems are those that use CDX services to support the electronic reporting or exchange of information among trading partners or from the regulated entities to EPA. Enabling Faster Receipt, Processing, and Quality Checking of Data: This terminology means the services used to ensure quality data entering the data and that they are submitted in a much faster way than the previous legacy methods, e.g., electronic and Internet-based as opposed to a paper or other method that involves mailing to the Agency. CDX: Central Data Exchange. CDX is the point of entry on the Environmental Information Exchange Network (Exchange Network) for environmental data submissions to the Agency. CDX assembles the registration/submission requirements of many different data exchanges with EPA and the States, Tribes, local governments and the regulated community into a centralized environment. This system improves performance tracking of external customers and overall management by making those processes more consistent and comprehensive. The creation of a centralized registration system, coupled with the use of web forms and web-based approaches to submitting the data, invite opportunities to introduce additional automated quality assurance procedures for the system and reduce human error. For more information, visit: http://www.epa.gov/cdx/index.htm
2a. Original Data Source:
Users of CDX from the Private sector, State, local, and Tribal government; entered into the CDX Customer Registration Subsystem

CDX Users at EPA program offices include the:
· Office of Air and Radiation (OAR)
· Office of Enforcement and Compliance Assurance (OECA)
· Office of Environmental Information (OEI)
· Office of Prevention, Pesticides and Toxic Substances (OPPTS)
· Office of Solid Waste and Emergency Response (OSWER)
· Office of Water (OW)

2b. Source Data Collection:

Source Data Collection Methods:

Reports are routinely generated from log files on CDX servers that support user registration and identity management.

EPA QA Requirements/Guidance Governing Collection:

QA/QC is performed in accordance with a CDX Quality Assurance Plan ["Quality Assurance Project Plan for the Central Data Exchange," 10/8/2004] and the CDX Design Document v.3, Appendix K registration procedures[Central Data Exchange Electronic Reporting Prototype System Requirements: Version 3; Document number: EP005S3; December 2000]. Specifically, data are reviewed for authenticity and integrity. Automated edit checking routines are performed in accordance with program specifications and the CDX Quality Assurance Plan. EPA currently has a draft plan developed in August 2007. In FY 2011, CDX will develop robust quality criteria, which will include performance metric results and align with the schedule for the upcoming CDX contract recompete.

Spatial Detail Covered By the Source Data: This is not applicable other than a user's address.

2c. Source Data Reporting:

Form/Mechanism for Receiving data and entering into EPA System:

CDX manages the collection of data and documents in a secure way either by users entering data onto web forms or via a batch file transfer, both of which are completed using the CDX environment. These data are then transported to the appropriate EPA system.

Timing and Frequency of Reporting: Annual

3a. Relevant Information Systems:

CDX Customer Registration Subsystem. This subsystem is used to register external users for reporting or exchanging data with EPA via CDX.

CDX completed its last independent security risk assessment in June 2011, and all vulnerabilities are being reviewed or addressed.

Additional Information:

In addition, environmental data collected by CDX is delivered to National data systems in the Agency. Upon receipt, the National systems often conduct a more thorough data quality assurance procedure based on more intensive rules that can be continuously changing based on program requirements. As a result, CDX and these National systems appropriately share the responsibility for ensuring environmental data quality

3b. Data Quality Procedures:

The CDX system collects, reports, and tracks performance measures on data quality and customer service. While its automated routines are sufficient to screen systemic problems/issues, a more detailed assessment of data errors/problems generally requires a secondary level of analysis that takes time and human resources.

CDX incorporates a number of features to reduce errors in registration data and that contribute greatly to the quality of environmental data entering the Agency. These features include pre-populating data either from CDX or National systems, conducting web-form edit checks, implementing XML schemas for basic edit checking and providing extended quality assurance checks for selected Exchange Network Data flows using Schematron.

3c. Data Oversight:

Although not officially termed, CDX is a general support application that provides centralized services to a multitude of program offices in the Agency and data trading partners on the Exchange Network. The general answer is that EPA Program Office System Managers and their management chains are responsible for oversight of the data quality. The closest individual responsible for "data integrity purposes" is the Chief of the Information Technology Branch.

3d. Calculation Methodology:

Unit of analysis: Systems

No data transformations occur.

4a. Oversight and Timing of Final Results Reporting:

Oversight of Final Reporting: Reports on CDX quality and performance are conducted on an annual basis. The reports consist of both quantitative measures from system logs and qualitative measures from user and program office surveys.

Timing of Results Reporting:
Annually

4b. Data Limitations/Qualifications:

The potential error in registration data, under CDX responsibility has been assessed to be less than 1%. This is accomplished through as combination of automated edit checks in web form fields and processes in place to confirm the identity of individuals prior to approving access to CDX data flows.

4c. Third-Party Audits:

Third party security risk assessments are conducted every three years in accordance with FISMA requirements. Alternatives analysis reviews are also conducted in accordance with OMB CPIC requirements. Lastly, adhoc third party requirements are conducted internally.

Office of Environmental Information (OEI)

Goal Number and Title:
0 -

Objective Number and Title:
0 -

Sub-Objective Number and Title:
0 -

Strategic Target Code and Title:
0 -

Managing Office:
Office of Information Collection

1a. Performance Measure Term Definitions:
Able to exchange data: A trading partner has the programmatic and technical infrastructure in place to exchange data across the Exchange Network. Nodes: Nodes are points of presence on the Internet which are used to support the secure transport of data to trusted trading partners. Real-time: When the data is generated and approved, it is automatically transported to the destination of another trading partner. CDX: Central Data Exchange. CDX is the point of entry on the Environmental Information Exchange Network (Exchange Network) for environmental data submissions to the Agency. CDX assembles the registration/submission requirements of many different data exchanges with EPA and the States, Tribes, local governments and the regulated community into a centralized environment. This system improves performance tracking of external customers and overall management by making those processes more consistent and comprehensive. The creation of a centralized registration system, coupled with the use of web forms and web-based approaches to submitting the data, invite opportunities to introduce additional automated quality assurance procedures for the system and reduce human error. For more information, visit: http://www.epa.gov/cdx/index.htm

2a. Original Data Source:
Users of CDX from the Private sector, State, local, and Tribal government; entered into the CDX Customer Registration Subsystem CDX Users at EPA program offices include the: ·   Office of Air and Radiation (OAR) ·   Office of Enforcement and Compliance Assurance (OECA)

·	Office of Environmental Information (OEI)
·	Office of Prevention, Pesticides and Toxic Substances (OPPTS)
·	Office of Solid Waste and Emergency Response (OSWER)
·	Office of Water (OW)

2b. Source Data Collection:

Source Data Collection Methods:
Reports are routinely generated from log files on CDX servers that support user registration and identity management.

Tabulation of records. Collection is ongoing.

EPA QA Requirements/Guidance Governing Collection:
QA/QC is performed in accordance with a CDX Quality Assurance Plan ["Quality Assurance Project Plan for the Central Data Exchange," 10/8/2004] and the CDX Design Document v.3, Appendix K registration procedures[Central Data Exchange Electronic Reporting Prototype System Requirements: Version 3; Document number: EP005S3; December 2000]. Specifically, data are reviewed for authenticity and integrity. Automated edit checking routines are performed in accordance with program specifications and the CDX Quality Assurance Plan. EPA currently has a draft plan developed in August 2007. In FY 2011, CDX will develop robust quality criteria, which will include performance metric results and align with the schedule for the upcoming CDX contract recompete.

Spatial Detail Covered By the Source Data: This is not applicable other than a user's address.

2c. Source Data Reporting:

Form/Mechanism for Receiving Data and Entering into EPA System:
CDX manages the collection of data and documents in a secure way either by users entering data onto web forms or via a batch file transfer, both of which are completed using the CDX environment. These data are then transported to the appropriate EPA system.

Timing and Frequency of Reporting: Annual

3a. Relevant Information Systems:

CDX Customer Registration Subsystem. This subsystem is used to register external users for reporting or exchanging data with EPA via CDX.

CDX completed its last independent security risk assessment in June 2011, and all vulnerabilities are being reviewed or addressed.

Additional Information:
In addition, environmental data collected by CDX is delivered to National data systems in the Agency. Upon receipt, the National systems often conduct a more thorough data quality assurance procedure based on more intensive rules that can be continuously changing based on program requirements. As a result, CDX and these

National systems appropriately share the responsibility for ensuring environmental data quality

3b. Data Quality Procedures:

The CDX system collects, reports, and tracks performance measures on data quality and customer service. While its automated routines are sufficient to screen systemic problems/issues, a more detailed assessment of data errors/problems generally requires a secondary level of analysis that takes time and human resources.

CDX incorporates a number of features to reduce errors in registration data and that contribute greatly to the quality of environmental data entering the Agency. These features include pre-populating data either from CDX or National systems, conducting web-form edit checks, implementing XML schemas for basic edit checking and providing extended quality assurance checks for selected Exchange Network Data flows using Schematron.

3c. Data Oversight:

Although not officially termed, CDX is a general support application that provides centralized services to a multitude of program offices in the Agency and data trading partners on the Exchange Network. The general answer is that EPA Program Office System Managers and their management chains are responsible for oversight of the data quality. The closest individual responsible for "data integrity purposes" is the Chief of the Information Technology Branch.

3d. Calculation Methodology:

Unit of analysis: Users

No data transformations occur.

4a. Oversight and Timing of Final Results Reporting:

Oversight of Final Reporting: Reports on CDX quality and performance are conducted on an annual basis. The reports consist of both quantitative measures from system logs and qualitative measures from user and program office surveys.

Timing of Results Reporting:
Annually

4b. Data Limitations/Qualifications:

The potential error in registration data, under CDX responsibility has been assessed to be less than 1%. This is accomplished through a combination of automated edit checks in web form fields and processes in place to confirm the identity of individuals prior to approving access to CDX data flows.

4c. Third-Party Audits:

Third party security risk assessments are conducted every three years in accordance with FISMA requirements. Alternatives analysis reviews are also conducted in accordance with OMB CPIC requirements. Lastly, adhoc third party requirements are conducted internally

Measure Code: 999 - Total number of active unique users from states, tribes, laboratories, regulated facilities and other entities that electronically report environmental data to EPA through CDX.

Office of Environmental Information (OEI)

Goal Number and Title:
0 -

Objective Number and Title:
0 -

Sub-Objective Number and Title:
0 -

Strategic Target Code and Title:
0 -

Managing Office:
Office of Information Collection

1a. Performance Measure Term Definitions:
Active unique users: Active accounts include those who have logged in within the last two years in which the statistic is generated. In addition, users who have multiple accounts are only counted as one account (unique). Active unique users include: States, Tribes, laboratories, and regulated facilities.   CDX: Central Data Exchange. CDX is the point of entry on the Environmental Information Exchange Network (Exchange Network) for environmental data submissions to the Agency.   CDX assembles the registration/submission requirements of many different data exchanges with EPA and the States, Tribes, local governments and the regulated community into a centralized environment. This system improves performance tracking of external customers and overall management by making those processes more consistent and comprehensive. The creation of a centralized registration system, coupled with the use of web forms and web-based approaches to submitting the data, invite opportunities to introduce additional automated quality assurance procedures for the system and reduce human error. For more information, visit: http://www.epa.gov/cdx/index.htm

2a. Original Data Source:
Users of CDX from the Private sector, State, local, and Tribal government; entered into the CDX Customer Registration Subsystem   CDX Users at EPA program offices include the: ·   Office of Air and Radiation (OAR)   ·   Office of Enforcement and Compliance Assurance (OECA)   ·   Office of Environmental Information (OEI)   ·   Office of Prevention, Pesticides and Toxic Substances (OPPTS)   ·   Office of Solid Waste and Emergency Response (OSWER)   ·   Office of Water (OW)

2b. Source Data Collection:
Source Data Collection Methods: Reports are routinely generated from log files on CDX servers that support user registration and identity management.

Tabulation of records: The records of registration provide an up-to-date, accurate count of users.

Date/Time Intervals Covered by Source Data: Ongoing

EPA QA Requirements/Guidance Governing Collection: QA/QC is performed in accordance with a CDX Quality Assurance Plan ["Quality Assurance Project Plan for the Central Data Exchange," 10/8/2004] and the CDX Design Document v.3, Appendix K registration procedures[Central Data Exchange Electronic Reporting Prototype System Requirements: Version 3; Document number: EP005S3; December 2000]. Specifically, data are reviewed for authenticity and integrity. Automated edit checking routines are performed in accordance with program specifications and the CDX Quality Assurance Plan. EPA currently has a draft plan developed in August 2007. In FY 2012, CDX will develop robust quality criteria, which will include performance metric results and align with the schedule for the upcoming CDX contract recompete.

2c. Source Data Reporting:
Form/mechanism for receiving data and entering into EPA system: CDX manages the collection of data and documents in a secure way either by users entering data onto web forms or via a batch file transfer, both of which are completed using the CDX environment. These data are then transported to the appropriate EPA system.

Timing and frequency of reporting: Ongoing

3a. Relevant Information Systems:
CDX Customer Registration Subsystem. Users identify themselves with several descriptors and use a number of CDX security mechanisms for ensuring the integrity of individuals' identities

CDX completed its last independent security risk assessment in June 2011, and all vulnerabilities are being reviewed or addressed. CDX users register themselves via web forms on CDX to obtain access to data flows in which they receive privileges. This user information comes directly from the user and is not transformed.

Additional information:
In addition, environmental data collected by CDX is delivered to National data systems in the Agency. Upon receipt, the National systems often conduct a more thorough data quality assurance procedure based on more intensive rules that can be continuously changing based on program requirements. As a result, CDX and these National systems appropriately share the responsibility for ensuring environmental data quality.

3b. Data Quality Procedures:
The CDX system collects, reports, and tracks performance measures on data quality and customer service. While its automated routines are sufficient to screen systemic problems/issues, a more detailed assessment of

data errors/problems generally requires a secondary level of analysis that takes time and human resources.

CDX incorporates a number of features to reduce errors in registration data and that contribute greatly to the quality of environmental data entering the Agency. These features include pre-populating data either from CDX or National systems, conducting web-form edit checks, implementing XML schemas for basic edit checking and providing extended quality assurance checks for selected Exchange Network Data flows using Schematron.

3c. Data Oversight:

Although not officially termed, CDX is a general support application that provides centralized services to a multitude of program offices in the Agency and data trading partners on the Exchange Network. The general answer is that EPA Program Office System Managers and their management chains are responsible for oversight of the data quality. The closest individual responsible for "data integrity purposes" is the Chief of the Information Technology Branch.

3d. Calculation Methodology:

Unit of Analysis: Users
EPA counts users based on the above definition in 1a.

4a. Oversight and Timing of Final Results Reporting:

Oversight of Final Reporting: Reports on CDX quality and performance are conducted on an annual basis. The reports consist of both quantitative measures from system logs and qualitative measures from user and program office surveys.

Timing of Results Reporting: Annually

4b. Data Limitations/Qualifications:

The potential error in registration data, under CDX responsibility has been assessed to be less than 1%. This is accomplished through a combination of automated edit checks in web form fields and processes in place to confirm the identity of individuals prior to approving access to CDX data flows.

4c. Third-Party Audits:

Third party security risk assessments are conducted every three years in accordance with FISMA requirements. Alternatives analysis reviews are also conducted in accordance with OMB CPIC requirements. Lastly, adhoc third party requirements are conducted internally.

Office of Environmental Information (OEI)

Goal Number and Title:
0 -

Objective Number and Title:
0 -

Sub-Objective Number and Title:
0 -

Strategic Target Code and Title:
0 -

Managing Office:
Office of Information Analysis and Access

1a. Performance Measure Term Definitions:
TRI Program: Number of Data Quality Checks - the Regions and HQ will identify possible data quality issues and follow up with approximately 500 facilities annually to ensure accuracy of TRI data on HQ-generated lists of facilities.

2a. Original Data Source:
EPA receives this data from companies or entities required to report annually under EPCRA (see 2b.) The data quality checks are performed by EPA HQ and regional offices on the facility data submitted.

2b. Source Data Collection:
All covered facilities are required to annually submit toxic chemical release and other waste management quantities and facility-specific information for the previous calendar year on or before July 1 to EPA and the States if reporting threshold requirements [40 CFR Part 372] are exceeded. EPA makes the collected data available to the public through EPA's TRI National Analysis and various online tools (e.g., Envirofacts TRI Explorer, TRI.NET, and my RTK.

2c. Source Data Reporting:
Form/mechanism for receiving data and entering into EPA's system: More than 97 percent of covered facilities use EPA's web-based electronic reporting tool - TRI-MEweb - to report their releases and other waste management information on the TRI program. Timing and frequency of reporting: covered facilities are required to submit release and waste management information for previous calendar year on or before July 1 if they meet reporting requirements.

3a. Relevant Information Systems:
TRI-MEweb and TRIPS databases

3b. Data Quality Procedures:
• EPA provides guidance documents (general, chemical-specific and sector-specific), training modules and TRI hotline assistance. • EPA performs multiple quality control and quality assurance checks during reporting (TRI-MEweb DQ checks) and at the end of the reporting period (in-house DQ checks). Here are few examples:

- Facilities that reported large changes in release, disposal or waste management practices on sector-level for certain chemicals (e.g., PBT chemicals);
- Facilities that submit invalid Chemical Abstract Service (CAS) numbers that do not match the chemical name;
- Facilities that report invalid North American Industry Classification System (NAICs) codes;
- Facilities that report invalid/incorrect RCRA facility IDs when they send wastes to offsite locations for management;
- Facilities that did not report for the current reporting year but reported for the previous reporting year; and
- Facilities that reported incorrect quantities on Form R Schedule 1 for dioxin and dioxin-like compounds;

The TRI Program generates a list of facilities with potential data quality issues and sends the list to the 10 TRI Regional coordinators. The TRI Program HQ staff and Regional coordinators contact the facilities and discuss data quality issues. The facilities may revise their reports where errors are identified. Certain facilities may be referred to enforcement for further examination. For each annual TRI collection received on or before July 1, headquarters and regional personnel will identify potential data quality issues and work with the Regions to contact facility reporters and resolve the issues during the following fall and spring.

3c. Data Oversight:
EPA performs several data quality analyses to support the TRI National Analysis. For this measure, the Regions and the HQ staff annually identify potential data quality issues and contact approximately 500 facilities for follow up.

3d. Calculation Methodology:
Unit of Analysis: Number of facilities contacted

4a. Oversight and Timing of Final Results Reporting:
For TRI reports (due to EPA and the states annually on July 1), the TRI program will identify potential data quality issues and work with the Regions to contact facility reporters and resolve the issues during the following fall and spring.

4b. Data Limitations/Qualifications:
Over 97% of all TRI reporting facilities use TRI-MEweb.

4c. Third-Party Audits:
This program does not conduct third-party audits of the data quality data.

Office of Indian and Tribal Affairs (OITA) Record(s)

Measure Code: 5PR - Percent of Tribes conducting EPA approved environmental monitoring and assessment activities in Indian country (cumulative.)

Office of Indian and Tribal Affairs (OITA)

Goal Number and Title:
3 - Cleaning Up Communities and Advancing Sustainable Development
Objective Number and Title:
4 - Strengthen Human Health and Environmental Protection in Indian Country
Sub-Objective Number and Title:
1 - Improve Human Health and the Environment in Indian Country
Strategic Target Code and Title:
2 - By 2015, increase the percent of tribes conducting EPA-approved environmental monitoring
Managing Office:
AIEO
1a. Performance Measure Term Definitions:
A tribe is a governmental entity that is recognized by the federal government and eligible to receive federal funding.   The performance measure reports the number of active Quality Assurance Project Plans (QAPPs) for monitoring activities that have been approved by Regional Quality Assurance Officers. All ongoing environmental monitoring programs are required to have active QAPPs, which are used as a surrogate for the monitoring activities that occur in Indian country.   However, tribes often have more than one QAPP, so the count of total QAPPs is always higher than the number of tribes that have QAPPs as reported for this measure.   Environmental monitoring and assessment activities are those that measure biological, chemical, or physical measurements.   EPA-approved indicates a required QAPP for the activity has been approved by Regional Quality Assurance Officers.   Active QAPPs are those that have not expired.   This measure represents progression toward the goal of improving human health and the environment in Indian country by helping tribes plan, develop and establish environmental protection programs.
2a. Original Data Source:
Regional Quality Assurance Officers

217

2b. Source Data Collection:
Regional tribal program liaisons obtain information from Regional Quality Assurance Officers. Spatial Detail: Base unit is a tribe. Geographic coverage is national.

2c. Source Data Reporting:
Reports are in the form of tables with measures in the columns and years in the rows. The years can be compared. Data are input manually by regional Tribal Program Management System (TPMS) team members. The data are reported by the Regions into TPMS at the end of the year.

3a. Relevant Information Systems:
The Tribal Program Management System (TPMS) is a secure database that holds the performance information http://www.epa.gov/tribalportal/ The information is entered into standard query fields in the data system. Thus, there is no allowance for differences in reporting across EPA's Regional offices, and national reports can be assembled in a common framework. The assumption is that the authorized person who enters the data is knowledgeable about the performance status of the tribe and understands data definitions. Quality Management Plan (QMP) is being drafted by contractor

3b. Data Quality Procedures:
Standard Operating Procedures are detailed in the Data Definitions document. Each Regional Administrator, who has tribal activity in his regional area, is the EPA official who certifies information in TPMS prior to submission to EPA Headquarters American Indian Office (AIEO.) However, in some cases the Regional Administrator may wish to delegate the signatory authority to another official such as the Regional Indian Coordinator. This procedure generally follows guidance provided in EPA Information Quality Guidelines. (See http://www.epa.gov/quality/informationguidelines/ for more information.) Additionally, the data in TPMS are extracted by the regional TPMS team twice a year, and delivered by spreadsheet to the Regional TPMS Project Officers for review and verification. Attached Documents: TPMS Data Definitions.doc

3c. Data Oversight:
Regional Indian Coordinators

3d. Calculation Methodology:
Each row in the report is a fiscal year. Calculation methodology is: Count number of tribes with at least one active QAPP in a fiscal year. A tribe is counted once even if they have more than one QAPP. Calculation of Percentages: 572 is the number that is used to calculate percentage, and reflects tribal bands

that are independent entities and are eligible for EPA funding.

Attached Documents:

TPMS Data Definitions.doc

4a. Oversight and Timing of Final Results Reporting:

The procedures for collecting and reporting on the Goal 4 Objective 3 performance measures require that Regional program managers certify the accuracy of the data submitted by the regions to AIEO. This certification procedure is consistent with EPA Information Quality Guidelines and verified by AIEO personnel

4b. Data Limitations/Qualifications:

Because data are input by EPA's Regional Project Officers on an ongoing basis, there may be a time lag between when a tribal program status has been achieved and when the data are entered into the TPMS.

For the TPMS, errors could occur by mis-entering data or neglecting to enter data. However, the data from each region will be certified as accurate at the end of each reporting cycle; error is estimated to be low, about 1-2 percent.

4c. Third-Party Audits:

Not applicable

Measure Code: 5PQ - Percent of Tribes implementing federal regulatory environmental programs in Indian country (cumulative).

Office of Indian and Tribal Affairs (OITA)

Goal Number and Title:
3 - Cleaning Up Communities and Advancing Sustainable Development
Objective Number and Title:
4 - Strengthen Human Health and Environmental Protection in Indian Country
Sub-Objective Number and Title:
1 - Improve Human Health and the Environment in Indian Country
Strategic Target Code and Title:
1 - By 2015, increase the percent of tribes implementing federal regulatory environmental programs
Managing Office:
AIEO
1a. Performance Measure Term Definitions:
The performance measure tracks the number of "Treatment in a manner similar to a State" (TAS) program approvals or primacies and execution of "Direct Implementation Tribal Cooperative Agreements (DITCAs)."
TAS status grants a tribe eligibility to implement and administer the environmental statutes for a program within the tribe's boundaries comparable to the way States implement and administer the statutes outside of Indian country.
DITCAs are agreements negotiated between EPA and federally-recognized tribes and eligible intertribal consortia that enable the tribes to conduct agreed-upon activities and to help EPA implement federal environmental programs in Indian country in the absence of an acceptable tribal program.
The measure is based on a count of tribes, and a given tribe may have more than one TAS program, and may have DITCAs as well. Because of the tribes with multiple qualifying programs, the total number of TAS designations plus DITCAs in Indian country is higher than the number of tribes with regulatory environmental programs as reported for this measure.
This measure represents progression toward the goal of improving human health and the environment in Indian country by helping tribes plan, develop and establish environmental protection programs.
2a. Original Data Source:
Regions and Tribes
2b. Source Data Collection:
Data for the TPMS are input on an ongoing basis by Regional tribal programs and EPA headquarters.
2c. Source Data Reporting:
Reports are in the form of tables with measures in the columns and years in the rows. The years can be compared. Data are input manually by regional tribal Tribal Program Management System (TPMS) team

members. The data are reported by the Regions into TPMS and at the end of the year.
3a. Relevant Information Systems:
The Tribal Program Management System (TPMS) is a secure database that holds the performance information http://www.epa.gov/tribalportal/ The information is entered into standard query fields in the data system. Thus, there is no allowance for differences in reporting across EPA's Regional offices, and national reports can be assembled in a common framework. The assumption is that the authorized person who enters the data is knowledgeable about the performance status of the tribe and understands data definitions. Quality Management Plan (QMP) is being drafted by contractor
3b. Data Quality Procedures:
Standard Operating Procedures are detailed in the Data Definitions document. Each Regional Administrator, who has tribal activity in his regional area, is the EPA official who certifies information in TPMS prior to submission to EPA Headquarters American Indian Office (AIEO.) However, in some cases the Regional Administrator may wish to delegate the signatory authority to another official such as the Regional Indian Coordinator. This procedure generally follows guidance provided in EPA Information Quality Guidelines. (See http://www.epa.gov/quality/informationguidelines/ for more information.) Additionally, the data in TPMS are extracted by the regional TPMS team twice a year, and delivered by spreadsheet to the Regional TPMS Project Officers for review and verification.   Attached Documents: TPMS Data Definitions.doc
3c. Data Oversight:
Regional Indian Coordinators certify data and submit to AIEO
3d. Calculation Methodology:
Calculation methodology is: Count the number of active tribes with DITCA and TAS in a fiscal year. TAS do not have expiration dates and are cumulative. Calculation of Percentages: 572 is the number that is used to calculate percentage, and reflects tribal bands that are independent entities and are eligible for EPA funding. Because of the tribes with multiple qualifying programs, the total number of TAS designations plus DITCAs in Indian country is higher than the number of tribes with regulatory environmental programs as reported for this measure. Percent of Tribes implementing federal regulatory environmental programs in Indian country:

Attached Documents:

TPMS Data Definitions.doc

4a. Oversight and Timing of Final Results Reporting:
The procedures for collecting and reporting on the Goal 4 Objective 3 performance measures require that Regonal program managers certify the accuracy of the data submitted by the regions to AIEO. This certification procedure is consistent with EPA Information Quality Guidelines and is verified by AIEO personnel

4b. Data Limitations/Qualifications:
Because data are input by EPA's Regional Project Officers on an ongoing basis, there may be a time lag between when a tribal program status has been achieved and when the data are entered into the TPMS. For the TPMS, errors could occur by mis-entering data or neglecting to enter data. However, the data from each region will be certified as accurate at the end of each reporting cycle; error is estimated to be low, about 1-2 percent.

4c. Third-Party Audits:
Not Applicable

Office of Chemical Strategies and Pollution Prevention (OCSPP) Record(s)

Measure Code: 009 - Cumulative number of certified Renovation Repair and Painting firms

Office of Chemical Strategies and Pollution Prevention (OCSPP)

Goal Number and Title:
4 - Ensuring the Safety of Chemicals and Preventing Pollution
Objective Number and Title:
1 - Ensure Chemical Safety
Sub-Objective Number and Title:
1 - Protect Human Health from Chemical Risks
Strategic Target Code and Title:
2 - By 2014, reduce the percentage of children with blood lead levels above 5ug/dl to 1.0 percent or less
Managing Office:
Office of Pollution Prevention and Toxic
1a. Performance Measure Term Definitions:
Cumulative number: Number certified since October 1, 2009. Certified Renovation Repair and Painting firms: "Renovation, Repair and Painting" is generally defined as any activity that disturbs paint in housing and child-occupied facilities built before 1978, including remodeling, repair, maintenance, electrical work, plumbing, painting, carpentry and window replacement. Most minor repair and maintenance activities of less than six square feet per interior room or 20 square feet on the exterior or a home or building are exempt from the work practice requirements. However, this exemption does not apply to window replacements, demolitions or the use of prohibited practices. Background: On March 31, 2008, EPA issued a new rule (Renovation, Repair, and Painting Program Rule or RRP rule) aimed at protecting children from lead-based paint hazards. In October 2009, firms began to apply to EPA for certification to conduct renovations. As of April 2010, renovations in target (pre-1978) housing and child-occupied facilities must be conducted by certified Renovation, Repair and Painting firms, using renovators who have completed an accredited training course, and following the work practice requirements of the rule. Firm certifications are valid for five years.
2a. Original Data Source:
In states where EPA administers the RRP program, the agency tracks the number of certified firms through its Federal Lead-Based Paint Program (FLPP) database. Data are entered into the FLPP database either by an individual submitting an application via CDX or by a contractor who manually data enter information submitted via a paper application. In states that have received authorization from EPA to administer the program in lieu of the Federal program, state grantees collect data on the number of state certified Renovation, Repair and Painting firms.

2b. Source Data Collection:
In states where EPA administers the RRP program, the agency tracks the number of certified firms through its Federal Lead-Based Paint Program (FLPP) database. Data is entered into the FLPP database either by an individual submitting an application via CDX or by a contractor who manually data enter information submitted via a paper application. In states that have received authorization from EPA to administer the program in lieu of the Federal program, state grantees collect data on the number of state certified Renovation, Repair and Painting firms. In authorized states, EPA collects data on the numbers of firms certified in each state through quarterly reports from grantees as part of the Agency's oversight of authorized programs. Since the performance result is based on a simple count of certified firms by EPA and authorized states, there are no applicable quality assurance plans or procedures other than those described under section 3b below.

2c. Source Data Reporting:
Form/mechanism for receiving data and entering into EPA system: Firms seeking RRP certification submit applications in hard copy directly to EPA or electronically through the Agency's Central Data Exchange (CDX). Original hard copies are retained to augment the electronic records. Authorized states report data to EPA Regional Offices on the number of certified firms in the state. Timing and frequency of reporting: Application data are entered into the FLPP database continuously as applications to the Federal Program are received.

3a. Relevant Information Systems:
The Federal Lead-Based Paint Program (FLPP) database provides a record of all applications for the certification of Renovation Repair and Painting firms where EPA directly implements the program, the actions taken on those applications including final decisions, and the multiple steps in the process used for measurement. Thus, the number of certified firms can be obtained directly from the database. Documentation for the FLPP database is maintained internally at EPA and is available upon request. The database contains only source data as there is no need for data transformation in order to derive the performance result for this measure. The FLPP database has recently been upgraded to increase processing efficiency. The FLPP database was Certified and Accredited under the National Institute of Standards and Technology (NIST) Special Publication 800-53 Revision 3 requirements issued under the Federal Information Security Management Act (FISMA) in June 2013. The Certification and Accreditation stays in effect until June 2016 with continuous monitoring and performance testing of one third of FLPPs' security controls each year. FLPP is tracked in the Agency's XACTA database system for tracking IT security compliance with FISMA and is a reportable database system to the Office of Management and Budget (OMB).

3b. Data Quality Procedures:
The database is interactive, and operational usage in processing applications by Headquarters and the Regional offices provides ongoing internal quality reviews. Further, EPA periodically checks contractors' data entry quality. OPPT has in place a signed Quality Management Plan ("Quality Management Plan for the Office of Pollution

Prevention and Toxics; Office of Prevention, Pesticides and Toxic Substances", November 2008). Like the 2003 QMP, it will ensure the standards and procedures are applied to this effort. In addition, NPCD has an approved Quality Management Plan in place, dated July 2008. Applications and instructions for applying for certification and accreditation are documented and available at the Web site http://www2.epa.gov/lead/epa-lead-safe-certification-program Documentation for the FLPP database is maintained internally at EPA and is available upon request.

3c. Data Oversight:

Chief, Planning and Assessment Branch, Environmental Assistance Division, OPPT

3d. Calculation Methodology:

Since the measure simply tracks the number of firms currently certified to perform Lead RRP work, there is no need to transform the original data by any mathematical methods.

4a. Oversight and Timing of Final Results Reporting:

Planning and Accountability Lead in the Resource Management Staff in the Office of Program Management Operations. Reporting semiannually: mid-year and end-of-year.

4b. Data Limitations/Qualifications:

Data are obtained by totaling the number of firm certification applications received either directly by EPA or through EPA authorized State programs and reported to EPA Regional offices.

There is little or no sampling error in this performance measure because it is based on an evaluation of all applicable records for the Federal program. Data on firms certified in each authorized state are collected through quarterly reports from grantees as part of the Agency's oversight of authorized programs.

4c. Third-Party Audits:

Not applicable.

Office of Chemical Strategies and Pollution Prevention (OCSPP)

Goal Number and Title:
4 - Ensuring the Safety of Chemicals and Preventing Pollution
Objective Number and Title:
1 - Ensure Chemical Safety
Sub-Objective Number and Title:
1 - Protect Human Health from Chemical Risks
Strategic Target Code and Title:
5 - By 2014, reduce concentration of targeted chemicals in children
Managing Office:
Office of Pesticide Programs
1a. Performance Measure Term Definitions:

Reduce: The American Association of Poison Control Centers (AAPCC) maintains a national database of exposure incidents called the National Poison Data System (NPDS), which is a compilation of data collected by AAPCC's national network of 61 poison controls centers (PCCs). The incident data maintained in AAPCC's NPDS includes pesticide-related exposure incidents that may occur throughout the U.S. population, including all age groups and exposures occurring in both residential and occupational settings. Summary data on pesticide-related incident data is reported on an annual basis in AAPCC's Annual Report, including the number of incidents by age, reason for exposure, level of medical treatment, and medical severity.
 	
The performance measure is based on the annual number of rodenticide exposure incidents involving children less than six-years old, based on aggregated data reported in AAPCC's Annual Report. The baseline for the performance measure will be based on AAPCC's 2008 Annual Report.
 	
Exposure Incidents: Calls to Poison Control Centers are managed primarily by AAPCC-certified Specialists in Poison Information (SPIs). SPIs are required to complete detailed electronic medical records for both exposure and informational calls. Standardized definitions have been established to ensure database uniformity.
 	
For EPA's performance measure, all exposure incidents, regardless of medical severity, will be included in the performance measure calculation.
 	
Rodenticide insecticides: AAPCC's Annual Report reports the number of annual incidents stratified by chemical category. Particular rodenticide categories that will be used to identify incidents include:
 	
   ·   "ANTU (1-naphthalenylthiourea)"
   ·   "Bromethalin Rodenticides"
   ·   "Cholecalciferol Rodenticides"
   ·   "Cyanide Rodenticides"

- "Long-Acting Anticoagulant Rodenticides"
- "Other Types of Rodenticide"
- "PNU (n-3-pyridylmethyln1-p-nitrophenyl urea)"
- "Strychnine Rodenticides"
- "Unknown Types of Rodenticide"
- "Warfarin Type Anticoagulant Rodenticides"
- "Zinc Phosphide Rodenticides"

Children: The performance measure will focus on exposure incidents reported to AAPCC than involved children less than six-years old. This age category is standardized by AAPCC and included as a data field in AAPCC's annual report.

Background:
- The reduction in rodenticide incidents is expected to result from EPA's risk mitigation decision that requires consumer use rodenticides be used in protective bait stations that limit direct contact by young children. As part of this risk mitigation decision, EPA is taking action to cancel and remove from the consumer market 12 D-Con brand mouse and rat poison products. These products fail to comply with safety measures and are commonly reported to U.S. poison control centers. Further information on EPA's risk mitigation is available at: http://www.epa.gov/pesticides/mice-and-rats/

2a. Original Data Source:
NPDS is a comprehensive source of surveillance data on poisonings in the United States. NPDS is a uniform database of 61 PCCs, which are members of the American Association of Poison Control Centers (AAPCC), and are distributed throughout the United States. The database was established in 1985 and now includes information on more than 36 million exposure cases. In 2006, 61 PCCs received more than 4 million cases, including more than 2.4 million human exposure cases and 1.4 million informational calls. NPDS is a valuable public health resource and has been utilized to identify hazards, develop education priorities, guide clinical research, and identify chemical and bioterrorism incidents. As a result, NPDS has helped prompt product reformulations, recalls, and bans, support regulatory actions, and provide post-marketing surveillance of new drugs.

2b. Source Data Collection:
Individual PCC provides 24-hour emergency medical information on the diagnosis and treatment of poisonings. Calls are routed from a single, nationally-available phone number to the PCC generally in closest proximity to the caller. Since the service is provided on a national scale, even though PCCs may not be located in every state, aggregate PCC data is generally considered to be national in scope. The calls are managed primarily by AAPCC-certified Specialists in Poison Information (SPIs), who are typically pharmacists and nurses. SPIs are required to complete detailed electronic medical records for both exposure and informational calls. The electronic medical records include general demographic information, including age, gender, location of exposure, and more detailed information if an exposure may have occurred, including suspected substance,

reason for exposure, route of exposure, management site, symptoms, and medical outcome. To assist SPIs and ensure database uniformity, many of the fields included in the electronic medical records use categories that have been defined by the AAPCC. For example, SPIs characterize the medical severity of possible exposures using the medical outcome field, which includes the AAPCC-defined categories "None," "Minor," "Moderate," "Major," or "Death." Additionally, the records may also contain several open fields, which allow SPIs to record additional information that may be relevant to the treatment and diagnosis of each case

2c. Source Data Reporting:

AAPCC produces the NPDS Annual Report giving statistics and information on all the poisonings in a calendar year. The NPDS Annual Report has three basic sections of information: general charts and statistics, a section of individual fatality listings, and a section listing demographic profile of single-substance exposure cases by generic category. The report is available to the general public to be downloaded for free and is usually made public the December following the close of a calendar year. This means the 2010 NPDS Annual Report was released around December of 2011. The report is typically published in the peer-reviewed journal Clinical Toxicology and is also publically available through AAPCC's website at:

http://www.aapcc.org/dnn/NPDSPoisonData/NPDSAnnualreports.aspx

3a. Relevant Information Systems:

EPA does not require specialized information systems for the purposes of collecting, calculating, and/or reporting the results for this measure. Rather, AAPCC maintains standardized reporting procedures and is responsible for aggregating the summary data that is available in AAPCC's annual report and utilized in the performance measure. Following the publication of AAPCC's annual report, EPA uses MS-Excel to further summarize aggregated data on moderate to severe exposure incidents associated with organophosphate and carbamate insecticides

System Description: Not Applicable

Source/Transformed Data: Not Applicable

Information System Integrity Standards: Not Applicable

3b. Data Quality Procedures:

AAPCC's annual report reflects only those cases that are not duplicates and classified by the regional PC as CLOSED. A case is closed when the PC has determined that no further follow-up/recommendations are required or no further information is available. Exposure cases are followed to obtain the most precise medical outcome possible. Depending on the case specifics, most calls are "closed" within the first hours of the initial call. Some calls regarding complex hospitalized patients or cases resulting in death may remain open for weeks or months while data continues to be collected. Follow-up calls provide a proven mechanism for monitoring the appropriateness of management recommendations, augmenting patient guidelines, and providing poison prevention education, enabling continual updates of case information as well as obtaining final/known medical outcome status to make the data collected as accurate and complete as possible.

3c. Data Oversight:

Source Data Reporting Oversight Personnel: Not applicable.

Source Data Reporting Oversight Responsibilities: Not applicable.

Information Systems Oversight Personnel: Appointed Measures Representative(s) for Health Effects Division, in conjunction with the Division Director and Associate Division Director

Information Systems Oversight Responsibilities: To review and analyze data and report it to the OPP measures representative for reporting

3d. Calculation Methodology:

Decision Rules for Selecting Data: The performance measure uses summary data from AAPCC's Annual Report. Specific incident data that will be selected will involve children less than six-years old and involve the following AAPCC-defined rodenticide categories:

- "ANTU (1-naphthalenylthiourea)"
- "Bromethalin Rodenticides"
- "Cholecalciferol Rodenticides"
- "Cyanide Rodenticides"
- "Long-Acting Anticoagulant Rodenticides"
- "Other Types of Rodenticide"
- "PNU (n-3-pyridylmethyln1-p-nitrophenyl urea)"
- "Strychnine Rodenticides"
- "Unknown Types of Rodenticide"
- "Warfarin Type Anticoagulant Rodenticides"
- "Zinc Phosphide Rodenticides"

Explanation of Calculations:

Annual performance will be evaluated using the equation below:

Where:

Baselinecount = Total number of exposure incidents that meet the case definition during the baseline period.

Performancecount = Total number of exposure incidents that meet the case definition during performance period.

Explanation of Assumptions: The performance measure is based on summary data published in AAPCC's Annual Report. The data is used without making any additional transformations, so no assumption will be made to transform the data.

Unit of Measure: Incident Count

Timeframe of Result: AAPCC's Annual Report is usually made public the December following the close of a calendar year. This means the 2010 NPDS Annual Report was released around December of 2011. Each report provides a summary of the total number of exposure incidents during the complete calendar year.Units:

4a. Oversight and Timing of Final Results Reporting:

Branch Chief, Financial Management and Planning Branch.
Planning and Accountability Lead in the Resource Management Staff in the Office of Program Management Operations. Reporting semiannually: mid-year and end-of-year.

4b. Data Limitations/Qualifications:

General Limitations/Qualifications:
· EPA has issue its risk mitigation decision and issued a Notice of Intent to Cancel in order to cancel the registration of 12 non-compliant rodenticide products. The registrant Reckitt Benckiser, however, has requested an administrative hearing to challenge EPA's decision to cancel the registrations of 12 D-Con mouse and rat poison products. Until the hearing before an EPA Administrative Law Judge is completed, the registrant may continue to market the 12 non-complying products. As such, there is uncertainty in how the upcoming administrative law hearing will impact EPA's efforts to reduce rodenticide exposure incidents involving young children.
· In general, PCC's provide medical management services through their response hotline and do not perform active surveillance of pesticide exposure incidents as part of NPDS. Due to this limitation, NPDS may be subject to reporting bias because of underreporting and differences in utilization rates among difference segments of the U.S. population.
· Because the incidents are self-reported, there is a potential bias in the data. However, there is no reason to believe that the bias will change from year to year.

Data Lag Length and Explanation: AAPCC's Annual Report is published December of every year and made publicly available. For example, 2010 Annual Report was available to EPA in December 2011 and the 2011 Annual Report is expected to be available to EPA in December 2012.

Methodological Changes: Not Applicable

4c. Third-Party Audits:

AAPCC is an independent organization and not subject to third-party audits by the U.S. Government. AAPCC'S Annual Report is publically available (http://www.aapcc.org/dnn/NPDSPoisonData/NPDSAnnualReports.aspx and published in the peer-reviewed journal Clinical Toxicology.

Measure Code: 091 - Percent of decisions completed on time (on or before PRIA or negotiated due date).

Office of Chemical Strategies and Pollution Prevention (OCSPP)

Goal Number and Title:
4 - Ensuring the Safety of Chemicals and Preventing Pollution

Objective Number and Title:
1 - Ensure Chemical Safety

Sub-Objective Number and Title:
1 - Protect Human Health from Chemical Risks

Strategic Target Code and Title:
5 - By 2014, reduce concentration of targeted chemicals in children

Managing Office:
Office of Pesticide Programs

1a. Performance Measure Term Definitions:
Decisions: Each action is assigned a decision number when it is received and with time, actions and decisions have come to mean about the same. A decision may be an application to register a new pesticide product, to amend a registered product's label, to review a protocol, to establish a tolerance or to make a decision on a request to waive a study requirement.

Completed: An action or decision is completed when OPP makes a decision on the application, i.e. the product is registered, a label is stamped, protocol reviewed, or the action is denied, the label not approved, etc. A decision memorandum is issued describing the decision made and the date that the delegated official signs the memo is the date that the decision is completed. In the case of a label, the date that the label is stamped as approved is the date that the application to register or amend a label is completed.

PRIA: The Pesticide Registration Improvement Act (PRIA) of 2003 established pesticide registration service fees for registration actions. The Pesticide Registration Improvement Renewal Act (PRIA 2), effective October 1, 2007, reauthorized the PRIA for five more years until 2012. The PRIA 2 legislation increased the number of actions covered by fees, modified the payment process and application in-processing. The category of action, the amount of pesticide registration service fee, and the corresponding decision review periods by year are prescribed in these statutes. Their goal is to create a more predictable evaluation process for affected pesticide decisions, and couple the collection of individual fees with specific decision review periods. They also promote shorter decision review periods for reduced-risk applications.

On time (on or before PRIA or negotiated due date): Each PRIA 2 fee category has an associated period of time in which the Agency must make a determination, which has been called a decision review period or PRIA 2 timeframe, or "PRIA due date." The PRIA 2 due date may be extended by a mutual agreement between the applicant and the Agency. The new due date is called a negotiated due date. Negotiated due dates occur predominately as a result of missing information or data or data deficiencies identified during an in-depth review of the application. The due date then is extended to allow the applicant the time to submit the data or information and for the Agency to review the data and make a determination.

Background:

This measure is a program output which represents the program's statutory requirements to ensure that pesticides entering the marketplace are safe for human health and the environment, and when used in accordance with the packaging label present a reasonable certainty of no harm. In addition, under PRIA and PRIA 2, there are specific timelines, based on the type of registration action, by which the Agency must make a decision. These laws do allow the decision due date under PRIA to be negotiated to a later date, after consultation with and agreement by the submitter of the application. The timeliness measure represents the Agency's effectiveness in meeting these PRIA timelines.

For more information, see
- http://www.epa.gov/pesticides/fees/
- FIFRA Sec 3(c)(5)
- FFDCA Sec 408(a)(2).

2a. Original Data Source:
EPA senior managers.

2b. Source Data Collection:

Source Data Collection Methods: EPA senior managers review justifications and make final decisions to extend or negotiate a PRIA due date and whether or not to issue a "PRIA Determination to Not Grant" a registration. The Agency employs continuous monitoring of the status of PRIA decisions. Numerous internal Agency meetings continue to monitor workload and compliance with PRIA due dates. Throughout the pesticide registration program, weekly meetings are held to review the status of pending decisions, due date extensions, and refunds; to identify potential issues and target their resolution; to resolve fee category questions; and to coordinate schedules with science support organizations.

EPA QA requirements/guidance governing collection: All risk assessments are subject to public and scientific peer review. All registration actions must employ sound science and meet the Food Quality Protection Act (FQPA) safety standards. The office adheres to its Quality Management Plan (Nov. 2006) in ensuring data quality and that procedures are properly applied.

2c. Source Data Reporting:

All registration actions received under the PRIA and PRIA 2 are entered and tracked in the Pesticide Registration Information System (PRISM). Reports developed in Business Objects (using PRISM as the data source) allow senior management to more effectively track the workload (e.g., pending actions with upcoming PRIA due dates, actions for which the PRIA date appears to have passed etc.) and ensure that PRIA or negotiated due dates are met.

OPP uses several internal controls within the OPPIN/PRISM system. First of all, users must be FIFRA CBI cleared in order to access the system. Within the system, security measures are taken to allow only authorized users to perform certain operations, which are managed by our Data Base Administrator (DBA).

For example, only Branch Chiefs can enter a negotiated due date in the Registration Division. The DBA must receive an Access Form from users wanting to use the system and their supervisor must sign the Access Form.

Applications are pin punched upon receipt by a NOWCC in ISB/ITRMD/OPP and the pin punch date is entered into OPPIN by another NOWCC in ISB. The pin punch date is the receipt date in OPPIN. The EPA team leader performs periodic/random checks of their work. Experts from the three registering divisions review each application and place it in a PRIA fee category generally on the date of receipt.

PRIA 2 requires that certification of payment be submitted together with the application. . Beginning January 2, 2008, ISB started to hold any application that did not contain certification of payment. ISB contacts the submitter to request certification of payment. When the certification is received, ISB generates an acknowledgement and sends it to the submitter. If no certification of payment is received within 14 days, ISB prepares a rejection letter for the Deputy Office Director's signature. After the rejection letter is signed, ISB posts the rejection to OPPIN, and invoices the submitter for 25% of the appropriate PRIA fee.

Any issues related to assigning a fee category are discussed with divisional management and may be elevated. If a full fee is being paid, the date that begins the PRIA timeframe or start date is the latest of 21 days after receipt of the application or the day payment is received by the Washington Finance Center/ OCFO. Staff in OCFO enter the amount and date of receipt of the payment into IFMS. OPP downloads IFMS and electronically transfers the data into OPPIN.

Once the IFMS data is transferred to OPPIN, OPPIN automatically calculates due dates from the start date using the time frames in the FR Notice on the fee schedule. Due dates can be extended through negotiations with the registrant or applicant. Negotiated due dates are manually entered and the rights to enter a negotiated due date belong to only branch chiefs, the Division Directors and other individuals designated such rights by a Division Director. In BPPD, negotiated PRIA due dates are entered in OPPIN by the branch chiefs, branch team leaders, or its Administrative Specialist while in RD, only a branch chief enters the date. According to OPP's procedures, a negotiated due date cannot be entered into the system until the Deputy Office Director or Office Director approves the negotiated date by signing the negotiated due date form. A copy of the negotiated due date form and documentation of the applicant's agreement with the due date are filed. Beginning July 2011, OPP transition to using Webforms for processing negotiated due date forms. Forms are routed, approved, and retained electronically.

The date that an action is completed is entered by staff in RD, AD, and BPPD according to their internal procedures. Documentation of the date of completion is filed in the product's file. Once data is entered into OPPIN, start dates and due dates cannot be changed by staff in the regulatory divisions. Changes are made by staff programming OPPIN in ITRMD. "Data fixes" must be requested by generating a SCR (Systems Change Request). These requests are reviewed by ITRMD staff and management and representatives of the regulatory divisions. Questions and issues are elevated to the PRIA Senior Advisor and if needed to OPP management. OPP management holds a Bi-weekly PRIA meeting in which these issues are discussed and resolved. The OPP Immediate Office uses a number of monitoring reports to identify actions that are past their due date or appear to have been logged out past their due date. An issue is then resolved with the appropriate division and generally involves an action that needs to be logged out as completed or a negotiated due date that needs to be entered. OPPIN software issues have also been identified through this oversight effort and an SCR is developed to make the necessary programming corrections.

PRIA data is an internally generated tracking data base with data entries being made during normal business hours.

Annually, the Office of the Inspector General conducts an audit that includes verifying the accurate entry of the date an action is received, extended and completed.

3a. Relevant Information Systems:

All registration actions received under the PRIA and PRIA 2 are entered and tracked in the Pesticide Registration Information System (PRISM).

The Office of Pesticide Programs (OPP) has migrated all of its major data systems including regulatory and scientific data, workflow tracking and electronic document management into one integrated system, the Pesticide Registration Information System (PRISM). PRISM provides a centralized source of information on all registered pesticide products, including chemical composition, toxicity, name and address of registrant, brand names, registration actions, and related data. It is maintained by the EPA and tracks regulatory data submissions and studies, organized by scientific discipline, which are submitted by the registrant in support of a pesticide's registration. All registration actions received under the PRIA and PRIA 2 are entered and tracked in PRISM.

PRISM is the successor to the Office of Pesticide Programs Information System Network (OPPIN). Data has been migrated from the following databases: Chemical Vocabulary (CV), Company Name and Address (CNAD), Pesticide Document Management System (PDMS), Pesticide Product Information System (PPIS), Chemical Review Management System (CRMS), FIFRA CBI Access (FAS), Jackets, Product Data Call-In (PDCI), Phones, Pesticide Regulatory Action Tracking (PRAT), Reference System (REFS), Tolerance Indexes (TIS and TOTS). Sources of the input are paper copy and electronic data. EPA's Central Data Exchange (CDX), scheduled as EPA 097, is the gateway for electronic submissions. It consolidates information stored on the mainframe, the OPP LAN, on stand-alone computers and in paper copy. PRISM (Pesticide Registration Information System) consolidates various pesticides program databases.

EPA recently constructed a module in PRISM tracking major Registration Review milestones. This module enhances tracking capabilities and is an important management tool.

For information on disposition of records in this database, please see EPA Records Schedule 329, http://www.epa.gov/records/policy/schedule/sched/329.htm

OPP adheres to its Quality Management Plan (Nov. 2006) in ensuring data quality and that procedures are properly applied.

PRISM was developed between 1997 and 2003 and has been operational since June 2, 2003. PRISM provides e-government capabilities to share pesticide information with OPP stakeholders. PRISM supports OPP's responsibilities under a variety of regulatory requirements including FIFRA, FQPA, PRIA, PRIA II, Pesticide Registration Review and for the Endocrine Disrupter Screening Program and will standardize the structure of a chemical case where appropriate to define the key tasks and documents used in a number of pesticide review processes. EDSP components are used to order, monitor, track and manage scientific tests associated with pesticide chemicals. Pesticide Registration Improvement Renewal Act (PRIA II).

PRISM was developed in response to the requirements of the following laws and regulations:

· The Title III of the E-Government Act of 2002 - Federal Information Security Management Act (FISMA) – Public Law 107-347: A security plan must be developed and practiced throughout all life cycles of the agency's information systems.

· Office of Management and Budget (OMB) Circular A-130, Management of Federal Information Resources: A System Security Plan (SSP) is to be developed and documented for each GSS and Major Application (MA) consistent with guidance issued by the National Institute of Standards and Technology (NIST).

· Federal Information Processing Standards (FIPS) Publication 199, Standards for Security Categorization of Federal Information and Information Systems: This document defines standards for the security categorization of information and information systems. System security categorization must be included in SSPs.

· FIPS Publication 200, Minimum Security Requirements for Federal Information and Information Systems: This document contains information regarding specifications for minimum security control requirements for federal information and information systems. Minimum security controls must be documented in SSPs.

· NIST Special Publication (SP) 800-18 Revision 1, Guide for Developing Security Plans for Federal Information Systems: The minimum standards for an SSP are provided in this NIST document.

· NIST SP 800-53, Revision 3, Recommended Security Controls for Federal Information Systems and Organizations: This document contains a list of security controls that are to be implemented into federal information systems based on their FIPS 199 categorization. This document is used in conjunction with FIPS 200 to define minimum security controls, which must be documented in SSPs.

· EPA Information Security Planning Policy. A system security plan shall be developed for each system cited on the EPA Inventory of Major Information Systems, including major applications and general support systems

Most, if not all, of PRISM data should be considered "source" data. This means that these data originate from primary data providers, particularly pesticide product registrants, submitting information sent to EPA directly in response to FIFRA regulatory requirements.

PRISM contains source data and from this source data, certain dates, such as the date due are calculated automatically.

3b. Data Quality Procedures:

OPP adheres to its Quality Management Plan (Nov. 2006) in ensuring data quality and that procedures are properly applied.

3c. Data Oversight:

Peter Caulkins, PRIA Coordinator and Special Assistant to the Deputy Office Director, Office of Pesticide Programs. Handles all aspects of data collection and verification.

3d. Calculation Methodology:

Unit of analysis: Percent

The percent completed on time is calculated by taking the total number of decisions or actions completed and withdrawn on or before their due date and dividing by the total number decisions or actions completed and withdrawn within the date range specified.

Total PRIA actions completed for the FY less PRIA actions completed late for the FY divided by total PRIA actions completed for the FY equals the percent of PRIA actions completed on time for the FY, where total PRIA actions completed includes actions completed, actions withdrawn, and actions rejected.

4a. Oversight and Timing of Final Results Reporting:

Vickie Richardson, Branch Chief Resource Management Staff in the Office of Program Management Operations. Reporting semiannually: mid-year and end-of-year.

4b. Data Limitations/Qualifications:

No Data Limitations.

4c. Third-Party Audits:

Not applicable.

Measure Code: 164 - Number of pesticide registration review dockets opened.

Office of Chemical Strategies and Pollution Prevention (OCSPP)

Goal Number and Title:
4 - Ensuring the Safety of Chemicals and Preventing Pollution

Objective Number and Title:
1 - Ensure Chemical Safety

Sub-Objective Number and Title:
1 - Protect Human Health from Chemical Risks

Strategic Target Code and Title:
5 - By 2014, reduce concentration of targeted chemicals in children

Managing Office:
Office of Pesticide Programs

1a. Performance Measure Term Definitions:

Registration Review dockets: EPA initiates a registration review by establishing a docket for a pesticide registration review case and opening the docket for public review and comment. Each docket contains a Summary Document that explains what information EPA has on the pesticide and the anticipated path forward. The Summary Document includes:

·   A Preliminary Work Plan highlighting anticipated risk assessment and data needs, providing an anticipated timeline for completing the pesticide's review, and identifying the types of information that would be especially useful to the Agency in conducting the review;

·   A fact sheet providing general background information and summarizing the current status of the pesticide;

·   Ecological risk assessment problem formulation and human health scoping sections describing the data and scientific analyses expected to be necessary to complete the pesticide's registration review.

Opened: EPA initiates a registration review by establishing a docket for a pesticide registration review case and opening the docket for public review and comment. The Agency publishes a Federal Register notice that announces the availability of the docket and provides a comment period of at least 60 days. See http://www.epa.gov/oppsrrd1/registration_review/reg_review_process.htm for more information.

Background:

The Food Quality Protection Act of 1996 directed EPA to establish a Registration Review program with the goal of reviewing all registered pesticides, AIs and products, on a 15-year cycle to ensure that they continue to meet the standards of registration. EPA issued the final rule in 2006 and began implementing the program in 2007. Under the rule, EPA posts registration review schedules and these will provide a baseline for expected AI case dockets that will be opened for the next three year cycle and for decisions expected over the next several years. The first step of Registration Review is to open a public docket for each pesticide case entering the process to show the public what the Agency knows about the AI and seek comment. When comments are evaluated and data needs are finalized, OPP posts a Final Work Plan (FWP) for each AI case. Although the docket openings and the FWPs are tracked, both steps require notable resources to complete.

All registrations must be based on sound science and meet the Food Quality Protection Act (FQPA) safety standard. All risk assessments are subject to public and scientific peer review. In addition, OPP management reviews and signs new documents before being placed in the docket or posted on EPA's website.

For more information, see:
http://www.epa.gov/oppsrrd1/registration_review/

2a. Original Data Source:

OPP staff, working collaboratively across the program, develop the draft preliminary work plan taking into account existing policies, data requirements, and standard operating procedures.

2b. Source Data Collection:

Each preliminary work plan is approved by Director of the appropriate OPP division (Antimicrobial Division, Biopesticides and Pollution Prevention Division, and Pesticide Re-evaluation Division). All preliminary work plans are included in the docket for that registration review case and are available via the pesticide program website at http://www.epa.gov/pesticides

Data collected are for national actions taken on an annual basis. There is no spatial component.

EPA QA requirements/guidance governing collection: The office adheres to its Quality Management Plan (Nov. 2006) in ensuring data quality and that procedures are properly applied.

2c. Source Data Reporting:

Form/mechanism for receiving data and entering into EPA system: As described in 2b, all preliminary work plans are posted to the docket for that registration review case and are available via the pesticide program website. Counts for preliminary work plans completed are tracked and tabulated in a master spreadsheet maintained by the Pesticide Re-evaluation Division.

Timing and frequency of reporting: Preliminary work plans are developed on a quarterly basis. Counts of actions completed are available at the end of each quarter.

EPA QA requirements/guidance governing collection: The office adheres to its Quality Management Plan (Nov. 2006) in ensuring data quality and that procedures are properly applied.

3a. Relevant Information Systems:

The Office of Pesticide Programs (OPP) has migrated all of its major data systems including regulatory and scientific data, workflow tracking and electronic document management into one integrated system, the Pesticide Registration Information System (PRISM). PRISM provides a centralized source of information on all registered pesticide products, including chemical composition, toxicity, name and address of registrant, brand names, registration actions, and related data. It is maintained by the EPA and tracks regulatory data submissions and studies, organized by scientific discipline, which are submitted by the registrant in support of a pesticide's registration. All registration actions received under the PRIA and PRIA 2 are entered and tracked in PRISM.

PRISM is the successor to the Office of Pesticide Programs Information System Network (OPPIN). Data have been migrated from the following databases: Chemical Vocabulary (CV), Company Name and Address (CNAD), Pesticide Document Management System (PDMS), Pesticide Product Information System (PPIS), Chemical Review Management System (CRMS), FIFRA CBI Access (FAS), Jackets, Product Data Call-In (PDCI), Phones, Pesticide Regulatory Action Tracking (PRAT), Reference System (REFS), Tolerance Indexes (TIS and TOTS). Sources of the input are paper copy and electronic data. EPA's Central Data Exchange (CDX), scheduled as EPA 097, is the gateway for electronic submissions. It consolidates information stored on the mainframe, the OPP LAN, on stand-alone computers and in paper copy. PRISM (Pesticide Registration Information System) consolidates various pesticides program databases.

EPA recently constructed a module in PRISM tracking major Registration Review milestones. This module enhances tracking capabilities and is an important management tool.

For information on disposition of records in this database, please see EPA Records Schedule 329, http://www.epa.gov/records/policy/schedule/sched/329.htm

PRISM was developed between 1997 and 2003 and has been operational since June 2, 2003. PRISM provides e-government capabilities to share pesticide information with OPP stakeholders. PRISM supports OPP's responsibilities under a variety of regulatory requirements including FIFRA, FQPA, PRIA, PRIA II, Pesticide Registration Review and for the Endocrine Disrupter Screening Program and will standardize the structure of a chemical case where appropriate to define the key tasks and documents used in a number of pesticide review processes. EDSP components are used to order, monitor, track and manage scientific tests associated with pesticide chemicals. Pesticide Registration Improvement Renewal Act (PRIA II).

PRISM was developed in response to the requirements of the following laws and regulations:

·   The Title III of the E-Government Act of 2002 - Federal Information Security Management Act (FISMA) – Public Law 107-347: A security plan must be developed and practiced throughout all life cycles of the agency's information systems.

·   Office of Management and Budget (OMB) Circular A-130, Management of Federal Information Resources: A System Security Plan (SSP) is to be developed and documented for each GSS and Major Application (MA) consistent with guidance issued by the National Institute of Standards and Technology (NIST).

·   Federal Information Processing Standards (FIPS) Publication 199, Standards for Security Categorization of Federal Information and Information Systems: This document defines standards for the security categorization of information and information systems. System security categorization must be included in SSPs.

·   FIPS Publication 200, Minimum Security Requirements for Federal Information and Information Systems: This document contains information regarding specifications for minimum security control requirements for federal information and information systems. Minimum security controls must be documented in SSPs.

- NIST Special Publication (SP) 800-18 Revision 1, Guide for Developing Security Plans for Federal Information Systems: The minimum standards for an SSP are provided in this NIST document.

- NIST SP 800-53, Revision 3, Recommended Security Controls for Federal Information Systems and Organizations: This document contains a list of security controls that are to be implemented into federal information systems based on their FIPS 199 categorization. This document is used in conjunction with FIPS 200 to define minimum security controls, which must be documented in SSPs.

- EPA Information Security Planning Policy. A system security plan shall be developed for each system cited on the EPA Inventory of Major Information Systems, including major applications and general support systems

Most, if not all, of PRISM data should be considered "source" data. This means that these data originate from primary data providers, particularly pesticide product registrants, submitting information sent to EPA directly in response to FIFRA regulatory requirements.

3b. Data Quality Procedures:
OPP adheres to its Quality Management Plan (Nov. 2006) in ensuring data quality and that procedures are properly applied. The Quality Management Plan is updated periodically, with the most recent plan approved on April 26, 2012.
3c. Data Oversight:
Planning and Accountability Lead in the Resource Management Staff in the Office of Program Management Operations. Reporting semiannually: mid-year and end-of-year. Rick Keigwin (Director, Pesticide Re-evaluation Division, OPP), Keith Matthew (Director, Biospesticides and Pollution Prevention Division, OPP), and Joan Harrigan-Farrelly (Director, Antimicrobials Division, OPP)-
3d. Calculation Methodology:
Identification of Unit of Measure and Timeframe: Timeframe is the fiscal year. Unit of measure is the number of preliminary work plans completed each year. The Agency develops a preliminary workplan for each pesticide subject to the registration review program. To be counted under this measure, each preliminary workplan must be signed by the appropriate division director and a docket is established to allow for public comment on the preliminary workplan. Workplans are only counted when signed by the division director. There are no other variables or assumptions. Calculations are conducted by summing the number or preliminary workplans issued each fiscal year.
4a. Oversight and Timing of Final Results Reporting:
Vickie Richardson, Branch Chief, Financial Management and Planning Branch Reporting is done twice a year
4b. Data Limitations/Qualifications:
No data limitations.
4c. Third-Party Audits:
Not applicable.

Measure Code: 247 - Percent of new chemicals or organisms introduced into commerce that do not pose unreasonable risks to workers, consumers, or the environment.

Office of Chemical Strategies and Pollution Prevention (OCSPP)

Goal Number and Title:
4 - Ensuring the Safety of Chemicals and Preventing Pollution

Objective Number and Title:
1 - Ensure Chemical Safety

Sub-Objective Number and Title:
1 - Protect Human Health from Chemical Risks

Strategic Target Code and Title:
0 -

Managing Office:
Office of Pollution Prevention and Toxics

1a. Performance Measure Term Definitions:
New chemicals or organisms: The term "new chemical substance" (which includes microorganisms) means any chemical substance which is not included in the chemical substance list compiled and published under TSCA section 8(b) (i.e., the TSCA Inventory).   Introduced (or "distributed")into commerce: The terms "distribute in commerce" and "distribution in commerce" when used to describe an action taken with respect to a chemical substance or mixture or article containing a substance or mixture mean to sell, or the sale of, the substance, mixture, or article in commerce; to introduce or deliver for introduction into commerce, or the introduction or delivery for introduction into commerce of, the substance, mixture, or article; or to hold, or the holding of, the substance, mixture, or article after its introduction into commerce.   Unreasonable risk: The term "unreasonable risk" is not defined in TSCA. The legislative history, however, indicates that unreasonable risk involves the balancing of the probability that harm will occur and the magnitude and severity of that harm against the effect of a proposed regulatory action on the availability to society of the expected benefits of the chemical substance. In the context of the New Chemicals Program, EPA's determination that manufacture, processing, use, distribution in commerce, or disposal of an individual substance which has been the subject of a notice under section 5 of the TSCA may present an unreasonable risk of injury to human health or the environment is based on consideration of (i) the size of the risks identified by EPA; (ii) limitations on risk that would result from specific safeguards (generally, exposure and release controls) sought based on Agency review and (iii) the benefits to industry and the public expected to be provided by new chemical substances intended to be manufactured after Agency review. In considering risk, EPA considers factors including environmental effects, distribution, and fate of the chemical substance in the environment, disposal methods, waste water treatment, use of protective equipment and engineering controls, use patterns, and market potential of the chemical substance.

2a. Original Data Source:
The original data source is EPA. The agency maintains records of all TSCA Section 5 PMN submissions and

Section 8(e) submissions. The Section 5 and Section 8(e) submissions are provided to EPA by external parties, typically chemical manufacturers in the case of PMN submissions and chemical manufacturers, processors and distributors in the case of Section 8(e) notices.

2b. Source Data Collection:

The agency tabulates data submitted under TSCA Section 5 and Section 8(e) on a daily basis and maintains the data in the various databases described in subsection 3(a) below. The individual submitting notices to EPA under TSCA Section 5 must certify by signature that "All information provided in the notice is complete and truthful as of the date of submission." Please see subsection 3(b) for information on the data quality procedures followed by the agency.

2c. Source Data Reporting:

Not applicable. Since the original data source is EPA, the source data are not transmitted to the agency by any independent entity.   As noted above, TSCA Section 5 and Section 8(e) submissions are provided to EPA by external parties, typically chemical manufacturers in the case of PMN submissions and chemical manufacturers, processors and distributors in the case of Section 8(e) notices.

3a. Relevant Information Systems:

Implementation of this measure requires the use of several EPA databases: Chemical Information System (CIS), the legacy Management Information Tracking System (MITS), Pre-manufacture Notice (PMN) Lotus Notes, PMN CBI Local Area Network (LAN), 8(e) ISIS database for new chemicals, and the Focus database.

The following information from these databases is used collectively in applying this measure:
•CIS: Tracking information on ePMNs received;
• MITS: Legacy database that contains NCP regulatory dispositions for section 5 cases since 1979 and which stopped being an active database in 2013. • PMN Lotus Notes: Records PMN review and decision, assessment reports on chemicals submitted for review. New workflow system for new chemicals submitted since August 2008.
• PMN CBI LAN: Records documenting PMN review and decision, assessment reports on chemicals submitted for review before August 2008. In addition, the information developed for each PMN is kept in hard copy in the Confidential Business Information Center (CBIC);
•8(e) ISIS Database: Data submitted by industry under the Toxic Substances Control Act (TSCA) Section 8(e). TSCA 8(e) requires that chemical manufacturers, processors, and distributors notify EPA immediately of new (e.g. not already reported), unpublished chemical information that reasonably supportsaconclusion of substantial risk. TSCA 8(e) substantial risk information notices most often contain toxicity data but may also contain information on exposure, environmental persistence, or actions being taken to reduce human health and environmental risks. It is an important information-gathering tool that serves as an early warning mechanism;
• Focus Database: Rationale for decisions emerging from Focus meeting, including decisions on whether or not to drop chemicals from further review.

3b. Data Quality Procedures:

OPPT has in place a signed Quality Management Plan ("Quality Management Plan for the Office of Pollution Prevention and Toxics; Office of Prevention, Pesticides and Toxic Substances," November 2008). Like the 2003

QMP, it ensures the standards and procedures are applied to this effort.

3c. Data Oversight:

Source data reporting oversight: Not applicable for reasons set out above.

Information systems oversight: Primary responsibility resides with the Director of OPPT's Information Management Division for most databases. Responsibility for the 8(e) ISIS Database resides with the Director of OPPT's Risk Assessment Division.

3d. Calculation Methodology:

EPA's methods for implementing this measure involve determining whether EPA's current PMN review practices would have failed to prevent the introduction of chemicals or microorganisms into commerce that pose an unreasonable risk to workers, consumers or the environment, based on comparisons of 8(e) and previously-submitted new chemical review data. The "unreasonable risk" determination is based on consideration of (1) the magnitude of risks identified by EPA, (2) limitations on risk that result from specific safeguards applied, and (3) the benefits to industry and the public expected to be provided by the new chemical substance. In considering risk, EPA looks at anticipated environmental effects, distribution and fate of the chemical substance in the environment, patterns of use, expected degree of exposure, the use of protective equipment and engineering controls, and other factors that affect or mitigate risk. The following are the steps OPPT will follow in comparing the 8(e) data with the previously-submitted new chemical review data:

1. Match all 8(e) submissions in the 8(e) database with associated TSCA Section 5 notices. TSCA Section 5 requires manufacturers to give EPA a 90-day advance notice (via a pre-manufacture notice or PMN) of their intent to manufacture and/or import a new chemical. The PMN includes information such as specific chemistry identity, use, anticipated production volume, exposure and release information, and existing available test data. The information is reviewed through
the New Chemicals Program to determine whether action is needed to prohibit or limit manufacturing, processing, or use of a chemical.
2. Characterize the resulting 8(e) submissions based on the PMN review phase. For example, were the 8(e) submissions received: a) before the PMN notice was received by EPA, b) during the PMN review process, or c) after the PMN review was completed?
3. Review 8(e) data focusing on 8(e)s received after the PMN review period was completed.
4. Compare hazard evaluation developed during PMN review with the associated 8(e) submission.
5. Report on the accuracy of the initial hazard determination.
6. Revise risk assessment to determine if there was an unreasonable risk based on established risk assessment and risk management guidelines and whether current PMN Review practices would have detected and prevented that risk.
7. Measurement results are calculated on a fiscal-year basis and draw on relevant information received over the 12-month fiscal year

4a. Oversight and Timing of Final Results Reporting:

Planning and Accountability Lead in the Resource Management Staff in the Office of Program Management

Operations. Reporting annually at end-of-year.

4b. Data Limitations/Qualifications:
There are some limitations of EPA's review which result from differences in the quality and completeness of 8(e) data provided by industry; for example, OPPT cannot evaluate submissions that do not contain adequate information on chemical identity. The review is also affected in some cases by a lack of available electronic information. In particular the pre-1996 PMN cases are only retrievable in hard copy and may have to be requested from the Federal Document Storage Center. This may introduce some delays to the review process.

4c. Third-Party Audits:
None.

Measure Code: 230 - Number of pesticide registration review final work plans completed.

Office of Chemical Strategies and Pollution Prevention (OCSPP)

Goal Number and Title:
4 - Ensuring the Safety of Chemicals and Preventing Pollution

Objective Number and Title:
1 - Ensure Chemical Safety

Sub-Objective Number and Title:
2 - Protect Ecosystems from Chemical Risks

Strategic Target Code and Title:
1 - By 2015, no watersheds will exceed aquatic life benchmarks for targeted pesticides

Managing Office:
Office of Pesticide Programs

1a. Performance Measure Term Definitions:
Registration Review dockets: EPA initiates a registration review by establishing a docket for a pesticide registration review case and opening the docket for public review and comment. Each docket contains a Summary Document that explains what information EPA has on the pesticide and the anticipated path forward. The Summary Document includes: · A Preliminary Work Plan highlighting anticipated risk assessment and data needs, providing an anticipated timeline for completing the pesticide's review, and identifying the types of information that would be especially useful to the Agency in conducting the review; · A fact sheet providing general background information and summarizing the current status of the pesticide; · Ecological risk assessment problem formulation and human health scoping sections describing the data and scientific analyses expected to be necessary to complete the pesticide's registration review. Completed: After the closure of the public comment period for the preliminary work plan, EPA reviews those comments and revises (as necessary) the work plan, resulting in the issuance of a final work plan. See http://www.epa.gov/oppsrrd1/registration_review/reg_review_process.htm for more information.   Background: The Food Quality Protection Act of 1996 directed EPA to establish a Registration Review program with the goal of reviewing all registered pesticides, AIs and products, on a 15-year cycle to ensure that they continue to meet the standards of registration. EPA issued the final rule in 2006 and began implementing the program in 2007. Under the rule, EPA posts registration review schedules and these will provide a baseline for expected AI case dockets that will be opened for the next three year cycle and for decisions expected over the next several years. The first step of Registration Review is to open a public docket for each pesticide case entering the process to show the public what the Agency knows about the AI and seek comment. When comments are evaluated and data needs are finalized, OPP posts a Final Work Plan (FWP) for each AI case. Although the docket openings and the FWPs are tracked, both steps require notable resources to complete. All registrations must be based on sound science and meet the Food Quality Protection Act (FQPA) safety

standard. All risk assessments are subject to public and scientific peer review. In addition, OPP management reviews and signs new documents before being placed in the docket or posted on EPA's website.

For more information, see:
http://www.epa.gov/oppsrrd1/registration_review/

2a. Original Data Source:

OPP staff, working collaboratively across the program, review the public comments and develop the draft final work plan taking into account existing policies, data requirements, and standard operating procedures.

2b. Source Data Collection:

Each final work plan is approved by Director of the appropriate OPP division (Antimicrobial Division, Biopesticides and Pollution Prevention Division, and Pesticide Re-evaluation Division). All final work plans are included in the docket for that registration review case and are available via the pesticide program website at http://www.epa.gov/pesticides

Data collected are for national actions taken on an annual basis. There is no spatial component.

EPA QA requirements/guidance governing collection: The office adheres to its Quality Management Plan (Nov. 2006) in ensuring data quality and that procedures are properly applied.

2c. Source Data Reporting:

Form/mechanism for receiving data and entering into EPA system: As described in 2b, all final work plans are posted to the docket for that registration review case and are available via the pesticide program website. Counts for final work plans completed are tracked and tabulated in a master spreadsheet maintained by the Pesticide Re-evaluation Division.

Timing and frequency of reporting: Final work plans are developed on a quarterly basis. Counts of actions completed are available at the end of each quarter.

EPA QA requirements/guidance governing collection: The office adheres to its quality Management Plan (Nov. 2006) in ensuring data quality and that procedures are properly applied.

3a. Relevant Information Systems:

The Office of Pesticide Programs (OPP) has migrated all of its major data systems including regulatory and scientific data, workflow tracking and electronic document management into one integrated system, the Pesticide Registration Information System (PRISM). PRISM provides a centralized source of information on all registered pesticide products, including chemical composition, toxicity, name and address of registrant, brand names, registration actions, and related data. It is maintained by the EPA and tracks regulatory data submissions and studies, organized by scientific discipline, which are submitted by the registrant in support of a pesticide's registration. All registration actions received under the PRIA and PRIA 2 are entered and tracked in PRISM.

PRISM is the successor to the Office of Pesticide Programs Information System Network (OPPIN). Data has been migrated from the following databases: Chemical Vocabulary (CV), Company Name and Address (CNAD),

Pesticide Document Management System (PDMS), Pesticide Product Information System (PPIS), Chemical Review Management System (CRMS), FIFRA CBI Access (FAS), Jackets, Product Data Call-In (PDCI), Phones, Pesticide Regulatory Action Tracking (PRAT), Reference System (REFS), Tolerance Indexes (TIS and TOTS). Sources of the input are paper copy and electronic data. EPA's Central Data Exchange (CDX), scheduled as EPA 097, is the gateway for electronic submissions. It consolidates information stored on the mainframe, the OPP LAN, on stand-alone computers and in paper copy. PRISM (Pesticide Registration Information System) consolidates various pesticides program databases.

EPA recently constructed a module in PRISM tracking major Registration Review milestones. This module enhances tracking capabilities and is an important management tool.

For information on disposition of records in this database, please see EPA Records Schedule 329, http://www.epa.gov/records/policy/schedule/sched/329.htm

PRISM was developed between 1997 and 2003 and has been operational since June 2, 2003. PRISM provides e-government capabilities to share pesticide information with OPP stakeholders. PRISM supports OPP's responsibilities under a variety of regulatory requirements including FIFRA, FQPA, PRIA, PRIA II, Pesticide Registration Review and for the Endocrine Disrupter Screening Program and will standardize the structure of a chemical case where appropriate to define the key tasks and documents used in a number of pesticide review processes. EDSP components are used to order, monitor, track and manage scientific tests associated with pesticide chemicals. Pesticide Registration Improvement Renewal Act (PRIA II).

PRISM was developed in response to the requirements of the following laws and regulations:

·   The Title III of the E-Government Act of 2002 - Federal Information Security Management Act (FISMA) – Public Law 107-347: A security plan must be developed and practiced throughout all life cycles of the agency's information systems.

·   Office of Management and Budget (OMB) Circular A-130, Management of Federal Information Resources: A System Security Plan (SSP) is to be developed and documented for each GSS and Major Application (MA) consistent with guidance issued by the National Institute of Standards and Technology (NIST).

·   Federal Information Processing Standards (FIPS) Publication 199, Standards for Security Categorization of Federal Information and Information Systems: This document defines standards for the security categorization of information and information systems. System security categorization must be included in SSPs.

·   FIPS Publication 200, Minimum Security Requirements for Federal Information and Information Systems: This document contains information regarding specifications for minimum security control requirements for federal information and information systems. Minimum security controls must be documented in SSPs.

·   NIST Special Publication (SP) 800-18 Revision 1, Guide for Developing Security Plans for Federal Information Systems: The minimum standards for an SSP are provided in this NIST document.

·   NIST SP 800-53, Revision 3, Recommended Security Controls for Federal Information Systems and Organizations: This document contains a list of security controls that are to be implemented into federal information systems based on their FIPS 199 categorization. This document is used in conjunction with FIPS 200 to define minimum security controls, which must be documented in SSPs.

·   EPA Information Security Planning Policy. A system security plan shall be developed for each system cited on the EPA Inventory of Major Information Systems, including major applications and general support systems

Most, if not all, of PRISM data should be considered "source" data. This means that these data originate from primary data providers, particularly pesticide product registrants, submitting information sent to EPA directly in response to FIFRA regulatory requirements.

3b. Data Quality Procedures:

OPP adheres to its Quality Management Plan (Nov. 2006) in ensuring data quality and that procedures are properly applied. The Quality Management Plan is updated periodically, with the most recent plan approved on April 26, 2012.

3c. Data Oversight:

Rick Keigwin (director, Pesticide Re-evaluation Division, OPP), Keith Matthews (Director, Biopesticides and Pollution Prevention Division, OPP), and Joan Harrigan-Farrelly (Director, Antimicrobials Division, OPP)

3d. Calculation Methodology:

Timeframe is the fiscal year. Unit of measure is the number of final work plans completed each year.The Agency develops a final workplan for each pesticide subject to the registration review program. To be counted under this measure, each final workplan must be signed by the appropriate division director and placed in the docket established for that pesticide. Workplans are only counted when signed by the division director. There are no other variables or assumptions. Calculations are conducted by summing the number or final workplans issued each fiscal year.

4a. Oversight and Timing of Final Results Reporting:

Branch Chief, Financial Management and Planning Branch, OPP - accountable for oversight of data gathering, confirmation of data accuracy and final reporting of measure results.Results are reported twice a year.

4b. Data Limitations/Qualifications:

No data limitations.

4c. Third-Party Audits:

Not applicable.

Measure Code: 269 - Percent of agricultural watersheds that do not exceed EPA aquatic life benchmarks for two key pesticides of concern (azinphos-methyl and chlorpyrifos).

Office of Chemical Strategies and Pollution Prevention (OCSPP)

Goal Number and Title:
4 - Ensuring the Safety of Chemicals and Preventing Pollution
Objective Number and Title:
1 - Ensure Chemical Safety
Sub-Objective Number and Title:
2 - Protect Ecosystems from Chemical Risks
Strategic Target Code and Title:
1 - By 2015, no watersheds will exceed aquatic life benchmarks for targeted pesticides
Managing Office:
Office of Pesticide Programs
1a. Performance Measure Term Definitions:
Agricultural watersheds: Agricultural Site is a site that has less than or equal to 5 percent urban land and greater than 50 percent agricultural area. Watershed : is the portion of the surface of the Earth that contributes water to a stream through overland run-off, including tributaries and impoundments. EPA aquatic life benchmarks: The aquatic life benchmarks (for freshwater species) are based on toxicity values reviewed by EPA and used in the Agency's most recent risk assessments developed as part of the decision-making process for pesticide registration. The Office of Pesticide Programs (OPP) in EPA relies on studies required under the Federal Insecticide, Fungicide, and Rodenticide Act (FIFRA), as specified at 40 CFR Part 158, as well as a wide range of environmental laboratory and field studies available in the public scientific literature to assess environmental risk. Each Aquatic Life Benchmark is based on the most sensitive, scientifically acceptable toxicity endpoint available to EPA for a given taxon (for example, freshwater fish) of all scientifically acceptable toxicity data available to EPA. For more information, please see information from OPP at http://www.epa.gov/oppefed1/ecorisk_ders/aquatic_life_benchmark.htm Key pesticides of concern: Azinphos-methyl and chlorpyrifos were selected for this measure because EPA anticipates ongoing registration activity will have a direct effect on reducing exceedences of aquatic life benchmarks. Where ongoing registration activity may be mitigation to labels, a phase out of a chemical registration etc. Background: ·   Water quality is a critical endpoint for measuring exposure and risk to the environment. It is a high-level measure of our ability to reduce exposure from key pesticides of concern. This measure evaluates the reduction in water concentrations of pesticides as a means to protect aquatic life. Reduced water column concentration is a major indicator of the efficacy of risk assessment, risk management, risk mitigation and risk communication actions. It will illuminate program progress in meeting the Agency's strategic pesticide and water quality goals. The goal is to develop long-term consistent and comparable information on the amount of

pesticides in streams, ground water, and aquatic ecosystems to support sound management and policy decisions. USGS-NAWQA data, used for this measure, can help inform EPA of the long-term results of its risk management decisions based on trends in pesticide concentrations.

· EPA will request that USGS add additional insecticides to their sampling protocols to establish base line information for newer products (e.g., the synthetic pyrethroids) that have been replacing the organophosphates. Although the USGS has performed a reconnaissance of pyrethoids occurrence is bed sediment, there is not currently a comprehensive monitoring strategy.

2a. Original Data Source:

USGS National Water-Quality Assessment program.

Since 1991, the USGS NAWQA program has been collecting and analyzing data and information in major river basins and aquifers across the Nation.

2b. Source Data Collection:

Collection Methodology:

Monitoring plans call for yearly monitoring in 8 agricultural watersheds; biennial sampling in 3 agricultural dominated watersheds; and sampling every four years in a second set of 25 agricultural watersheds.

The sampling frequency for these sites will range from approximately 13 to 26 samples per year depending on the size of the watershed and the extent of pesticide use period. Sampling frequency is seasonally weighted so more samples are collected when pesticide use is expected to be highest.

The USGS database provides estimates of analytical methods and associated variability estimates (http://water.usgs.gov/nawqa/pnsp/pubs/circ1291/appendix8/8a.html

Quality Procedures:

The data that will be used for the outcome measure are subject to well-established QA-QC procedures in the USGS-NAWQA program (http://water.usgs.gov/nawqa/pnsp/pubs/qcsummary.html and http://water.usgs.gov/owq/FieldManual/index.html

Geographical Extent: NAWQA Study-Units cover a variety of hydrologic and ecological resources; critical sources of contaminants, including agricultural, urban, and natural sources; and a high percentage of population served by municipal water supply and irrigated agriculture. Study Unit boundaries frequently cross State boundaries and usually encompass more than 10,000 square kilometers (about 3,900 square miles). (http://water.usgs.gov/nawqa/studies/study_units.html

Spatial Detail: The Study-Unit design uses a rotational sampling scheme; therefore, sampling intensity varies year to year at the different sites. In general, about one-third of the Study Units are intensively investigated at any given time for 3-4 years, followed by low-intensity monitoring. Trends are assessed about every 10 years. During the first decade, 20 investigations began in 1991; 16 in 1994; and 15 in 1997.

During the second decade (2001-2012), monitoring continues in 42 of the 51 Study Units completed in the first decade, following a rotational scheme of 14 investigations beginning in 2001, 2004, and 2007. Findings will help to establish trends at selected surface-water and ground-water sites that have been consistently monitored and characterize water-quality conditions. (http://water.usgs.gov/nawqa/studies/study_units.html

Dates Covered by Source Data: Baseline data are derived from the USGS National Water-Quality Assessment (NAWQA) program's 2006 report: Pesticides in the Nation's Streams and Ground Water, 1992-2001. USGS is currently developing the report on its second cycle (cycle II) from 2002-2012. Data are available to the public on the USGS-NAWQA website from the (http://water.usgs.gov/nawqa/ USGS is currently developing sampling plans for 2013 – 2022. Future data will be available from USGS as it is made available on public websites.

2c. Source Data Reporting:

Data Submission Instrument: Baseline data are derived from the USGS National Water-Quality Assessment (NAWQA) program's 2006 report: Pesticides in the Nation's Streams and Ground Water, 1992-2001. USGS is currently developing the report on its second cycle (cycle II) from 2002-2012. Data are available to the public on the USGS-NAWQA website from the (http://water.usgs.gov/nawqa/ USGS is currently developing sampling plans for 2013 – 2022. Future data will be available from USGS as it is made available on public websites.

EPA does not rely on the production of the new report to receive the data. The report is when the data is available to the public. However, since this measure is reported every other year and requires two years worth of data, the USGS NAWQA program collects and analyzes data and information in major river basins and aquifers across the Nation every year, taking samples multiple times throughout the year. Then, two years' worth of data are sent to EPA biennially and entered into the EPA information system.

Frequency of Data Transmission to EPA: New results are available biennially.

Timing of Data Transmission to EPA: Data is provided to EPA from USGA NAQWA biennially. The data for the previous two years are received the reporting year for analysis and submission.

Data Entry Mechanism: All data are received in an excel spreadsheet from USGS-NAQWA. The data are analyzed within the spreadsheet and reported to the OPP measures representative. The data are then tabulated in a master spreadsheet for all OPP measures.

3a. Relevant Information Systems:

All data are received in an excel spreadsheet from USGS-NAQWA. The data are analyzed within the spreadsheet and reported to the OPP measures representative. The data are then tabulated in a master spreadsheet for OPP measures

Source/Transformed Data: There is one excel spreadsheet kept by EFED for each reporting cycle and for either agricultural orurban watershed. The spreadsheet has a tab containing source data, and a tab containing the analysis of the data along with the reported results.

Information System Integrity Standards: Standard not applicable.

3b. Data Quality Procedures:

EPA adheres to its approved Quality Management Plan in ensuring the quality of the data obtained from USGS. The data that will be used for the outcome measure are subject to well-established QA-QC procedures

in the USGS-NAWQA program (http://ca.water.usgs.gov/pnsp/rep/qcsummary/ and
http://water.usgs.gov/owq/FieldManual/index.html

Since 1991, the USGS NAWQA program has been collecting and analyzing data and information in major river basins and aquifers across the Nation. The program has undergone periodic external peer-review (http://dels.nas.edu/Report/Opportunities-Improve-USGS-National/10267

3c. Data Oversight:

Source Data Reporting Oversight Personnel: Not applicable.

Source Data Reporting Oversight Responsibilities: Not applicable.

Information Systems Oversight Personnel: Appointed Measures Representative(s) for Environmental Fate and Effects Division, in conjunction with the Division Director and Associate Division director(s).

Information Systems Oversight Responsibilities: To look over the source data, analyze the data, and report it to the OPP measures representative for reporting. The information systems oversight personnel keep a copy of all data spreadsheets.

3d. Calculation Methodology:

Decision Rules for Selecting Data: The data selected are completely provided by USGS and was determined by the 2006 USGS National Water-Quality Assessment (NAWQA) program's report. This report was used to determine the baseline for this measure and as a result, determined where data will be obtained. Moreover, all data provided to EPA from USGS are used in determining the analysis of the measure.

Definitions of Variables: Definitions of variables for the source data can be found in the documentation for the Pesticide National Synthesis Project http://water.usgs.gov/nawqa/pnsp/data/ The variables used during calculating the measure is represented as a percentile (the percent of watersheds that had an exceedance when compared to the watersheds sampled without exceedances). Please see the explanation of calculations for how the calculations are performed.

Explanation of Calculations: For each site within the two-year reporting timeframe provided by USGS, the monitoring data are compared to aquatic life benchmarks for each pesticide of concern. Acute aquatic life benchmarks are compared to each measured concentration for the representative year for each site. Chronic benchmarks for invertebrates and fish are compared to 21-day and 60-day moving averages, respectively. Moving average concentrations for 21- and 60-day periods are computed for each day of the year for each stream site from hourly concentration estimates determined by straight-line interpolation between samples.

Explanation of Assumptions: Not applicable.

Unit of Measure: Percentage of watersheds

Timeframe of Result: Source data is received biennially and contains two years' worth of data. The data is then evaluated the reporting year. From receipt of source data until reporting data results to the OPP measures representative is about one month.

Documentation of Methodological Changes: Not applicable at this time.

4a. Oversight and Timing of Final Results Reporting:

The Environmental Fate and Effects Division Director, Don Brady, will oversee the final reporting from the appointed measures representative. The EFED measures representative then reports to the central measures representative, Vickie Richardson - Branch Chief, Financial Management and Planning Branch, for all of OPP who reports all measures

Final Reporting Oversight Responsibilities: The EFED Division Director and measures representative meet to discuss the results in order to be able to explain any deviations, positive or negative, in reporting goals. This is done to also see if goals should be updated, are the chemicals being looked at still applicable (are they even being found, discontinued, etc.). Once this meeting occurs, the final results with explanations are sent to the OPP measures representative who maintains a log of all of the OPP measures for reporting

Final Reporting Timing: This measure is reported on a biennial basis.

4b. Data Limitations/Qualifications:

General Limitations/Qualifications:

These data continue to be evaluated and data limitations will be characterized during developmental stages of the measure and a complete evaluation will be provided in the NAWQA "Cycle II" Study Report. EPA has requested that USGS add additional insecticides to their sampling protocols to establish base line information for newer products that have been replacing the organophosphates (e.g., the synthetic pyrethroids). Although the USGS has performed a reconnaissance of pyrethoids occurrence in bed sediment, there is not currently a comprehensive monitoring strategy.

References: USGS National Water-Quality Assessment (NAWQA) program's 2006 report: Pesticides in the Nation's Streams and Ground Water, 1992-2001.

The NAWQA 2011 "Cycle II" Study Report is still being completed , thus there is no citation at this time.

The USGS database provides estimates of analytical methods and associated variability estimates (http://water.usgs.gov/nawqa/pnsp/pubs/circ1291/appendix8/8a.html

Data Lag Length and Explanation: Source data covers a two-year period (example: 2010-2011), this data is then received in 2012 once USGS has compiled the data for EPA into either agricultural or urban watershed data. This data are usually received by September of the reporting year (i.e. 2012 for this example), and then analyzed and reported by end of Septemeber (i.e. 2012). As a result, from the date the data collection is completed, there is a lag of 9 months until EPA receives the data, and 1 month from receipt until the data are reported.

Methodological Changes: Not applicable at this time. How the measure is calculated and data are collected remains the same at this time.

4c. Third-Party Audits:

The USGS NAWQA program has undergone periodic external peer review. For information on evaluation conducted by the National Research Council please see: http://www.nap.edu/catalog.php?record_id=10267 and http://water.usgs.gov/nawqa/opportunities.html

EPA's pesticide registration program, including this performance measure, was evaluated by OMB as part of the PART process. For more information, see:
http://georgewbush-whitehouse.archives.gov/omb/expectmore/detail/10000234.2003.html

Measure Code: 008 - Percent of children (aged 1-5 years) with blood lead levels (>5 ug/dl).

Office of Chemical Strategies and Pollution Prevention (OCSPP)

Goal Number and Title:
4 - Ensuring the Safety of Chemicals and Preventing Pollution
Objective Number and Title:
1 - Ensure Chemical Safety
Sub-Objective Number and Title:
1 - Protect Human Health from Chemical Risks
Strategic Target Code and Title:
2 - By 2014,reduce the percentage of children with blood lead levels above 5ug/dl to 1.0 percent or less
Managing Office:
Office of Pollution Prevention and Toxics;National Program Chemicals Division.

1a. Performance Measure Term Definitions:

Blood lead level: Blood lead level measures the amount of lead in the blood expressed in micrograms per deciliter (μg/dL). Until recently, the Centers for Disease Control and Prevention (CDC) identified children as having a blood lead level of concern if the test result was 10 or more micrograms of lead in a deciliter of blood. CDC experts now use a new level based on the U.S. population of children ages 1-5 years who are in the top 2.5% (the 97.5th percentile) of children tested for lead in their blood. According to CDC, the 97.5th percentile of the NHANES-generated blood lead level distribution in children 1-5 years old is 5 μg/dL.

http://www.cdc.gov/nceh/lead/ACCLPP/blood_lead_levels.htm

μg/dL: Micrograms of lead per deciliter of blood.

Background: This performance measure supports EPA's long-term goal of eliminating childhood lead poisoning as a public health concern and continuing to maintain the elimination of childhood lead poisoning over time. EPA's Lead Risk Reduction program contributes to the goal of eliminating childhood lead poisoning by: (1) establishing standards governing lead hazard identification and abatement practices and maintaining a national pool of professionals trained and certified to implement those standards; (2) providing information to housing occupants so they can make informed decisions and take actions about lead hazards in their homes; and (3) establishing a national pool of certified firms and individuals who are trained to carry out renovation and repair and painting projects while adhering to the lead-safe work practice standards and to minimize lead dust hazards created in the course of such projects.

Recent data show significant progress in the continuing effort to eliminate childhood lead poisoning as a public health concern. However, results of recent studies indicate adverse health effects to children at low blood levels, below 10μg/dL. In response to this new information and the fact that approximately three-quarters of the nation's housing stock built before 1978 still contains some lead-based paint , the EPA is now targeting reductions in the number of children with blood lead levels of 5 μg/dL or higher, as reflected in this performance measure.

2a. Original Data Source:

The original data source is the CDC's National Health and Nutrition Examination Survey (NHANES), which is recognized as the primary database in the United States for national blood lead statistics, http://www.cdc.gov/nchs/nhanes/about_nhanes.htm NHANES is a probability sample of the non-institutionalized population of the United States. The survey examines a nationally representative sample of approximately 5,000 men, women, and children each year located across the U.S.

2b. Source Data Collection:

Methods of data collection (by original data source): Data are obtained by analysis of blood and urine samples collected from survey participants. Health status is assessed by physical examination. Demographic and other survey data regarding health status, nutrition, and health-related behaviors are collected by personal interview, either by self-reporting or, for children under 16 and some others, as reported by an informant. Detailed interview questions cover areas related to demographic, socio-economic, dietary, and health-related questions. The survey also includes an extensive medical and dental examination of participants, physiological measurements, and laboratory tests. NHANES is unique in that it links laboratory-derived biological markers (e.g. blood, urine etc.) to questionnaire responses and results of physical exams.

Quality procedures followed (by original data source): According to the CDC, the process of preparing NHANES data sets for release is as rigorous as other aspects of the survey. After a CDC contractor performs basic data cleanup, the CDC NHANES staff ensure that the data are edited and cleaned prior to release. NHANES staff devotes at least a full year after the completion of data collection to careful data preparation. Additionally, NHANES data are published in a wide array of peer-reviewed professional journals.

Background documentation is available at the NHANES Web site at: http://www.cdc.gov/nchs/nhanes.htm
The analytical guidelines are available at the Web site:

http://www.cdc.gov/nchs/about/major/nhanes/nhanes2003-2004/analytical_guidelines.htm

Geographical extent of source data, if relevant: Data are collected to be representative of the U.S. population. The population data are extrapolated from sample data by the application of standard statistical procedures.

Spatial detail of source data, if relevant: NHANES sampling procedures provide nationally representative data.

2c. Source Data Reporting:

Form/mechanism for receiving data and entering into EPA system: EPA monitors the periodic issuance of NHANES reports and other data releases to obtain the data relevant to this measure.

Timing and frequency of reporting: NHANES is a continuous survey and examines a nationally representative sample of about 5,000 persons each year. These persons are located in counties across the country, 15 of which are visited each year.

Files of raw data, containing measured blood lead levels in NHANES participants, are currently released to the public in two-year sets. CDC also periodically publishes reports containing summary statistics for lead and more than 200 other chemicals measured in NHANES, at www.cdc.gov/exposurereport

3a. Relevant Information Systems:

There are no EPA systems utilized in collecting data for this measure as the Agency is able to secure the necessary data directly from NHANES reports and data releases.

3b. Data Quality Procedures:

EPA does not have any procedures for quality assurance of the underlying data as this function is performed by the CDC itself. CDC has periodically reviewed and confirmed EPA's calculation of NHANES summary statistics from the raw data files. The Agency determines the performance result for this measure either directly from the NHANES data or by performing simple arithmetical calculations on the data.

3c. Data Oversight:

Chief, Planning and Assessment Branch, Environmental Assistance Division, Office of Pollution Prevention and Toxics

3d. Calculation Methodology:

Decision rules for selecting data: EPA uses the blood lead level values generated by the NHANES surveys. EPA however, limits the age of the child to under six, based on the most sensitive receptor age group noted in Section 401 of TSCA.

Definitions of variables: Key terms are defined in 1(a) above.

Explanation of the calculations: Not applicable. Performance results obtained from NHANES.

Explanation of assumptions: Not applicable for the same reason as above.

Identification of unit of measure: Micrograms per deciliter (μg/dL)

Identification of timeframe of result: The performance result is computed from data released by the CDC in sets covering the particular time period over which sampling occurs. Thus, the timeframe that applies to the measured result is the same period for which the NHANES data are released. It is not a simple snapshot at a specific moment in time.

4a. Oversight and Timing of Final Results Reporting:

Planning and Accountability Lead in the Resource Management Staff in the Office of Program Management Operations. Reporting semiannually: mid-year and end-of-year, but subject to a data lag due to the periodic nature of NHANES reporting.

4b. Data Limitations/Qualifications:

NHANES is a voluntary survey and selected persons may refuse to participate. In addition, the NHANES survey uses two steps, a questionnaire and a physical exam. There are sometimes different numbers of subjects in the interview and examinations because some participants only complete one step of the survey. Participants may answer the questionnaire but not provide the more invasive blood sample. Special weighting techniques are used to adjust for non-response. NHANES is not designed to provide detailed estimates for populations that are highly exposed to lead.

4c. Third-Party Audits:

Report of the NHANES Review Panel to the NCHS Board of Scientific Counselors.

Cover letter can be accessed at: http://www.cdc.gov/nchs/data/bsc/bscletterjune8.pdf
Report can be accessed at: http://www.cdc.gov/nchs/data/bsc/NHANESReviewPanelReportrapril09.pdf

Measure Code: 297 - Metric Tons of Carbon Dioxide Equivalent (MTCO2e) reduced, conserved, or offset through pollution prevention.

Office of Chemical Strategies and Pollution Prevention (OCSPP)

Goal Number and Title:
4 - Ensuring the Safety of Chemicals and Preventing Pollution

Objective Number and Title:
2 - Promote Pollution Prevention

Sub-Objective Number and Title:
1 - Prevent Pollution and Promote Environmental Stewardship

Strategic Target Code and Title:
2 - By 2015, reduce 9 million MTof carbon dioxide equivalent (MMTCO2Eq.) through pollution prevention

Managing Office:
Office of Pollution Prevention and Toxics

1a. Performance Measure Term Definitions:
Carbon Dioxide Equivalent: A measure of reductions in emissions from greenhouse gases based upon their global warming potential (GWP). The measurement unit is "million metric tons of carbon dioxide equivalents (MMTCO2Eq), which is equal to (million metric tons of a gas) multiplied by (GWP of the gas). See greenhouse gas, global warming potential, metric ton.

Offset: Emission savings or storage that cancel out emissions that would otherwise have occurred. For example, electricity produced from burning landfill gas replaces electricity from the grid. This creates a carbon offset because landfill gas production and combustion produces fewer GHG emissions than fossil-fuel grid electricity does. http://epa.gov/climatechange/wycd/waste/downloads/warm-definitions-and-acronyms.pdf

P2 Programs related to this measure include:

Green Suppliers Network (GSN) and Economy, Energy, and the Environment (E3) are related programs-
o Green Suppliers Network is a coordinated effort of two federal agencies to help large manufacturers engage their small and medium-sized suppliers in undergoing low-cost technical reviews to improve their processes and minimize their wastes.
o Economy, Energy, and the Environment is a coordinated federal and local technical assistance initiative to help manufacturers become more sustainable. More federal agencies contribute to E3 technical assessments than to GSN assessments. The agencies provide technical production-process assessments and training to help manufacturers increase the energy efficiency and sustainability of their manufacturing processes, and reduce their environmental wastes, carbon emissions, and business costs.

Environmentally Preferable Purchasing (EPP) uses the federal buying power to stimulate market demand for green products and services.
o The Federal Electronics Challenge (FEC) encourages federal partners to purchase greener electronic products, reduce impacts of electronic products during use, and manage obsolete electronics in an

environmentally safe way.

o The Electronic Product Environmental Assessment Tool (EPEAT) program facilitates the development of Institute of Electrical and Electronics Engineers (IEEE) standards that address the environmental performance of electronics equipment. The program also facilitates the use of these IEEE-standard products by maintaining a registry of products that meet these standards.

Green Chemistry (GC)

o The Green Chemistry Program promotes the research, development, and implementation of technologies that reduce or eliminate the use or generation of hazardous substances.

o Through annual recognition, the Presidential Green Chemistry Challenge (PGCC) awards demonstrates the human health and environmental benefits and market competitiveness that green chemistry technologies offer.

Design for the Environment (DfE)/ Green Engineering GE

o Design for the Environment: DfE does not contribute to this measure

o Green Engineering:

§ The Green Engineering program has provided support to the DSW rule, resulting in a 'remanufacturing' exclusion in facilitating the recovery / extension of life of solvents commonly used in pharmaceutical, basic organic, coating and plastic industries. This work builds on the GE program's industrial activities which involve working with industry to implement greener solvent management methods and other process changes. GE continues development of its Sustainable Materials Management Tool (SMMT), which assess emissions to all media and raw material usage.   The SMMT tool estimates the benefits ((GHG, dollars, water, etc.) of extending the life of chemical substances..

§ The Green Engineering program continues its decade long effort in the development and dissemination of green / sustainable engineering materials, to incorporate sustainable materials management approaches in the engineering curriculum.

· Technical Assistance – The P2 Program issues two kinds of grants – P2 Grants to States and Tribes and Source Reduction Assistance Grants. States and Tribes are eligible for both kinds; localities, non-profits, universities and community groups are eligible for one kind – Source Reduction Assistance Grants. The purpose of these grants is to help small and medium businesses adopt sustainable P2 technologies and practices. Grantees provide technical assistance and implement award programs to achieve results.

Additional information about the P2 programs listed here can be found at:
http://www.epa.gov/p2/pubs/partnerships.htm

2a. Original Data Source:

GSN/E3:

The entities providing the data that EPA uses are EPA and US Department of Energy (DOE) DOE for modeled/aggregated Energy Star data, US Department of Commerce (DOC) for modeled/aggregated industrial process data, and industrial facilities for facility-level utility and materials-management data. National program staff's use of modeled data has been based on the assumption that many partner facilities will not submit project-completion data to maintain confidentiality or accept DOC sharing any non-aggregated potential or project-completed data with EPA. The national program is transitioning to finding ways to use more project-completed data as source data.

EPP:

EPEAT: The entities providing the data that EPA uses are the Institute of Electrical and Electronics Engineers (IEEE) and the Information Technology Industry Council (ITI). IEEE is a leading standards development organization for the development of industrial standards. IEEE provides EPA with the environmental standards-specifications for the manufacturing and performance of electronics equipment that is EPEAT-registered. ITI provides EPA with a tabulation of global equipment sales. Computers, laptops, monitors, imaging equipment and TVs are covered.

FEC: The entities providing the data that EPA uses are IEEE and the federal agencies that are FEC partners. IEEE provides EPA with environmental standards-specifications for the manufacturing and performance of electronics equipment that is EPEAT-registered. FEC-partner federal agencies provide EPA with a tabulation of their purchases and user practices for electronics equipment. Computers, laptops, monitors, imaging equipment and TVs are covered.

Green Chemistry (GC): Participants in the PGCC awards self-nominate. The awards are public, so confidential business information for nominated technologies is not accepted.

GE: The GE program has adapted and expanded the GlaxoSmithKline (GSK) partial life cycle modules which generate the multimedia emissions based on the life stages of a chemical substance: creation, reuse and disposal.[1] GSK provided extensive support during this adaptation process. The program expanded on the GSK's model unique contribution of 'manufacturing emissions' in assessing the benefits of extending solvent life. These manufacuting emissions were expanded further utilizing work from the ETH Zurich Institute.[2]

Other sources of information used for this work include: Green Engineering Textbook, US DOE OIT's Energy and Enviromental Profile of the US Chemical Industry, TRI Production Related Waste Data (transfers / inventories), Chemical Trees for the studied solvents supplied by GSK, data obtained from working with industry, and Kirk-Othmer Encyclopedia of chemical processes. The 2008 DSW RIA provided additional information on end of life impacts, including disposal and waste management costs.

Technical Assistance: The entities providing the data that EPA uses are facilities that received State P2 technical assistance and facilities that applied to a State to receive a P2 environmental award. These facilities provided the States with data taken from their facility utility bills and to a lesser degree from their facility materials-management records.

1/ Waste treatment modules – a Partial Life Cycle Inventory, Concepcion Jimenez-Gonzalez, Michael R Overcash and Alan Curzons, Journal of Chemical Technology and Biotechnology 76: 707-716, 2001.

2/ What is a green solvent? A comprehensive framework for the environmental assessment of solvents. Capello, C., Fischer, U., and Hungerbuhler, K. Green Chemistry, 9, 927-934, 2007.

2b. Source Data Collection:

GSN/E3: DOC grantees, DOE grantees, and EPA grantees collect the source data. Manufacturing Extension Partnership Centers (MEPs) are grantees of DOC's National Institute for Standards and Technology (NIST), and they collect energy-performance data on products and practices they recommend to businesses, such as modeled EnergyStar product ratings developed by EPA and DOE. MEPs record potential energy savings associated with each set of MEP E3 business-review recommendations, plus any utility-based data in facilities responses to voluntary MEP questionnaires on implemented E3 projects, and then DOC aggregates all data before sharing data with EPA. DOE grantees and EPA grantees likewise collect utility and any materials-management data on energy savings from facilities that implemented their respective grantees' E3 recommendations. DOC, DOE, and EPA grantees all send data to their grantor agencies, and the agencies input the data into their respective databases. All grantees follow their respective agencies' QA/QC requirements. Federal agencies participating in E3 are developing a second complementary database for their collective use.

EPP:

EPEAT: EPA and the Information Technology Industry Council (ITI) collect the source data. EPA collects equipment specifications from IEEE for computers, laptops, monitors, imaging equipment and TVs. ITI tabulates manufacturer records to compile annual worldwide sales. Manufacturers of EPEAT-registered products sign a Memorandum of Understanding in which they warrant the accuracy of the data they provide.

FEC: EPA and other Federal agencies collect the source data. EPA collects equipment manufacturing and performance specifications for computers, laptops, monitors, imaging equipment and TVs. Federal agency partners tabulate their purchasing records and extra-power-management activities (that they employ beyond their computers' default setting) and end-of-life recycling. National program staff uses the reported EOY data that relate to enabling ENERGY STAR features and reusing electronics and the GHG benefits of ancillary electronics recycling. Staff makes documented assumptions for how many of each class of electronics are ENERGY STAR 4 or ENERGY STAR 5.0.

Green Chemistry: PGCC awards nominations are provided to EPA. EPA prescreens the nominations, and then provides those that meet scope criteria to an external peer review expert panel for judging. Suggested winners are returned to EPA for final verification and validation.

GE: The GE program has, to date, has not reported GPRA measurements. It is envisioned, though, that industry, upon implementation of GE approaches, will be able to use the SMMT to generate the metrics, which would be used for reporting purposes.

Technical Assistance: EPA grantees collect the source data. P2 Program grantees (or their sub-grantees) collect utility bill data and any relevant materials-management records from facilities who have completed project implementation or who have applied for State environmental awards. Grantees follow EPA guidelines, and QAPP requirements, as appropriate, in collecting data. Grantees sometimes transform utility and materials-management records into MTCO2e; sometimes the EPA P2 Regional Coordinator needs to transform the data. Grantees transforming the data must be transparent about the methodological tool they used to make the transformation. The P2 Program provides grantees its GHG Reduction Calculator for their voluntary use, the same tool the P2 Regional Coordinators use.

2c. Source Data Reporting:

Data Submission Instrument

GSN/E3: NIST/DOC submits data to EPA in a document. State grantees submit data electronically or by mail to EPA in grant reports.

EPP:

EPEAT: The Information Technology Industry council (ITI) submits data to the non-profit Green Electronics Council, who then submits the data to EPA.

FEC: FEC partners submit data to EPA in an online FEC Annual Reporting Form.

GC: PGCC awards nominations are provided to EPA as an electronic report.

GE: The GE program has explored development of a boiler plate template which industry could use in reporting metrics, with linkages to the SMMT. Please also see 2b above.

Technical Assistance: State grantees submit data to EPA electronically or by mail in grant reports.

Data Entry Mechanism

GSN/E3: NIST sends a document of data to EPA program staff, who keep the data in a file. DOE sends to DOC and EPA; EPA program staff keeps the data in a file. State grantees submit data in grant reports to Regional project officers, who then manually enter the data into GrantsPlus, our program database.

EPP:

EPEAT: ITI sends the data to the Green Electronics Council which sends them to EPA. EPA program staff enters the data into EPA's Electronics Environmental Benefits Calculator (EEBC) to transform them into environmental and economic benefits.

FEC: FEC partners send data in the Annual FEC Reporting Form to EPA.

GC: Benefits data in PGCC awards nominations provided to EPA are entered into an internal database.

GE: The GE program envisions a central database in which information submitted by industry will be entered manually.

Technical Assistance: State grantees submit data in grant reports to Regional project officers, who then manually enter the data into GrantsPlus, our program database.

Frequency of Data Transmission to EPA

GSN/E3:

GSN: monthly

E3: quarterly

EPP: annually.

GC: annually

GE: annually

Technical Assistance: semi-annually.

Timing of Data Transmission to EPA

GSN/E3:

GSN: monthly

E3: every fiscal quarter

EPP:

EPEAT: the Green Electronics Council must submit data for the prior fiscal year by September 30th.

FEC: Partners must submit data for the prior fiscal year by January 31st.

GC: End of Fiscal Year

GE: End of Fiscal Year

Technical Assistance: Grantees must submit data if possible by the close of the fiscal year, and any amendments for the prior fiscal year by March 1st.

3a. Relevant Information Systems:

GSN/E3: DOC's current information system -- its Customer Relationship Management (CRM) database – is an information system used for storing source data and data transformed by modeling and aggregation, although EPA does not have direct access. EPA program staff uses its own desktop data storage, such as Excel spreadsheets. Regional project officers issuing E3 grants use P2 GrantsPlus as their information system; see the description of that system under Technical Assistance below. Another multi-agency database (for all agencies involved in E3 projects) is under development, which would contain a combination of secondary source data and source data transformed by modeling. EPA's Information System Integrity Standards are not applicable to DOC and DOE databases. EPA's P2 GrantsPlus database meets EPA's IT security policy.

EPP:

EPEAT: National program staff uses EPA's peer-reviewed Electronics Environmental Benefits Calculator (EEBC) as its data transformation system. All assumptions underlying calculation methods in EEBC were peer reviewed, and are peer reviewed for each version upgrade. Information on the EEBC can be found at http://www.epa.gov/fec/publications.html - calculator The link to the EEBC is found at http://www.federalelectronicschallenge.net/resources/bencalc.htm

FEC: National program staff uses the FEC Administrative Database to store modeled source data and annual reporting information from FEC government partners. Staff also uses the EEBC as its data transformation system (see description above).

GC: internal database

GE: Most relevant information system will be the SMMT. Other relevant information systems which will help with focused targeting and application is the TRI Production Related Waste (PRW) database. Technical Assistance: Regional project officers use P2 GrantsPlus as the information system to store source data. This system contains grant-specific data that have not been normalized. P2 GrantsPlus satisfies EPA's IT security policy. An extensive description of the system is available from EPA's contractor, and will be available on line in the future.

3b. Data Quality Procedures:

OPPT: All OPPT programs operate under the Information Quality Guidelines as found at http://www.epa.gov/quality/informationguidelines as well as under the Pollution Prevention and Toxics Quality Management Plan (QMP) ("Quality Management Plan for the Office of Pollution Prevention and Toxics; Office of Prevention, Pesticides and Toxic Substances," November 2008), and the programs will ensure that those standards and procedures are applied to this effort. The Quality Management Plan is for internal use only. EPP: EPEAT: EPEAT manufacturers sign a Memorandum of Understanding in which they warrant accuracy of the data they provide. EPA/EPEAT Program Managers review all data before using them in the EEBC. FEC: National program staff provides guidelines and instructions on how Federal partners should report data. National program staff and managers review all data before entering them into the EEBC. GC: The GC program operates under the Information Quality Guidelines found at http://www.epa.gov/quality/informationguidelines as well as under the Pollution Prevention and Toxics Quality Management Plan (QMP) ("Quality Management Plan for the Office of Pollution Prevention and Toxics; Office of Prevention, Pesticides and Toxic Substances," November 2008), and the program will ensure that those standards and procedures are applied to this effort. The Quality Management Plan is for internal use only. GE: As the GE work assignment includes both outreach and modeling work, a Quality Assurance Plan has been developed which contains the data quality procedures. EPA also performed quality review of the models and information used from both GSK (reference paper 1) and Zurich (reference paper 2), which have been used and adapted by industry, academia and government. Technical Assistance: Regional project officers determine whether a grantee is generating source data and, if so, require a Quality Assurance Project Plan; otherwise they provide other data quality procedures. Regional and national program staff reviews data before entering them into P2 Grants Plus. To ensure consistency in data reporting standards across all regions, regional and national program staff periodically updates the Regional P2 Measurement Guidance (last update, beginning of FY 2013).

3c. Data Oversight:

The Branch Chief of the Planning and Assessment Branch (PAB) in OPPT oversees source data reporting and information systems through periodic updates and discussions with the national program staff members and managers who monitor their own source data reporting. This oversight is also accomplished through written

protocols developed by PAB and national program managers.

3d. Calculation Methodology:

VariVariables:

Unit of analysis: MTCO2e (see Section 1a for definition)

GSN/E3: The national program had been relying primarily on a projected implementation rate to apply to aggregated modeled potential data (all actions that facilities could implement) to achieve reportable MTCO2e, and supplemented those with project-completed facility data. The national program is shifting to using more project-completed facility data as the basis for reporting MTCO2e results. For the projected implementation rate, the national program was relying on the following E3 assumptions: (2010, 30%; 2011, 32%; 2012, 34%; 2013, 36%, 2014, 38%; and, 2015, 40%) and GSN assumptions (2010, 35%; 2011, 37%; 2012, 39%; 2013, 41%; 2014, 43%; and 2015, 45%).The calculation methodology for project-complete facility annual data collected by EPA grantees is embodied in the GHG Reductions Calculator (see description under Technical Assistance). The national program also calculates the results that are expected to recur for six years. EPA is using an average lifetime of equipment or process change as the rationale for counting six years of results.

EPP: The calculation methodology for EPEAT and FEC is in the national program's Electronics Environmental Benefits Calculator (EEBC). All assumptions underlying calculation methods in EEBC were peer reviewed, and are peer reviewed for each version upgrade. Information on the EEBC can be found at http://www.epa.gov/fec/publications.html - calculator The link to the EEBC is found at http://www.federalelectronicschallenge.net/resources/bencalc.htm

GC: Simple summation of realized quantitative pounds reported from valid PGCC award nominations received.

GE: The SMMT generates CO2eq. The program used the SMMT to derive GE GPRA goals for 2014. In addition, the DSW rule, finalized at the end of the year, and implementation of the remanufacturing exclusion will result in additional GHG savings.

Technical Assistance: Regional project officers use the national program's Greenhouse Gas Reductions Calculator as their calculation methodology. The GHG tool was reviewed by an expert panel in 2009 and 2010. Assumptions as well as justifications as to data sources are transparent and clearly identified in the tools. End users such as grantees, regions, states, academia, businesses and others have completed extensive training on the suite of P2 tools. Live webinar training is held twice a year, and training materials/tools can be downloaded at: http://www.p2.org/general-resources/p2-data-calculators/ The latest update to the GHG tool is FY 2013.

4a. Oversight and Timing of Final Results Reporting:

Planning and Accountability Lead in the Resource Management Staff in the Office of Program Management Operations. Reporting semiannually: mid-year and end-of-year.

4b. Data Limitations/Qualifications:

Green Suppliers Network (GSN) and Economy, Energy, and the Environment (E3): To a degree, EPA assumes that partner facilities report actual data accurately to NIST Manufacturing Extension Partnership (NIST MEP) headquarters, that MEP and State technical assistance providers make accurate estimates of potential P2 results if projects are implemented, and that NIST MEP headquarters accurately aggregates the data before sharing them with EPA.

The program assumes that many partner facilities will choose not to submit actual P2 outcome data to maintain confidentiality and that facility partners will not accept NIST MEP headquarters sharing any non-aggregated potential or actual P2 data with EPA.

Facilities reviewed by NIST MEP and State technical assistance providers are often reluctant to have their individual facility opportunity assessments shared with EPA or to share proprietary information on quantitative benefits with NIST or EPA. MEP programs can also vary in the level of detail they report from the facility-level opportunity assessments (potential results) to MEP Headquarters, where data are aggregated and then sent to EPA. To address these limitations, EPA has strengthened the Request for Proposals requirements for the grantee MEP centers eligible to perform GSN and E3 reviews.

EPP: FEC has a built-in reliance on partners for data reporting. EPEAT relies on manufacturers of EPEAT-registered products, and the GEC, for data reporting.

GC: Because the PGCC awards are public, companies cannot submit confidential business information. As such, data provided can be qualitative rather than quantitative; qualitative data is not counted towards measures, so the data that is reported is conservative.

GE: The Program has used the TRI PRW data to obtain chemical substance, amount of chemical substance, and the unit operations. It has always been a goal to be able to use the actual production data from the facilities. As EPA adapted and developed the tool, the Program will work directly with industrial facilities and expect more accurate data will be available.

4c. Third-Party Audits:

EPP: The Electronics Environmental Benefits Calculator (EEBC) underwent internal and external review during its development phases. It was also reviewed and beta-tested during development of version 2.0.

GC: PGCC award nominations are reviewed by an external peer review expert panel.

GE: The method used in SMMT has gone through external review during its stages of development. The method was published in a peer reviewed journal. Note that the tool is still under development, with anticipation of completing this year.

Measure Code: 268 - Percent of urban watersheds that do not exceed EPA aquatic life benchmarks for three key pesticides of concern (diazinon, chlorpyrifos and carbaryl).

Office of Chemical Strategies and Pollution Prevention (OCSPP)

Goal Number and Title:
4 - Ensuring the Safety of Chemicals and Preventing Pollution
Objective Number and Title:
1 - Ensure Chemical Safety
Sub-Objective Number and Title:
2 - Protect Ecosystems from Chemical Risks
Strategic Target Code and Title:
1 - By 2015, no watersheds will exceed aquatic life benchmarks for targeted pesticides
Managing Office:
Office of Pesticide Programs
1a. Performance Measure Term Definitions:
Urban watersheds (as per USGS NAWQA glossary): Urban Site is a site that has greater than 25 percent urbanized and less than or equal to 25 percent agricultural area. Watershed : is the portion of the surface of the Earth that contributes water to a stream through overland run-off, including tributaries and impoundments.

EPA aquatic life benchmarks: The aquatic life benchmarks (for freshwater species) are based on toxicity values reviewed by EPA and used in the Agency's most recent risk assessments developed as part of the decision-making process for pesticide registration. The Office of Pesticide Programs (OPP) in EPA relies on studies required under the Federal Insecticide, Fungicide, and Rodenticide Act (FIFRA), as specified at 40 CFR Part 158, as well as a wide range of environmental laboratory and field studies available in the public scientific literature to assess environmental risk. Each Aquatic Life Benchmark is based on the most sensitive, scientifically acceptable toxicity endpoint available to EPA for a given taxon (for example, freshwater fish) of all scientifically acceptable toxicity data available to EPA. For more information, please see information from OPP at http://www.epa.gov/oppefed1/ecorisk_ders/aquatic_life_benchmark.htm

Key pesticides of concern: The pesticides diazinon, chlorpyrifos, and carbaryl were selected for measurement because of recent registration activity that is expected to reduce exceedences of aquatic life benchmarks. Where ongoing registration activity may be mitigation to labels, a phase out of a chemical registration etc.

Background:
·   Water quality is a critical endpoint for measuring exposure and risk to the environment. It is a high-level measure of our ability to reduce exposure from key pesticides of concern. This measure evaluates the reduction in water concentrations of pesticides as a means to protect aquatic life. Reduced water column concentration is a major indicator of the efficacy of risk assessment, risk management, risk mitigation and risk communication actions. It will illuminate program progress in meeting the Agency's strategic pesticide and water quality goals. The goal is to develop long-term consistent and comparable information on the amount of |

pesticides in streams, ground water, and aquatic ecosystems to support sound management and policy decisions. USGS-NAWQA data, used for this measure, can help inform EPA of the long-term results of its risk management decisions based on trends in pesticide concentrations.

·   EPA will request that USGS add additional insecticides to their sampling protocols to establish base line information for newer products (e.g., the synthetic pyrethroids) that have been replacing the organophosphates. Although the USGS has performed a reconnaissance of pyrethoids occurrence is bed sediment, there is not currently a comprehensive monitoring strategy.

2a. Original Data Source:

Since 1991, the USGS NAWQA program has been collecting and analyzing data and information in major river basins and aquifers across the Nation.

2b. Source Data Collection:

Collection Methodology:

Monitoring plans call for biennial sampling in 8 urban watersheds; and sampling every four years in a second set of 9 urban watersheds.

The sampling frequency for these sites will range from approximately 13 to 26 samples per year depending on the size of the watershed and the extent of pesticide use period. Sampling frequency is seasonally weighted so more samples are collected when pesticide use is expected to be highest.

The USGS database provides estimates of analytical methods and associated variability estimates
http://water.usgs.gov/nawqa/pnsp/pubs/circ1291/appendix8/8a.html

Quality Procedures:

The data that will be used for the outcome measure are subject to well-established QA-QC procedures in the USGS-NAWQA program (http://water.usgs.gov/nawqa/pnsp/pubs/qcsummary.html and http://water.usgs.gov/owq/FieldManual/index.html

Geographical Extent: NAWQA Study-Units cover a variety of hydrologic and ecological resources; critical sources of contaminants, including agricultural, urban, and natural sources; and a high percentage of population served by municipal water supply and irrigated agriculture. Study Unit boundaries frequently cross State boundaries and usually encompasses more than 10,000 square kilometers (about 3,900 square miles). (http://water.usgs.gov/nawqa/studies/study_units.html

Spatial Detail: The Study-Unit design uses a rotational sampling scheme; therefore, sampling intensity varies year to year at the different sites. In general, about one-third of the Study Units are intensively investigated at any given time for 3-4 years, followed by low-intensity monitoring. Trends are assessed about every 10 years. During the first decade, 20 investigations began in 1991; 16 in 1994; and 15 in 1997.

During the second decade (2001-2012), monitoring continues in 42 of the 51 Study Units completed in the first decade, following a rotational scheme of 14 investigations beginning in 2001, 2004, and 2007. Findings will help to establish trends at selected surface-water and ground-water sites that have been consistently monitored and characterize water-quality conditions. (http://water.usgs.gov/nawqa/studies/study_units.html

Dates Covered by Source Data: Baseline data are derived from the USGS National Water-Quality Assessment

(NAWQA) program's 2006 report: Pesticides in the Nation's Streams and Ground Water, 1992-2001. USGS is currently developing the report on its second cycle (cycle II) from 2002-2012. Data are available to the public on the USGS-NAWQA website from the (http://water.usgs.gov/nawqa/ USGS is currently developing sampling plans for 2013 – 2022. Future data will be available from USGS as it is made available on public websites.

2c. Source Data Reporting:

Data Submission Instrument: Baseline data are derived from the USGS National Water-Quality Assessment (NAWQA) program's 2006 report: Pesticides in the Nation's Streams and Ground Water, 1992-2001. USGS is currently developing the report on its second cycle (cycle II) from 2002-2012. Data are available to the public on the USGS-NAWQA website from the (http://water.usgs.gov/nawqa/ USGS is currently developing sampling plans for 2013 – 2022. Future data will be available from USGS as it is made available on public websites.

EPA does not rely on the production of the new report to receive the data. The report is when the data is available to the public. However, since this measure is reported every other year and requires two years worth of data, the USGS NAWQA program collects and analyzes data and information in major river basins and aquifers across the Nation every year, taking samples multiple times throughout the year. Then, two years worth of data are sent to EPA biennially and entered into the EPA information system.

Frequency of Data Transmission to EPA: New results are available biennially.

Timing of Data Transmission to EPA: Data is provided to EPA from USGA NAQWA biennially. The data for the previous two years are received the reporting year for analysis and submission.

Data Entry Mechanism: All data is received in an excel spreadsheet from USGS-NAQWA. The data is analyzed within the spreadsheet and reported to the OPP measures representative. The data is then tabulated in a master spreadsheet for all OPP measures.

3a. Relevant Information Systems:

System Description: All data is received in an excel spreadsheet from USGS-NAQWA. The data is analyzed within the spreadsheet and reported to the OPP measures representative. The data is then tabulated in a master spreadsheet for all OPP measures.

Source/Transformed Data: There is one excel spreadsheet kept by EFED for each reporting cycle and for either agricultural or urban watershed. The spreadsheet has a tab containing source data and a tab containing the analysis of the data along with the reported results.

Information System Integrity Standards: Standard not applicable.

3b. Data Quality Procedures:

EPA adheres to its approved Quality Management Plan in ensuring the quality of the data obtained from USGS. The data that will be used for the outcome measure is based on well-established QA-QC procedures in the USGS-NAWQA program (http://ca.water.usgs.gov/pnsp/rep/qcsummary/ and http://water.usgs.gov/owq/FieldManual/index.html

Since 1991, the USGS NAWQA program has been collecting and analyzing data and information in major river basins and aquifers across the Nation. The program has undergone periodic external peer-review (http://dels.nas.edu/Report/Opportunities-Improve-USGS-National/10267

3c. Data Oversight:

Source Data Reporting Oversight Personnel: Not applicable.

Source Data Reporting Oversight Responsibilities: Not applicable.

Information Systems Oversight Personnel: Appointed Measures Representative(s) for Environmental Fate and Effects Division, in conjunction with the Division Director and Associate Division director(s).

Information Systems Oversight Responsibilities: To look over the source data, analyze the data, and report it to the OPP measures representative for reporting. The information systems oversight personnel keep a copy of all data spreadsheets.

3d. Calculation Methodology:

Decision Rules for Selecting Data: The data selected are completely provided by USGS and was determined by the 2006 USGS National Water-Quality Assessment (NAWQA) program's report. This report was used to determine the baseline for this measure and as a result, determined where data will be obtained. Moreover, all data provided to EPA from USGS are used in determining the analysis of the measure.

Definitions of Variables: Definitions of variables for the source data can be found in the documentation for the Pesticide National Synthesis Project http://water.usgs.gov/nawqa/pnsp/data/ The variables used during calculating the measure is represented as a percentile (the percent of watersheds that had an exceedance when compared to the watersheds sampled without exceedances). Please see the explanation of calculations for how the calculations are performed.

Explanation of Calculations: For each site within the two-year reporting timeframe provided by USGS, the monitoring data is compared to aquatic life benchmarks for each pesticide of concern. Acute aquatic life benchmarks are compared to each measured concentration for the representative year for each site. Chronic benchmarks for invertebrates and fish are compared to 21-day and 60-day moving averages, respectively. Moving average concentrations for 21- and 60-day periods are computed for each day of the year for each stream site from hourly concentration estimates determined by straight-line interpolation between samples.

Explanation of Assumptions: Not applicable.

Unit of Measure: Percentage of watersheds

Timeframe of Result: Source data is received biennially and contains two years worth of data. The data is then evaluated the reporting year. From receipt of source data until reporting data results to the OPP measures representative is about one month.

Documentation of Methodological Changes: Not applicable at this time.

4a. Oversight and Timing of Final Results Reporting:

Final Reporting Oversight Personnel: The Environmental Fate and Effects Division Director, Don Brady will oversee the final reporting from the appointed measures representative. The EFED measures representative then reports to a central measures representative, Vickie Richardson - Branch Chief, Financial Management and Planning Branch, for all of OPP who reports all measures.

Final Reporting Oversight Responsibilities: The EFED Division Director and measures representative meet to discuss the results in order to be able to explain any deviations, positive or negative, in reporting goals. This is done to also see if goals should be updated, are the chemicals being looked at still applicable (are they even being found, discontinued, etc.). Once this meeting occurs, the final results with explanations are sent to the OPP measures representative who maintains a log of all of the OPP measures for reporting.

Final Reporting Timing: This measure is reported on a biennial basis.

4b. Data Limitations/Qualifications:

General Limitations/Qualifications:

These data continue to be evaluated and data limitations will be characterized during developmental stages of the measure and a complete evaluation will be provided in the NAWQA "Cycle II" Study Report. EPA has requested that USGS add additional insecticides to their sampling protocols to establish base line information for newer products that have been replacing the organophosphates (e.g., the synthetic pyrethroids). Although the USGS has performed a reconnaissance of pyrethoids occurrence in bed sediment, there is not currently a comprehensive monitoring strategy.

References: USGS National Water-Quality Assessment (NAWQA) program's 2006 report: Pesticides in the Nation's Streams and Ground Water, 1992-2001.

The NAWQA 2011 "Cycle II" Study Report is still being completed , thus there is no citation at this time.

The USGS database provides estimates of analytical methods and associated variability estimates (http://water.usgs.gov/nawqa/pnsp/pubs/circ1291/appendix8/8a.html

Data Lag Length and Explanation: Source data covers a two-year period (example: 2010-2011), this data is then received in 2012 once USGS has compiled the data for EPA into either agricultural or urban watershed data. This data is usually received by September of the reporting year (i.e. 2012 for this example), and then analyzed and reported by end of Septemeber (i.e. 2012). As a result, from the date the data collection is completed, there is a lag of 9 months until EPA receives the data, and 1 month from receipt until the data is reported.

Methodological Changes: Not applicable at this time. How the measure is calculated and data is collected

remains the same at this time.
4c. Third-Party Audits:
The USGS NAWQA program has undergone periodic external peer-review. For information on evaluation conducted by the National Research Council please see: http://www.nap.edu/catalog.php?record_id=10267 and http://water.usgs.gov/nawqa/opportunities.html EPA's pesticide registration program, including this performance measure, was evaluated by OMB as part of the PART process. For more information, see: http://georgewbush-whitehouse.archives.gov/omb/expectmore/detail/10000234.2003.html

Measure Code: 266 - Reduction in concentration of targeted pesticide analytes in the general population.

Office of Chemical Strategies and Pollution Prevention (OCSPP)

Goal Number and Title:
4 - Ensuring the Safety of Chemicals and Preventing Pollution
Objective Number and Title:
1 - Ensure Chemical Safety
Sub-Objective Number and Title:
1 - Protect Human Health from Chemical Risks
Strategic Target Code and Title:
3 - By 2014, reduce concentration of targeted chemicals in the general population
Managing Office:
Office of Pesticide Programs
1a. Performance Measure Term Definitions:
Reduction: Each fiscal year, EPA compares the most recent biomonitoring data available on the analytes of targeted organophosphate pesticides in urine samples from the general public that have been analyzed by the Centers for Disease Control and Prevention (CDC) to the baseline concentrations. The baseline years (corresponding to the NHANES sampling period) chosen for this measure are 2001-2002. The percent for which the population's 95th percentile concentration changed between the baseline year and the latest measurements will be calculated. The result of these calculations is then compared to the target set for the year in which performance is being measured. Concentration: 95th percentile concentration measured in the micrograms per liter (ug/L), at standard detection limits. Targeted pesticide analytes: The pesticides targeted by this measure are organophosphate pesticides. The measure is based on levels of the following metabolites that CDC measures in urine samples: six non-specific organophosphate dialkyl phosphate metabolites – and the chlorpyrifos-specific metabolite 3,5,6-Trichloro-2-pyridinol. The dialkyl phosphate and 3,5,6-Trichloro-2-pyridinol metabolites can be present in urine after low level exposures to organophosphorus insecticides that do not cause clinical symptoms or inhibition of cholinesterase activity, and measurement of these metabolites reflects recent exposure, predominantly in the previous few days. The metabolites may also occur in the environment as a result of degradation of organophosphorus insecticides, and therefore, the presence in a person's urine may reflect exposure to the metabolite itself. General population: the non-institutionalized population of the United States. This measure focuses on all age groups included in NHANES. Background: NHANES is a major program of the National Center for Health Statistics (NCHS). NCHS is part of the Centers for

Disease Control and Prevention (CDC), U.S. Public Health Service, and has the responsibility for producing vital and health statistics for the Nation. NCHS is one of the Federal statistical agencies belonging to the Interagency Council on Statistical Policy (ICSP). The ICSP, which is led by the Office of Management and Budget (OMB), is composed of the heads of the Nation's 10 principal statistical agencies plus the heads of the statistical units of four non-statistical agencies. The ICSP coordinates statistical work across organizations, enabling the exchange of information about organization programs and activities, and provides advice and counsel to OMB on statistical activities. The statistical activities of these agencies are predominantly the collection, compilation, processing or analysis of information for statistical purposes. Within this framework, NCHS functions as the Federal agency responsible for the collection and dissemination of the Nation's vital and health statistics. Its mission is to provide statistical information that will guide actions and policies to improve the health of the American people.

· 	 To carry out its mission, NCHS conducts a wide range of annual, periodic, and longitudinal sample surveys and administers the national vital statistics systems.

· 	 As the Nation's principal health statistics agency, NCHS leads the way with accurate, relevant, and timely data. To assure the accuracy, relevance, and timeliness of its statistical products, NCHS assumes responsibility for determining sources of data, measurement methods, methods of data collection and processing while minimizing respondent burden; employing appropriate methods of analysis, and ensuring the public availability of the data and documentation of the methods used to obtain the data. Within the constraints of resource availability, NCHS continually works to improve its data systems to provide information necessary for the formulation of sound public policy. As appropriate, NCHS seeks advice on its statistical program as a whole, including the setting of statistical priorities and on the statistical methodologies it uses. NCHS strives to meet the needs for access to its data while maintaining appropriate safeguards for the confidentiality of individual responses.

· 	 The Centers for Disease and Prevention's (CDC) National Health and Nutrition Examination Survey (NHANES) program is a survey designed to assess the health and nutritional status of adults and children in the U.S. NHANES was selected as the performance database because it is an ongoing program that is statistically designed to be nationally representative of the U.S. civilian, non-institutionalized population.

· 	 Baseline for this measure was established using existing NHANES biomonitoring data. During each fiscal year, performance will then be evaluated by comparing subsequent NHANES biomonitoring data with the established baseline.

· 	 This measure supports the long-term goal of reducing the risk and ensuring the safety of chemicals and preventing pollution at the source by enabling EPA to better assess progress in reducing exposure to targeted chemicals, as reflected in concentration levels among the general population and key subpopulations.

· 	 Analytes for organophosphate pesticides were selected for this measure because EPA anticipates recent registration activity will have a direct effect on reducing exposure in the general population.

For more information on the pesticides, visit http://www.epa.gov/pesticides/about/types.htm

2a. Original Data Source:

NHANES: CDC's NHANES survey program began in the early 1960s as a periodic study and continues as an annual survey (http://www.cdc.gov/nchs/NHANES.htm The survey examines a nationally representative sample of approximately 5,000 men, women, and children each year located across the U.S. CDC's National Center for Health Statistics (NCHS) is responsible for the conduct of the survey and the release of the data to the public through their website at: http://www.cdc.gov/nchs/nhanes/nhanes_questionnaires.htm NHANES is designed to collect data on the health and nutritional status of the U.S. population. NHANES collects information about a wide range of health-related behaviors, performs physical examinations, and collects samples for laboratory tests. NHANES is unique in its ability to examine public health issues in the U.S. population, such as risk factors for cardiovascular disease. Beginning in 1999, NHANES became a continuous survey, sampling the U.S. population annually and releasing the data in 2-year cycles.

2b. Source Data Collection:

Collection Methods: The sampling plan follows a complex, stratified, multistage, probability-cluster design to select a representative sample of the civilian, noninstitutionalized population in the United States based on age, gender, and race/ethnicity. The NHANES survey contains detailed interview questions covering areas related to demographic, socio-economic, dietary, and health-related subjects. It also includes an extensive medical and dental examination of participants, physiological measurements, and laboratory tests. NHANES is unique in that it links laboratory-derived biological markers (e.g. blood, urine etc.) to questionnaire responses and results of physical exams. Analytical guidelines issued by NCHS provide guidance on how many years of data should be combined for an analysis. NHANES measures blood levels in the same units (i.e., ug/dL) and at standard detection limits.

Environmental chemicals are measured in blood, serum, or urine specimens collected as part of the examination component of NHANES. The participant ages for which a chemical was measured varied by chemical group. Most of the environmental chemicals were measured in randomly selected subsamples within specific age groups. Randomization of subsample selection is built into the NHANES design before sample collection begins. Different random subsamples include different participants. This subsampling was needed to ensure an adequate quantity of sample for analysis and to accommodate the throughput of the mass spectrometry analytical methods.

Geographical Extent: NHANES is designed to be a representative sample of the civilian, noninstitutionalized population in the United States based on age, gender, and race/ethnicity.

Quality Procedures: NCHS assures the security of its statistical and analytic information products through the enforcement of rigorous controls that protect against unauthorized access to the data, revision or corruption of the data, or unauthorized use of the data. Some of the major controls used at NCHS include access control, user authentication, encryption, access monitoring, provision of unalterable electronic content, and audit trails. All NCHS statistical and analytic information products undergo a formal clearance process before dissemination. Publications and reports, whether in electronic or paper form, are reviewed by a designated official within the author's office or division and by the NCHS Associate Director for Science (ADS). These reviews cover the clarity of descriptive text, the appropriateness of the methodology, the soundness of the analysis, the adherence to confidentiality and disclosure avoidance restrictions, the readability of tabular and graphic presentations of data, etc. Finally, all products undergo editorial review (e.g., formatting, proofreading, spell checks, proper punctuation, etc.). In addition, all public-use tapes are reviewed for accuracy and appropriate confidentiality protections. Oral presentations are subject to appropriate supervisory review.

NCHS statistical and analytic information products are derived using generally acceptable statistical practices and methodologies, which are well documented and available to the public. These procedures enable responsible statisticians and analysts outside of NCHS to replicate the NCHS statistical methods and obtain results consistent with those obtained by NCHS.

References:

CDC (2009a). National Health and Nutrition Examination Survey, 2007-2008 Overview. Available at: <http://www.cdc.gov/nchs/data/nhanes/nhanes_07_08/overviewbrochure_0708.pdf>

CDC (2009b) Fourth National Report on Human Exposure to Environmental Chemicals. Available at: <http://www.cdc.gov/exposurereport/pdf/FourthReport.pdf>

CDC (2009c). NCHS Guidelines for Ensuring the Quality of Information Disseminated to the Public. Available at: < http://www.cdc.gov/nchs/about/policy/quality.htm>

2c. Source Data Reporting:

Data Submission Instrument: CDC's National Center for Health Statistics (NCHS) is responsible for the release of the data to the public through their website at: http://www.cdc.gov/nchs/nhanes/nhanes_questionnaires.htm The data utilized for the performance measure is released as part of the NHANES laboratory files. The naming convention and organization of the laboratory data files may change between survey cycles, so NHANES laboratory documentation should be reviewed to identify the correct data fields for analysis. In 2001-2002, the SAS Transport File containing the targeted pesticide data ("l26PP_B.xpt") can be identified through the "2001-2002 Laboratory Variable List" at: http://www.cdc.gov/nchs/nhanes/nhanes2001-2002/varlab_b.htm

In recent years, CDC has published a national exposure report based on the data from the NHANES. CDC has scheduled release of data, and scheduled release of national exposure reports through NHANES. The most current update of the National Report on Human Exposure to Environmental Chemicals was released February 2012 and is available at the Web site http://www.cdc.gov/exposurereport/ Performance results will be updated as new peer reviewed NHANES data are published either in the official CDC report on human exposure to environmental chemicals or other journal articles as the data becomes available.

3a. Relevant Information Systems:

CDC is responsible for all NHANES data collection and reporting. As such, no EPA information systems are involved in the process of collecting, calculating and/or reporting the results for this measure. In order to calculate the results for the performance measure, EPA accesses and downloads the NHANES data files that are publically available through CDC/NCSH at: http://www.cdc.gov/nchs/nhanes/nhanes_questionnaires.htm The NHANES data files are downloaded as SAS Transport files and uploaded into SAS for statistical analysis.

System Description: Not Applicable

Source/Transformed Data: Not Applicable

Information System Integrity Standards: Not Applicable

3b. Data Quality Procedures:

NCHS assures the security of its statistical and analytic information products through the enforcement of rigorous controls that protect against unauthorized access to the data, revision or corruption of the data, or unauthorized use of the data. Some of the major controls used at NCHS include access control, user authentication, encryption, access monitoring, provision of unalterable electronic content, and audit trails.

All NCHS statistical and analytic information products undergo a formal clearance process before dissemination. Publications and reports, whether in electronic or paper form, are reviewed by a designated official within the author's office or division and by the NCHS Associate Director for Science (ADS). These reviews cover the clarity of descriptive text, the appropriateness of the methodology, the soundness of the analysis, the adherence to confidentiality and disclosure avoidance restrictions, the readability of tabular and graphic presentations of data, etc. NCHS statistical and analytic information products are derived using generally acceptable statistical practices and methodologies, which are well documented and available to the public. These procedures enable responsible statisticians and analysts outside of NCHS to replicate the NCHS statistical methods and obtain results consistent with those obtained by NCHS.

References:
CDC (2009c). NCHS Guidelines for Ensuring the Quality of Information Disseminated to the Public. Available at: <http://www.cdc.gov/nchs/about/policy/quality.htm>

3c. Data Oversight:

Appointed measures representative(s) for the Health Effects Division, in conjunction with the Division Director and Associate Division Director, to look over, review, analyze the data and report it to the OPP measures representative for reporting.

3d. Calculation Methodology:

Decision Rules for Selecting Data: The performance measure uses NHANES pesticide biomonitoring data published by CDC/NCHS. No pesticide biomonitoring data will be excluded from the performance measure calculations.

Definitions of Variables: Key data fields related to the NHANES survey design and targeted pesticide analytes are defined below:

- WTSPP2YR   Pesticides Subsample 2 year Mec Weight
- SEQN   Respondent sequence number
- URXCPM   3,5,6-trichloropyridinol (ug/L) result
- URXOP1   Dimethylphosphate (ug/L) result
- URXOP2   Diethylphosphate (ug/L) result
- URXOP3   Dimethylthiophosphate (ug/L) result
- URXOP4   Diethylthiophosphate (ug/L) result
- URXOP5   Dimethyldithiophosphate (ug/L) result

·   URXOP6   Diethyldithiophosphate (ug/L) result

Explanation of the Calculations:

Annual performance will be evaluated using the equation below:

Where:

Baseline95th = 95th percentile urinary concentration during the baseline period.
Performance95th = 95th percentile urinary concentration during performance period.

Explanation of Assumptions: The performance measure is based on NHANES pesticide biomonitoring data published by CDC/NCHS. The data is used without making any additional transformations, so no assumption will be made to transform the data.

Timeframe of Result: NHANES is a continuous survey, sampling the U.S. population annually and releasing the data in 2-year cycles. As such, the span of time represented by the results represents a 2-year timeframe.

Unit of Measure: micrograms per liter (ug/L), at standard detection limits

Documentation of Methodological Changes: Not applicable.

4a. Oversight and Timing of Final Results Reporting:

Branch Chief, Financial Management and Planning Branch.

Measure is reported on a biennially

4b. Data Limitations/Qualifications:

NHANES provides the most comprehensive biomonitoring data on the U.S. population. While it provides most comprehensive data for evaluating national-level trends, there are some limitations that should be considered when evaluating the biomonitoring results. With regard to the general interpretation of biomonitoring data, CDC highlights that there are a number of factors that can influence the concentration of chemicals in urine. Some examples described by CDC include :

·   Chemical half-life (i.e., persistence of chemical in blood or urine);
·   Route of exposure;
·   Genetic susceptibility;
·   Demographic characteristics (e.g., age or gender);
·   Health status and nutrition(e.g., reduced kidney function, iron deficiency);
·   Lifestyle or behavioral factors (e.g. , smoker versus non-smoker, or occupation); and

·   Geography (e.g., proximity to environmental chemical sources, or climate).

In addition to these interpretive considerations, an important design limitation of the NHANES survey is that data is not publically available to evaluate seasonal and geographic trends. While this seasonal and geographic data is not available, EPA believes the data are suitable for the performance measure because EPA is interested in evaluating national-level trends between years.

References:

CDC (2010). Important Analytic Considerations and Limitations Regarding Environmental Chemical Data Analyses. Available at: <http://www.cdc.gov/nchs/tutorials/environmental/critical_issues/limitations/index.htm>

DATA LAG: Data lags may prevent performance results from being determined for every reporting year. Performance results will be updated as NHANES data are published either in the official CDC report on human exposure to environmental chemicals or other journal articles or as the data becomes available. There can be a substantial lag between CDC sampling and publication of data. For instance, in 2012, the most recently available data were from the sampling period of 2007-2008.

Methodological Changes: Not Applicable

4c. Third-Party Audits:

In 2009, the Board of Scientific Counselors (BSC) of NCHS commissioned a panel to review the NHANES as part of an ongoing program review process and to report its findings to the BSC (Available at: www.cdc.gov/nchs/data/bsc/NHANESReviewPanelReportrapril09.pdf The Panel concluded that NHANES simply must continue, and in a form that will fully sustain its historical importance as a key source of health information. With regard to the biomonitoring data that is the focus of EPA's performance measure, BSC did not make any specific recommendations that should impact the the methodology(ies), model(s), data, and information system(s) used to measure/collect/report performance.   BSC emphasized the importance of the biomonitoring component of NHANES, stating:

Environmental monitoring is another major responsibility folded into the on-going NHANES. The survey's biomonitoring component documents exposure to environmental toxicants by direct measure of chemicals in blood or other biological specimens from individuals. The collection of blood and urine specimens in NHANES provides a unique opportunity for monitoring environmental exposure in the U.S. population....

Measure Code: 264 - Pounds of hazardous materials reduced through pollution prevention.

Office of Chemical Strategies and Pollution Prevention (OCSPP)

Goal Number and Title:
4 - Ensuring the Safety of Chemicals and Preventing Pollution
Objective Number and Title:
2 - Promote Pollution Prevention
Sub-Objective Number and Title:
1 - Prevent Pollution and Promote Environmental Stewardship
Strategic Target Code and Title:
1 - By 2015, reduce 15 billion pounds of hazardous materials cumulatively through pollution prevention
Managing Office:
Office of Pollution Prevention and Toxics
1a. Performance Measure Term Definitions:
Pounds of Hazardous Materials reduced: A measure of hazardous materials (hazardous waste, hazardous inputs, air emissions, and water effluent) reduced from the implementation of P2 practices by P2 program participants. P2 Programs related to this measure include: Green Suppliers Network (GSN) and Economy, Energy, and the Environment (E3) are related programs- o GSN: a coordinated effort of EPA and the US Department of Commerce to help large manufacturers engage their suppliers in undergoing low-cost technical reviews to improve their processes from an environmental and energy perspective and, specific to this measure, to achieve related reductions in pounds of hazardous materials used or released. o E3: a coordinated federal and local technical assistance initiative to help manufacturers become more sustainable. More federal agencies contribute to E3 technical assessments than to GSN assessments. Agencies provide technical assessments and training to help manufacturers increase their energy efficiency/sustainability and, specific to this measure, to achieve related reductions in pounds of hazardous materials used or released. Environmentally Preferable Purchasing (EPP) uses the federal buying power to stimulate market demand for green products and services. o The Federal Electronics Challenge (FEC) encourages federal partners to purchase greener electronic products, reduce impacts of electronic products during use, and manage obsolete electronics in an environmentally safe way. o The Electronic Product Environmental Assessment Tool (EPEAT) program facilitates the development of Institute of Electrical and Electronics Engineers (IEEE) standards that address the environmental performance of electronics equipment. The program also facilitates the use of these IEEE-standard products by maintaining a registry of products that meet these standards.

Green Chemistry (GC)

o The Green Chemistry Program promotes the research, development, and implementation of technologies that reduce or eliminate the use or generation of hazardous substances.

o Through annual recognition, the Presidential Green Chemistry Challenge (PGCC) awards demonstrate the human health and environmental benefits and market competitiveness that green chemistry technologies offer.

Design for the Environment (DfE)/ Green Engineering GE

Design for the Environment:

§EPA allows safer products to carry the Design for the Environment (DfE) label. This mark enables consumers to quickly identify and choose products that can help protect the environment and are safer for families.

§The DfE logo on a product means that the DfE scientific review team has screened each ingredient for potential human health and environmental effects and that—based on currently available information, EPA predictive models, and expert judgment—the product contains only those ingredients that pose the least concern among chemicals in their class.

§Product manufacturers who become DfE partners, and earn the right to display the DfE logo on recognized products, have invested heavily in research, development and reformulation to ensure that their ingredients and finished product line up on the green end of the health and environmental spectrum while maintaining or improving product performance.

Green Engineering:

§The Green Engineering program has provided support to the DSW rule, resulting in a 'remanufacturing' exclusion in facilitating the recovery / extension of life of solvents commonly used in pharmaceutical, basic organic, coating and plastic industries. This work builds on the GE program's industrial activities which involve working with industry to implement greener solvent management methods and other process changes. GE continues development of its Sustainable Materials Management Tool (SMMT), which assesses emissions to all media and raw material usage.   The SMMT tool assesses reductions in pounds of chemical substance by conserving / extending the life of chemical substances

§The Green Engineering program continues its decade long effort in the development and dissemination of green / sustainable engineering materials, to incorporate sustainable materials management approaches in the engineering curriculum.

Technical Assistance – The P2 Program issues two kinds of grants – P2 Grants to States and Tribes and Source Reduction Assistance Grants. States and Tribes are eligible for both kinds; localities, non-profits, universities and community groups are eligible for one kind – Source Reduction Assistance Grants. The purpose of these grants is to help small and medium businesses adopt sustainable P2 technologies and practices. Grantees provide technical assistance and implement award programs to achieve results.

2a. Original Data Source:

GSN/E3: EPA grantees provide hazardous-material reduction data from facilities that have implemented E3 projects, and the US Department of Commerce (DOC) grantees provide aggregated facility-level hazardous-material reduction data from facilities willing to provide these data. EPA had been relying on modeled data (assuming many partner facilities will maintain confidentiality to the extent they will neither submit project-completion data nor accept DOC sharing any non-aggregated potential or project-completed data with EPA), but is transitioning to using project-completed data as source data.

EPP:
EPEAT: The Institute of Electrical and Electronics Engineers (IEEE), a standards-development organization, provides environmental specifications for the manufacturing and performance of electronics equipment that is EPEAT-registered. The Information Technology Industry Council (ITI) tabulates global equipment sales of computers, laptops, monitors, imaging equipment and TVs.
FEC: The IEEE provides the source data listed above. FEC-partner federal agencies tabulate their purchases and user practices for computers, laptops, monitors, imaging equipment and TVs.

Green Chemistry (GC): Participants in the PGCC awards self-nominate. The awards are public, so confidential business information for nominated technologies is not accepted.

DfE/GE:
DfE: Partners, i.e., those companies who have DfE-labeled products, provide proprietary information on the production volume of their safer formulations. These data are aggregated and averaged to protect the confidential nature of this information.
GE: The GE program has adapted and expanded the GlaxoSmithKline (GSK) partial life cycle modules which generate the multimedia emissions based on the life stages of a chemical substance: creation, reuse and disposal. Waste treatment modules – a Partial Life Cycle Inventory, Concepcion Jimenez-Gonzalez, Michael R Overcash and Alan Curzons, Journal of Chemical Technology and Biotechnology 76: 707-716, 2001. GSK provided extensive support during this adaptation process. The program expanded on the GSK's model unique contribution of 'manufacturing emissions' in assessing the benefits of extending solvent life. These manufacturing emissions were expanded further utilizing work from the ETH Zurich Institute. What is a green solvent? A comprehensive framework for the environmental assessment of solvents. Capello, C., Fischer, U., and Hungerbuhler, K. Green Chemistry, 9, 927-934, 2007.

Other sources of information used for this work include: Green Engineering Textbook, US DOE OIT's Energy and Environmental Profile of the US Chemical Industry, TRI Production Related Waste Data (transfers / inventories), Chemical Trees for the studied solvents supplied by GSK, data obtained from working with industry, and Kirk-Othmer Encyclopedia of chemical processes. The 2008 DSW RIA provided additional information on end of life impacts, including disposal and waste management costs.

Technical Assistance: Facilities receiving State P2 technical assistance or applying for a State P2 award provide grantees (States) with data from their materials-management records.

2b. Source Data Collection:

GSN/E3: Grantees of DOC and EPA collect source data. Grantees of DOC (Manufacturing Extension Partnership Centers (MEPs)) collect hazardous pollution/input reduction data associated with technologies and practices they recommend to businesses. MEPs collect any hazardous pollution/ input reduction data that facilities supply directly to MEPs or through follow-up voluntary MEP questionnaires on implemented E3 projects. DOC aggregates these data and then shares them with EPA. EPA grantees collect hazardous pollution/input reduction data from facilities that implemented their E3 recommendations. DOC and EPA grantees send data to their grantor agencies, and the agencies input the data into their respective databases. All grantees follow their respective agencies' QA/QC requirements. Federal agencies participating in E3 are developing a second complementary database for their collective use.

EPP:

EPEAT: EPA collects IEEE manufacturing specifications for computers, laptops, monitors, imaging equipment and TVs, from which EPA can calculate reductions in hazardous materials. ITI tabulates manufacturer records to compile annual worldwide sales. EPEAT-registered product manufacturers sign Memoranda of Understanding to warrant the accuracy of data they provide.

FEC: EPA collects IEEE specifications as stated above. Federal partners tabulate their purchasing records, extra-power-management practices (employed beyond their computers' default setting), and end-of-life recycling. EPA uses their EOY data on reusing electronics and ancillary electronics recycling. Staff documents assumptions for how many of each class of electronics are ENERGY STAR 4.0 or 5.0.

Green Chemistry: PGCC awards nominations are provided to EPA. EPA prescreens the nominations, and then provides those that meet scope criteria to an external peer review expert panel for judging. Suggested winners are returned to EPA for final verification and validation.

DfE/GE:

DfE: Production volume is a standard measure for companies who sell chemical products such as cleaning formulations. This information is directly related to the sales of their product, and therefore they have the incentive to track it accurately. Production volumes reported to EPA include the total volume of each DfE-labeled product sold in the US or internationally.

GE: The GE program has, to date, has not reported GPRA measurements. It is envisioned, though, that industry, upon implementation of GE approaches, will be able to use the SMMT to generate the metrics, which would be used for reporting purposes.

Technical Assistance: P2 Program grantees (or sub-grantees) collect records relating to hazardous pollution and inputs from facilities who have completed P2 implementation or who have applied for State environmental awards. Grantees follow EPA data collection guidelines and submit QAPPs if required. Grantees must be transparent about any methodology to transform gallons to pounds. Grantees are provided the same GHG Reduction Calculator that P2 Regional Coordinators use.

2c. Source Data Reporting:

Data Submission Instrument

GSN/E3: NIST/DOC submits data to EPA in a document. State grantees submit data electronically or by mail to EPA in grant reports.

EPP:
EPEAT: The Information Technology Industry council (ITI) submits data to the non-profit Green Electronics Council, who then submits the data to EPA.
FEC: FEC partners submit data to EPA in an online FEC Annual Reporting Form.

GC: PGCC awards nominations are provided to EPA as an electronic report.

DfE/GE:
DfE: Production volume data are submitted to EPA along with a company's application to receive the DfE label. These data are captured on an as needed basis in a database maintained by EPA contractors. EPA receives production volume data when a company chooses to submit an application for recognition under the program.

GE: The GE program has explored development of a boiler plate template which industry could use in reporting metrics, with linkages to the SMMT. Please also see 2b above.

Technical Assistance: State grantees submit data to EPA electronically or by mail in grant reports.

Data Entry Mechanism
GSN/E3: NIST sends a document of data to EPA program staff, who keep the data in a file. DOE sends to DOC and EPA; EPA program staff keeps the data in a file. State grantees submit data in grant reports to Regional project officers, who then manually enter the data into GrantsPlus, our program database.

EPP:
EPEAT: ITI sends the data to the Green Electronics Council which sends them to EPA. EPA program staff enters the data into EPA's Electronics Environmental Benefits Calculator (EEBC) to transform them into environmental and economic benefits.
FEC: FEC partners send data in the Annual FEC Reporting Form to EPA.

GC: Benefits data in PGCC awards nominations provided to EPA are entered into an internal database.

DfE/GE:
DfE: Production volume is a standard measure for companies who sell chemical products such as cleaning formulations. This information is directly related to the sales of their product, and therefore they have the incentive to track it accurately. Production volumes reported to EPA include the total volume of each DfE-labeled product sold in the US or internationally. DfE includes this information when tracking partnership status for partner companies.
GE: The GE program envisions a central database in which information submitted by industry will be entered

manually.

Technical Assistance: State grantees submit data in grant reports to Regional project officers, who then manually enter the data into GrantsPlus, our program database.

Frequency of Data Transmission to EPA
GSN/E3:
GSN: monthly
E3: quarterly
EPP: annually.
GC: annually
DfE/GE:

      DfE: annually

      GE: annually

Technical Assistance: semi-annually.

Timing of Data Transmission to EPA
GSN/E3:
GSN: monthly
E3: every fiscal quarter
EPP:
EPEAT: the Green Electronics Council must submit data for the prior fiscal year by September 30th.
FEC: Partners must submit data for the prior fiscal year by January 31st.
GC: End of Fiscal Year
DfE/GE:

      DfE: End of Fiscal Year

      GE: End of Fiscal Year

Technical Assistance: Grantees must submit data if possible by the close of the fiscal year, and any amendments for the prior fiscal year by March 1st.

3a. Relevant Information Systems:

GSN/E3: DOC's Customer Relationship Management database is a system used to store source data and data transformed by modeling and aggregation; EPA does not have direct access. EPA program staff uses its own desktop data storage, such as Excel spreadsheets. Regional project officers issuing E3 grants use P2 GrantsPlus as their information system, described below under Technical Assistance. Another multi-agency database (for all agencies in E3) is under development to house secondary source data and source data transformed by modeling. EPA's Information System Integrity Standards are not applicable to DOC and DOE databases. EPA's P2 GrantsPlus database meets EPA's IT security policy.

EPP:
EPEAT: EPA uses its Electronics Environmental Benefits Calculator (EEBC) as a system to transform IEEE data and ITI sales data into quantities of hazardous materials reduced. The EEBC (and upgrades) are peer reviewed.

See http://www.epa.gov/fec/publications.html - calculator and
http://www.federalelectronicschallenge.net/resources/bencalc.htm

FEC: EPA stores modeled and federal-partner source data in an FEC Administrative Database and transforms the data using the EEBC (described above).

GC: internal database

DfE/GE:

DfE: Because the collection and analysis of this data is a relatively simply process, DfE uses its internal Access database to capture this information along with other pieces of information about the company manufacturing a DfE-labeled product. The source of the data is always the company who manufacturers the labeled product. As described above, data from these companies are aggregated and averaged across all DfE-labeled products in order to protect the confidentiality of the data.

GE: Most relevant information system will be the SMMT. Other relevant information systems which will help with focused targeting and application is the TRI Production Related Waste (PRW) database.

Technical Assistance: Regional project officers use a web-based P2 GrantsPlus information system to store source data and some transformed data. The system houses non-normalized grant-specific data, and satisfies EPA's IT security policy. EPA's contractor has an extensive systems description; a description will soon be available on line. Regional project officers use the program's Gallons to Pounds Converter Tool to convert any grantee records of hazardous-material gallons into hazardous-material pounds. Regional project officers check to see that water-use reduction value and water-effluent reduction value are not the same; if they are the same, a P.O. would divide the value of water effluent by 10,000 before entering this sub-value of hazardous materials reduced into P2 GrantsPlus.

3b. Data Quality Procedures:

OPPT: All OPPT programs operate under the Information Quality Guidelines as found at http://www.epa.gov/quality/informationguidelines as well as under the Pollution Prevention and Toxics Quality Management Plan (QMP) ("Quality Management Plan for the Office of Pollution Prevention and Toxics; Office of Prevention, Pesticides and Toxic Substances," November 2008), and the programs will ensure that those standards and procedures are applied to this effort. The Quality Management Plan is for internal use only.

EPP:

EPEAT: EPEAT manufacturers sign a Memorandum of Understanding in which they warrant accuracy of the data they provide. EPA/EPEAT Program Managers review all data before using them in the EEBC.

FEC: National program staff provides guidelines and instructions on how Federal partners should report data. National program staff and managers review all data before entering them into the EEBC.

GC: The GC program operates under the Information Quality Guidelines found at http://www.epa.gov/quality/informationguidelines as well as under the Pollution Prevention and Toxics Quality Management Plan (QMP) ("Quality Management Plan for the Office of Pollution Prevention and Toxics; Office of Prevention, Pesticides and Toxic Substances," November 2008), and the program will ensure that those standards and procedures are applied to this effort. The Quality Management Plan is for internal use only.

DfE/GE:

DfE: The DfE program operates under the Information Quality Guidelines found at http://www.epa.gov/quality/informationguidelines as well as under the Pollution Prevention and Toxics Quality Management Plan (QMP) ("Quality Management Plan for the Office of Pollution Prevention and Toxics; Office of Prevention, Pesticides and Toxic Substances," November 2008), and the program will ensure that those standards and procedures are applied to this effort. The Quality Management Plan is for internal use only.

Data undergo a technical screening review by DfE before being added to the data collection spreadsheet. DfE determines whether data submitted adequately support the environmental benefits described. In addition, the DfE Program maintains Quality Assurance Project Plans (QAPPs) and Information Collection Requests (ICRs) for the collection of technical and performance data.

In response to recommendations in the OIG's report "Measuring and Reporting Performance Results for the Pollution Prevention Program Need Improvement" (January 2009), EPA established Standard Operating Procedures to govern its collection, tracking, analyzing, and public reporting of data on environmental and other performance parameters. These SOPs pertain to the type, format and quality of data to be submitted to the Agency by partners, contractors, and program beneficiaries for use in reporting P2 Program performance.

GE: As the GE work assignment includes both outreach and modeling work, a Quality Assurance Plan has been developed which contains the data quality procedures. EPA also performed quality review of the models and information used from both GSK (reference paper 1) and Zurich (reference paper 2), which have been used and adapted by industry, academia and government.

Technical Assistance: If Regional project officers determine a grantee is generating source data, they require a Quality Assurance Project Plan; in all cases they supply data quality procedures to grantees. Regional project officers follow Regional P2 Measurement Guidance (revised October 2012). Regional and national program staff reviews data before entry into P2 Grants Plus. Regional project officers check to see that the water-use reduction and water-effluent reduction values are not the same; if they are the same, the P.O. would divide the water-effluent value by 10,000 to get a more accurate version of this sub-value of hazardous materials reduced.

3c. Data Oversight:

The Branch Chief of the Planning and Assessment Branch (PAB) in OPPT oversees source data reporting and information systems through periodic updates and discussions with the national program staff members and managers who monitor their own source data reporting. This oversight is also accomplished through written protocols developed by PAB and national program managers.

3d. Calculation Methodology:

Variables:

Unit of analysis: MTCO2e (see Section 1a for definition)

GSN/E3: The national program is shifting away from applying a projected implementation rate to aggregated modeled potential data (all potential actions a facility could take) to report reductions in pounds of hazardous materials. EPA is shifting to using project-completed facility data collected from DOC or EPA grantees as the basis for reporting pound results. EPA calculates results will recur for four years. EPA considers average equipment/product lifetime and the quality of data records to arrive at a four-year recurring formula (counting results in year 1 when new and in three additional years).

EPP: EPEAT and FEC use the Electronics Environmental Benefits Calculator (EEBC) to calculate. Records are chosen in relation to hazardous materials; variables are for the equipment manufacturing process and disposal. The EEBC attributes hazardous material reductions to the year of manufacture and the year of disposal. EPA enters records associated with the phase of product life in the year it occurs. 3.a. lists EEBC references.

GC: Simple summation of realized quantitative pounds reported from valid PGCC award nominations received.

DFE: Data are included if the product has received the DfE label. Not all companies provide production volume information, so DfE has summed the total production volumes provided, and calculated an average production volume per recognized product. The production volume represents the pounds of chemicals in the product. Each chemical in the product is screened by DfE and must be a safer chemical (according to EPA criteria) in order for the product to be recognized. Because the measure is reported as a percent increase each year, a baseline year was chosen. That baseline year is 2009, the year preceding the development of the measure.

GE: The SMMT generates pounds. The program used the SMMT to derive GE GPRA goals for 2014. In addition, the DSW rule, finalized at the end of the year and implementation of the remanufacturing exclusion will result in additional pounds saved.

Technical Assistance: Regional project officers use EPA's Gallons to Pounds Converter tool to assist their calculations; otherwise, the record data submitted by grantees suffice. Records are chosen according to the subcomponents of hazardous materials, namely, hazardous waste, hazardous inputs, air emissions, and water effluent. Measurement guidance to grantees and internal quality reviews apply a 10,000-to-1 formula to segregate gallons of water from pounds of water effluent so that water effluent is not over-represented.

4a. Oversight and Timing of Final Results Reporting:

Planning and Accountability Lead in the Resource Management Staff in the Office of Program Management Operations. Reporting semiannually: mid-year and end-of-year.

4b. Data Limitations/Qualifications:

GSN/E3: EPA relies on partner facilities for data reporting to NIST, DOE, and EPA grantees, and on NIST for aggregating data before sharing them with EPA.

EPP: FEC has a built-in reliance on partners for data reporting. EPEAT relies on manufacturers of EPEAT-registered products, and the GEC, for data reporting.

GC: Because the PGCC awards are public, companies cannot submit confidential business information. As such, data provided can be qualitative rather than quantitative; qualitative data is not counted towards measures, so the data that is reported is conservative.

DFE:

DfE: As mentioned previously, the measure is an average based on production volumes provided to DfE. We believe the measure is conservative, because the average is derived primarily from products sold by small to medium-sized businesses. Consumer product companies, with larger production volumes, are not factored into the current calculation.

GE: The Program has used the TRI PRW data to obtain chemical substance, amount of chemical substance, and the unit operations. It has always been a goal to be able to use the actual production data from the facilities. As EPA adapted and developed the tool, the Program will work directly with industrial facilities and expect more accurate data will be available.

Technical Assistance: Regions rely primarily on grantees for data reporting.

4c. Third-Party Audits:

EPP: The Electronics Environmental Benefits Calculator (EEBC) underwent internal and external review during its development phases. It was also reviewed and beta-tested during development of version 2.0.

GC: PGCC award nominations are reviewed by an external peer review expert panel.

DfE/GE:

DfE: OMB previously reviewed this measure as part of its analysis of Pollution Prevention programs. This measure was one of the primary contributions to OMB's high rating of the pollution prevention programs.

GE: The method used in SMMT has gone through external review during its stages of development. The method was published in a peer reviewed journal. (Footnote 1) Note that the tool is still under development, with anticipation of completing this year.

Measure Code: 262 - Gallons of water reduced through pollution prevention.

Office of Chemical Strategies and Pollution Prevention (OCSPP)

Goal Number and Title:
4 - Ensuring the Safety of Chemicals and Preventing Pollution
Objective Number and Title:
2 - Promote Pollution Prevention
Sub-Objective Number and Title:
1 - Prevent Pollution and Promote Environmental Stewardship
Strategic Target Code and Title:
3 - By 2015, reduce water use by an additional 24billion gallons
Managing Office:
Office of Pollution Prevention and Toxics
1a. Performance Measure Term Definitions:
Gallons of Water Conserved: A measure of water conserved from the implementation of P2 practices by P2 program participants. P2 Programs related to this measure include: Green Suppliers Network (GSN) and Economy, Energy, and the Environment (E3) are related programs- o GSN: a coordinated effort of EPA and the US Department of Commerce to help large manufacturers engage their suppliers in undergoing low-cost technical reviews to improve their processes from an environmental and energy perspective, and to reduce their use of water. o E3: a coordinated federal and local technical assistance initiative to help manufacturers become more sustainable. More federal agencies contribute to E3 technical assessments than to GSN assessments. Agencies provide technical assessments and training to help manufacturers increase their energy efficiency/sustainability and reduce their use of water. Green Chemistry (GC) o The Green Chemistry Program promotes the research, development, and implementation of technologies that reduce or eliminate the use or generation of hazardous substances. o Through annual recognition, the Presidential Green Chemistry Challenge (PGCC) awards demonstrate the human health and environmental benefits and market competitiveness that green chemistry technologies offer. Design for the Environment (DfE)/ Green Engineering GE o Design for the Environment: DfE does not contribute to this measure o Green Engineering: · The Green Engineering program has provided support to the DSW rule, resulting in a 'remanufacturing' exclusion in facilitating the recovery / extension of life of solvents commonly used in pharmaceutical, basic organic, coating and plastic industries. This work builds on the GE program's industrial activities which involve

working with industry to implement greener solvent management methods and other process changes. GE continues development of its Sustainable Materials Management Tool (SMMT), which assess emissions to all media and raw material usage. The SMMT tool assesses reductions in gallons water by conserving / extending the life of chemical substances. The Green Engineering program continues its decade long effort in the development and dissemination of green / sustainable engineering materials, to incorporate sustainable materials management approaches in the engineering curriculum.

· Technical Assistance – The P2 Program issues two kinds of grants – P2 Grants to States and Tribes and Source Reduction Assistance Grants. States and Tribes are eligible for both kinds; localities, non-profits, universities and community groups are eligible for one kind – Source Reduction Assistance Grants. The purpose of these grants is to help small and medium businesses adopt sustainable P2 technologies and practices. Grantees provide technical assistance and implement award programs to achieve results.

2a. Original Data Source:

GSN/E3: EPA grantees provide utility data from facilities that have implemented E3 projects, and US Department of Commerce (DOC) grantees provide aggregated facility-level utility data from facilities willing to provide these data. EPA had been relying on modeled data (assuming many partner facilities will maintain confidentiality to the extent they will neither submit project-completion data nor accept DOC sharing any non-aggregated potential or project-completed data with EPA), but EPA is transitioning to using project-completed data as source data.

Green Chemistry (GC): Participants in the PGCC awards self-nominate. The awards are public, so confidential business information for nominated technologies is not accepted.

GE: The GE program has adapted and expanded the GlaxoSmithKline (GSK) partial life cycle modules which generate the multimedia emissions based on the life stages of a chemical substance: creation, reuse and disposal.[1] GSK provided extensive support during this adaptation process. The program expanded on the GSK's model unique contribution of 'manufacturing emissions' in assessing the benefits of extending solvent life. These manufacturing emissions were expanded further utilizing work from the ETH Zurich Institute.[2]

Other sources of information used for this work include: Green Engineering Textbook, US DOE OIT's Energy and Environmental Profile of the US Chemical Industry, TRI Production Related Waste Data (transfers / inventories), Chemical Trees for the studied solvents supplied by GSK, data obtained from working with industry, and Kirk-Othmer Encyclopedia of chemical processes. The 2008 DSW RIA provided additional information on end of life impacts, including disposal and waste management costs.

Technical Assistance: Facilities receiving State P2 technical assistance or applying for a State P2 award provide grantees (States) with data from their water utility bills.

------

1 - Waste treatment modules – a Partial Life Cycle Inventory, Concepcion Jimenez-Gonzalez, Michael R Overcash and Alan Curzons, Journal of Chemical Technology and Biotechnology 76: 707-716, 2001.
2 - What is a green solvent? A comprehensive framework for the environmental assessment of solvents. Capello, C., Fischer, U., and Hungerbuhler, K. Green Chemistry, 9, 927-934, 2007.

2b. Source Data Collection:
GSN/E3: Grantees of DOC and EPA collect source data. Grantees of DOC (Manufacturing Extension Partnership Centers (MEPs)) collect water conservation data associated with technologies and practices they recommend to businesses. MEPs collect any water conservation data that facilities supply directly to MEPs or through follow-up voluntary MEP questionnaires on implemented E3 projects. DOC aggregates these data and then shares them with EPA. EPA grantees collect water conservation data from facilities that implemented their E3 recommendations. DOC and EPA grantees send data to their grantor agencies, and the agencies input the data into their respective databases. All grantees follow their respective agencies' QA/QC requirements. Federal agencies participating in E3 are developing a second complementary database for their collective use. Green Chemistry: PGCC awards nominations are provided to EPA. EPA prescreens the nominations, then provides those that meet scope criteria to an external peer review expert panel for judging. Suggested winners are returned to EPA for final verification and validation. GE: The GE program has, to date, has not reported GPRA measurements. It is envisioned, though, that industry, upon implementation of GE approaches, will be able to use the SMMT to generate the metrics, which would be used for reporting purposes. Technical Assistance: P2 Program grantees (or sub-grantees) collect records relating to water-use reductions from facilities who have completed P2 implementation or who have applied for State environmental awards. Grantees follow EPA data collection guidelines and submit QAPPs if required.
2c. Source Data Reporting:
Data Submission Instrument GSN/E3: NIST/DOC submits data to EPA in a document. State grantees submit data electronically or by mail to EPA in grant reports. GC: PGCC awards nominations are provided to EPA as an electronic report. GE: The GE program has explored development of a boiler plate template which industry could use in reporting metrics, with linkages to the SMMT. Please also see 2b above. Technical Assistance: State grantees submit data to EPA electronically or by mail in grant reports. Data Entry Mechanism GSN/E3: NIST sends a document of data to EPA program staff, who keep the data in a file. DOE sends to DOC and EPA; EPA program staff keeps the data in a file. State grantees submit data in grant reports to Regional project officers, who then manually enter the data into GrantsPlus, our program database. GC: Benefits data in PGCC awards nominations provided to EPA are entered into an internal database.

GE: The GE program envisions a central database in which information submitted by industry will be entered manually.

Technical Assistance: State grantees submit data in grant reports to Regional project officers, who then manually enter the data into GrantsPlus, our program database.

Frequency of Data Transmission to EPA
GSN/E3:
GSN: monthly
E3: quarterly
GC: annually
GE: annually
Technical Assistance: semi-annually.

Timing of Data Transmission to EPA
GSN/E3:
GSN: monthly
E3: every fiscal quarter
GC: see above
GE: See above
Technical Assistance: Grantees must submit data if possible by the close of the fiscal year, and any amendments for the prior fiscal year by March 1st.

3a. Relevant Information Systems:

GSN/E3: DOC's Customer Relationship Management database is a system used to store source data and data transformed by modeling and aggregation; EPA does not have direct access. EPA program staff uses its own desktop data storage, such as Excel spreadsheets. Regional project officers issuing E3 grants use P2 GrantsPlus as their information system, described below under Technical Assistance. Another multi-agency database (for all agencies in E3) is under development to house secondary source data and source data transformed by modeling. EPA's Information System Integrity Standards are not applicable to DOC and DOE databases. EPA's P2 GrantsPlus database meets EPA's IT security policy.

GC: internal database

GE: Most relevant information system will be the SMMT. Other relevant information systems which will help with focused targeting and application is the TRI Production Related Waste (PRW) database.

Technical Assistance: Regional project officers use a web-based P2 GrantsPlus information system to store source data. The system houses non-normalized grant-specific data, and satisfies EPA's IT security policy. EPA's contractor has an extensive systems description; a description will soon be available on line.

3b. Data Quality Procedures:

OPPT: All OPPT programs operate under the Information Quality Guidelines as found at http://www.epa.gov/quality/informationguidelines as well as under the Pollution Prevention and Toxics Quality Management Plan (QMP) ("Quality Management Plan for the Office of Pollution Prevention and Toxics; Office of Prevention, Pesticides and Toxic Substances," November 2008), and the programs will ensure that those standards and procedures are applied to this effort. The Quality Management Plan is for internal use only.

GC: The GC program operates under the Information Quality Guidelines found at http://www.epa.gov/quality/informationguidelines as well as under the Pollution Prevention and Toxics Quality Management Plan (QMP) ("Quality Management Plan for the Office of Pollution Prevention and Toxics; Office of Prevention, Pesticides and Toxic Substances," November 2008), and the program will ensure that those standards and procedures are applied to this effort. The Quality Management Plan is for internal use only.

GE: As the GE work assignment includes both outreach and modeling work, a Quality Assurance Plan has been developed which contains the data quality procedures. EPA also performed quality review of the models and information used from both GSK (reference paper 1) and Zurich (reference paper 2), which have been used and adapted by industry, academia and government.

Technical Assistance: Technical Assistance: If Regional project officers determine a grantee is generating source data, they require a Quality Assurance Project Plan; in all cases they supply data quality procedures to grantees. Regional project officers follow Regional P2 Measurement Guidance (revised October 2012). Regional and national program staff reviews data before entry into P2 Grants Plus.

3c. Data Oversight:

The Branch Chief of the Planning and Assessment Branch (PAB) in OPPT oversees source data reporting and information systems through periodic updates and discussions with the national program staff members and managers who monitor their own source data reporting. This oversight is also accomplished through written protocols developed by PAB and national program managers.

3d. Calculation Methodology:

Variables:
Unit of analysis: MTCO2e (see Section 1a for definition)

GSN/E3: The national program is shifting away from applying a projected implementation rate to aggregated modeled potential data (all potential actions a facility could take) to report reductions in water use. EPA is shifting to using project-completed facility data collected from DOC or EPA grantees as the basis for reporting gallons of water conserved. EPA calculates results will recur for four years.3 EPA considers average equipment/P2 practice lifetime and the quality of data records to arrive at a four-year recurring formula (counting results in year 1 when new and in three additional years).

GC: Simple summation of realized quantitative pounds reported from valid PGCC award nominations received.

GE: The SMMT generates emissions to water; i.e., COD, TDS, as well as water used as a raw material.   The planned addition of a water balance will also take into account process water used for washing, quenching, etc. The program used the SMMT to derive GE GPRA goals for 2014. These targets result in significant gallons of water saved. Water is used not only as a primary raw material, but also throughout the chemical manufacturing process in heating, cooling and washing. Any gaseous substance leaving a reactor is washed with water and shifted to this media. In addition, the DSW rule, finalized at the end of the year and implementation of the remanufacturing exclusion will result in additional water savings.

Technical Assistance: Regional project officers use the water saving values reported by grantees in their grant reports. EPA calculates results will recur for four years. EPA considers average equipment/P2 practice lifetime and the quality of data records to arrive at a four-year recurring formula (counting results in year 1 when new and in three additional years).

—

3 Allen, D. and Shonnard, D. Green Engineering: Environmentally Conscious Design of Chemical Processes, Chapter 10.3 Process Mass Integration, p 317 – 346, 2001.

4a. Oversight and Timing of Final Results Reporting:

Planning and Accountability Lead in the Resource Management Staff in the Office of Program Management Operations. Reporting semiannually: mid-year and end-of-year.

4b. Data Limitations/Qualifications:

GSN/E3: EPA relies on partner facilities for data reporting to NIST, DOE, and EPA grantees, and on NIST for aggregating data before sharing them with EPA.

GC: Because the PGCC awards are public, companies cannot submit confidential business information. As such, data provided can be qualitative rather than quantitative; qualitative data is not counted towards measures, so the data that is reported is conservative.

GE: The Program has used the TRI PRW data to obtain chemical substance, amount of chemical substance, and the unit operations. It has always been a goal to be able to use the actual production data from the facilities. As EPA adapted and developed the tool, the Program will work directly with industrial facilities and expect more accurate data will be available.

Technical Assistance: Regions rely primarily on grantees for data reporting.

4c. Third-Party Audits:

GC: PGCC award nominations are reviewed by an external peer review expert panel.

GE: The method used in SMMT has gone through external review during its stages of development. The method was published in a peer reviewed journal. (Footnote 1) Note that the tool is still under development, with anticipation of completing this year.

Measure Code: 10D - Percent difference in the geometric mean blood level in low-income children 1-5 years old as compared to the geometric mean for non-low income children 1-5 years old.

Office of Chemical Strategies and Pollution Prevention (OCSPP)

Goal Number and Title:
4 - Ensuring the Safety of Chemicals and Preventing Pollution
Objective Number and Title:
1 - Ensure Chemical Safety
Sub-Objective Number and Title:
1 - Protect Human Health from Chemical Risks
Strategic Target Code and Title:
2 - By 2014,reduce the percentage of children with blood lead levels above 5ug/dl to 1.0 percent or less
Managing Office:
Office of Pollution Prevention and Toxics;National Program Chemicals Division
1a. Performance Measure Term Definitions:
Geometric mean blood lead level: This term refers to a type of average which indicates the central tendency or typical value of a set of numbers. As used in this measure, it represents the central tendency of reported blood lead levels (micrograms of lead per deciliter of blood, or µg/dL) of children ages 1-5.
Low-income children: As used in this measure, this term means children whose families are below the poverty income ratio (PIR) of 1.0. The poverty income ratio is a measure of income to the poverty threshold.
Non-low-income children: Children whose families have a PIR above 1.0
Background: This performance measure examines the disparities of blood lead levels in low-income children as compared to non-low-income children so that EPA can track progress toward its long-term goal of eliminating childhood lead poisoning in harder to reach vulnerable populations. EPA's Lead Risk Reduction program contributes to the goal of eliminating childhood lead poisoning by: (1) establishing standards governing lead hazard identification and abatement practices and maintaining a national pool of professionals trained and certified to implement those standards; (2) providing information to housing occupants so they can make informed decisions and take actions about lead hazards in their homes; and (3) establishing a national pool of certified firms and individuals who are trained to carry out renovation and repair and painting projects while adhering to the lead-safe work practice standards and to minimize lead dust hazards created in the course of such projects.
Recent data show significant progress in the continuing effort to eliminate childhood lead poisoning as a public health concern. However, results of recent studies indicate adverse health effects to children at low blood levels, below 10 µg/dL. In response to this new information and the fact that approximately three-quarters of the nation's housing stock built before 1978 still contains some lead-based paint, the EPA is now targeting reductions in the number of children with blood lead levels of 5 µg/dL or higher, as reflected in this performance measure.

2a. Original Data Source:

The original data source is the Centers for Disease Control and Prevention's (CDC) National Health and Nutrition Examination Survey (NHANES), which is recognized as the primary database in the United States for national blood lead statistics, http://www.cdc.gov/nchs/nhanes/about_nhanes.htm NHANES is a probability sample of the non-institutionalized population of the United States. The survey examines a nationally representative sample of approximately 5,000 men, women, and children each year located across the U.S.

2b. Source Data Collection:

Methods of data collection (by original data source): Data are obtained by analysis of blood and urine samples collected from survey participants. Health status is assessed by physical examination. Demographic and other survey data regarding health status, nutrition, and health-related behaviors are collected by personal interview, either by self-reporting or, for children under 16 and some others, as reported by an informant. Detailed interview questions cover areas related to demographic, socio-economic, dietary, and health-related questions. The survey also includes an extensive medical and dental examination of participants, physiological measurements, and laboratory tests. NHANES is unique in that it links laboratory-derived biological markers (e.g. blood, urine etc.) to questionnaire responses and results of physical exams.

Quality procedures followed (by original data source): According to the CDC, the process of preparing NHANES data sets for release is as rigorous as other aspects of the survey. After a CDC contractor performs basic data cleanup, the CDC NHANES staff ensure that the data are edited and cleaned prior to release. NHANES staff devotes at least a full year after the completion of data collection to careful data preparation. Additionally, NHANES data are published in a wide array of peer-reviewed professional journals.

 Background documentation is available at the NHANES Web site at: http://www.cdc.gov/nchs/nhanes.htm
The analytical guidelines are available at the Web site:
http://www.cdc.gov/nchs/about/major/nhanes/nhanes2003-2004/analytical_guidelines.htm

Geographical extent of source data, if relevant: Data are collected to be representative of the U.S. population. The population data are extrapolated from sample data by the application of standard statistical procedures.

Spatial detail of source data, if relevant: NHANES sampling procedures provide nationally representative data.

2c. Source Data Reporting:

Form/mechanism for receiving data and entering into EPA system: EPA monitors the periodic issuance of NHANES reports and other data releases to obtain the data relevant to this measure.

Timing and frequency of reporting: NHANES is a continuous survey and examines a nationally representative sample of about 5,000 persons each year. These persons are located in counties across the country, 15 of which are visited each year.

Files of raw data, containing measured blood lead levels in NHANES participants, are currently released to the public in two-year sets. CDC also periodically publishes reports containing summary statistics for lead and more than 200 other chemicals measured in NHANES, at www.cdc.gov/exposurereport

3a. Relevant Information Systems:

There are no EPA systems utilized in collecting data for this measure as the Agency is able to secure the necessary data directly from NHANES reports and data releases.

3b. Data Quality Procedures:

EPA does not have any procedures for quality assurance of the underlying data as this function is performed by the CDC itself. CDC has periodically reviewed and confirmed EPA's calculation of NHANES summary statistics from the raw data files. The Agency determines the performance result for this measure by performing standard mathematical operations on reported NHANES data to derive geometric mean blood lead levels by income group and to estimate the disparity in those levels between low-income and non-low-income children.

3c. Data Oversight:

Chief, Planning and Assessment Branch, Environmental Assistance Division, Office of Pollution Prevention and Toxics

3d. Calculation Methodology:

Decision rules for selecting data: Not applicable. EPA simply uses the geometric mean blood lead level values for low-income and non-low-income children that are generated from NHANES survey data, as described below. EPA however, limits the age of the child to under six, based on the most sensitive receptor age group noted in Section 401 of TSCA.

Definitions of variables: Key terms are defined in 1(a) above.

Explanation of the calculations: EPA performs standard mathematical operations on the published NHANES survey data. After calculating geometric mean blood lead levels by income group from the public use data files, EPA (1) determines the absolute disparity in blood lead level values between the two groups of children by subtracting the lower value from the higher; (2) averages the values for the two groups; and (3) divides the absolute disparity (i.e., the result of calculation (1)) by the average of the values (i.e., the result of calculation (2)), to express the disparity as a percent difference between the blood lead levels of the two groups.

Explanation of assumptions: Not applicable.

Identification of unit of measure: Percent difference in blood lead levels as determined by the methods described under "Explanation of the calculations" above.

Identification of timeframe of result: The performance result is computed from data released by the CDC in sets covering the particular time period over which sampling occurs. Thus, the timeframe that applies to the measured result is the same period for which the NHANES data are released. It is not a simple snapshot at a specific moment in time.

4a. Oversight and Timing of Final Results Reporting:

Planning and Accountability Lead in the Resource Management Staff in the Office of Program Management Operations. Reporting semiannually: mid-year and end-of-year, but subject to a data lag due to the periodic nature of NHANES reporting.

4b. Data Limitations/Qualifications:

NHANES is a voluntary survey and selected persons may refuse to participate. In addition, the NHANES survey uses two steps, a questionnaire and a physical exam. There sometimes are different numbers of subjects in the interview and examinations because some participants only complete one step of the survey. Participants may answer the questionnaire but not provide the more invasive blood sample. Special weighting techniques are used to adjust for non-response. NHANES is not designed to provide detailed estimates for populations that are highly exposed to lead.

4c. Third-Party Audits:

Report of the NHANES Review Panel to the NCHS Board of Scientific Counselors.

Cover letter can be accessed at: http://www.cdc.gov/nchs/data/bsc/bscletterjune8.pdf
Report can be accessed at: http://www.cdc.gov/nchs/data/bsc/NHANESReviewPanelReportrapril09.pdf

Office of Chemical Strategies and Pollution Prevention (OCSPP)

Goal Number and Title:
4 - Ensuring the Safety of Chemicals and Preventing Pollution

Objective Number and Title:
2 - Promote Pollution Prevention

Sub-Objective Number and Title:
1 - Prevent Pollution and Promote Environmental Stewardship

Strategic Target Code and Title:
5 - Through 2015, increase the use of safer chemicals by 40 percent

Managing Office:
Office of Pollution Prevention and Toxics

1a. Performance Measure Term Definitions:
Safer chemicals: A safer chemical is a substance which meets DfE Safer Ingredient Criteria. Background: The EPA allows safer products to carry the Design for the Environment (DfE) label. This mark enables consumers to quickly identify and choose products that can help protect the environment and are safer for families. The DfE logo on a product means that the DfE scientific review team has screened each ingredient for potential human health and environmental effects and that—based on currently available information, EPA predictive models, and expert judgment—the product contains only those ingredients that pose the least concern among chemicals in their class. Product manufacturers who become DfE partners, and earn the right to display the DfE logo on recognized products, have invested heavily in research, development and reformulation to ensure that their ingredients and finished product line up on the green end of the health and environmental spectrum while maintaining or improving product performance.

2a. Original Data Source:
Partners, i.e., those companies who have DfE-labeled products, provide proprietary information on the production volume of their safer formulations. These data are aggregated and averaged to protect the confidential nature of this information.

2b. Source Data Collection:
Production volume is a standard measure for companies who sell chemical products such as cleaning formulations. This information is directly related to the sales of their product, and therefore they have the incentive to track it accurately. Production volumes reported to the EPA include the total volume of each DfE-labeled product sold in the US or internationally.

2c. Source Data Reporting:
Production volume data are submitted to the EPA along with a company's application to receive the DfE label. These data are captured on an as needed basis in a database maintained by EPA contractors. The EPA receives

production volume data when a company chooses to submit an application for recognition under the program.

3a. Relevant Information Systems:

Because the collection and analysis of this data is a relatively simple process, DfE uses its internal Access database to capture this information along with other pieces of information about the company manufacturing a DfE-labeled product. The source of the data is always the company who manufacturers the labeled product. As described above, data from these companies are aggregated and averaged across all DfE-labeled products in order to protect the confidentiality of the data.

3b. Data Quality Procedures:

The DfE program operates under the Information Quality Guidelines found at http://www.epa.gov/quality/informationguidelines as well as under the Pollution Prevention and Toxics Quality Management Plan (QMP) ("Quality Management Plan for the Office of Pollution Prevention and Toxics; Office of Prevention, Pesticides and Toxic Substances," November 2008), and the program will ensure that those standards and procedures are applied to this effort. The Quality Management Plan is for internal use only.

Data undergo a technical screening review by DfE before being added to the data collection spreadsheet. DfE determines whether data submitted adequately support the environmental benefits described. In addition, the DfE Program maintains Quality Assurance Project Plans (QAPPs) and Information Collection Requests (ICRs) for the collection of technical and performance data.

In response to recommendations in the OIG's report "Measuring and Reporting Performance Results for the Pollution Prevention Program Need Improvement" (January 2009), EPA established Standard Operating Procedures to govern its collection, tracking, analyzing, and public reporting of data on environmental and other performance parameters. These SOPs pertain to the type, format and quality of data to be submitted to the Agency by partners, contractors, and program beneficiaries for use in reporting P2 Program performance.

3c. Data Oversight:

Chief, Planning and Assessment Branch, Environmental Assistance Division, Office of Pollution Prevention and Toxics

3d. Calculation Methodology:

Data are included if the product has received the DfE label. Not all companies provide production volume information, so DfE has summed the total production volumes provided, and calculated an average production volume per recognized product. The production volume represents the pounds of chemicals in the product. Each chemical in the product is screened by DfE and must be a safer chemical (according to EPA criteria) in order for the product to be recognized. Because the measure is reported as a percent increase each year, a baseline year was chosen. That baseline year is 2009, the year preceding the development of the measure.

4a. Oversight and Timing of Final Results Reporting:

Planning and Accountability Lead in the Resource Management Staff in the Office of Program Management Operations. Reporting semiannually: mid-year and end-of-year.

4b. Data Limitations/Qualifications:

As mentioned previously, the measure is an average based on production volumes provided to DfE. We believe the measure is conservative, because the average is derived primarily from products sold by small to medium-sized businesses. Consumer product companies, with larger production volumes, are not factored into the current calculation.

4c. Third-Party Audits:

OMB previously reviewed this measure as part of its analysis of Pollution Prevention programs. This measure was one of the primary contributions to OMB's high rating of the pollution prevention programs.

Measure Code: J15 - Reduction in concentration of targeted pesticide analytes in children.

Office of Chemical Strategies and Pollution Prevention (OCSPP)

Goal Number and Title:	
4 - Ensuring the Safety of Chemicals and Preventing Pollution	
Objective Number and Title:	
1 - Ensure Chemical Safety	
Sub-Objective Number and Title:	
1 - Protect Human Health from Chemical Risks	
Strategic Target Code and Title:	
5 - By 2014, reduce concentration of targeted chemicals in children	
Managing Office:	
Office of Pesticide Programs	
1a. Performance Measure Term Definitions:	

Reduction: Each fiscal year, EPA compares the most recent biomonitoring data available on the analytes of targeted organophosphate pesticides in urine samples from the general public that have been analyzed by the Centers for Disease Control and Prevention (CDC) to the baseline concentrations. The baseline years (corresponding to the NHANES sampling period) chosen for this measure are 2001-2002. The percent for which the population's 95th percentile concentration changed between the baseline year and the latest measurements will be calculated. The result of these calculations is then compared to the target set for the year in which performance is being measured.

Concentration: 95th percentile concentration measured in the micrograms per liter (ug/L), at standard detection limits.

Targeted pesticide analytes: The pesticides targeted by this measure are organophosphate pesticides. The measure is based on levels of the following metabolites that CDC measures in urine samples: six non-specific organophosphate dialkyl phosphate metabolites – and the chlorpyrifos-specific metabolite 3,5,6-Trichloro-2-pyridinol. The dialkyl phosphate and 3,5,6-Trichloro-2-pyridinol metabolites can be present in urine after low level exposures to organophosphorus insecticides that do not cause clinical symptoms or inhibition of cholinesterase activity, and measurement of these metabolites reflects recent exposure, predominantly in the previous few days. The metabolites may also occur in the environment as a result of degradation of organophosphorus insecticides, and therefore, the presence in a person's urine may reflect exposure to the metabolite itself.

Children: The measure is intended to evaluate exposure trends in the U.S. population and will target the youngest available child age group in which pesticide analytes are measured, 6 - 11 year olds.

Background:
·   NHANES is a major program of the National Center for Health Statistics (NCHS). NCHS is part of the Centers for Disease Control and Prevention (CDC), U.S. Public Health Service, and has the responsibility for

producing vital and health statistics for the Nation. NCHS is one of the Federal statistical agencies belonging to the Interagency Council on Statistical Policy (ICSP). The ICSP, which is led by the Office of Management and Budget (OMB), is composed of the heads of the Nation's 10 principal statistical agencies plus the heads of the statistical units of four non-statistical agencies. The ICSP coordinates statistical work across organizations, enabling the exchange of information about organization programs and activities, and provides advice and counsel to OMB on statistical activities. The statistical activities of these agencies are predominantly the collection, compilation, processing or analysis of information for statistical purposes. Within this framework, NCHS functions as the Federal agency responsible for the collection and dissemination of the Nation's vital and health statistics. Its mission is to provide statistical information that will guide actions and policies to improve the health of the American people.

·   To carry out its mission, NCHS conducts a wide range of annual, periodic, and longitudinal sample surveys and administers the national vital statistics systems.

·   As the Nation's principal health statistics agency, NCHS leads the way with accurate, relevant, and timely data. To assure the accuracy, relevance, and timeliness of its statistical products, NCHS assumes responsibility for determining sources of data, measurement methods, methods of data collection and processing while minimizing respondent burden; employing appropriate methods of analysis, and ensuring the public availability of the data and documentation of the methods used to obtain the data. Within the constraints of resource availability, NCHS continually works to improve its data systems to provide information necessary for the formulation of sound public policy. As appropriate, NCHS seeks advice on its statistical program as a whole, including the setting of statistical priorities and on the statistical methodologies it uses. NCHS strives to meet the needs for access to its data while maintaining appropriate safeguards for the confidentiality of individual responses.

·   The Centers for Disease and Prevention's (CDC) National Health and Nutrition Examination Survey (NHANES) program is a survey designed to assess the health and nutritional status of adults and children in the U.S. NHANES was selected as the performance database because it is an ongoing program that is statistically designed to be nationally representative of the U.S. civilian, non-institutionalized population.

·   Baseline for this measure was established using existing NHANES biomonitoring data. During each fiscal year, performance will then be evaluated by comparing subsequent NHANES biomonitoring data with the established baseline.

·   This measure supports the long-term goal of reducing the risk and ensuring the safety of chemicals and preventing pollution at the source by enabling EPA to better assess progress in reducing exposure to targeted chemicals, as reflected in concentration levels among the general population and key subpopulations.

·   Analytes for organophosphate pesticides were selected for this measure because EPA anticipates recent registration activity will have a direct effect on reducing exposure in the general population.

For more information on the pesticides, visit http://www.epa.gov/pesticides/about/types.htm

2a. Original Data Source:
NHANES: CDC's NHANES survey program began in the early 1960s as a periodic study and continues as an annual survey (http://www.cdc.gov/nchs/NHANES.htm The survey examines a nationally representative sample of approximately 5,000 men, women, and children each year located across the U.S. CDC's National Center for Health Statistics (NCHS) is responsible for the conduct of the survey and the release of the data to the public through their website at: http://www.cdc.gov/nchs/nhanes/nhanes_questionnaires.htm NHANES is designed to

collect data on the health and nutritional status of the U.S. population. NHANES collects information about a wide range of health-related behaviors, performs physical examinations, and collects samples for laboratory tests. NHANES is unique in its ability to examine public health issues in the U.S. population, such as risk factors for cardiovascular disease. Beginning in 1999, NHANES became a continuous survey, sampling the U.S. population annually and releasing the data in 2-year cycles.

2b. Source Data Collection:

Collection Methods: The sampling plan follows a complex, stratified, multistage, probability-cluster design to select a representative sample of the civilian, noninstitutionalized population in the United States based on age, gender, and race/ethnicity. The NHANES survey contains detailed interview questions covering areas related to demographic, socio-economic, dietary, and health-related subjects. It also includes an extensive medical and dental examination of participants, physiological measurements, and laboratory tests. NHANES is unique in that it links laboratory-derived biological markers (e.g. blood, urine etc.) to questionnaire responses and results of physical exams. Analytical guidelines issued by NCHS provide guidance on how many years of data should be combined for an analysis. NHANES measures blood levels in the same units (i.e., ug/dL) and at standard detection limits.

Environmental chemicals are measured in blood, serum, or urine specimens collected as part of the examination component of NHANES. The participant ages for which a chemical was measured varied by chemical group. Most of the environmental chemicals were measured in randomly selected subsamples within specific age groups. Randomization of subsample selection is built into the NHANES design before sample collection begins. Different random subsamples include different participants. This subsampling was needed to ensure an adequate quantity of sample for analysis and to accommodate the throughput of the mass spectrometry analytical methods.

Geographical Extent: NHANES is designed to be a representative sample of the civilian, noninstitutionalized population in the United States based on age, gender, and race/ethnicity.

Quality Procedures: NCHS assures the security of its statistical and analytic information products through the enforcement of rigorous controls that protect against unauthorized access to the data, revision or corruption of the data, or unauthorized use of the data. Some of the major controls used at NCHS include access control, user authentication, encryption, access monitoring, provision of unalterable electronic content, and audit trails. All NCHS statistical and analytic information products undergo a formal clearance process before dissemination. Publications and reports, whether in electronic or paper form, are reviewed by a designated official within the author's office or division and by the NCHS Associate Director for Science (ADS). These reviews cover the clarity of descriptive text, the appropriateness of the methodology, the soundness of the analysis, the adherence to confidentiality and disclosure avoidance restrictions, the readability of tabular and graphic presentations of data, etc. Finally, all products undergo editorial review (e.g., formatting, proofreading, spell checks, proper punctuation, etc.). In addition, all public-use tapes are reviewed for accuracy and appropriate confidentiality protections. Oral presentations are subject to appropriate supervisory review.

NCHS statistical and analytic information products are derived using generally acceptable statistical practices and methodologies, which are well documented and available to the public. These procedures enable responsible statisticians and analysts outside of NCHS to replicate the NCHS statistical methods and obtain results consistent with those obtained by NCHS.

References:

CDC (2009a). National Health and Nutrition Examination Survey, 2007-2008 Overview. Available at: <http://www.cdc.gov/nchs/data/nhanes/nhanes_07_08/overviewbrochure_0708.pdf>

CDC (2009b) Fourth National Report on Human Exposure to Environmental Chemicals. Available at: <http://www.cdc.gov/exposurereport/pdf/FourthReport.pdf>

.CDC (2009c). NCHS Guidelines for Ensuring the Quality of Information Disseminated to the Public. Available at: < http://www.cdc.gov/nchs/about/policy/quality.htm>

2c. Source Data Reporting:

Data Submission Instrument: CDC's National Center for Health Statistics (NCHS) is responsible for the release of the data to the public through their website at: http://www.cdc.gov/nchs/nhanes/nhanes_questionnaires.htm The data utilized for the performance measure is released as part of the NHANES laboratory files. The naming convention and organization of the laboratory data files may change between survey cycles, so NHANES laboratory documentation should be reviewed to identify the correct data fields for analysis. In 2001-2002, the SAS Transport File containing the targeted pesticide data ("l26PP_B.xpt") can be identified through the "2001-2002 Laboratory Variable List" at: http://www.cdc.gov/nchs/nhanes/nhanes2001-2002/varlab_b.htm

In recent years, CDC has published a national exposure report based on the data from the NHANES. CDC has scheduled release of data, and scheduled release of national exposure reports through NHANES. The most current update of the National Report on Human Exposure to Environmental Chemicals was released February 2012 and is available at the Web site http://www.cdc.gov/exposurereport/ Performance results will be updated as new peer reviewed NHANES data are published either in the official CDC report on human exposure to environmental chemicals or other journal articles as the data become available.

3a. Relevant Information Systems:

CDC is responsible for all NHANES data collection and reporting. As such, no EPA information systems are involved in the process of collecting, calculating and/or reporting the results for this measure. In order to calculate the results for the performance measure, EPA accesses and downloads the NHANES data files that are publically available through CDC/NCSH at: http://www.cdc.gov/nchs/nhanes/nhanes_questionnaires.htm The NHANES data files are downloaded as SAS Transport files and uploaded into SAS for statistical analysis.

System Description: Not Applicable

Source/Transformed Data: Not Applicable

Information System Integrity Standards: Not Applicable

3b. Data Quality Procedures:

NCHS assures the security of its statistical and analytic information products through the enforcement of rigorous controls that protect against unauthorized access to the data, revision or corruption of the data, or unauthorized use of the data. Some of the major controls used at NCHS include access control, user authentication, encryption, access monitoring, provision of unalterable electronic content, and audit trails.

All NCHS statistical and analytic information products undergo a formal clearance process before dissemination. Publications and reports, whether in electronic or paper form, are reviewed by a designated official within the author's office or division and by the NCHS Associate Director for Science (ADS). These reviews cover the clarity of descriptive text, the appropriateness of the methodology, the soundness of the analysis, the adherence to confidentiality and disclosure avoidance restrictions, the readability of tabular and graphic presentations of data, etc. NCHS statistical and analytic information products are derived using generally acceptable statistical practices and methodologies, which are well documented and available to the public. These procedures enable responsible statisticians and analysts outside of NCHS to replicate the NCHS statistical methods and obtain results consistent with those obtained by NCHS.

References:
CDC (2009c). NCHS Guidelines for Ensuring the Quality of Information Disseminated to the Public. Available at: <http://www.cdc.gov/nchs/about/policy/quality.htm>

3c. Data Oversight:

Source Data Reporting Oversight Personnel: Not applicable; Data is from CDC.

Source Data Reporting Oversight Responsibilities: Not applicable; Data is from CDC.

Information Systems Oversight Personnel: Not applicable; Data is from CDC.

Information Systems Oversight Responsibilities: Not applicable; Data is from CDC.

3d. Calculation Methodology:

Decision Rules for Selecting Data: The performance measure uses NHANES pesticide biomonitoring data published by CDC/NCHS. EPA's measure focuses on 6 – 11 year old children, so the only selection criteria is survey respondents aged 6 to 11 years at time of screening. This selection criteria is applied during data analysis using the age at screening data field "RIDAGEYR."

Definitions of Variables: Key data fields related to the NHANES survey design and targeted pesticide analytes are defined below:

- WTSPP2YR   Pesticides Subsample 2 year Mec Weight
- SEQN   Respondent sequence number
- RIDAGEYR   Age at Screening Adjudicated - Recode

- URXCPM   3,5,6-trichloropyridinol (ug/L) result
- URXOP1   Dimethylphosphate (ug/L) result
- URXOP2   Diethylphosphate (ug/L) result
- URXOP3   Dimethylthiophosphate (ug/L) result
- URXOP4   Diethylthiophosphate (ug/L) result
- URXOP5   Dimethyldithiophosphate (ug/L) result
- URXOP6   Diethyldithiophosphate (ug/L) result

Explanation of the Calculations:

Annual performance will be evaluated using the equation below:

Where:

Baseline95th = 95th percentile urinary concentration during the baseline period.
Performance95th = 95th percentile urinary concentration during performance period.

Explanation of Assumptions: The performance measure is based on NHANES pesticide biomonitoring data published by CDC/NCHS. The data is used without making any additional transformations, so no assumption will be made to transform the data.

Timeframe of Result: NHANES is a continuous survey, sampling the U.S. population annually and releasing the data in 2-year cycles. As such, the span of time represented by the results represents a 2-year timeframe.

Unit of Measure: micrograms per liter (ug/L), at standard detection limits

Documentation of Methodological Changes: Not applicable.

4a. Oversight and Timing of Final Results Reporting:

Final Reporting Oversight Personnel: Branch Chief, Financial Management and Planning Branch

Final Reporting Timing: Performance results will be updated as NHANES data are published either in the official CDC report on human exposure to environmental chemicals or other journal articles or as the data become available.

4b. Data Limitations/Qualifications:

NHANES provides the most comprehensive biomonitoring data on the U.S. population. While it provides the most comprehensive data for evaluating national-level trends, there are some limitations that should be considered when evaluating the biomonitoring results. With regard to the general interpretation of biomonitoring data, CDC highlights that there are a number of factors that can influence the concentration of chemicals in urine. Some examples described by CDC include :

- Chemical half-life (i.e., persistence of chemical in blood or urine);
- Route of exposure;
- Genetic susceptibility;
- Demographic characteristics (e.g., age or gender);
- Health status and nutrition(e.g., reduced kidney function, iron deficiency);
- Lifestyle or behavioral factors (e.g., smoker versus non-smoker, or occupation); and
- Geography (e.g., proximity to environmental chemical sources, or climate).

In addition to these interpretive considerations, an important design limitation of the NHANES survey is that data are not publically available to evaluate seasonal and geographic trends. While this seasonal and geographic data are not available, EPA believes the data are suitable for the performance measure because EPA is interested in evaluating national-level trends between years.

References:

CDC (2010). Important Analytic Considerations and Limitations Regarding Environmental Chemical Data Analyses. Available at: <http://www.cdc.gov/nchs/tutorials/environmental/critical_issues/limitations/index.htm>

DATA LAG: Data lags may prevent performance results from being determined for every reporting year. Performance results will be updated as NHANES data are published either in the official CDC report on human exposure to environmental chemicals or other journal articles or as the data become available. There can be a substantial lag between CDC sampling and publication of data. For instance, in 2012, the most recently available data were from the sampling period of 2007-2008.

Methodological Changes: Not Applicable

4c. Third-Party Audits:

In 2009, the Board of Scientific Counselors (BSC) of NCHS commissioned a panel to review the NHANES as part of an ongoing program review process and to report its findings to the BSC (Available at: www.cdc.gov/nchs/data/bsc/NHANESReviewPanelReportrapril09.pdf The Panel concluded that NHANES simply must continue, and in a form that will fully sustain its historical importance as a key source of health information. With regard to the biomonitoring data that is the focus of EPA's performance measure, BSC did not make any specific recommendations that should impact the the methodology(ies), model(s), data, and information system(s) used to measure/collect/report performance. BSC emphasized the importance of the biomonitoring component of NHANES, stating:

Environmental monitoring is another major responsibility folded into the on-going NHANES. The survey's biomonitoring component documents exposure to environmental toxicants by direct measure of chemicals in blood or other biological specimens from individuals. The collection of blood and urine specimens in NHANES provides a unique opportunity for monitoring environmental exposure in the U.S. population....

Measure Code: J11 - Reduction in moderate to severe exposure incidents associated with organophosphates and carbamate insecticides in the general population.

Office of Chemical Strategies and Pollution Prevention (OCSPP)

Goal Number and Title:
4 - Ensuring the Safety of Chemicals and Preventing Pollution
Objective Number and Title:
1 - Ensure Chemical Safety
Sub-Objective Number and Title:
1 - Protect Human Health from Chemical Risks
Strategic Target Code and Title:
1 - By 2015, reduce percent of moderate to severe incidents affecting workers exposed to pesticides
Managing Office:
Office of Pesticide Programs
1a. Performance Measure Term Definitions:
Reduce: The American Association of Poison Control Centers (AAPCC) maintains a national database of exposure incidents called the National Poison Data System (NPDS), which is a compilation of data collected by AAPCC's national network of 61 poison controls centers (PCCs). The incident data maintained in AAPCC's NPDS includes pesticide-related exposure incidents that may occur throughout the U.S. population, including all age groups and exposures occurring in both residential and occupational settings. Summary data on pesticide-related incident data is reported on an annual basis in AAPCC's Annual Report, including the number of incidents by age, reason for exposure, level of medical treatment, and medical severity.   The performance measure is based on the annual number of moderate to severe organophosphate and carbamate exposure incidents, based on aggregated data reported in AAPCC's Annual Report. The baseline for the performance measure will be based on AAPCC's 2008 Annual Report.   Moderate to severe exposure incidents: Calls to Poison Control Centers are managed primarily by AAPCC-certified Specialists in Poison Information (SPIs). SPIs are required to complete detailed electronic medical records for both exposure and informational calls. Standardized definitions have been established to ensure database uniformity.   For EPA's performance measure, the determination of medical outcome will be used as an exclusion category for selecting incidents. All organophosphate and carbamate incidents designated as "Moderate" or "Major" will be included in the performance measure calculations.   Organophosphate and carbamate insecticides: AAPCC's Annual Report reports the number of annual incidents stratified by chemical category. Particular organophosphate and carbamate categories that will be used to identify incidents include: ·   "Organophosphate/ Carbamate/Chlorinated Hydrocarbon (Fixed- Combo)"

- "Carbamate Insecticides Alone"
- "Carbamate Insecticides in Combination with Other Insecticides"
- "Organophosphate   Insecticides Alone"
- "Organophosphate   Insecticides in Combination with Carbamate Insecticides"
- "Organophosphate   Insecticides in Combination with Non-Carbamate Insecticides"

General population: The general population means that the performance measure will focus on all exposure incidents reported to AAPCC, regardless of age.

Background:
- The reduction in poisoning incidents (i.e., moderate to severe exposure incidents) is expected to result from mitigation measures made during reregistration, from increased availability of lower risk alternative products resulting from the Agency's reduced risk registration process, and from the continued implementation of worker protection enforcement and training. Carbamates and organophosphates were selected for measurement because EPA anticipates recent registration activity will have a direct effect on reducing exposure in the general population. http://www.epa.gov/pesticides/about/types.htm

2a. Original Data Source:
NPDS is a comprehensive source of surveillance data on poisonings in the United States. NPDS is a uniform database of 61 PCCs, which are members of the American Association of Poison Control Centers (AAPCC), and are distributed throughout the United States. The database was established in 1985 and now includes information on more than 36 million exposure cases. In 2006, 61 PCCs received more than 4 million cases, including more than 2.4 million human exposure cases and 1.4 million informational calls. NPDS is a valuable public health resource and has been utilized to identify hazards, develop education priorities, guide clinical research, and identify chemical and bioterrorism incidents. As a result, NPDS has helped prompt product reformulations, recalls, and bans, support regulatory actions, and provide post-marketing surveillance of new drugs.

2b. Source Data Collection:
Individual PCC provides 24-hour emergency medical information on the diagnosis and treatment of poisonings. Calls are routed from a single, nationally-available phone number to the PCC generally in closest proximity to the caller. Since the service is provided on a national scale, even though PCCs may not be located in every state, aggregate PCC data is generally considered to be national in scope. The calls are managed primarily by AAPCC-certified Specialists in Poison Information (SPIs), who are typically pharmacists and nurses. SPIs are required to complete detailed electronic medical records for both exposure and informational calls. The electronic medical records include general demographic information, including age, gender, location of exposure, and more detailed information if an exposure may have occurred, including suspected substance, reason for exposure, route of exposure, management site, symptoms, and medical outcome. To assist SPIs and ensure database uniformity, many of the fields included in the electronic medical records use categories that have been defined by the AAPCC. For example, SPIs characterize the medical severity of possible exposures using the medical outcome field, which includes the AAPCC-defined categories "None," "Minor,"

"Moderate," "Major," or "Death." Additionally, the records may also contain several open fields, which allow SPIs to record additional information that may be relevant to the treatment and diagnosis of each case

2c. Source Data Reporting:
AAPCC produces the NPDS Annual Report giving statistics and information on all the poisonings in a calendar year. The NPDS Annual Report has three basic sections of information: general charts and statistics, a section of individual fatality listings, and a section listing demographic profile of single-substance exposure cases by generic category. The report is available to the general public to be downloaded for free and is usually made public the December following the close of a calendar year. This means the 2010 NPDS Annual Report was released around December of 2011. The report is typically published in the peer-reviewed journal Clinical Toxicology and is also publically available through AAPCC's website at: http://www.aapcc.org/dnn/NPDSPoisonData/NPDSAnnualreports.aspx

3a. Relevant Information Systems:
EPA does not require specialized information systems for the purposes of collecting, calculating, and/or reporting the results for this measure. Rather, AAPCC maintains standardized reporting procedures and is responsible for aggregating the summary data that is available in AAPCC's annual report and utilized in the performance measure. Following the publication of AAPCC's annual report, EPA uses MS-Excel to further summarize aggregated data on moderate to severe exposure incidents associated with organophosphate and carbamate insecticides   System Description: Not Applicable   Source/Transformed Data: Not Applicable   Information System Integrity Standards: Not Applicable

3b. Data Quality Procedures:
AAPCC's annual report reflects only those cases that are not duplicates and classified by the regional PC as CLOSED. A case is closed when the PC has determined that no further follow-up/recommendations are required or no further information is available. Exposure cases are followed to obtain the most precise medical outcome possible. Depending on the case specifics, most calls are "closed" within the first hours of the initial call. Some calls regarding complex hospitalized patients or cases resulting in death may remain open for weeks or months while data continues to be collected. Follow-up calls provide a proven mechanism for monitoring the appropriateness of management recommendations, augmenting patient guidelines, and providing poison prevention education, enabling continual updates of case information as well as obtaining final/known medical outcome status to make the data collected as accurate and complete as possible.

3c. Data Oversight:
Source Data Reporting Oversight Personnel: Not applicable.   Source Data Reporting Oversight Responsibilities: Not applicable.   Information Systems Oversight Personnel: Appointed Measures Representative(s) for Health Effects Division, in conjunction with the Division Director and Associate Division Director

Information Systems Oversight Responsibilities: To review and analyze data and report it to the OPP measures representative for reporting .

3d. Calculation Methodology:

Decision Rules for Selecting Data: The performance measure uses summary data from AAPCC's Annual Report. Specific incident data that will be selected for the performance measure "Moderate" and "Major" medical outcome incidents that involved the following AAPCC-defined chemical include:

- "Organophosphate/ Carbamate/Chlorinated Hydrocarbon (Fixed- Combo)"
- "Carbamate Insecticides Alone"
- "Carbamate Insecticides in Combination with Other Insecticides"
- "Organophosphate Insecticides Alone"
- "Organophosphate Insecticides in Combination with Carbamate Insecticides"
- "Organophosphate Insecticides in Combination with Non-Carbamate Insecticides"

Definitions of Variables: For EPA's performance measure, the determination of medical outcome will be used as an exclusion category for selecting incidents. All organophosphate and carbamate incidents designated as "Moderate" or "Major" will be included in the performance measure calculations. The AAPCC definitions for these medical outcome categories are defined below:

- Moderate effect: The patient exhibited signs or symptoms as a result of the exposure that were more pronounced, more prolonged, or more systemic in nature than minor symptoms. Usually, some form of treatment is indicated. Symptoms were not life-threatening, and the patient had no residual disability or disfigurement (e.g., corneal abrasion, acid-base disturbance, high fever, disorientation, hypotension that is rapidly responsive to treatment, and isolated brief seizures that respond readily to treatment).
- Major effect: The patient exhibited signs or symptoms as a result of the exposure that were life-threatening or resulted in significant residual disability or disfigurement (e.g., repeated seizures or status epilepticus, respiratory compromise requiring intubation, ventricular tachycardia with hypotension, cardiac or respiratory arrest, esophageal stricture, and disseminated intravascular coagulation).

Explanation of Calculations:

Annual performance will be evaluated using the equation below:

Where:

Baselinecount = Total number of exposure incidents that meet the case definition during the baseline period.

Performancecount = Total number of exposure incidents that meet the case definition during performance period.

Explanation of Assumptions: The performance measure is based on summary data published in AAPCC's Annual Report. The data is used without making any additional transformations, so no assumption will be made to transform the data.

Unit of Measure: Incident Count

Timeframe of Result: AAPCC's Annual Report is usually made public the December following the close of a calendar year. This means the 2010 NPDS Annual Report was released around December of 2011. Each report provides a summary of the total number of exposure incidents during the complete calendar year.

Documentation of Methodological Changes: Not Applicable

4a. Oversight and Timing of Final Results Reporting:

Branch Chief, Financial Management and Planning Branch

Measure is reported on a biennial basis

4b. Data Limitations/Qualifications:

General Limitations/Qualifications:

·   In general, PCC's provide medical management services through their response hotline and do not perform active surveillance of pesticide exposure incidents as part of NPDS. Due to this limitation, NPDS may be subject to reporting bias because of underreporting and differences in utilization rates among difference segments of the U.S. population.

·   Because the incidents are self-reported, there is a potential bias in the data. However, there is no reason to believe that the bias will change from year to year.

Data Lag Length and Explanation: AAPCC's Annual Report is published December of every year and made publicly available. For example, 2010 Annual Report was available to EPA in December 2011 and the 2011 Annual Report is expected to be available to EPA in December 2012.

Methodological Changes: Not Applicable

4c. Third-Party Audits:

AAPCC is an independent organization and not subject to third-party audits by the U.S. Government. AAPCC'S Annual Report is publically available (http://www.aapcc.org/dnn/NPDSPoisonData/NPDSAnnualReports.aspx and published in the peer-reviewed journal Clinical Toxicology.

Measure Code: E02 - Number of chemicals for which EDSP Tier 1 test orders have been issued

Office of Chemical Strategies and Pollution Prevention (OCSPP)

Goal Number and Title:
4 - Ensuring the Safety of Chemicals and Preventing Pollution

Objective Number and Title:
1 - Ensure Chemical Safety

Sub-Objective Number and Title:
1 - Protect Human Health from Chemical Risks

Strategic Target Code and Title:
0 -

Managing Office:
Office of Science Coordination and Policy

1a. Performance Measure Term Definitions:
Chemicals: The initial pesticide chemicals to be screened in the EDSP. EDSP Tier 1: The Endocrine Disruptor Screening Program's (EDSP) Tier 1 screening assays, which are designed to identify compounds that have the potential to interact with the body's endocrine system. Test orders: The initial issuance of orders to conduct EDSP Tier 1 screening tests to entities initially identified by EPA as being the producers of specific chemicals that may have the potential to interact with the estrogen, androgen, or thyroid hormone systems, all of which are part of the endocrine system. Issued: Issuance of EDSP Tier 1 test orders follows the policies and procedures that are described in detail in the Federal Register at 74FR17560. For the purpose of this measure, completing the issuance of Tier 1 test orders for a particular chemical will be defined as completing the initial issuance of orders to the order recipients initially identified by EPA. Subsequent issuance of orders to recipients who were not initially identified by EPA or to recipients who become subject to EDSP requirements after the initial issuance of test orders (referred to as "catch up" orders) will not be considered in this measure. As noted above, Background: ·   Tier 1 screening will include a battery of screening assays that would identify substances with the potential to interact with the estrogen, androgen, or thyroid hormone systems, according to text at http://www.epa.gov/endo/pubs/edspoverview/components.htm - 2 ·   Consistent with EPA plans to integrate the EDSP Tier 1 test orders into the pesticide registration review process, issuance of test orders for additional chemicals (including industrial chemicals that are water contaminants) is expected to continue in FY 2012 and beyond. ·   EPA anticipates that an increasing proportion of the resources allocated to the EDSP will be used for review of EDSP submissions of Tier 1 screening battery data in FY 2012. As a result, a measure based on the number of chemicals for which EDSP decisions have been completed captures an important shift in resource utilization for the program. ·   Given the dynamic nature of chemical markets, some companies may be missed in EPA's analysis or

companies may enter new markets subjecting them to the EDSP requirements for a chemical after the initial test orders for that chemical have been issued. EPA's policies and procedures allow for "catch up" orders to address these situations. Given that the time horizon for "catch up" orders is 15 years after the initial test orders are issued for a chemical, for purposes of this measure, a chemical will be counted as completed after initial test orders are issued.

· With EPA plans to integrate EDSP Tier 1 test orders into the pesticide registration review process and as EPA develops subsequent lists of chemicals, EPA anticipates that an increasing proportion of the EDSP resources will be used for the issuance of Tier 1 test orders and data review. Therefore, a measure based on the number of Tier 1 test orders issued captures performance of activities on which the program will be spending a larger proportion of its future resources.

· In general, it is anticipated that the EDSP decisions will vary from chemical to chemical with respect to complexity and timing. Therefore, careful analysis will be needed in setting performance targets each year. It is anticipated that annual performance targets will be established by considering (to the extent practicable) the number of chemicals for which EDSP Tier 1 test orders have been issued, the identity of the chemicals, the number of Tier 1 test order recipients, any other available chemical specific information and EPA resources available to complete data evaluations. However, several factors remain unpredictable and will impact the schedule for completing EDSP decisions. These include, for example, the number of pesticide cancellations and other regulatory actions that may remove a chemical from commerce and/or discontinue manufacture and import (voluntary and enforced), unforeseen laboratory capacity limits, and unforeseen technical problems with completing the Tier 1 assays for a particular chemical. Each of these factors can move the timeline for completing an EDSP decision for a particular chemical beyond the fiscal year in which the decision was originally anticipated.

· Annual performance targets for this measure will be subject to obtaining an approved Information Collection Request and the EPA resources available for issuing EDSP Tier 1 test orders.

· Annual performance targets may be influenced by a number of factors including OCSPP's identification of manufacturers of chemicals and the corresponding issuance of Information Collection Requests. Therefore, careful analysis will be needed in setting performance targets each year.

· The results from this performance measure, together with additional chemical specific information, will help set performance targets for another EDSP measure: the number of chemicals for which Endocrine Disruptor Screening Program (EDSP) decisions have been completed.

· Endocrine Disruptor Screening Program; Second List of Chemicals for Tier 1 Screening [Federal Register Notice: November 17, 2010 (Volume 75, Number 221, pages 70248-70254)]

· http://www.epa.gov/endo/ (including Highlights box on right side of page)

· http://www.epa.gov/endo/pubs/edspoverview/background.htm

http://www.epa.gov/endo/pubs/edsp_orders_status.pdf

2a. Original Data Source:
EPA staff, including scientists and regulatory managers from relevant program offices, are responsible for issuing and documenting the test orders.

2b. Source Data Collection:
Source Data Collection Methods: Using several databases, EPA initially completed a comprehensive analysis to identify companies that are

potential test order recipients because of their association with specific chemicals. These chemicals may have the potential to interact with the estrogen, androgen, or thyroid hormone systems.

The policies and procedures regarding issuance of EDSP Tier 1 test orders are described in detail in the Federal Register (74FR17560). The policies and procedures regarding issuance of EDSP Tier 1 test orders that are described in detail in the Federal Register (74FR17560) are being adapted to address additional chemicals (including water contaminants). EPA completes a comprehensive analysis using several databases to identify companies that are potential order recipients for each chemical.

2c. Source Data Reporting:

EPA has created and is maintaining an on-line report for tracking the status of chemicals screened in the EDSP (see Highlights box at http://www.epa.gov/endo The report includes for each chemical: the date a test order was issued, to whom the test order was issued, the due date for completing and submitting the data, the recipient's response to the order, and regulatory status (e.g., pesticide registration cancelled), as appropriate. In addition, the report will include information on EDSP Tier 1 decisions.

3a. Relevant Information Systems:

EPA has created and is maintaining an on-line report for tracking the number of chemicals for which EDSP Tier 1 test orders have been issued.

EPA's on-line report for tracking the status of chemicals screened in the EDSP includes for each chemical: the date a test order was issued, to whom the test order was issued, the due date for completing and submitting the data, the recipient's response to the order, regulatory status (e.g., pesticide registration cancelled), as appropriate, and other information.

Additional information:
Since the data generated for this measure will correspond to the on-line reporting on the status of chemicals in the EDSP, the public and other interested parties will be able to easily determine the accuracy of the reported results.

3b. Data Quality Procedures:

The number of chemicals for which Tier 1 test orders have been issued can be checked against order related documentation.

Data on the number of orders issued that are related to this measure will be reviewed for accuracy before submitting.

3c. Data Oversight:

Deputy Director, Office of Science Coordination and Policy

3d. Calculation Methodology:

Unit of analysis: Number of chemicals for which Endocrine Disruptor Screening Program (EDSP) Tier 1 test orders have been issued.

4a. Oversight and Timing of Final Results Reporting:

Planning and Accountability Lead in the Resource Management Staff in the Office of Program Management

Operations. Reporting semiannually: mid-year and end-of-year.

4b. Data Limitations/Qualifications:

Issuance of test orders is based largely on EPA actions, thus minimal error is anticipated with this estimate.

4c. Third-Party Audits:

The American Society of Human Genetics, the American Society for Reproductive Medicine, the Endocrine Society, the Genetics Society of America, the Society for Developmental Biology, the Society for Pediatric Urology, the Society for the Study of Reproduction, and the Society for Gynecologic Investigation. Assessing Chemical Risk: Societies Offer Expertise. Science, March 3, 2011 DOI: 10.1126/science.331.6021.1136-a or http://www.sciencemag.org/content/331/6021/1136.1

Measure Code: E01 - Number of chemicals for which Endocrine Disruptor Screening Program (EDSP) decisions have been completed

Office of Chemical Strategies and Pollution Prevention (OCSPP)

Goal Number and Title:
4 - Ensuring the Safety of Chemicals and Preventing Pollution

Objective Number and Title:
1 - Ensure Chemical Safety

Sub-Objective Number and Title:
1 - Protect Human Health from Chemical Risks

Strategic Target Code and Title:
6 - By 2015, complete Endocrine Disruptor Screening Program (EDSP)

Managing Office:
Office of Science Coordination and Policy

1a. Performance Measure Term Definitions:
Chemicals: Universe of chemicals to be screened in the EDSP. E01 EDSP Decisions Endocrine Disruptor Screening Program Decisions: EPA will measure the number of chemicals for which Endocrine Disruptor Screening Program decisions have been completed. EDSP chemical decisions span a broad range of decisions, inclusive of determining the potential to interact with the estrogen, androgen, or thyroid hormone systems to otherwise determining whether endocrine related testing is necessary. These decisions will take into consideration Tier 1 screening battery data reviews, other scientifically relevant information and/or the regulatory status of a chemical, as applicable. In addition, cancellations of all chemical manufacture or all pesticide chemical registered uses will be counted as regulatory decisions, as long as the chemical resides in the EDSP chemical universe, as defined under FFDCA 408 (p) and SDWA section 1457. The inclusion of these types of decisions for this measure will be counted irrespective of whether it is aligned with the formal listing (e.g., list 1, 2, etc.) of that chemical for EDSP screening and the temporal decision of cancellation or cease of all manufacturing decision(s). This measure is a count of completed decisions that include: ·   EDSP decisions for a particular chemical (in Tier 1) can be organized into two broad categories: (1) regulatory actions and (2) determinations regarding potential to interact with E, A, or T. In both cases, the decisions will determine whether further endocrine related testing is necessary for that chemical. The number of chemicals which EPA judges to have been fully assessed using the Endocrine Disruptor Screening Program's (EDSP) Tier 1 screening assays and other scientifically relevant information (as applicable) for their potential to interact with E, A, or T, will be counted as completed for this measure. ·   There are several regulatory actions that will remove a chemical from further consideration for endocrine related testing in the EDSP. These include, but would not be limited to, cancellation of pesticide registrations, ceasing sales of the chemical for use in pesticide products, and discontinuing the manufacture

and import of the chemical. These actions may be voluntary on the part of a Tier 1 test order recipient or the result of an EPA regulatory determination. In either case, when such regulatory decisions have been completed for a chemical that decision will be counted for this measure.

·   Chemicals are also counted for this measure when EPA makes a regulatory determination to remove or exempt a chemical from further consideration for endocrine related testing in the EDSP. The decision to exempt a biologic substance or other substance under FFDCA Section 408 (p), section 4 from EDSP screening will be determined on a case by case basis, dependent on the weight of evidence supporting "...the determination that the substance is anticipated not to produce any effect in humans similar to an effect produced by a naturally occurring estrogen."

Background:
·   EPA anticipates that an increasing proportion of the resources allocated to the EDSP will be used for technical and scientific review of EDSP submissions of Tier 1 screening battery data in FY 2013. As a result, a measure based on the number of chemicals for which EDSP decisions have been completed captures an important shift in resource utilization for the program.

·   In general, it is anticipated that the EDSP decisions will vary from chemical to chemical with respect to scientific complexity and timing. Therefore, careful analysis will be needed in setting performance targets each year. It is anticipated that annual performance targets will be established by considering (to the extent practicable) the number of chemicals for which EDSP Tier 1 test orders have been issued, the identity of the chemicals, the number of Tier 1 test order recipients, volume and type of available chemical specific information and EPA resources available to complete the extensive and rigorous data evaluations. However, several factors remain unpredictable and will likely impact the schedule for completing EDSP decisions. These include, for example, the number chemicals and timing for which test orders will be issued, number of pesticide cancellations and other regulatory actions that may remove a chemical from commerce and/or discontinue manufacture and import (voluntary and enforced), unforeseen laboratory capacity limits, and unforeseen technical problems with completing the Tier 1 assays for a particular chemical. Each of these factors can move the timeline for completing an EDSP decision for a particular chemical beyond the fiscal year in which the decision was originally anticipated.

·   This performance measure is best used in conjunction with another EDSP annual performance measure (Number of chemicals for which EDSP scientific weight of evidence reviews have been completed). Measuring the number of completed EDSP scientific weight of evidence reviews in combination with the number of EDSP Tier 1 assay data reviews, will help anchor the performance targets for the number of chemicals for which EDSP decisions have been completed.

·   Endocrine Disruptor Screening Program; Second List of Chemicals for Tier 1 Screening [Federal Register Notice: November 17, 2010 (Volume 75, Number 221, pages 70248-70254)]
http://www.regulations.gov/contentStreamer?disposition=attachment&objectId=0900006480b954bf&contentType=ht

ml
2a. Original Data Source:
EPA staff, including scientists and regulatory managers from relevant program offices, are responsible for making and documenting the decisions, as described in the Comprehensive Management Plan (see below).
2b. Source Data Collection:
Source Data Collection Methods: The decisions will take into consideration Tier 1 screening battery data, other scientifically relevant information (OSRI), and/or the regulatory status of a chemical, as applicable. EPA has developed guidance on how to conduct the Weight of Evidence (WoE) analysis that will lead to decisions about whether chemicals have the potential to interact with E, A, or T and whether further testing will be required (see Highlights box at http://www.epa.gov/endo EPA has also recently developed and issued the agency's Endocrine Disruptor Screening Program Comprehensive Management Plan in June 2012 that describes the strategic guidance for how the agency plans to implement the program over the next five years and includes key activities described and captured in the performance measures.(http://www.epa.gov/endo Date/time intervals covered by source data: FY 2013-present
2c. Source Data Reporting:
Form/mechanism for receiving data and entering into EPA system: EPA has created and is maintaining an on-line report for tracking the status of chemicals screened in the EDSP (see Highlights box at http://www.epa.gov/endo The report includes for each chemical: the date a test order was issued, to whom the test order was issued, the due date for completing and submitting the data, the recipient's response to the order, and regulatory status (e.g., pesticide registration cancelled), as appropriate. In addition, the report will include information on EDSP decisions. Decisions will be counted once EPA announces them via updates to the EDSP website (http://www.epa.gov/endo/pubs/EDSP_OSRI_Response_Table.pdf Timing and frequency of reporting: Annual
3a. Relevant Information Systems:
EPA has created and is maintaining an on-line report for tracking the status of chemicals screened in the EDSP (see Highlights box at http://www.epa.gov/endo The report includes for each chemical: the date a test order was issued, to whom the test order was issued, the due date for completing and submitting the data, the recipient's response to the order, and regulatory status (e.g., pesticide registration cancelled), as appropriate. In addition, the report will include information on EDSP decisions. EPA anticipates expanding this report to include chemicals other than pesticides. Additional information: Since the data will correspond to the on-line reporting on the status of chemicals in the EDSP, the public and other interested parties will be able to easily determine the accuracy of the reported results.

3b. Data Quality Procedures:	
Data on the number of decisions generated for this measure will be reviewed for accuracy before submitting. The number of chemicals for which EDSP Tier 1 decisions have been completed can be checked against supporting records documenting the decisions.	
3c. Data Oversight:	
Deputy Director, Office of Science Coordination and Policy	
3d. Calculation Methodology:	
Unit of analysis: Number of chemicals for which Endocrine Disruptor Screening Program (EDSP) Tier 1 decisions have been completed.	
4a. Oversight and Timing of Final Results Reporting:	
Planning and Accountability Lead in the Resource Management Staff in the Office of Program Management Operations. Reporting semiannually: mid-year and end-of-year.	
4b. Data Limitations/Qualifications:	
Decisions are based on EPA regulatory actions and data review once data are received, thus minimal error is anticipated with this estimate.	
4c. Third-Party Audits:	
The American Society of Human Genetics, the American Society for Reproductive Medicine, the Endocrine Society, the Genetics Society of America, the Society for Developmental Biology, the Society for Pediatric Urology, the Society for the Study of Reproduction, and the Society for Gynecologic Investigation. Assessing Chemical Risk: Societies Offer Expertise. Science, March 3, 2011 DOI: 10.1126/science.331.6021.1136-a or http://www.sciencemag.org/content/331/6021/1136.1	

Measure Code: C18 - Percentage of existing CBI claims for chemical identity in health and safety studies reviewed and challenged, as appropriate.

Office of Chemical Strategies and Pollution Prevention (OCSPP)

Goal Number and Title:
4 - Ensuring the Safety of Chemicals and Preventing Pollution

Objective Number and Title:
1 - Ensure Chemical Safety

Sub-Objective Number and Title:
3 - Ensure Transparency of Chemical Health and Safety Information

Strategic Target Code and Title:
1 - 2015, make all health and safety information available to the public for chemicals in commerce

Managing Office:
Office of Pollution Prevention and Toxics

1a. Performance Measure Term Definitions:
Existing CBI claims: Under TSCA, companies may claim that information they submit to EPA should be treated as "confidential business information" (CBI) and not be disclosed to the public. "Existing" CBI claims are CBI claims in TSCA filings in the possession of the Agency as of August 21, 2010. Health and safety studies: EPA has initiated a general practice of reviewing confidentiality claims for chemical identities in health and safety studies, and in data from health and safety studies, submitted under the Toxic Substances Control Act (TSCA) in accordance with Agency regulations at 40 CFR part 2, subpart B. The term "Health and safety study or study" is defined in various regulatory provisions promulgated under TSCA. For example, in 40 C.F.R. 720.3(k), health and safety study is defined as "any study of any effect of a chemical substance or mixture on health or the environment or on both, including underlying data and epidemiological studies, studies of occupational exposure to a chemical substance or mixture, toxicological, clinical, and ecological, or other studies of a chemical substance or mixture, and any test performed under the Act. Chemical identity is always part of a health and safety study." CBI cases: Each filing submitted to EPA under TSCA is assigned a case number by EPA which indicates the type of filing, the fiscal year in which the submission was made, and a numeric identifier. Any subsequent submissions received relating to the original submission are assigned the same case number. For purposes of the Performance Measure, the term "CBI cases" refers to all documents submitted that have been assigned a unique case number identifying the original filing containing a CBI claim for the chemical identity of the chemical substance or mixture addressed in that filing. Reviewed and, as appropriate, challenged: To achieve this measure, EPA will complete the following actions for all new and historical submissions by the end of 2015: 1) determine if a challenge to the CBI claim is warranted; 2) execute the challenge if warranted; and 3) where legally defensible, declassify the information claimed as CBI. Section 14(b) of TSCA does not extend confidential treatment to health and safety studies, or data from health and safety studies, which, if made public, would not disclose processes used in the manufacturing or processing of a chemical substance or mixture or, in the case of a mixture, would not

disclose the portion of the mixture comprised by any of the chemical substances in the mixture. Where a chemical identity does not explicitly contain process information or reveal portions of a mixture, EPA expects to find that the information would clearly not be entitled to confidential treatment. Where EPA determines that the information is not eligible for confidential treatment, the Agency will notify companies, and in those instances where the company will not voluntarily relinquish the claims, EPA may initiate administrative action consistent with Section 14 of TSCA.

Background:

This performance measure supports EPA's strategic measure through 2015 to make all health and safety studies available to the public for chemicals in commerce, to the extent allowed by law. A similar strategic measure appeared in EPA's Strategic Plan for FY 2011-2015. For pesticides, EPA will continue to make risk assessments and supporting information available through its long standing Public Participation Process.

The effort has involved a renewed focus on companies' practices in submitting CBI claims, first announced on January 21, 2010, when EPA said it planned to reject CBI claims for chemicals submitted to EPA with studies that show a substantial risk to people's health and the environment, where the subject chemicals were previously disclosed on the TSCA Chemical Inventory. In a follow-up communication issued May 27, 2010, EPA said it planned to generally deny confidentiality claims for the identity of chemicals in health and safety studies filed under TSCA, except in specified circumstances.

For more information, please see:
(1) http://www.epa.gov/oppt/tsca8e/pubs/confidentialbusinessinformation.html
(2) http://www.epa.gov/oppt/existingchemicals/pubs/transparency.html
(3) http://www.regulations.gov/ - !documentDetail;D=EPA-HQ-OPPT-2010-0446-0001

2a. Original Data Source:

Data are provided by EPA Headquarters Staff.

2b. Source Data Collection:

Historical data used to identify existing CBI health and safety data will come from staff and contractor maintained internal databases. The Agency relies on existing databases that track TSCA filings and identify data elements for each document, including CBI claims. These databases are used for a wide variety of purposes relating to TSCA implementation. Quality controls include standard operating procedures related to data processing. This process is enhanced, in the CBI review context, by reviews of the actual data (hard copy, microfiche and pdf) to ensure that what is tracked is consistent with the submitted filings.

2c. Source Data Reporting:

EPA receives information by paper or electronic submission under the authority of TSCA. CBI reviews are initiated under specific regulatory authority (e.g. 40 CFR 720, 40 CFR part 2 etc). TSCA submitters (e.g., companies that manufacture industrial chemicals) transmit various types of information required under TSCA electronically or by mail to the TSCA Confidential Business Information Center (CBIC) at EPA Headquarters. These filings are submitted according to the various provisions of TSCA and applicable regulations. There are no specific reporting requirements directly related to the CBI review initiatives nor is there a regular reporting period with beginning and ending dates as EPA receives materials pursuant to TSCA on a daily basis throughout the year.

3a. Relevant Information Systems:

Information about TSCA submissions is entered into internal agency data tracking databases. For purposes of the performance measure, OPPT retrieves information from these databases regarding cases that meet relevant criteria for review. A listing of cases identified for review is provided to various divisions within OPPT to complete particular elements of the case reviews. The results of these reviews are entered into various informal data repositories, including Excel spreadsheets and Access databases. The information from these sources is compiled into two separate Access databases from which various reports are generated. The data stored in these databases include the records identified for review, date of receipt, review status, claim validation, letter or call sent, 2.204(d)(2) Action, and declassification status. For chemicals in 8(e) filings the database will also track if the chemical name has process or portion of mixture information and if it is claimed as research and development (R&D) or as a pesticide. Data elements used to track the cases for which CBI claims are reviewed consist of: (1) new process-specific elements entered by reviewers and (2) elements associated with health and safety studies and previously entered into OPPT databases. The databases are limited to tracking filings, CBI review status and other data element consistent with internal management of the program. The databases do not contain any transformed data.

Given the nature of the information processes used for this measure, there are no formal Information System Lifecycle Management policies or other Information System Integrity Standards that apply.

3b. Data Quality Procedures:

EPA reviews all subject filings with CBI claims to ensure that such claims are consistent with provisions of statute and Agency policy. Participants in the review include legal and technical staff persons who rely on advice from EPA's Office of General Counsel.

3c. Data Oversight:

Chief, Planning and Assessment Branch, Environmental Assistance Division, OPPT

3d. Calculation Methodology:

Between the date when TSCA was enacted and August 21, 2010, a total of 22,483 CBI cases potentially containing TSCA health and safety information were submitted for chemicals potentially in commerce. In recent years, hundreds of such cases have been submitted annually.

EPA has identified all filings and cases subject to review and placed these in a tracking database. As filings and cases are reviewed and appropriate actions undertaken, the database is updated to capture the new status. Thus, the number of CBI cases that are reviewed and challenged can be readily obtained from the database and expressed as a percentage of all existing CBI health and safety study cases.

The unit of measure in which the performance result is expressed is CBI cases and the timeframe represented in the performance result extends from the first day through the last day of the relevant fiscal year. Variables used in this measure are defined in Section 1a above.

4a. Oversight and Timing of Final Results Reporting:
Planning and Accountability Lead in the Resource Management Staff in the Office of Program Management Operations. Reporting semiannually: mid-year and end-of-year.
4b. Data Limitations/Qualifications:
Data limitations include: ·   Some archived data may have been lost or damaged. ·   The DTS database does not differentiate between types of CBI claims, so some studies tracked in the DTS system may, in theory, already be public. ·   Some submissions may be redundant due to overlap in processing. ·   The number of errors that could have been made during data entry is not known.
4c. Third-Party Audits:
OIG published a report in 2010 finding that CBI claims were excessive, and encouraging EPA to increase public access to information filed under TSCA. For more information, see http://www.epa.gov/oig/reports/2010/20100217-10-P-0066.pdf

Measure Code: C19 - Percentage of CBI claims for chemical identity in health and safety studies reviewed and challenged, as appropriate, as they are submitted.

Office of Chemical Strategies and Pollution Prevention (OCSPP)

Goal Number and Title:
4 - Ensuring the Safety of Chemicals and Preventing Pollution
Objective Number and Title:
1 - Ensure Chemical Safety
Sub-Objective Number and Title:
3 - Ensure Transparency of Chemical Health and Safety Information
Strategic Target Code and Title:
1 - 2015, make all health and safety information available to the public for chemicals in commerce
Managing Office:
Office of Pollution Prevention and Toxics
1a. Performance Measure Term Definitions:
CBI claims: Summarized, under TSCA, companies may claim that information they submit to EPA should be treated as "confidential business information" (CBI) and not be disclosed to the public except as authorized under the law. This performance measure addresses EPA's review of "new" CBI claims for chemical name in health and safety studies. Such claims are sometimes made in TSCA section 5 and TSCA section 8 filings.   Health and safety studies: EPA has initiated a general practice of reviewing confidentiality claims for chemical identities in health and safety studies, and health and safety data from health and safety studies, submitted pursuant to the Toxic Substances Control Act (TSCA). The term "Health and safety study or study" is defined in various regulatory provisions promulgated under TSCA. For example, in 40 C.F.R. 720.3(k), health and safety study is defined as "any study of any effect of a chemical substance or mixture on health or the environment or on both, including underlying data and epidemiological studies, studies of occupational exposure to a chemical substance or mixture, toxicological, clinical, and ecological, or other studies of a chemical substance or mixture, and any test performed under the Act. Chemical identity is always part of a health and safety study."   CBI cases: Each filing submitted to EPA under TSCA is assigned a case number by EPA which indicates the type of filing, the fiscal year in which the submission was made, and a numeric identifier. Any subsequent submissions received relating to the original submission are assigned the same case number. For purposes of the Performance Measure, the term "CBI cases" refers to all documents submitted that have been assigned a unique case number identifying the original filing containing a CBI claim for the chemical identity of the chemical substance or mixture addressed in that filing.
Reviewed and, as appropriate, challenged: To achieve this measure, EPA must complete the following actions for all new and historical submissions by the end of 2015: 1) determine if a challenge to the CBI claim is warranted; 2) execute the challenge if warranted; and 3) where legally authorized, declassify the information claimed as CBI. Section 14(b) of TSCA does not extend confidential treatment to health and safety studies, or data from health and safety studies, which, if made public, would not disclose processes used in the

manufacturing or processing of a chemical substance or mixture or, in the case of a mixture, would not disclose the portion of the mixture comprised by any of the chemical substances in the mixture. Where a chemical identity does not explicitly contain process information or reveal portions of a mixture, EPA expects to find that the information would clearly not be entitled to confidential treatment. Where EPA determines that the information is not eligible for confidential treatment, the Agency will notify companies, and in those instances where the company will not voluntarily relinquish the claims, EPA may initiate administrative action consistent with Section 14 of TSCA and applicable regulations.

Background:

This performance measure supports EPA's strategic measure through 2015 to make all health and safety studies available to the public for chemicals in commerce, to the extent allowed by law. A similar strategic measure appeared in EPA's Strategic Plan for FY 2011-2015. For pesticides, EPA will continue to make risk assessments and supporting information available through its long standing Public Participation Process.

The effort has involved the renewed focus on companies' practices in submitting CBI claims, first announced on January 21, 2010, when EPA said it planned to reject CBI claims for chemicals submitted to EPA accompanied by studies that show a substantial risk to people's health and the environment, where the subject chemicals were previously disclosed on the TSCA Chemical Inventory. In a follow-up communication of May 27, 2010, EPA said it planned to generally deny confidentiality claims for the identity of chemicals in health and safety studies filed under TSCA, except in specified circumstances.

For more information, please see:
(1) http://www.epa.gov/oppt/tsca8e/pubs/confidentialbusinessinformation.html
(2) http://www.epa.gov/oppt/existingchemicals/pubs/transparency.html
(3) http://www.regulations.gov/ - !documentDetail;D=EPA-HQ-OPPT-2010-0446-0001

2a. Original Data Source:
Data are provided by EPA Headquarters Staff.

2b. Source Data Collection:
The Agency relies on existing databases that track TSCA filings and identify data elements for each document, including CBI claims. These databases are used for a wide variety of purposes relating to TSCA implementation. Quality controls include standard operating procedures related to data processing. This process is enhanced, in the CBI review context, by reviews of the actual data to ensure that what is tracked is consistent with the submitted filings.

2c. Source Data Reporting:
EPA receives information by paper or electronic submission under the authority of TSCA. CBI reviews are initiated under specific regulatory authority (e.g. 40 CFR part 2 etc). TSCA submitters (e.g., companies that manufacture industrial chemicals) transmit various types of information required under TSCA electronically or by mail to the TSCA Confidential Business Information Center (CBIC) at EPA Headquarters. These filings are submitted according to the various provisions of TSCA and applicable regulations. There are no specific

reporting requirements directly related to the CBI review initiatives nor is there a regular reporting period with beginning and ending dates as EPA receives materials pursuant to TSCA on a daily basis throughout the year.

3a. Relevant Information Systems:

Information about TSCA submissions is entered into the Chemical Information System (CIS) by Agency staff and contractors. For purposes of the performance measure, OPPT retrieves information from CIS regarding cases that meet relevant criteria for review. A listing of cases identified for review is provided to various divisions within OPPT to complete particular elements of the case reviews. The results of these reviews are entered into various data repositories, including Excel spreadsheets and Access databases. The material from these sources is compiled into two separate Access databases from which various reports are generated. The data stored in these databases include the records identified for review, date of receipt, review status, claim validation, letter or call sent, 2.204(d)(2) Action, and declassification status. For chemicals in 8(e) filings the database will also track if the chemical name has process or portion of mixture information and if it is claimed as research and development (R&D) or as a pesticide. Data elements used to track the cases for which CBI claims are reviewed consist of: (1) new process-specific elements entered by reviewers and (2) elements associated with health and safety studies and previously entered into OPPT databases. The databases are limited to tracking filings, CBI review status and other data element consistent with internal management of the program. The databases do not contain any transformed data.

Given the nature of the information processes used for this measure, there are no formal Information System Lifecycle Management policies or other Information System Integrity Standards that apply.

3b. Data Quality Procedures:

EPA reviews all subject filings with CBI claims to ensure that such claims are consistent with provisions of statute and Agency policy. Participants in the review include legal and technical staff persons who rely on advice from EPA's Office of General Counsel.

3c. Data Oversight:

Chief, Planning and Assessment Branch, Environmental Assistance Division, OPPT

3d. Calculation Methodology:

EPA's experience in recent years has been that hundreds of CBI cases potentially containing TSCA health and safety information are submitted annually. The agency identifies all filings and cases subject to review and places them in an internal tracking database. As filings and cases are reviewed and appropriate actions undertaken, the database is updated to capture the new status. Thus, the number of CBI cases that are reviewed and, as appropriate, challenged can be readily obtained from the database and expressed as a percentage of all existing CBI health and safety study cases.

The unit of measure in which the performance result is expressed is CBI cases and the timeframe represented in the performance result extends from the first day through the last day of the relevant fiscal year. Variables used in this measure are defined in Section 1a above.

4a. Oversight and Timing of Final Results Reporting:

Planning and Accountability Lead in the Resource Management Staff in the Office of Program Management

Operations. Reporting semiannually: mid-year and end-of-year.
4b. Data Limitations/Qualifications:
Data limitations include: ·   Some submissions may be redundant due to overlap in processing. ·   The number of errors that could have been made during data entry is not known.
4c. Third-Party Audits:
OIG published a report in 2010 finding that CBI claims were excessive, and encouraging EPA to increase public access to information filed under TSCA. For more information, see http://www.epa.gov/oig/reports/2010/20100217-10-P-0066.pdf

Measure Code: D6A - Reduction in concentration of PFOA in serum in the general population.

Office of Chemical Strategies and Pollution Prevention (OCSPP)

Goal Number and Title:
4 - Ensuring the Safety of Chemicals and Preventing Pollution

Objective Number and Title:
1 - Ensure Chemical Safety

Sub-Objective Number and Title:
1 - Protect Human Health from Chemical Risks

Strategic Target Code and Title:
4 - By 2014, reduce the disparity of concentration of chemicals in low income populations

Managing Office:

1a. Performance Measure Term Definitions:
PFOA: Perfluorooctanoic Acid

PFOA: Perfluorooctanoic Acid

Reduction in serum concentration: The percent reduction in the blood serum concentration of PFOA among the U.S. general population from the baseline year to a subsequent reporting year, as estimated from sample data obtained from the periodic National Health and Nutrition Examination Surveys (NHANES) administered by the Centers for Disease Control and Prevention (CDC).

General population: The population of all adults and children ages 12 years and older living in the United States, excluding those who are institutionalized.

Background: Perfluorooctanoic acid (PFOA) is a synthetic chemical that does not occur naturally in the environment. Companies use PFOA to make fluoropolymers, substances with special properties that have hundreds of important manufacturing and industrial applications. They are used to provide non-stick surfaces on cookware and waterproof, breathable membranes for clothing, and are used in many industry segments, including the aerospace, automotive, building/construction, chemical processing, electronics, semiconductors and textile industries. PFOA can also be produced by the breakdown of some fluorinated telomers, substances that are used in surface treatment products to impart soil, stain, grease and water resistance.

The EPA began investigating PFOA in the 1990s, following an investigation of perfluorooctyl sulfonates (PFOS), a similar group of chemicals found to present concerns for persistence, bioaccumulation and toxicity. The research concluded that PFOA, too, is very persistent in the environment, is found at very low levels in the environment and in the blood of the general U.S. population, and causes developmental and other adverse effects in laboratory animals. In response to these findings, the agency is pursuing several complementary approaches to encourage reduction in PFOA use. For instance, an action plan to address Long-Chain Perfluorinated Chemicals (LCPFCs), including PFOA, is being implemented and a number of potential chemical substitutes are being explored.

This performance measure tracks the extent to which reductions in the serum concentration of PFOA among U.S. residents are actually being achieved. Progress is measured as the percent reduction from the 2005-2006

baseline serum concentration as estimated through NHANES.

2a. Original Data Source:

The original data source is the Centers for Disease Control and Prevention's (CDC) National Health and Nutrition Examination Survey (NHANES), which is recognized as the primary database in the United States for national biomonitoring statistics on blood and urine concentrations of potentially harmful chemicals such as PFOA, http://www.cdc.gov/nchs/nhanes/about_nhanes.htm NHANES is a probability sample of the non-institutionalized population of the United States. The survey examines a nationally representative sample of approximately 5,000 men, women, and children each year located across the U.S.

2b. Source Data Collection:

Methods of data collection (by original data source): Data are obtained by analysis of blood and urine samples collected from survey participants. Health status is assessed by physical examination. Demographic and other survey data regarding health status, nutrition, and health-related behaviors are collected by personal interview, either by self-reporting or, for children under 16 and some others, as reported by an informant. Detailed interview questions cover areas related to demographic, socio-economic, dietary, and health-related questions. The survey also includes an extensive medical and dental examination of participants, physiological measurements, and laboratory tests. NHANES is unique in that it links laboratory-derived biological markers (e.g. blood, urine etc.) to questionnaire responses and results of physical exams.

Quality procedures followed (by original data source): According to the CDC, the process of preparing NHANES data sets for release is as rigorous as other aspects of the survey. After a CDC contractor performs basic data cleanup, the CDC NHANES staff ensure that the data are edited and cleaned prior to release. NHANES staff devotes at least a full year after the completion of data collection to careful data preparation. Additionally, NHANES data are published in a wide array of peer-reviewed professional journals.

Background documentation is available at the NHANES Web site at http://www.cdc.gov/nchs/nhanes.htm The analytical guidelines are available at the Web site http://www.cdc.gov/nchs/about/major/nhanes/nhanes2003-2004/analytical_guidelines.htm

Geographical extent of source data, if relevant: Data are collected to be representative of the U.S. population. The population data are extrapolated from sample data by the application of standard statistical procedures.

Spatial detail of source data, if relevant: NHANES sampling procedures provide nationally representative data.

2c. Source Data Reporting:

Form/mechanism for receiving data and entering into EPA system: EPA monitors the periodic issuance of NHANES reports and other data releases to obtain the data relevant to this measure.

Timing and frequency of reporting: NHANES is a continuous survey and examines a nationally representative sample of about 5,000 persons each year. These persons are located in counties across the country, 15 of which are visited each year.

Files of raw data, containing measured serum concentrations of PFOA among NHANES participants, are currently released to the public in two-year sets. CDC also periodically publishes reports containing summary statistics for PFOA and more than 200 other chemicals measured in NHANES, at www.cdc.gov/exposurereport

3a. Relevant Information Systems:

There are no EPA systems utilized in collecting data for this measure as the Agency is able to secure the necessary data directly from NHANES reports and data releases.

3b. Data Quality Procedures:

EPA does not have any procedures for quality assurance of the underlying data as this function is performed by the CDC itself. CDC has periodically reviewed and confirmed EPA's calculation of NHANES summary statistics from the raw data files. The Agency determines the performance result for this measure either directly from the NHANES data or by performing simple arithmetical calculations on the data.

3c. Data Oversight:

Chief, Planning and Assessment Branch, Environmental Assistance Division, Office of Pollution Prevention and Toxics

3d. Calculation Methodology:

Decision rules for selecting data: EPA uses the blood serum concentration values for PFOA that are generated by the NHANES surveys. Values from all NHANES participants with serum PFOA measurements in a two-year NHANES cycle are used (along with the appropriate NHANES sample weights) in calculating the geometric mean concentrations.

Definitions of variables: Key terms are defined in 1(a) above.

Explanation of the calculations: Not applicable. Performance results obtained from NHANES.

Explanation of assumptions: Not applicable for the same reason as above.

Identification of unit of measure: Geometric mean serum concentration (in ug/L)

Identification of timeframe of result: The performance result is computed from data released by the CDC in sets covering the particular time period over which sampling occurs. Thus, the timeframe that applies to the measured result is the same period for which the NHANES data are released. It is not a simple snapshot at a specific moment in time.

4a. Oversight and Timing of Final Results Reporting:

Planning and Accountability Lead in the Resource Management Staff in the Office of Program Management Operations. Reporting semiannually: mid-year and end-of-year, but subject to a data lag due to the periodic nature of NHANES reporting.

4b. Data Limitations/Qualifications:

NHANES is a voluntary survey and selected persons may refuse to participate. In addition, the NHANES survey uses two steps, a questionnaire and a physical exam. There are sometimes different numbers of subjects in the interview and examinations because some participants only complete one step of the survey. Participants may answer the questionnaire but not provide the more invasive blood sample. Special weighting techniques are used to adjust for non-response. NHANES is not designed to provide detailed estimates for populations that are highly exposed to PFOA.

4c. Third-Party Audits:

Report of the NHANES Review Panel to the NCHS Board of Scientific Counselors.

Cover letter can be accessed at: http://www.cdc.gov/nchs/data/bsc/bscletterjune8.pdf
Report can be accessed at: http://www.cdc.gov/nchs/data/bsc/NHANESReviewPanelReportrapril09.pdf

Measure Code: HC1 - Annual number of hazard characterizations completed for HPV chemicals

Office of Chemical Strategies and Pollution Prevention (OCSPP)

Goal Number and Title:
4 - Ensuring the Safety of Chemicals and Preventing Pollution
Objective Number and Title:
1 - Ensure Chemical Safety
Sub-Objective Number and Title:
1 - Protect Human Health from Chemical Risks
Strategic Target Code and Title:
0 -
Managing Office:
Office of Pollution Prevention and Toxics
1a. Performance Measure Term Definitions:
Hazard characterizations: "Hazard characterizations" refers to Screening Level Hazard Characterization Reports prepared by EPA staff based on information submitted by the companies that make the chemicals, as well as on data identified from a targeted search of publicly available sources of information specifically relevant to characterizing hazards. Completed: Screening Level Hazard Characterization Reports are deemed "completed" once they are deemed by senior Agency scientists and OPPT management to be suitable for posting on the program's website. In order for reports to be completed, the source Screening Information Data Set data submissions must be judged by the Agency to be adequate. HPV chemicals: High Production Volume chemicals produced or imported in the United States in quantities of 1 million pounds or more per year. Background: · EPA's High Production Volume Challenge (HPV Challenge) program has inspired chemical manufacturers and users to deliver health and environmental effects data on many of the most heavily used chemicals in U.S. commerce to the agency. More information is available at: http://www.epa.gov/hpv/ · EPA is investigating the hazard characteristics of heavily used chemicals in conjunction with the Organization for Economic Cooperation and Development (OECD). The OECD's criteria for including chemicals in its Screening Information Data Sets (SIDS) program are production in one OECD Member country in quantities above 10,000 metric tons (22 million lbs) per annum or above 1,000 metric tons (2.2 million lbs) in two or more OECD countries. More information is available at http://www.epa.gov/opptintr/sids/pubs/overview.htm Screening Level Hazard Characterization Reports are supplemented and aligned twice a year with the international database of chemicals sponsored internationally through Screening Information Data Sets (SIDs) Initial Assessment Meetings. Hazard characterizations are made publicly available through OPPT's High

Production Volume Information System (HPVIS): http://www.epa.gov/hpvis/
2a. Original Data Source:
Submissions from chemical sponsors, for both U.S. HPVs and international Screening Information Data Sets (SIDs) chemicals.
2b. Source Data Collection:
Tabulation of records or activities: Screening Level Hazard Characterization Reports are prepared by EPA staff based on submissions from chemical sponsors and are reviewed by senior scientists and management to determine whether they are complete. Each screening level hazard characterization document represents a thorough review by qualified EPA personnel of the information provided by the submitter, as well as other targeted sources of information. For more information about sources utilized, please visit: http://www.epa.gov/hpvis/hazardinfo.htm This measure analyzes and supplements data received through EPA's High Production Volume (HPV) challenge, the EPA program that has inspired companies to deliver health and environmental effects data on many of the most heavily used chemicals in U.S. commerce to the agency. An assessment of adequacy is made for HPV chemicals, defined as approximately 2,450 chemicals (1400 US Sponsored chemicals, 850 International sponsored chemicals, and 200 Original Organization for Economic Cooperation and Development (OECD) SIDS Initial Assessment Reports (SIARs)). The measure is a count of completed reports from all of these sources, which are then posted on EPA's website. http://www.epa.gov/hpvis/abouthc.htm EPA QA requirements/guidance governing collection: OPPT has in place a signed Quality Management Plan (Quality Management Plan for the Office of Pollution Prevention and Toxics; Office of Prevention, Pesticides and Toxic Substances, November 2008).
2c. Source Data Reporting:
Form/mechanism for receiving data and entering into EPA system: EPA staff complete Screening Level Hazard Characterization Reports based on submissions from chemical sponsors. Once a report is completed, as determined by senior scientist and management review, an internal reporting spreadsheet called HPV HC Tracking Data is updated with the chemical name and date of completion. The HPV tracking system is updated by EPA staff upon posting of final documents to the EPA web site at the end of each quarter. The number of chemicals reviewed and posted is then recorded in the internal reporting spread sheet. Timing and frequency of reporting: As new HCs are posted at the end of each quarter, the number of chemicals posted is recorded in the internal tracking spreadsheet.
3a. Relevant Information Systems:
EPA uses a reporting spreadsheet called HPV HC Tracking Data to track the number of completed Screening Level Hazard Characterization Reports. There are no transformed data in this spreadsheet as this is a simple tracking measure.

3b. Data Quality Procedures:
Not Available
3c. Data Oversight:
Branch Chief, Planning and Assessment Branch
3d. Calculation Methodology:
The performance result is simply a count of Screening Level Hazard Characterization Reports completed by EPA either quarterly or over the fiscal year.
4a. Oversight and Timing of Final Results Reporting:
Planning and Accountability Lead in the Resource Management Staff in the Office of Program Management Operations. Reporting semiannually: mid-year and end-of-year.
4b. Data Limitations/Qualifications:
Not Available
4c. Third-Party Audits:
Recent GAO reviews found that EPA does not routinely assess the risks of all existing chemicals and faces challenges in obtaining the information necessary to do so. EPA has taken several steps to respond to these reviews including more aggressive efforts to collect data, continued efforts to assess data through hazard characterizations, and increased emphasis on risk management activities for chemicals of concern. GAO-05-458: Chemical Regulation: Options Exist to Improve EPA's Ability to Assess Health Risks and Manage Its Chemical Review Program, June 2005. GAO-06-1032T: Chemical Regulation: Actions Are Needed to Improve the Effectiveness of EPA's Chemical Review Program, August 2006. GAO-09-271: High Risk Series-An update. Transforming EPA's Processes for Assessing and Controlling Toxic Chemicals, January 2009.

Measure Code: 10A - Annual percentage of lead-based paint certification and refund applications that require less than 20 days of EPA effort to process.

Office of Chemical Strategies and Pollution Prevention (OCSPP)

Goal Number and Title:
4 - Ensuring the Safety of Chemicals and Preventing Pollution

Objective Number and Title:
1 - Ensure Chemical Safety

Sub-Objective Number and Title:
1 - Protect Human Health from Chemical Risks

Strategic Target Code and Title:
2 - By 2014,reduce the percentage of children with blood lead levels above 5ug/dl to 1.0 percent or less

Managing Office:
Office of Pollution Prevention and Toxics

1a. Performance Measure Term Definitions:
Lead-based Paint Certification and Refund Applications: Applicants interested in receiving certification and/or accreditation must submit a complete application package to EPA and pay the required fees. In some instances, applicants for certification or accreditation either submit payment in excess of what is required or they decide to withdraw their application. In those cases, the Agency prepares a refund to return those funds to the applicant. 20 Days of EPA Effort: Because the Agency relies in part on contract employees to perform data entry of applications, this measure only counts the portion of the application approval process that is handled by EPA Regional and Headquarters staff. EPA Regional Offices are measured on the number and percentage of individual certification applications processed in less than 20 calendar days. This measure is calculated by using two timeframes. Timeframe 1 is the number of days elapsed from the "Sent to Regional Office" date (when the Contractor sends the application to the Regional Office) to the "Regional Office Review" date (when the Regional Office enters its recommendation to approve/disapprove.) Timeframe 2 is the number of days from the "Approval or Disapproval Letter Generated" date entered by the Regional Office to the "Final Package Sent" date entered by the Regional Office. Timeframes 1 and 2 are added together to give the total processing time. These two timeframes do not include time from any other Federal Lead-based Paint Program (FLPP) process and specifically exclude any time associated with fee confirmation. All of the dates discussed are only valid if recorded in FLPP, and the date recorded in FLPP is the date that these activities are checked off in the database.

2a. Original Data Source:
The original data source is EPA Headquarters and Regional offices. Applications are submitted by external entities but the agency itself generates the data on number of applications and processing time. The original data source is EPA Headquarters and Regional offices. Applications are submitted by external entities but the agency itself generates the data on number of applications and processing time.

2b. Source Data Collection:

Data are entered initially into the FLPP database either by an individual submitting an application via CDX or by a contractor who performs manual entry of information submitted via a paper application. When a paper application is submitted, the contractor is required to contact the applicant if any information is missing or unclear on the application. The CDX system is designed so that all fields must be completed before the application will be accepted.

2c. Source Data Reporting:

Since the original data source is the agency itself, EPA does not rely on source data reporting by any external entity. The original certification and refund applications – from which the number of applications and average processing time are determined – are submitted by outside parties as noted above.

3a. Relevant Information Systems:

The National Program Chemicals Division (NPCD) in the Office of Pollution Prevention and Toxics (OPPT) maintains the Federal Lead-Based Paint Program (FLPP) database.

The FLPP electronic database contains applications for certification by individuals and firms and applications for accreditation by training providers in states and tribal lands administered by the Federal lead abatement program. The database provides a record of all applications the actions on those applications including final decisions and the multiple steps in the process used for measurement. Thus, the database contains only source data. The database is augmented by hard copy records of the original applications. EPA uses an Oracle Discoverer application to query the database to collect measurable performance data.

The FLPP database is available internally to EPA Headquarters, the federal program contractors and Regional lead program staff who process the applications or oversee the processing. The database is maintained on EPA servers at the National Computer Center (NCC) located in Research Triangle Park (RTP), North Carolina. Access to the database is granted by the Lead, Heavy Metals, and Inorganics Branch (LHMIB) in NPCD. Retention and disposal of records in accordance with the EPA Records Schedule 089 and the National Archives and Records Administration General Records Schedule 23/8. Application records maintained in the system are deleted/destroyed two years after the date of the last entry.

In FY 2013, NPCD had the FLPP database certified and accredited under the National Institute of Standards and Technology's (NIST's) Special Publication (SP) 800-53 Revision 3 requirements issued under the Federal Information Security Management Act (FISMA). The certification and accreditation stays in effect until June 2016 with continuous monitoring and performance testing of one third of FLPP's security controls each year. FLPP's Risk Assessment impact values for confidentiality, integrity, and Availability -- the overall security categorization (severity of impact) -- was determined to be moderate for confidentiality, integrity, and Availability. Of the 205 security controls examined and tested there were no high weaknesses for the system; nine medium weaknesses; and 37 low weaknesses. Under the Office of Management and Budget (OMB) guidance on implementing FISMA requirements, these 46 material weaknesses have been documented in FLPP's Plan of Action and Milestones (POA&M) and entered into the Agency XACTA database for tracking. In FY 2014, NPCD will incorporate the NIST SP 800-53 Revision 4 standards for identifying information security controls. Also, NPCD will review and test one-third of FLPP's security controls and perform continuous

monitoring of the database.

3b. Data Quality Procedures:
The FLPP database is an internal EPA database, maintained for the purpose of processing and tracking applications. The database is interactive, and operational usage in processing applications by Headquarters and the Regional offices provides ongoing internal quality reviews. The contractors perform quality review of data entered into the database and update incorrect data. Further, EPA periodically checks contractors' data entry quality when performing HQ reviews.

OPPT has in place a signed Quality Management Plan ("Quality Management Plan for the Office of Pollution Prevention and Toxics; Office of Prevention, Pesticides and Toxic Substances," November 2008). Like the 2003 QMP, it will ensure the standards and procedures are applied to this effort. In addition, NPCD has an approved Quality Management Plan in place, dated July 2008. Applications and instructions for applying for certification and accreditation are documented and available at the Web site http:/ /www.epa.gov/lead/pubs/traincert.htm Documentation for the FLPP database is maintained internally at EPA and is available upon request. |

3c. Data Oversight:
Chief, Planning and Assessment Branch, Environmental Assistance Division,OPPT

3d. Calculation Methodology:
Each complete application for certification is processed (approximately 3000 per year). Certification is issued if all criteria are met. Some applications may be returned to the applicant or withdrawn by the applicant. For the applications that are fully processed, the length of time for EPA processing can be determined from data fields in the FLPP database. Accordingly, a census of all the fully processed applications for certification is conducted monthly , and the percentage of applications that took more than the prescribed number of days (e.g., 20) of EPA effort to process is computed based on this census. The data used to estimate this performance measure directly reflect all information that has been recorded pertaining to certification applications and are the most acceptable for this requirement.

4a. Oversight and Timing of Final Results Reporting:
Planning and Accountability Lead in the Resource Management Staff in the Office of Program Management Operations. Reporting semiannually: mid-year and end-of-year.

4b. Data Limitations/Qualifications:
The agency is not aware of any significant data limitations for this measure.

4c. Third-Party Audits:
None.

Measure Code: RA1 - Annual number of chemicals for which risk assessments are finalized through EPA's TSCA Existing Chemicals Program.

Office of Chemical Strategies and Pollution Prevention (OCSPP)

Goal Number and Title:
4 - Ensuring the Safety of Chemicals and Preventing Pollution

Objective Number and Title:
1 - Ensure Chemical Safety

Sub-Objective Number and Title:
1 - Protect Human Health from Chemical Risks

Strategic Target Code and Title:
0 -

Managing Office:
OPPT

1a. Performance Measure Term Definitions:
Risk Assessment: As EPA uses the term, "risk" is the chance of harmful effects to human health or to ecological systems resulting from exposure to an environmental stressor such as a toxic chemical. The objective of a "risk assessment" is to characterize the nature and magnitude of risks to human health and/or ecological receptors from chemical substances and other stressors that may be present in the environment. Risk managers can use the information gained from risk assessment to help them decide how to protect humans and the environment from such stressors (e.g., chemicals). (http://www.epa.gov/riskassessment/basicinformation.htm - arisk   Finalized: For purposes of this measure, a risk assessment is considered to have been finalized if the risk assessment has been issued in final form following completion of peer review.   Through EPA's TSCA Existing Chemicals Program: For purposes of this measure, a risk assessment is considered to have been finalized "through EPA's TSCA Existing Chemicals Program" if the chemical for which the finalized risk assessment was conducted is on the list of TSCA Work Plan chemicals released by EPA on March 1, 2012, as updated periodically; or, for other chemicals, if the finalized risk assessment was publicly issued after FY 2012 by EPA's TSCA Existing Chemicals Program.   Background: Under TSCA, the EPA has significant responsibilities for ensuring that commercial chemicals do not present unreasonable risk to human health or the environment. The TSCA Existing Chemicals Program focuses on assessing and managing the potential risks of chemicals that entered commerce before TSCA took effect (i.e., "existing chemicals"). For selected chemicals identified by EPA for priority review, the Agency is considering data collection, risk assessment and risk management actions as may be needed.   In March 2012, EPA identified a work plan of 83 chemicals for further risk assessment, beginning with seven to be assessed in FY 2012. The initial list of Work Plan Chemicals may be updated periodically as new information becomes available. EPA intends to use this list to focus and direct the activities of the Existing

Chemicals Program over the next several years. From time to time, EPA may also conduct risk assessments of other existing chemicals pursuant to TSCA.

2a. Original Data Source:

Since all risk assessments tracked through this measure are performed by EPA, the original data source is the Agency itself. All such risk assessments are managed by EPA Headquarters so there is no need for EPA Regions or states to report accomplishments toward the target.

2b. Source Data Collection:

Process managers in OPPT responsible for developing the risk assessments obtain status information from developers of specific components of the assessments and provide periodic progress reports to the Office Director.

2c. Source Data Reporting:

Process managers in OPPT responsible for developing the risk assessments obtain status information from developers of specific components of the assessments and provide periodic progress reports to the Office Director.

Form/mechanism for receiving data and entering into EPA system: Not applicable since there is no data source outside EPA that submits data to the Agency. As noted above, the process of conducting risk assessments is managed by EPA.

Timing and frequency of reporting: Not applicable for the reasons given above.

3a. Relevant Information Systems:

Not applicable due to small number of deliverables.

3b. Data Quality Procedures:

OPPT complies with established EPA or Office guidance, protocols and standard operating procedures for developing chemical risk assessments.

3c. Data Oversight:

Process managers in OPPT responsible for developing the risk assessments obtain status information from developers of specific components of the assessments and provide periodic progress reports to the Office Director.

3d. Calculation Methodology:

Since the measure simply counts the number of chemicals for which risk assessments are completed, there is no need to transform the original data by any mathematical methods. Data are not selected by any formal decision rules except insofar as risk assessments are identified as being subject to the performance measure based on the definitional criteria set out in 1a above.

For definitions of variables and key terms, please see section 1a above. The unit of measure is simply the number of risk assessments completed, and the timeframe of the performance result extends from the first day of the fiscal year through the last day. Explanations of calculations and assumptions are not applicable to this measure as this is a simple output measure involving counting activity completions.

4a. Oversight and Timing of Final Results Reporting:
Planning and Accountability Lead in the Resource Management Staff in the Office of Program Management Operations. Reporting semiannually: mid-year and end-of-year.

4b. Data Limitations/Qualifications:
Since all risk assessments tracked under this measure are conducted by EPA Headquarters (OPPT), there is little chance of incomplete reporting or significant data inconsistencies. There is no sampling error because statistical sampling is not employed. The measure is not subject to a data lag. As the measure is new, there have not been any methodological changes over the life of the measure.

4c. Third-Party Audits:
Not applicable.

Measure Code: P26 - Number of safer chemicals and safer chemical products

Office of Chemical Strategies and Pollution Prevention (OCSPP)

Goal Number and Title:
-

Objective Number and Title:
-

Sub-Objective Number and Title:
-

Strategic Target Code and Title:
-

Managing Office:
Office of Pollution Prevention and Toxics

1a. Performance Measure Term Definitions:
"Safer Chemical Products" are products recognized under the Design for the Environment's (DfE) Safer Products Labeling Program. For more information please see http://www.epa.gov/dfe/ "Safer Chemicals" are chemicals listed on the Safer Chemicals Ingredients List. For more information please see http://www.epa.gov/dfe/saferingredients.htm

2a. Original Data Source:
Safer Products Labeling Program & Safer Chemical Ingredients List: Product manufactures with DfE-labeled products, provided information on the chemical composition of their safer formulations.

2b. Source Data Collection:
Safer Products Labeling Program & Safer Chemical Ingredient List: Product and chemical manufacturers submit chemical information to EPA for review against DfE's Standard for Safer Products.[1] Products that meet the Standard are qualified to carry the DfE label. Chemicals that meet DfE's Criteria for Safer Chemical Ingredients qualify for listing on the Safer Chemical Ingredients List.[2]   The Design for the Environment Program receives chemical content and production volume, which are standard measures for companies who sell chemical products such as cleaning formulations. This information is directly related to the sales of their product, and therefore the companies have the incentive to track it accurately. Production volumes reported to the EPA include the total volume of each DfE-labeled product sold in the US or internationally. [1] http://www.epa.gov/dfe/pubs/projects/gfcp/standard-for-safer-products.pdf [2] http://www.epa.gov/dfe/pubs/projects/gfcp/index.htm - GeneralScreen

2c. Source Data Reporting:
Safer Products Labeling Program: Participating companies sign Partnership Agreements with the EPA, detailed here: http://www.epa.gov/dfe/pubs/projects/formulat/about.htm All products recognized by the Design for the Environment Program are listed on the EPA website, available here:

http://www.epa.gov/dfe/pubs/projects/formulat/formpart.htm Partnership agreements are in effect for three years, partners must renew their agreements every three years or the products may not carry the Design for the Environment logo.

Safer Chemical Ingredients List: Chemicals that meet DfE's Criteria for Safer Chemical Ingredients are drawn from DfE-labeled products and listed on the EPA website, available here: http://www.epa.gov/dfe/saferingredients.htm

3a. Relevant Information Systems:

Safer Products Labeling Program & Safer Chemical Ingredient List: DfE uses an internal Access database to capture relevant information along with other pieces of information about the company manufacturing a DfE-labeled product. The source of the data is always the company who manufacturers the labeled product. The Design for the Environment program is developing a Sales Force-based data system that will replace the existing Access system.

3b. Data Quality Procedures:

Safer Products Labeling Program & Safer Chemical Ingredient List: The DfE program operates under the Information Quality Guidelines found at http://www.epa.gov/quality/informationguidelines as well as under the Pollution Prevention and Toxics Quality Management Plan (QMP) ("Quality Management Plan for the Office of Pollution Prevention and Toxics; Office of Prevention, Pesticides and Toxic Substances," November 2008), and the program will ensure that those standards and procedures are applied to this effort. The Quality Management Plan is for internal use only.

3c. Data Oversight:

Branch Chief, Planning and Assessment Branch (PAB), Environmental Assistance Division, Office of Pollution Prevention Division. The Branch Chief oversees source data reporting and information systems through periodic updates and discussions with the national program staff members and managers who monitor their own source data reporting. This oversight is also accomplished through written protocols developed by PAB and national program managers.

3d. Calculation Methodology:

Safer Products Labeling Program: The number of additional products that qualify to carry the DfE label and are added to the DfE Safer Products Labeling Program website in a given fiscal year will be one contribution to the measure. Additionally, the information gathered on production volume may be used to estimate the volume and weight of safer chemicals used in these products.

Safer Chemical Ingredients List: The number of additional chemicals that meet DfE's Criteria for Safer Chemical Ingredients and are added to the Safer Chemicals Ingredients List in a given fiscal year will be the other contribution to the measure.

These two contributions will be added for the total contribution to this measure.

4a. Oversight and Timing of Final Results Reporting:

Planning and Accountability Lead, Resource Management Staff, Office of Program Management Operations, Office of Chemical Safety and Pollution Prevention. Reporting is conducted semi-annually, at mid-year and end

of year.

4b. Data Limitations/Qualifications:

Safer Products Labeling Program: All product manufactures with DfE-labeled products must go through a renewal process every three years to maintain the DfE label on their products. During this renewal, products are re-reviewed against the DfE Standard for Safer Products to ensure continuing compliance.

Safer Chemical Ingredient List: All chemicals listed on the Safer Chemical Ingredients List are re-reviewed every 3 years to maintain listing as a safer chemical. Not all chemicals that qualify may be listed due to confidentiality concerns related to product formulations, these chemicals will not be counted towards the annual number of Safer Chemicals

4c. Third-Party Audits:

DfE Partners are audited each year of the three-year partnership cycle, undergoing each of the following: a desk (paper) audit, on-site audit (if manufacturing occurs at more than one facility, then two audits are conducted), and renewal (paper) audit.

Office of Research and Development (ORD) Record(s)

Measure Code: SW1 - Percentage of planned research products completed on time by the Safe and Sustainable Water Resources research program.

Office of Research and Development (ORD)

Goal Number and Title:
2 - Protecting America's Waters

Objective Number and Title:
0 -

Sub-Objective Number and Title:
0 -

Strategic Target Code and Title:
0 -

Managing Office:
Office of Program Accountability and Resource Management- Planning;Budget and Performance Analysis Branch

1a. Performance Measure Term Definitions:
A research product is "a deliverable that results from a specific research project or task. Research products may require translation or synthesis before integration into an output ready for partner use." This secondary performance measure tracks the timely completion of research products. Sustainability Research Strategy, available from: http://epa.gov/sciencematters/april2011/truenorth.htm http://www.epa.gov/risk_assessment/health-risk.htm

2a. Original Data Source:
EPA and its partners confirm the schedule for completing research outputs and products that are transformed or synthesized into outputs. ORD tracks progress toward delivering the outputs; clients are notified of progress. Scheduled milestones are compared to actual progress on a quarterly basis. At the end of the fiscal year, outputs are either classified as "met" or "not met" to determine the overall percentage of planned products that have been met by the research program. The actual product completion date is self-reported.

2b. Source Data Collection:
Each output is assigned to a Lab or Center representative before the start of the fiscal year. This individual provides quarterly status updates via ORD's Resource Management System. Status reports are reviewed by senior management, including the Lab or Center Director and National Program Director. Overall status data is generated and reviewed by ORD's Office of Program Accountability and Resource Management.

2c. Source Data Reporting:
Quarterly status updates are provided via ORD's Resource Management System.

3a. Relevant Information Systems:

Internal database or internal tracking system such as the Resources Management System (RMS).
3b. Data Quality Procedures:
EPA and its partners confirm the schedule for completing research outputs and products that are transformed or synthesized into outputs. ORD tracks progress toward delivering the outputs; clients are notified of progress. Scheduled milestones are compared to actual progress on a quarterly basis. At the end of the fiscal year, outputs are either classified as "met" or "not met" to determine the overall percentage of planned products that have been met by the program.
3c. Data Oversight:
The National Program Director oversees the source data reporting, specifically, the process of establishing agreement with program stakeholders and senior ORD managers on the list and content of the planned products, and subsequent progress, completion, and delivery of these products.
3d. Calculation Methodology:
At the end of the fiscal year, outputs are either classified as "met" or "not met". An overall percentage of planned products met by the program is reported.
4a. Oversight and Timing of Final Results Reporting:
The Office of Program Accountability and Resource Management is responsible for reporting program progress in meeting its target of completion of 100% of program planned products.
4b. Data Limitations/Qualifications:
This measure does not capture directly the quality or impact of the research products.
4c. Third-Party Audits:
Not applicable

Measure Code: HS1 - Percentage of planned research products completed on time by the Homeland Security research program.

Office of Research and Development (ORD)

Goal Number and Title:
4 - Ensuring the Safety of Chemicals and Preventing Pollution

Objective Number and Title:
0 -

Sub-Objective Number and Title:
0 -

Strategic Target Code and Title:
0 -

Managing Office:
Office of Program Accountability and Resource Management- Planning;Budget and Performance Analysis Branch

1a. Performance Measure Term Definitions:
A research product is "a deliverable that results from a specific research project or task. Research products may require translation or synthesis before integration into an output ready for partner use."   This secondary performance measure tracks the timely completion of research products.   Sustainability Research Strategy, available from: http://epa.gov/sciencematters/april2011/truenorth.htm   http://www.epa.gov/risk_assessment/health-risk.htm

2a. Original Data Source:
EPA and its partners confirm the schedule for completing research outputs and products that are transformed or synthesized into outputs. ORD tracks progress toward delivering the outputs; clients are notified of progress. Scheduled milestones are compared to actual progress on a quarterly basis. At the end of the fiscal year, outputs are either classified as "met" or "not met" to determine the overall percentage of planned products that have been met by the research program. The actual product completion date is self-reported.

2b. Source Data Collection:
Each output is assigned to a Lab or Center representative before the start of the fiscal year. This individual provides quarterly status updates via ORD's Resource Management System. Status reports are reviewed by senior management, including the Lab or Center Director and National Program Director. Overall status data is generated and reviewed by ORD's Office of Program Accountability and Resource Management.

2c. Source Data Reporting:
Quarterly status updates are provided via ORD's Resource Management System.

3a. Relevant Information Systems:
Internal database or internal tracking system such as the Resources Management System (RMS).

3b. Data Quality Procedures:

EPA and its partners confirm the schedule for completing research outputs and products that are transformed or synthesized into outputs. ORD tracks progress toward delivering the outputs; clients are notified of progress. Scheduled milestones are compared to actual progress on a quarterly basis. At the end of the fiscal year, outputs are either classified as "met" or "not met" to determine the overall percentage of planned products that have been met by the program.

3c. Data Oversight:

The National Program Director oversees the source data reporting, specifically, the process of establishing agreement with program stakeholders and senior ORD managers on the list and content of the planned products, and subsequent progress, completion, and delivery of these products.

3d. Calculation Methodology:

At the end of the fiscal year, outputs are either classified as "met" or "not met". An overall percentage of planned products met by the program is reported.

4a. Oversight and Timing of Final Results Reporting:

The Office of Program Accountability and Resource Management is responsible for reporting program progress in meeting its target of completion of 100% of program planned products.

4b. Data Limitations/Qualifications:

This measure does not capture directly the quality or impact of the research products.

4c. Third-Party Audits:

Not applicable

Measure Code: RA1 - Percentage of planned research products completed on time by the Human Health Risk Assessment research program.

Office of Research and Development (ORD)

Goal Number and Title:
4 - Ensuring the Safety of Chemicals and Preventing Pollution

Objective Number and Title:
1 - Ensure Chemical Safety

Sub-Objective Number and Title:
0 -

Strategic Target Code and Title:
0 -

Managing Office:
Office of Program Accountability and Resource Management- Planning;Budget and Performance Analysis Branch

1a. Performance Measure Term Definitions:
A research product is "a deliverable that results from a specific research project or task. Research products may require translation or synthesis before integration into an output ready for partner use." This secondary performance measure tracks the timely completion of research products. Sustainability Research Strategy, available from:http://epa.gov/sciencematters/april2011/truenorth.htm http://www.epa.gov/risk_assessment/health-risk.htm

2a. Original Data Source:
EPA and its partners confirm the schedule for completing research outputs and products that are transformed or synthesized into outputs. ORD tracks progress toward delivering the outputs; clients are notified of progress. Scheduled milestones are compared to actual progress on a quarterly basis. At the end of the fiscal year, outputs are either classified as "met" or "not met" to determine the overall percentage of planned products that have been met by the research program. The actual product completion date is self-reported.

2b. Source Data Collection:
Each output is assigned to a Lab or Center representative before the start of the fiscal year. This individual provides quarterly status updates via ORD's Resource Management System. Status reports are reviewed by senior management, including the Lab or Center Director and National Program Director. Overall status data is generated and reviewed by ORD's Office of Program Accountability and Resource Management.

2c. Source Data Reporting:
Quarterly status updates are provided via ORD's Resource Management System.

3a. Relevant Information Systems:
Internal database or internal tracking system such as the Resources Management System (RMS).

3b. Data Quality Procedures:

EPA and its partners confirm the schedule for completing research outputs and products that are transformed or synthesized into outputs. ORD tracks progress toward delivering the outputs; clients are notified of progress. Scheduled milestones are compared to actual progress on a quarterly basis. At the end of the fiscal year, outputs are either classified as "met" or "not met" to determine the overall percentage of planned products that have been met by the program.

3c. Data Oversight:

The National Program Director oversees the source data reporting, specifically, the process of establishing agreement with program stakeholders and senior ORD managers on the list and content of the planned products, and subsequent progress, completion, and delivery of these products.

3d. Calculation Methodology:

At the end of the fiscal year, outputs are either classified as "met" or "not met". An overall percentage of planned products met by the program is reported.

4a. Oversight and Timing of Final Results Reporting:

The Office of Program Accountability and Resource Management is responsible for reporting program progress in meeting its target of completion of 100% of program planned products.

4b. Data Limitations/Qualifications:

This measure does not capture directly the quality or impact of the research products.

4c. Third-Party Audits:

Not applicable

Measure Code: CS1 - Percentage of planned research products completed on time by the Chemical Safety for Sustainability research program.

Office of Research and Development (ORD)

Goal Number and Title:
4 - Ensuring the Safety of Chemicals and Preventing Pollution
Objective Number and Title:
0 -
Sub-Objective Number and Title:
0 -
Strategic Target Code and Title:
0 -
Managing Office:
Office of Program Accountability and Resource Management- Planning;Budget and Performance Analysis Branch
1a. Performance Measure Term Definitions:
A research product is "a deliverable that results from a specific research project or task. Research products may require translation or synthesis before integration into an output ready for partner use."   This secondary performance measure tracks the timely completion of research products.   Sustainability Research Strategy, available from: http://epa.gov/sciencematters/april2011/truenorth.htm   http://www.epa.gov/risk_assessment/health-risk.htm
2a. Original Data Source:
EPA and its partners confirm the schedule for completing research outputs and products that are transformed or synthesized into outputs. ORD tracks progress toward delivering the outputs; clients are notified of progress. Scheduled milestones are compared to actual progress on a quarterly basis. At the end of the fiscal year, outputs are either classified as "met" or "not met" to determine the overall percentage of planned products that have been met by the research program. The actual product completion date is self-reported.
2b. Source Data Collection:
Each output is assigned to a Lab or Center representative before the start of the fiscal year. This individual provides quarterly status updates via ORD's Resource Management System. Status reports are reviewed by senior management, including the Lab or Center Director and National Program Director. Overall status data is generated and reviewed by ORD's Office of Program Accountability and Resource Management.
2c. Source Data Reporting:
Quarterly status updates are provided via ORD's Resource Management System.
3a. Relevant Information Systems:
Internal database or internal tracking system such as the Resources Management System (RMS).
3b. Data Quality Procedures:

EPA and its partners confirm the schedule for completing research outputs and products that are transformed or synthesized into outputs. ORD tracks progress toward delivering the outputs; clients are notified of progress. Scheduled milestones are compared to actual progress on a quarterly basis. At the end of the fiscal year, outputs are either classified as "met" or "not met" to determine the overall percentage of planned products that have been met by the program.
3c. Data Oversight:
The National Program Director oversees the source data reporting, specifically, the process of establishing agreement with program stakeholders and senior ORD managers on the list and content of the planned products, and subsequent progress, completion, and delivery of these products.
3d. Calculation Methodology:
At the end of the fiscal year, outputs are either classified as "met" or "not met". An overall percentage of planned products met by the program is reported.
4a. Oversight and Timing of Final Results Reporting:
The Office of Program Accountability and Resource Management is responsible for reporting program progress in meeting its target of completion of 100% of program planned products.
4b. Data Limitations/Qualifications:
This measure does not capture directly the quality or impact of the research products.
4c. Third-Party Audits:
Not applicable

Measure Code: HC1 - Percentage of planned research products completed on time by the Safe and Healthy Communities research program.

Office of Research and Development (ORD)

Goal Number and Title:
3 - Cleaning Up Communities and Advancing Sustainable Development
Objective Number and Title:
0 -
Sub-Objective Number and Title:
0 -
Strategic Target Code and Title:
0 -
Managing Office:
Office of Program Accountability and Resource Management- Planning;Budget and Performance Analysis Branch
1a. Performance Measure Term Definitions:
A research product is "a deliverable that results from a specific research project or task. Research products may require translation or synthesis before integration into an output ready for partner use."   This secondary performance measure tracks the timely completion of research products.   Sustainability Research Strategy, available from: http://epa.gov/sciencematters/april2011/truenorth.htm   http://www.epa.gov/risk_assessment/health-risk.htm
2a. Original Data Source:
EPA and its partners confirm the schedule for completing research outputs and products that are transformed or synthesized into outputs. ORD tracks progress toward delivering the outputs; clients are notified of progress. Scheduled milestones are compared to actual progress on a quarterly basis. At the end of the fiscal year, outputs are either classified as "met" or "not met" to determine the overall percentage of planned products that have been met by the research program. The actual product completion date is self-reported.
2b. Source Data Collection:
Each output is assigned to a Lab or Center representative before the start of the fiscal year. This individual provides quarterly status updates via ORD's Resource Management System. Status reports are reviewed by senior management, including the Lab or Center Director and National Program Director. Overall status data is generated and reviewed by ORD's Office of Program Accountability and Resource Management.
2c. Source Data Reporting:
Quarterly status updates are provided via ORD's Resource Management System.
3a. Relevant Information Systems:
Internal database or internal tracking system such as the Resources Management System (RMS).
3b. Data Quality Procedures:

EPA and its partners confirm the schedule for completing research outputs and products that are transformed or synthesized into outputs. ORD tracks progress toward delivering the outputs; clients are notified of progress. Scheduled milestones are compared to actual progress on a quarterly basis. At the end of the fiscal year, outputs are either classified as "met" or "not met" to determine the overall percentage of planned products that have been met by the program.

3c. Data Oversight:

The National Program Director oversees the source data reporting, specifically, the process of establishing agreement with program stakeholders and senior ORD managers on the list and content of the planned products, and subsequent progress, completion, and delivery of these products.

3d. Calculation Methodology:

At the end of the fiscal year, outputs are either classified as "met" or "not met". An overall percentage of planned products met by the program is reported.

4a. Oversight and Timing of Final Results Reporting:

The Office of Program Accountability and Resource Management is responsible for reporting program progress in meeting its target of completion of 100% of program planned products.

4b. Data Limitations/Qualifications:

This measure does not capture directly the quality or impact of the research products.

4c. Third-Party Audits:

Not applicable

Measure Code: AC1 - Percentage of products completed by Air, Climate, and Energy.

Office of Research and Development (ORD)

Goal Number and Title:
1 - Taking Action on Climate Change and Improving Air Quality

Objective Number and Title:
0 -

Sub-Objective Number and Title:
0 -

Strategic Target Code and Title:
0 -

Managing Office:
Office of Program Accountability and Resource Management- Planning;Budget and Performance Analysis Branch

1a. Performance Measure Term Definitions:
A research product is "a deliverable that results from a specific research project or task. Research products may require translation or synthesis before integration into an output ready for partner use."   This secondary performance measure tracks the timely completion of research products.   Sustainability Research Strategy, available from: http://epa.gov/sciencematters/april2011/truenorth.htm   http://www.epa.gov/risk_assessment/health-risk.htm

2a. Original Data Source:
EPA and its partners confirm the schedule for completing research outputs and products that are transformed or synthesized into outputs. ORD tracks progress toward delivering the outputs; clients are notified of progress. Scheduled milestones are compared to actual progress on a quarterly basis. At the end of the fiscal year, outputs are either classified as "met" or "not met" to determine the overall percentage of planned products that have been met by the research program. The actual product completion date is self-reported.

2b. Source Data Collection:
Each output is assigned to a Lab or Center representative before the start of the fiscal year. This individual provides quarterly status updates via ORD's Resource Management System. Status reports are reviewed by senior management, including the Lab or Center Director and National Program Director. Overall status data is generated and reviewed by ORD's Office of Program Accountability and Resource Management.

2c. Source Data Reporting:
Quarterly status updates are provided via ORD's Resource Management System.

3a. Relevant Information Systems:
Internal database or internal tracking system such as the Resources Management System (RMS).

3b. Data Quality Procedures:

EPA and its partners confirm the schedule for completing research outputs and products that are transformed or synthesized into outputs. ORD tracks progress toward delivering the outputs; clients are notified of progress. Scheduled milestones are compared to actual progress on a quarterly basis. At the end of the fiscal year, outputs are either classified as "met" or "not met" to determine the overall percentage of planned products that have been met by the ACE program.

3c. Data Oversight:

The National Program Director oversees the source data reporting, specifically, the process of establishing agreement with program stakeholders and senior ORD managers on the list and content of the planned products, and subsequent progress, completion, and delivery of these products.

3d. Calculation Methodology:

At the end of the fiscal year, outputs are either classified as "met" or "not met". An overall percentage of planned products met by the ACE program is reported.

4a. Oversight and Timing of Final Results Reporting:

The Office of Program Accountability and Resource Management is responsible for reporting program progress in meeting its target of completion of 100% of Ace, Climate, and Energy program planned products.

4b. Data Limitations/Qualifications:

This measure does not capture directly the quality or impact of the research products.

4c. Third-Party Audits:

Not applicable

Office of Solid Waste and Emergency Response (OSWER) Record(s)

Measure Code: 113 - Number of LUST cleanups completed that meet risk-based standards for human exposure and groundwater migration in Indian Country.

Office of Solid Waste and Emergency Response (OSWER)

Goal Number and Title:
3 - Cleaning Up Communities and Advancing Sustainable Development
Objective Number and Title:
3 - Restore Land
Sub-Objective Number and Title:
3 - Cleanup Contaminated Land
Strategic Target Code and Title:
6 - Through 2015, reduce the backlog of LUST cleanups in Indian Country
Managing Office:
Office of Underground Storage Tanks (OUST)
1a. Performance Measure Term Definitions:
Cleanups Completed –The number of cleanups completed is the cumulative number of confirmed releases where cleanup has been initiated and where EPA has determined that no further actions are currently necessary to protect human health and the environment. This number includes sites with post-closure monitoring as long as site-specific (e.g., risk-based) cleanup goals have been met. Site characterization, monitoring plans, and site-specific cleanup goals must be established and cleanup goals must be attained for sites being remediated by natural attenuation to be counted in this category. Clarification: "Cleanups Completed" is a cumulative category–sites should never be deleted from this category. A "no further action" determination made by the Region that satisfies the "cleanups initiated" measure above, also satisfies this "cleanups completed" measure. This determination will allow a confirmed release that does not require further action to meet the definition of both an initiated and completed cleanup. For complete definition see EPA OUST's UST And LUST Performance Measures Definitions. January 18, 2008. http://www.epa.gov/OUST/cat/PMDefinitions.pdf Risk-based standards for human exposure and groundwater migration. Reference: Semi-annual Report of UST Performance Measures, End Of Fiscal Year 2013 – as of September 30, 2013 , dated December 2013;http://epa.gov/oust/cat/ca-13-34.pdf
2a. Original Data Source:
The original data source is EPA which is responsible for the implementation of the UST Program in Indian Country. For more information:

1.	For complete definitions see EPA OUST's UST and LUST Performance Measures Definitions, January 18, 2008 - http://epa.gov/oust/cat/PMDefinitions.pdf

2b. Source Data Collection:

Determination of cleanup completion requires consideration of environmental data, such as field sampling, which can vary by project. The overall measure requires tabulation of the number LUST clean-ups completed. Spatial Coverage: National

For contracts: EPA Regions determine which quality requirements are applicable. Contracts must be current and specify: QA roles and responsibilities for EPA and national LUST contractors; and quality requirements including responsibilities for final review and approval. Default quality requirements include: organization-level QA documentation (i.e. QMP) for the primary contractors; and project-level QAPPs for each Tribal LUST remedial Work Assignment. Sample EPA contract language: "the Contractor shall comply with the higher-level quality standard selected below: Specifications and Guidelines for Quality Systems for Environmental Data Collection and Environmental Technology Programs (ANSI/ASQC E4, 1994). As authorized by FAR 52.246-11, the higher-level quality standard ANSI/ASQC E4 is tailored as follows: The solicitation and contract require the offerors/contractor to demonstrate conformance to ANSI/ASQC E4 by submitting the quality documentation described below. [Specifically,...] ... The Contractor shall not commence actual field work until until the Government has approved the quality documentation (i.e., QAPP)."

Note: Regions keep copies of individual QAPPs associated with contracts.

2c. Source Data Reporting:

LUST4 also allows for bulk (batch) uploading by EPA that already have the location & measures-related data captured in a data system or have the technical expertise to create flat files through another method in exactly the format and layout specified. This batch uploading is supported by OUST; data providers not comfortable with this approach are encouraged to use the interactive online features of the Locations Subsystem and Measures Subsystem. Access to the LUST4 Locations and Measures Subsystems is available online via the EPA portal at http://portal.epa.gov/ under the My Communities/Underground Storage Tank menu page. Reports are submitted biannually (e.g., Mid-Year and End-of-Year reports).

3a. Relevant Information Systems:

LUST4. This database is the master database of all LUST program-related data. EPA reports data for activity and measures directly into LUST4. LUST4 includes both source data and transformed data (e.g., data aggregated into Regional totals).

The program's Oracle web-based system-- LUST4-- accessed through EPA's portal.

OSWER Performance Assessment Tool (PAT). This tool serves as the primary external servicing resource for organizing and reporting OSWER's performance data. PAT collects information from OSWER program systems, and conforms it for uniform reporting and data provisioning. PAT captures data from LUST4; replicates business logic used by LUST4 for calculating measures; can deliver that data to EPA staff and managers via a business intelligence dashboard interface for analytic and reporting use; enables LUST point of contact to document status and provide explanation for each measure; and transmits data to the Budget Automation System.

Budget Automation System (BAS). BAS is the final repository of the performance values.

3b. Data Quality Procedures:

.EPA's regional program managers provide first-level data quality reviews and oversight of their program performance measure results.

OUST uses a combination of automated validation along with manual QA/QC review.

QA/QC REVIEW BY REGIONS. EPA/OUST oversees the use of the QA/QC checklist, which is incorporated into the LUST4 oracle web-based system. Regions complete the QA/QC checklist, sign it electronically and submit it to EPA/OUST for review, comment and approval of each record.
NOTE: This QA/QC checklist was last updated 10/1/2009 and is accessed through the user interface of LUST4.

Regional QA/QC Evaluation Checklist –
Note: Checklist is to be completed by Regional reviewer and will appear "shaded" to others.
1. Previous Totals Column
-- Verify the previous total number is correct by comparing it to the total from the last reporting period. If there is a discrepancy, report the information in the "Correction To Previous Data" column. Please add comments in the "Comments" column for any corrections that are made to the applicable performance measure.
2. Actions This Reporting Period
For each performance measure, if the "Reported" number deviates by more than 10% from the last period's number, the Region must include an explanation of the deviation.
3. Corrections to Previous Data Column
-- Ensure that any corrections to measures are documented at the Regional office and in the LUST4 system.
-- Evaluate if the corrections will impact other performance measures (e.g., if the number of cleanups completed is adjusted downward by a correction, does this also result in a commensurate downward adjustment of cleanups initiated?) Include any additional comments in the "Comments" column as necessary.
4. Totals (Cumulative, if applicable)
-- Verify accuracy of all cumulative totals
-- Include any additional comments in the "Comments" column as necessary

AUTOMATED VALIDATION. This feature is no longer applicable. The feature ONLY applied to the American Recovery and Reinvestment Act (ARRA) funded projects.

EPA/OUST provides second-level data quality reviews of all data

LUST4. LUST4 operates under OSWER's QMP, including the security policy specified in that QMP. LUST4 does not have any stand-alone certifications related to the EPA security policy or the Systems Life Cycle Management policy. The LUST4 system is built upon Oracle Business Intelligence tools provided by the EPA Business Intelligence Analytics Center, which ensures that a stand-alone security certification is not necessary.

PAT. PAT operates under the OSWER Quality Management Plan (QMP). PAT has a security certification confirming that a security policy is not necessary because no sensitive data are handled and PAT is built upon the Oracle-based business intelligence system. PAT's security certification indicates that it follows all security guidelines for EPA's Oracle Portal and that PAT is (1) not defined as a "Major Application" according to NIST Special Publication 800-18, Guide for Developing Security Plans for Information Technology Systems, section 2.3.1; (2) does not store, process, or transmit information that the degree of sensitivity is assessed as high by considering the requirements for availability, integrity, and confidentiality according to NIST Special Publication 800-18, Guide for Developing Security Plans for Information Technology Systems, section 3.7.2. (3) is not covered by EPA Order 2100.2A1 Information Technology Capital Planning and Investment Control (CPIC). Data Flow:

Step 1. Performance measure are entered into LUST4 by Regions (for EPA contractors).

Step 2. Each Region conducts Regional level review of their data from the LUST4 system.

Step 3. Headquarters' staff perform performs National Program Review, using data from the LUST4 system. Rejected data must be corrected by the Region (Step 2).

Step 4. PAT pulls data from LUST4. Headquarters staff compare PAT results to LUST4 results. If PAT does not match LUST4 then there was an error with the upload and data is reloaded. Headquarters staff enter into PAT the ACS status information of "Indicator" for each measure and, if desired, explanation.(Note: PAT allows for programs to identify status other than "Indicator." When programs select a status of "no status," "data not available," or "target not met," PAT requires that an explanation be provided. LUST program policy is to resolve all reporting issues prior to ACS reporting, so "Indicator" is the only status chosen and explanations for that status are optional.)

Step 5. Headquarters approves PAT results, and PAT pushes results into BAS/Measures Central.

Step 6. Measures Central aggregates Regional data into a national total. OUST reporting lead reviews and certifies results.

3c. Data Oversight:

An EPA Headquarters primary contact maintains a list of the HQ (OUST and OEI) and Regional users; a record of changes to the list is also maintained. The primary HQ contact ensures that Regional reporting is on track, conducts QA on LUST performance measures, ensures QA issues are resolved and/or documented, and oversees final reporting to BAS.

Regional Program Managers are ultimately responsible for regional-level data. They conduct their review

based upon a national QA/QC checklist.
3d. Calculation Methodology:
The cumulative number of confirmed releases where cleanup has been initiated and where the region has determined that no further actions are currently necessary to protect human health and the environment, includes sites where post-closure monitoring is necessary as long as site specific (e.g., risk based) cleanup goals have been met. Site characterization, monitoring plans and site-specific cleanup goals must be established and cleanup goals must be attained for sites being remediated by natural attenuation to be counted in this category. (See http://www.epa.gov/OUST/cat/PMDefinitions.pdf. The unit of analysis is site cleanup
4a. Oversight and Timing of Final Results Reporting:
Semiannual by Deputy Office Director. Responsible for final review to ensure LUST 4 System Manager has completed review, and numbers are accurate.
4b. Data Limitations/Qualifications:
Data quality depends on the accuracy and completeness of state records.
4c. Third-Party Audits:
Not applicable!

Measure Code: 337 - Percent of all FRP inspected facilities found to be non-compliant which are brought into compliance.

Office of Solid Waste and Emergency Response (OSWER)

Goal Number and Title:
3 - Cleaning Up Communities and Advancing Sustainable Development
Objective Number and Title:
3 - Restore Land
Sub-Objective Number and Title:
2 - Emergency Preparedness and Response
Strategic Target Code and Title:
3 - By 2015, no more than 1.5 million gallons will be spilled annually at FRP Facilities
Managing Office:
Office of Emergency Management
1a. Performance Measure Term Definitions:

1.   FRP facility: A facility which must submit a Facility Response Plan (FRP) to EPA to demonstrate its preparedness to respond to a worst-case discharge of oil under Clean Water Act, as amended by the Oil Pollution Act. FRP planholders typically represent facilities that pose a higher-risk to human health and the environment from a discharge of oil.

2.   Initially compliant: The tag used to describe facilities that are not in violation of the requirements in the FRP regulation of the Oil Pollution Act upon inspection in a given fiscal year.

3.   Non-compliant: The tag used to describe a facility that upon inspection is deemed to be in violation of the FRP regulation. Non-compliant facilities are brought into compliance through EPA follow-up activity.

4.   Brought into compliance: The tag used to describe a facility that was found non-compliant at some point during the measurement period and then was then found to be in compliance after a follow-up inspection activity.

5.   Not Subject: The tag used to describe a facility that, upon inspection, is not required to comply with the FRP regulation. Facility may have been subject previously but has made changes to the facility whereby it is no longer subject to the FRP regulation.

6.   Closed: The tag used to describe a facility that, upon inspection, is closed and no longer subject to the FRP rule. Facility may have been subject previously.

7.   Carry Over: The tag used to describe facilities that were found non-compliant at some point before the current FY in the measurement period and remain non-compliant as of the beginning of the current FY.

8.   Oil Database Application: The online database that is used to collect and store facility and inspection-related data for FRP facilities. The Oil Database is not a public database.

2a. Original Data Source:
The primary data used to calculate the oil measures are the number and outcomes of facility inspections and government initiated unannounced exercises (GIUEs) and the dates in which non-compliant facilities are brought into compliance. Specifically, calculation of the oil measures involves the recording of the inspection activity and the outcome (in compliance vs. out of compliance) and subsequent tracking of the dates when

non-compliant FRP facilities are brought into compliance. Starting with FY 2011, the reported value of facilities brought into compliance in any given FY must also capture inspections that occurred prior to the current FY (using FY 2010 as the base year).

2b. Source Data Collection:

In order to calculate the measure for each FY, the following data must be collected and entered by the regions for each inspection conducted:

1.   Facility Identifier – identifier assigned by the Oil Database Application used to track the activity for each unique facility.

2.   Inspection (Activity) Tracking Number – also assigned by the Oil Database Application, but prior to October 2011, these tracking numbers were Region-specific.

3.   Inspection Date (Activity Start/Schedule Date in the Oil Database Application) – date of the initial inspection or most recent compliance inspection.

4.   Inspection Outcome – either in compliance or not in compliance.

5.   Date of Confirmed Compliance (Brought into Compliance Date in the Oil Database Application) – date the facility is verified to be in compliance overall; could occur in subsequent FYs.

The business rules for entering data into the "Compliance Module" of the Oil Database Application used to calculate the end-of-year measures results are as follows:

·   Inspection Outcomes: When a facility is inspected for FRP compliance (including GIUE's), the outcome of the inspection should be recorded in the application. If the facility is found to be out of compliance, the outcome should be recorded as "out of compliance." If the facility is found to be compliant, then the outcome should be recorded as "in compliance."

·   "Not-Subject Facilities" Reporting: If an inspected facility is found to be not subject to FRP regulations, the outcome of the inspection should be recorded as "in compliance".

·   Tracking Multiple Inspections: If there are multiple inspections in the same FY for a facility, the SPCC inspection will be counted as a single inspection, and the FRP inspection/GIUE will be counted as a separate inspection/exercise. Multiple FRP inspections would not be added together by the Oil Database Application for the same facility. If there are multiple inspections and a GIUE in the same FY for a facility, in order to determine compliance status of the facility, the application will look at the activities to determine if the most recent activity has an "Inspection Outcome" or a "Brought into Compliance" date. If an "Inspection Outcome" or a "Brought into Compliance" date doesn't exist on the most recent activity, the Oil Database Application will look at the previous activity within the FY to determine the facility's compliance status.

·   Tracking Compliance vs. Enforcement: Once a facility is brought into compliance, it should be captured in the "Inspection Outcome" and "Brought into Compliance" fields in the application. If additional enforcement is pursued by the Region after a facility is brought into compliance, the facility record should retain the facility's status as "in compliance."

·   Treatment of Carryover Non-Compliant Facilities: The application automatically tracks the carryover of non-compliant facilities to the next FY, starting at the end of FY 2010. Specifically, the application will review the recorded inspections (FRP and GIUE's) performed since FY 2010 and for facilities that remain out of compliance as of the date the report is run, these facilities will represent the total population of non-compliant facilities, including carryover and newly-inspected, non-compliant facilities. For example, if you have a facility that was inspected in FY 2010 that was initially found to be "not in compliance" and is still showing as "not in compliance" at the end of FY 2011, the facility will be included in the carryover of non-compliant facilities to the next FY (i.e., FY 2012). This carryover count will be automatically added to the list of facilities found to be non-compliant in the current FY (or FY 2012 in this example).

·   Treatment of Follow-up Inspections: If a non-compliant facility is inspected for FRP or is exercised (i.e., GIUE) in a subsequent FY, this inspection should be identified as a "Follow-up Inspection". It should be noted that the application will identify this follow-up inspection as tied to the original inspection where the outcome as identified as "out of compliance." This follow-up inspection will not be recorded as new inspection in that FY to avoid double-counting in the "Facilities Inspected" column for the HQ Oil Measures Report. However, for programmatic tracking, this follow-up inspection will be retained in the facility record in the compliance module.

·   Treatment of Inspections with No Outcome: If the inspection or GIUE does not have an "Inspection Outcome" or a "Brought into Compliance" date populated in the compliance module of the application, the facility will count as "not in compliance" on the HQ Measures Report. When this occurs, the record should be updated with inspection outcome.

·   Tracking of Inspection and Plan Review Outcomes: For inspected facilities that are in compliance at the time of the inspection, but a subsequent Plan review reveals the Plan to be out of compliance, enter the two activities as different activities, field inspection vs. Plan review. Enter the inspection as "out of compliance" (using the date of the inspection), and then enter the Plan review activity as "not in compliance" (using the review date). When the Plan is deemed "in compliance", enter this compliance date in the "Brought into Compliance Field" for the Plan review activity and the inspection activity. Note that the "Brought into Compliance" date for the inspection activity is considered the overall compliance date for oil measures reporting.

2c. Source Data Reporting:

Data Submission Instrument: Data submission and data entry are handled via the Oil Database Application. The Oil Database Application can be accessed at http://emp.epa.gov/

Data Entry Mechanism: Data are entered by EPA personnel in the regions. Facility and inspection results data are entered for each initial inspection and follow-up activity.

Frequency of Data Transmission to EPA: Data are entered on an ongoing-basis as inspections are completed.

The initial measurement period for this measure starts with FY 2010 and concludes in FY 2015. The measure will then be extended to FY 2018 for the next iteration of the Agency's Strategic Plan.
A progressive rate (i.e., cumulative over the measurement period) of non-compliant facilities brought into compliance was selected starting with 15% in the base year leading up to 60% brought into compliance in the final year to account for carryover of non-compliant facilities to subsequent fiscal years. The reasoning behind this approach is based on program experience of the time-lag to bring facilities into compliance. For example, non-compliant facilities may need to install equipment or complete construction activities to achieve compliance. Another example is a facility may be subject to enforcement action as a result of non-compliance. Each or both of these scenarios could result in an extended timeframe to bring the facility into compliance.

Thus, a 60% target was selected for FY 2015 to account for the potential time-lag for achieving compliance. The goal over the FY 2016- 2018 measurement period is to maintain a cumulative 60% brought into compliance rate.

3a. Relevant Information Systems:

System Description: The Oil Database Application contains all of the necessary fields and data entry points for a facility inspection. The data are reported via the HQ Measures Report in the Oil Database Application. The Oil Database Application contains data on the following components of a facility and its inspection history:

1. General
2. Address
3. Contacts/Ownership
4. FRP
5. Compliance Module
6. Oil Capacity
7. Discharge History
8. Documents

Source/Transformed Data: The data in the system are source data about the facility and the inspection results.

Information System Integrity Standards: As an application module of the Emergency Management Portal (EMP) the Oil Database application contains Information System Integrity measures to secure its data. EMP has undergone an annual Continuous Monitoring Assessment (CMA) and SAISO Audit of its security posture and continuous monitoring activities. In addition, the CMA was to facilitate ongoing Security Authorization of the system in accordance with Office of Management & Budget Circular A-130, Appendix III, Security of Federal Automated Information Resource; NIST Special Publication 800-37, rev.1, Guide for Applying the Risk Management Framework to Federal Information Systems; and NIST Special Publication 800-53A Rev.1, a subset of the applicable Security Controls of EMP's System Security Plan (SSP) were assessed. EMP's SSP and its Security Controls get reviewed on an annual basis.

The EPA Annual Commitment System (ACS) in BAS is the database for the number of inspections/exercises at FRP facilities. Using data submitted directly by Regional staff as well as data in ACS, the Office of Emergency Management (OEM) tracks in a spreadsheet national information about Regional activities at FRP facilities. EPA will also be using its in-house SPCC/FRP Oil Database to pull data related to inspected facilities to assist measurement tracking.

3b. Data Quality Procedures:

Facility and Inspection data go through a multi-phase quality check. The Oil Database Application has built in data checks for common data entry problems (e.g., checks for missing data) for each facility and inspection outcome. Regions also engage in periodic (usually quarterly or bi-annual) manual QA/QC procedures to double check the data that are entered in the database.

OEM HQ also assists the regions with their manual data checks and resolves problems with missing or incorrect data. This happens on a bi-annual basis.

3c. Data Oversight:

Source Data Reporting Oversight Personnel:

- HQ Project Manager: Office of Emergency Management –Evaluation and Communications Division & Regulation and Policy Development Division
- HQ Project Support: Office of Emergency Management – Evaluation and Communications Division & Regulation and Policy Development Division
- Regional Oil Program Managers and Inspector Personnel

Source Data Reporting Oversight Responsibilities:

- Ensure Accuracy of source data entered into the Oil Database Application through manual QA/QC and data quality checks built into the application.

Information Systems Oversight Personnel:

- HQ Project Manager: Office of Emergency Management – Evaluation and Communications Division & Regulation and Policy Development Division
- HQ Project Support: Office of Emergency Management – Evaluation and Communications Division & Regulation and Policy Development Division

Information Systems Oversight Responsibilities:

- Maintain information systems and correct errors in report calculations.
- Provide upgrades to database capabilities.

3d. Calculation Methodology:

Explanation of Calculations: Data are subject to the following steps in order to determine the brought into compliance percentage (BIC%) that is the final reporting form for these measures.

1.   First, the Regional BIC% is calculated for each region using the following steps:

a.   For each Fiscal Year of the overall reporting period (initially FYs 2010 to 2015, extended to FY 2018) the number of facilities that are found non-compliant (NC#) is subtracted from the total number of inspections conducted in that FY, which includes those that were found compliant, not subject or closed (Total). NC# is that region's portion of the total denominator used to calculate the brought into compliance percentage (BIC %).

b.   Facilities that are brought into compliance during the same fiscal (BIC#) year make up that region's portion of the numerator of the BIC %.

c.   So Regional BIC% = BIC# divided by total NC# (facilities found non-compliant in current FY and carryover non-compliant facilities from prior FY)

2.   Then, the National BIC% is calculated by summing the regional BIC# and NC# and dividing them. So, National BIC% = (All Regional BIC# / All Regional NC#)

3. The Cumulative National BIC% is then calculated by dividing the National BIC# and the National NC# for all FYs from FY 2010 to the current FY. This Cumulative National BIC% accounts for Carry Over facilities from previous FYs, and it is the number that is reported as the final GPRA result for the current FY (i.e., it is not just the National BIC% for the current FY that is reported). So, Cumulative National BIC% = (National BIC# for All FYs / National NC# for All FYs).

The Unit of Measure is percentage of facilities brought into compliance (BIC%).

There are no Assumptions for this calculation.

The Timeframe is from FY 2010 – FY 2018. BIC% represents the percentage of facilities that were initially inspected and brought into compliance during this timeframe.

4a. Oversight and Timing of Final Results Reporting:

Final Reporting Oversight Personnel:
- HQ Project Manager: Office of Emergency Management – Evaluation and Communications Division & Regulation and Policy Development Division
- HQ Project Support: Office of Emergency Management – Evaluation and Communications Division & Regulation and Policy Development Division
- HQ Results Reporting Support: Office of Emergency Management - Evaluation and Communications Division & Regulation and Policy Development Division

Final Reporting Oversight Responsibilities:
- Ensure Accuracy of source data entered into the Oil Database Application through manual QA/QC and data quality checks built into the application.
- Calculate the Cumulative National BIC% (as outlined in Section 3d of this DQR).
- Enter final performance result into EPA's Budget Automation System (BAS).

Final Reporting Timing: The final cumulative brought into compliance percentage for FY 2010 through the current FY is calculated for FRP facilities for inclusion in OSWER reporting activities for the Government Performance and Results Act.

4b. Data Limitations/Qualifications:

OEM has not identified any systematic limitations or qualifications to the data beyond typical data entry and user error.

4c. Third-Party Audits:

There are no third-party audits for this measure.

Office of Solid Waste and Emergency Response (OSWER)

Goal Number and Title:
3 - Cleaning Up Communities and Advancing Sustainable Development
Objective Number and Title:
1 - Promote Sustainable and Livable Communities
Sub-Objective Number and Title:
2 - Assess and Cleanup Brownfields
Strategic Target Code and Title:
2 - By 2015, make an additional 17,800 acres of brownfield properties ready for reuse
Managing Office:
Brownfields
1a. Performance Measure Term Definitions:
Acres Made Ready for Reuse – Acres associated with properties benefiting from EPA Brownfields funding that have been assessed and determined not to require cleanup, or where cleanup has been completed and institutional controls are in place, if required, as reported by cooperative agreement recipients. This typically occurs when one of the following conditions applies: 1. A clean or no further action letter (or its equivalent) has been issued by the state or tribe under its voluntary response program (or its equivalent) for cleanup activities at the property; or 2. The cooperative agreement recipient or property owner, upon the recommendation of an environmental professional, has determined and documented that on-property work is finished. Ongoing operation and maintenance activities or monitoring may continue after a cleanup completion designation has been made. Note: a property can be counted under this measure if an assessment is completed and results in a determination of no further cleanup being required. A "property" is defined as a contiguous piece of land under unitary ownership. A property may contain several smaller components, parcels or areas. For additional information: http://www.epa.gov/brownfields/index.html
2a. Original Data Source:
Assessments and Cleanups are funded either through cooperative agreements, or through EPA contracts (for Targeted Brownfields Assessments (TBAs)). Cooperative agreement recipients (or sub-recipients) and contractors submit performance data to EPA in quarterly reports, and property profile reports. On a limited basis EPA personnel are allowed to update or supplement information when a cooperative agreement has been closed and outcomes have been reported to EPA.

2b. Source Data Collection:

Data collection may involve tabulation of records and review of field surveys that identify acreage. The program does not require or recommend a specific land surveying protocol for determining acreage. Data collection is ongoing as projects are implemented. Reporting instructions indicate that accomplishments are to be recorded as they occur.

Acres Made Ready for Reuse can be achieved by conducting an assessment and/or cleanup activities via an Assessment, Revolving Loan Fund or Cleanup (ARC) award, a Targeted Brownfield Assessment (TBA) or by 128(a) funding used for site specific activities.

Conditions for counting "ACRES Made Ready for Reuse" above and beyond the completion of the funded activity:

Under assessment activities:
-       If neither cleanup nor Institutional Controls (ICs) are required, then the acres are ready for resuse.
-       If ICs are required and they are in place, but cleanup is not required, then the acres are ready for resuse.
-       If cleanup is required and later conducted, (where EPA funds assessment activity, but does not fund cleanup) than the property associated with the original assessment is considered ready for reuse.

Under cleanup activities:
-       If cleanup is required and completed and ICs are not required, then acres are ready for resuse.
-       If cleanup is required and completed and ICs are required and they are in place, then acres are ready for reuse.

Geographic Detail: As of FY12 ACRES leverages a Google Maps application within the system to assign geocoordinates based on address information. Any deviation from these coordinated requires a manual override by the reporting party.

All the Brownfields cooperative agreements have a QA term and condition. Project-level QA documents (i.e. QAPPs) are a minimum requirement for EPA funding of Brownfields activities which include environmental data collection. The program prepares and provides the QA term and condition to the regional offices and requires them to include it in the cooperative agreements. The QA term and condition for Brownfields Assessment cooperative agreements reads as follows:

"B. Quality Assurance (QA) Requirement. 1. When environmental samples are collected as part of the brownfields assessment, the CAR shall comply with 40 CFR Part 31.45 requirements to develop and implement quality assurance practices sufficient to produce data adequate to meet project objectives and to minimize data loss. State law may impose additional QA requirements."

EPA contractors conducting Targeted Brownfiels Assessment should develop site-specific Quality Assurance Project Plans (QAPP) for environmental assessment activities or a site-specific QAPP addendum if a Generic

QAPP has already been approved for assessment activities. The EPA requires all environmental monitoring and measurement efforts be conducted in accordance with approved QAPPs. The purpose of the QAPP is to document the project planning process, enhance the credibility of sampling results, produce data of known quality, and potentially save time and money by gathering data that meets the needs of the project and intended use of the data. The QAPP is a formal document describing in comprehensive detail the necessary QA/QC and other technical activities that must be conducted to ensure the results of the work performed will satisfy performance criteria and can be used for their intended purposes. All QA/QC procedures shall be in accordance with applicable professional technical standards, EPA requirements, government regulations and guidelines, and specific project goals and requirements.

OSWER has available the following guidance: "Quality Assurance Guidance for Conducting Brownfields Assessments." EPA 540-R-98-038. 1998.

Attached Documents:
Quality Assurance Guidance for Conducting Brownfields Assessments 1998.pdf

2c. Source Data Reporting:

Cooperative agreement recipients (or sub-recipients) and contractors submit performance data to EPA in quarterly reports, and property profile reports. A Property Profile Form (PPF) collects information (environmental, historical, physical) from a property-specific investigation funded under the Brownfields Program.

Contract Agreement recipients have 3 submission options: complete and submit the Property Profile Form (PPF) in online format connected to the Assessment, Cleanup and Redevelopment Exchange System (ACRES) database; fill out a PPF version in Microsoft Excel format and submit it via e-mail or regular mail to the EPA Regional Representative; or for multiple properties (more than ten properties) fill out a multi-property Excel spreadsheet and submit it via email; or regular mail to the EPA Regional Representative. Any paper forms are entered into ACRES via EPA contractor.

The Property Profile Form is an approved OMB form - OMB No. 2050-0192. Online forms available to registered users here: http://www.epa.gov/brownfields/pubs/index.html
EPA contractors conducting TBAs provide the assessment report to the EPA Region, who in turn enters the data into ACRES. In some cases, the contractor will also provide a filled-out PPF.
In accordance with the Terms and Conditions of the Brownfields Cooperative Agreements, all Brownfields cooperative agreement recipients (CARs) must report accomplishments to EPA on a quarterly basis. Quarterly reports are due 30 days from the end of the federal fiscal quarter.

Attached Documents:
2009_multiple_ppf_template_external.xls
2009_property_profile_form_instructions.pdf
2009_property_profile_form.xls

3a. Relevant Information Systems:

Assessment, Cleanup, and Redevelopment Exchange System (ACRES). This database is the master database of all data supporting OBLR measures. Recipients and EPA report directly into ACRES. It includes source data and transformed data (e.g., data aggregated into Regional totals).

http://www.epa.gov/brownfields/pubs/acres/index.htm provides more information about this database.

ACRES quality is assured by adherence to a security plan and quality management plan:

-- Security plan. The latest version of the Security Plan for ACRES is dated 11/2009.

-- Quality Management Plan.ACRES operates under its own Quality Management Plan (Data Quality Management Plan for the Assessment, Cleanup, and Redevelopment Exchange System, Version 1.02), which is updated annually, has been updated as of 2/2010. Contact Ryan Smith for the most recent copy of the QMP.

OSWER Performance Assessment Tool (PAT). This tool serves as the primary external servicing resource for organizing and reporting OSWER's performance data, which collects information from OSWER program systems, and conforms it for uniform reporting and data provisioning. PAT captures data from CERCLIS; replicates business logic used by CERCLIS for calculating measures; delivers that data to EPA staff and managers via a business intelligence dashboard interface for analytic and reporting use; ; and transmits data to BAS. No current system specifications document is currently available for PAT, but will be provided when available. Contact Lisa Jenkins in OSWER regarding questions about PAT.

PAT operates under the OSWER Quality Management Plan (QMP), attached.

PAT has a security certification confirming that a security policy is not necessary because no sensitive data are handled and PAT is built upon the Oracle-based business intelligence system. PAT's security certification indicates that it follows all security guidelines for EPA's Oracle Portal and that PAT is (1) not defined as a "Major Application" according to NIST Special Publication 800-18, Guide for Developing Security Plans for Information Technology Systems, section 2.3.1; (2) does not store, process, or transmit information that the degree of sensitivity is assessed as high by considering the requirements for availability, integrity, and confidentiality according to NIST Special Publication 800-18, Guide for Developing Security Plans for Information Technology Systems, section 3.7.2. (3) is not covered by EPA Order 2100.2A1 Information Technology Capital Planning and Investment Control (CPIC). The security certification, attached, was submitted on 9/11/2008.

Budget Automation System (BAS). BAS is the final repository of the performance values.

Attached Documents:
OSWER QMP printed 2010-03-23.pdf
PAT SecurityCertificationNIST.doc

3b. Data Quality Procedures:

Data reported by cooperative award agreement recipients are reviewed by EPA Regional grant managers for accuracy, to verify activities and accomplishments, and to ensure appropriate interpretation of performance measure definitions.

Step 1. Performance measure data entered into ACRES by recipients and/or EPA HQ contractor (for data submitted by recipients in an alternate format, such as hard copy). For each cooperative agreement recipient,, all data entered are signed off by the EPA Regional Representative (Regional Project Officer) identified in the terms and conditions of the cooperative agreement. For contractors, the EPA Regional COR/WAM signs off on the data.

Step 2. Each Region conducts Regional level review of data from the ACRES system. Rejected data must be edited by the original data source. Approved data proceed to Step 3.

Step 3. HQ conducts National level review (EPA HQ contractors) of data approved by regions. Rejected data must be edited by the region (Step 2). Approved data is stored in ACRES.

Step 4. Each quarter, OSWER Performance Assessment Tool (PAT) database pulls the approved data (performance measure) from ACRES.

Step 5. Headquarters approves PAT results, and PAT pushes results into ACS/Measures Central.

Step 6. ACS/Measures Central aggregates Regional data into a national total. OBLR reporting lead reviews and confirms result

ACRES. ACRES quality is assured by adherence to a security plan and quality management plan:

3c. Data Oversight:

Headquarters-level oversight is provided by maintained by the EPA Contract Officer Technical Representative (COTR)

There is a Regional Project Officer assigned to each cooperative agreement. That Regional Project Officer is responsible for reviewing for completeness and correctness all data provided by cooperative agreement recipients and data related to Targeted Brownfields Assessment (TBA) contracts; their data is reviewed at the Headquarters level. A list of Regional Project Officers is maintained by the Regional Brownfields Coordinator in each region.

Each region also has a data manager (some Regions have a SEE Employee as their data manager). The responsibility of the data manager is to disseminate information about ACRES updates and accomplishments updates. This person serves as the regional point of contact for data related issues.

3d. Calculation Methodology:

"Acres of Brownfields property made ready for reuse" is an aggregate of "acreage assessed that does not require cleanup" and "acreage cleaned up as reported by Assessment Grantees, Regional Targeted Brownfields Assessments, Cleanup Grantees, RLF Grantees, and State and Tribal 128 Voluntary Response Program Grantees for which any required institutional controls are in place."

The unit of measure is acres.

4a. Oversight and Timing of Final Results Reporting:

The ACRES Project Manager is responsible for reporting accomplishments and program results recorded via ACRES.

4b. Data Limitations/Qualifications:

There are some known limitations related to the nature of much of the data being recipient-reported. Regional Project Officers review data to minimize errors (as described above), but some known quality issues remain. Most pertinent to this measure is that outcome data are sometimes not reported by recipients, in the event that the EPA funding expires before the work is complete (for instance, if EPA funding is only part of the funding used for an assessment for cleanup).

Given the reporting cycle and the data entry/QA period, there is typically a several month data lag getting reported data into ACRES.

4c. Third-Party Audits:

No external reviews.

Measure Code: B29 - Brownfield properties assessed.

Office of Solid Waste and Emergency Response (OSWER)

Goal Number and Title:
3 - Cleaning Up Communities and Advancing Sustainable Development

Objective Number and Title:
1 - Promote Sustainable and Livable Communities

Sub-Objective Number and Title:
2 - Assess and Cleanup Brownfields

Strategic Target Code and Title:
1 - By 2015, conduct environmental assessments at 20,600 (cumulative) brownfield properties

Managing Office:
Brownfields

1a. Performance Measure Term Definitions:
Properties Assessed -- Number of properties that have been environmentally assessed for the first time using EPA Brownfields funding. A property will be counted for this measure if the property has not previously been counted for this annual performance measure as a result of other assessments completed with regular EPA Brownfields funding. A "property" is defined as a contiguous piece of land under unitary ownership. A property may contain several smaller components, parcels or areas. "Assessments" can consist of a Phase I assessment, Phase II assessment, and/or supplemental assessments. Assessments are deemed complete when the reports for those assessments are deemed complete. A Phase I assessment report is final when an environmental professional or state official has signed and dated the report as required in the final rule (see 40 CFR 312.21 (c). For Phase II, the report is final when an environmental professional or state official has prepared an environmental assessment report that has been accepted by the grant recipient. For a supplemental assessment, the report is considered final when it has been accepted by the cooperative agreement recipient. For additional information: http://www.epa.gov/brownfields/index.html

2a. Original Data Source:
Assessments are funded either through cooperative agreements, or through EPA contracts (for Targeted Brownfields Assessments (TBAs)). Cooperative agreement recipients (or sub-recipients) and contractors submit performance data to EPA in quarterly reports, and property profile reports. On a limited basis EPA personnel are allowed to update or supplement information when a cooperative agreement has been closed and outcomes have been reported to EPA.

2b. Source Data Collection:

Field sampling is utilized during the assessment process to determine cleanup needs and to develop assessment reports. Formal completion of assessment reports is tabulated for this measure. Data collection is ongoing as projects are implemented. Reporting instructions indicate that accomplishments are to be recorded as they occur.

Assessment Pathways – Assessments meeting this definition can be completed by using funds via an Assessment Award, a Targeted Brownfields Assessment (TBA) or by completing activities funded by 128(a) awards.

Geographic Detail: As of FY12 ACRES leverages a Google Maps application within the system to assign geocoordinates based on address information. Any deviation from these coordinated requires a manual override by the reporting party.

All the Brownfields cooperative agreements have a QA term and condition. Project-level QA documents (i.e. QAPPs) are a minimum requirement for EPA funding of Brownfields activities which include environmental data collection. The program prepares and provides the QA term and condition to the regional offices and requires them to include it in the cooperative agreements. The QA term and condition for Brownfields Assessment cooperative agreements reads as follows:

"B. Quality Assurance (QA) Requirement. 1. When environmental samples are collected as part of the brownfields assessment, the CAR shall comply with 40 CFR Part 31.45 requirements to develop and implement quality assurance practices sufficient to produce data adequate to meet project objectives and to minimize data loss. State law may impose additional QA requirements."

EPA contractors conducting Targeted Brownfiels Assessment should develop site-specific Quality Assurance Project Plans (QAPP) for environmental assessment activities or a site-specific QAPP addendum if a Generic QAPP has already been approved for assessment activities. The EPA requires all environmental monitoring and measurement efforts be conducted in accordance with approved QAPPs. The purpose of the QAPP is to document the project planning process, enhance the credibility of sampling results, produce data of known quality, and potentially save time and money by gathering data that meets the needs of the project and intended use of the data. The QAPP is a formal document describing in comprehensive detail the necessary QA/QC and other technical activities that must be conducted to ensure the results of the work performed will satisfy performance criteria and can be used for their intended purposes. All QA/QC procedures shall be in accordance with applicable professional technical standards, EPA requirements, government regulations and guidelines, and specific project goals and requirements.

OSWER has available the following guidance: "Quality Assurance Guidance for Conducting Brownfields Assessments." EPA 540-R-98-038. 1998.

Attached Documents:

2c. Source Data Reporting:

Cooperative agreement recipients (or sub-recipients) and contractors submit performance data to EPA in quarterly reports, and property profile reports. A Property Profile Form (PPF) collects information (environmental, historical, physical) from a property-specific investigation funded under the Brownfields Program.

Contract Agreement recipients have 3 submission options: complete and submit the Property Profile Form (PPF) in online format connected to the Assessment, Cleanup and Redevelopment Exchange System (ACRES) database; fill out a PPF version in Microsoft Excel format and submit it via e-mail or regular mail to the EPA Regional Representative; or for multiple properties (more than ten properties) fill out a multi-property Excel spreadsheet and submit it via email; or regular mail to the EPA Regional Representative. Any paper forms are entered into ACRES via EPA contractor.

The Property Profile Form is an approved OMB form - OMB No. 2050-0192. Online forms available to registered users here: http://www.epa.gov/brownfields/pubs/index.html
EPA contractors conducting TBAs provide the assessment report to the EPA Region, who in turn enters the data into ACRES. In some cases, the contractor will also provide a filled-out PPF.
In accordance with the Terms and Conditions of the Brownfields Cooperative Agreements, all Brownfields cooperative agreement recipients (CARs) must report accomplishments to EPA on a quarterly basis. Quarterly reports are due 30 days from the end of the federal fiscal quarter.

Attached Documents:

2009_multiple_ppf_template_external.xls

2009_property_profile_form_instructions.pdf

2009_property_profile_form.xls

3a. Relevant Information Systems:

Assessment, Cleanup, and Redevelopment Exchange System (ACRES). This database is the master database of all data supporting OBLR measures. Recipients and EPA report directly into ACRES. It includes source data and transformed data (e.g., data aggregated into Regional totals).
http://www.epa.gov/brownfields/pubs/acres/index.htm provides more information about this database.

ACRES quality is assured by adherence to a security plan and quality management plan:
-- Security plan. The latest version of the security plan for ACRES is dated 11/2009. Contact Ryan Smith in OSWER for a copy of the security plan.

-- Quality Management Plan. ACRES operates under its own Quality Management Plan (Data Quality Management Plan for the Assessment, Cleanup, and Redevelopment Exchange System, Version 1.02), which is updated annually, has been updated as of 2/2010. Contact Ryan Smith for the most recent copy of the QMP.

OSWER Performance Assessment Tool (PAT). This tool serves as the primary external servicing resource for organizing and reporting OSWER's performance data, which collects information from OSWER program systems, and conforms it for uniform reporting and data provisioning. PAT captures data from CERCLIS; replicates business logic used by CERCLIS for calculating measures; delivers that data to EPA staff and managers via a business intelligence dashboard interface for analytic and reporting use; ; and transmits data to BAS. No current system specifications document is currently available for PAT, but will be provided when available. Contact Lisa Jenkins in OSWER regarding questions about PAT.

PAT operates under the OSWER Quality Management Plan (QMP), attached.

PAT has a security certification confirming that a security policy is not necessary because no sensitive data are handled and PAT is built upon the Oracle-based business intelligence system. PAT's security certification indicates that it follows all security guidelines for EPA's Oracle Portal and that PAT is (1) not defined as a "Major Application" according to NIST Special Publication 800-18, Guide for Developing Security Plans for Information Technology Systems, section 2.3.1; (2) does not store, process, or transmit information that the degree of sensitivity is assessed as high by considering the requirements for availability, integrity, and confidentiality according to NIST Special Publication 800-18, Guide for Developing Security Plans for Information Technology Systems, section 3.7.2. (3) is not covered by EPA Order 2100.2A1 Information Technology Capital Planning and Investment Control (CPIC). The security certification, attached, was submitted on 9/11/2008.

Budget Automation System (BAS). BAS is the final repository of the performance values.
Attached Documents:
OSWER QMP printed 2010-03-23.pdf
PAT SecurityCertificationNIST.doc

3b. Data Quality Procedures:

Data reported by cooperative award agreement recipients are reviewed by EPA Regional grant managers for accuracy, to verify activities and accomplishments, and to ensure appropriate interpretation of performance measure definitions.

Step 1. Performance measure data entered into ACRES by recipients and/or EPA HQ contractor (for data submitted by recipients in an alternate format, such as hard copy). For each cooperative agreement recipient,, all data entered are signed off by the EPA Regional Representative (Regional Project Officer) identified in the terms and conditions of the cooperative agreement. For contractors, the EPA Regional COR/WAM signs off on the data.

Step 2. Each Region conducts Regional level review of data from the ACRES system. Rejected data must be edited by the original data source. Approved data proceed to Step 3.

Step 3. HQ conducts National level review (EPA HQ contractors) of data approved by regions. Rejected data must be edited by the region (Step 2). Approved data is stored in ACRES.

Step 4. Each quarter, OSWER Performance Assessment Tool (PAT) database pulls the approved data (performance measure) from ACRES.

Step 5. Headquarters approves PAT results, and PAT pushes results into ACS/Measures Central.

Step 6. ACS/Measures Central aggregates Regional data into a national total. OBLR reporting lead reviews and confirms result

3c. Data Oversight:
Headquarters-level oversight is provided by maintained by the EPA Contract Officer Technical Representative (COTR) There is a Regional Project Officer assigned to each cooperative agreement. That Regional Project Officer is responsible for reviewing for completeness and correctness all data provided by cooperative agreement recipients and data related to Targeted Brownfields Assessment (TBA) contracts; their data is reviewed at the Headquarters level. A list of Regional Project Officers is maintained by the Regional Brownfields Coordinator in each region. Each region also has a data manager (some Regions have a SEE Employee as their data manager). The responsibility of the data manager is to disseminate information about ACRES updates and accomplishments updates. This person serves as the regional point of contact for data related issues.

3d. Calculation Methodology:
"Number of Brownfields properties assessed" is an aggregate of properties assessed using funding from Assessment Grants, Regional TBA funds, and State and Tribal 128 Voluntary Response Program funding. The unit of measure is "Properties"

4a. Oversight and Timing of Final Results Reporting:
The ACRES Project Manager is responsible for reporting accomplishments and program results recorded via ACRES.

4b. Data Limitations/Qualifications:
There are some known limitations related to the nature of much of the data being recipient-reported. Regional Project Officers review data to minimize errors (as described above), but some known quality issues remain. Most pertinent to this measure is that outcome data are sometimes not reported by recipients, in the event that the EPA funding expires before the work is complete (for instance, if EPA funding is only part of

the funding used for an assessment for cleanup).
Given the reporting cycle and the data entry/QA period, there is typically a several month data lag getting reported data into ACRES.

4c. Third-Party Audits:
No external reviews.

Office of Solid Waste and Emergency Response (OSWER)

Goal Number and Title:
3 - Cleaning Up Communities and Advancing Sustainable Development
Objective Number and Title:
2 - Preserve Land
Sub-Objective Number and Title:
2 - Minimize Releases of Hazardous Waste and Petroleum Products
Strategic Target Code and Title:
1 - By 2015,prevent releases at 500 hazardous waste management facilities with initial approved controls
Managing Office:
Office of Resource Conservation and Recovery
1a. Performance Measure Term Definitions:
Definition of "Hazardous Waste Facilities": This universe is composed of facilities that were subject to permits as of 10-1-1997 and subsequent years. EPA updates the list of units that need "updated controls" at the end of each Strategic Plan cycle. Those facilities that need updated controls are a smaller set within the larger permitting universe tracked for strategic and annual goals associated with the Government Performance and Results Act (GPRA). Definition of "New or Updated Controls": Facilities under control is an outcome-based measure as permits or similar mechanisms are not issued until facilities have met standards or permit conditions that are based on human health or environmental standards. Examples include sites cleaned up to a protective level; any groundwater releases controlled so no further attenuation is occurring; any remaining waste safely removed or capped (isolated); and long term controls in place to protect people and the environment at the site, if any contamination remains. New and updated controls ensure that the facilities are designed and operated to maintain continued safe management of hazardous wastes, thus minimizing the potential for releases and accidents, and protecting human health and the environment.
2a. Original Data Source:
States and EPA's Regional offices generate the data.
2b. Source Data Collection:
Facility data: The authorized states have ownership of their data and EPA has to rely on them to make changes. The data that determine if a facility has met its permit requirements are prioritized in update efforts. States and EPA's Regional offices manage data quality related to timeliness and accuracy.
2c. Source Data Reporting:
Data can be entered directly into RCRAInfo, submitted by the facility through the myRCRAid interface, or some use a different approach and then "translate" the information into RCRAInfo. Supporting documentation and reference materials are maintained in Regional and state files. Users log into the following URL: https://rcrainfo.epa.gov/

3a. Relevant Information Systems:
RCRAInfo, the national database which supports EPA's RCRA program, contains information on entities (generically referred to as "handlers") engaged in hazardous waste generation and management activities regulated under the portion of the Resource Conservation and Recovery Act (RCRA) that provides for regulation of hazardous waste. RCRAInfo has several different modules, and allows for tracking of information on the regulated universe of RCRA hazardous waste handlers, such as facility status, regulated activities, and compliance history. The system also captures detailed data on the generation of hazardous waste by large quantity generators and on waste management practices from treatment, storage, and disposal facilities. RCRAInfo is web accessible, providing a convenient user interface for Federal, state and local managers, encouraging development of in-house expertise for controlled cost, and states have the option to use commercial off-the-shelf software to develop reports from database tables. Access to RCRAInfo is open only to EPA Headquarters, Regional, and authorized state personnel. It is not available to the general public because the system contains enforcement sensitive data. The general public is referred to EPA's Envirofacts Data Warehouse to obtain information on RCRA-regulated hazardous waste sites. This non-sensitive information is supplied from RCRAInfo to Envirofacts.
3b. Data Quality Procedures:
Within RCRAInfo, the application software contains structural controls that promote the correct entry of the high-priority national components. Starting in January 2013, the RCRAInfo Team started an in-depth State-by-State data quality project. This entails taking a single state's data, processing it through the business rules, and identifying any errors. Once the errors are identified, they are sent to the State and the Regional representative for resolution. The RCRAInfo Team works closely with the State, in conjunction with the Region, to resolve all errors in RCRAInfo. Eleven (11) States have been processed thus far. This project will continue until all 57 entities that report to RCRAInfo have been processed and corrected.
3c. Data Oversight:
The Information Collection and Analysis Branch (ICAB) maintains a list of the Headquarters, Regional, and delegated state/territory users and controls access to the system. Branch members ensure data collection is on track, conduct QA reports, and work with Regional and state partners to resolve issues as they are discovered.
3d. Calculation Methodology:
Determination of whether or not the facility has approved controls in place is based primarily on the legal and operating status codes for each unit. Accomplishment of updated controls is based on the permit expiration date code and other related codes. The baseline is composed of facilities that can have multiple units. These units may consolidate, split or

undergo other activities that cause the number of units to change. There may be occasions where there needs to be minor baseline modifications. The larger permitting universe is carried over from one EPA strategic planning cycle to the next (starting with facilities subject to permits as of October 1, 1997) with minor changes made with each subsequent strategic planning cycle (e.g., facilities referred to Superfund are removed, or facilities never regulated are removed; facilities that applied for a permit within the last strategic cycle are added). EPA updates the list of units that need "updated controls" after the end of each strategic planning cycle. Those facilities that need updated controls are a smaller set within the larger permitting universe.

Complete data dictionary is available at: http://www.epa.gov/enviro/html/rcris/rcris_table.html

The unit of analysis for this measure is "facilities."

4a. Oversight and Timing of Final Results Reporting:

Program Implementation and Information Division (PIID) data analysts are responsible for the reporting.

4b. Data Limitations/Qualifications:

Even with the increasing emphasis on data quality, with roughly 10,000 units in the baseline (e.g., a facility can have more than one unit), there are problems with the number of facilities in the baseline and their supporting information, particularly with the older inactive facilities. EPA Headquarters works with the EPA Regional offices to resolve them.

Basic site data may become out-of-date because RCRA does not mandate the notification of all information changes. Nevertheless, EPA tracks the facilities by their ID numbers and those should not change even during ownership changes (RCRA Subtitle C EPA Identification Number, Site Status, and Site Tracking Guidance, March 21, 2005).

4c. Third-Party Audits:

The 1995 U.S. Government Accountability Office (GAO) report Hazardous Waste: Benefits of EPA's Information System Are Limited, AIMD-95-167, August 22, 1995, http://www.gao.gov/archive/1995/ai95167.pdf (accessed December 17, 2013) on EPA's Hazardous Waste Information System reviewed whether national RCRA information systems support EPA and the states in managing their hazardous waste programs. Those recommendations coincided with ongoing internal efforts to improve the definitions of data collected, and ensure that data collected provide critical information and minimize the burden on states. RCRAInfo, the current national database, has evolved in part as a response to this report. The "Permitting and Corrective Action Program Area Analysis" was the primary vehicle for the improvements made in the December 2008 release (V4).

EPA OIG report:

U.S. Environmental Protection Agency. "Permitting and Corrective Action Program Area Analysis". WIN/INFORMED Executive Steering Committee, July 28, 2005.

Measure Code: CA5 - Cumulative percentage of RCRA facilities with final remedies constructed.

Office of Solid Waste and Emergency Response (OSWER)

Goal Number and Title:	
3 - Cleaning Up Communities and Advancing Sustainable Development	
Objective Number and Title:	
3 - Restore Land	
Sub-Objective Number and Title:	
3 - Cleanup Contaminated Land	
Strategic Target Code and Title:	
4 - By 2015, , increase the number of RCRA facilities with final remedies constructed	
Managing Office:	
Office of Resource Conservation and Recovery	
1a. Performance Measure Term Definitions:	
The remedy construction measure tracks the RCRA Corrective Action Program's progress in moving sites towards final cleanup. For background information, please visit: http://www.epa.gov/osw/hazard/correctiveaction/index.htm Cumulative – made up of accumulated parts; increasing by successive additions RCRA Facilities – facilities subject to restriction or action from the Resource Conservation and Recovery Act; Definition of "final remedies constructed": The lead regulators (delegated state or EPA Region) for the facility select the remedy and determine when the facility has completed construction of that remedy. EPA collects the determinations as made by the lead regulator and this total is used for this measure.	
2a. Original Data Source:	
States and regions enter all data on determinations made.	
2b. Source Data Collection:	
Known and suspected facility-wide conditions are evaluated using a series of simple questions and flow-chart logic to arrive at a reasonable, defensible determination. These questions were issued as a memorandum titled: Interim Final Guidance for RCRA Corrective Action Environmental Indicators, Office of Solid Waste, February 5, 1999).The lead regulators (delegated state or EPA Region) for the facility select the remedy and determine when the facility has completed construction of that remedy. Construction completions are collected on both an area-wide and site-wide basis. States and regions generate the data and manage data quality related to timeliness and accuracy (i.e., the environmental conditions and determinations are correctly reflected by the data). EPA has provided guidance and training to states and regions to help ensure consistency in those determinations. RCRAInfo	

documentation, available to all users on-line, provides guidance to facilitate the generation and interpretation of data.

2c. Source Data Reporting:

The remedy construction measure tracks the RCRA Corrective Action Program's progress in moving sites towards final cleanup. The date of remedy construction is entered in the database. (EPA makes the same kind of entry related to facilities in non-delegated states.)

3a. Relevant Information Systems:

RCRAInfo, the national database which supports EPA's RCRA program, contains information on entities (generically referred to as "handlers") engaged in hazardous waste generation and management activities regulated under the portion of the Resource Conservation and Recovery Act (RCRA) that provides for regulation of hazardous waste.

RCRAInfo has several different modules, and allows for tracking of information on the regulated universe of RCRA hazardous waste handlers, such as facility status, regulated activities, and compliance history. The system also captures detailed data on the generation of hazardous waste by large quantity generators and on waste management practices from treatment, storage, and disposal facilities. Within RCRAInfo, the Corrective Action Module tracks the status of facilities that require, or may require, corrective actions, including the information related to the performance measure.

RCRAInfo is web accessible, providing a convenient user interface for Federal, state and local managers, encouraging development of in-house expertise for controlled cost, and states have the option to use commercial off-the-shelf software to develop reports from database tables.

 RCRAInfo is currently at Version 5 (V5), which was released in March 2010. V5 expanded on V4's capabilities and made updates to the Handler module to support two new rules that went into effect in 2009.

Access to RCRAInfo is open only to EPA Headquarters, Regional, and authorized state personnel. It is not available to the general public because the system contains enforcement sensitive data. The general public is referred to EPA's Envirofacts Data Warehouse to obtain information on RCRA-regulated hazardous waste sites. This non-sensitive information is supplied from RCRAInfo to Envirofacts. For more information, see:
http://www.epa.gov/enviro/index.html

Find more information about RCRAInfo at: http://www.epa.gov/enviro/html/rcris/index.html

3b. Data Quality Procedures:

Manual procedures: EPA Corrective Action sites are monitored on a facility-by-facility basis and QA/QC procedures are in place to ensure data validity.

Automated procedures: Within RCRAInfo, the application software enforces structural controls that ensure that high-priority national components of the data are properly entered. Training on use of RCRAInfo is provided on a regular basis, usually annually, depending on the nature of systems changes and user needs.

The latest version of RCRAInfo, Version 5 (V5), was released in March 2010 and has many added components that will help the user identify errors in the system.

3c. Data Oversight:

The Information Collection and Analysis Branch (ICAB) maintains a list of the Headquarters, Regional and delegated state/territory users and controls access to the system. Branch members ensure data collection is on track, conduct QA reports, and work with Regional and state partners to resolve issues as they are discovered.

3d. Calculation Methodology:

The remedy construction measure tracks the RCRA Corrective Action Program's progress in moving sites towards final cleanup. Like with the environmental indicators determination, the lead regulators for the facility select the remedy and determine when the facility has completed construction of that remedy. Construction completions are collected on both an area-wide and site-wide basis for sake of the efficiency measure.

4a. Oversight and Timing of Final Results Reporting:

Program Implementation and Information Division (PIID) data analysts are responsible for the reporting.

4b. Data Limitations/Qualifications:

With emphasis on data quality, EPA Headquarters works with the EPA Regional offices to ensure the national data pulls are consistent with the Regional data pulls.

4c. Third-Party Audits:

US Government Accountability Office (GAO) report: "Hazardous Waste: Early Goals Have Been Met in EPA's Corrective Action Program, but Resource and Technical Challenges Will Constrain Future Progress." (GAO-11-514, August 25, 2011, http://www.gao.gov/assets/330/321743.pdf

Measure Code: CA2 - Cumulative percentage of RCRA facilities with migration of contaminated groundwater under control.

Office of Solid Waste and Emergency Response (OSWER)

Goal Number and Title:
3 - Cleaning Up Communities and Advancing Sustainable Development

Objective Number and Title:
3 - Restore Land

Sub-Objective Number and Title:
3 - Cleanup Contaminated Land

Strategic Target Code and Title:
3 - By 2015, increase number of Resource Conservation and Recovery Act (RCRA) facilities

Managing Office:
Office of Resource Conservation and Recovery

1a. Performance Measure Term Definitions:
The performance measure is used to track the RCRA Corrective Action Program's progress in dealing with immediate threats to human health and groundwater resources. It is meant to summarize and report on facility-wide environmental conditions at RCRA Corrective Action Program's facilities nation-wide. For background information, please visit: http://www.epa.gov/osw/hazard/correctiveaction/index.htm   Cumulative – made up of accumulated parts; increasing by successive additions   RCRA Facilities – facilities subject to restriction or action from the Resource Conservation and Recovery Act;   Migration – to change position; movement from one location to another   Contaminated Groundwater – water in the subsurface which has become tainted with any number of dissolved contaminants at levels greater than the prescribed environmental standard levels   Definition of "under control": Known and suspected facility-wide conditions are evaluated using a series of simple questions and flow-chart logic to arrive at a reasonable, defensible determination. These questions were issued as a memorandum titled: Interim Final Guidance for RCRA Corrective Action Environmental Indicators, Office of Solid Waste, February 5, 1999). Lead regulators (delegated state or EPA Region) for the facility (authorized state or EPA) make the environmental indicator determination, but facilities or their consultants may assist EPA in the evaluation by providing information on the current environmental conditions. The determinations are entered directly into RCRAInfo. EPA collects the determinations as made by the lead regulator, and this total is used for this measure.

2a. Original Data Source:
States and regions enter all data on determinations made.

| 2b. Source Data Collection: |

Known and suspected facility-wide conditions are evaluated using a series of simple questions and flow-chart logic to arrive at a reasonable, defensible determination. These questions were issued as a memorandum titled: Interim Final Guidance for RCRA Corrective Action Environmental Indicators, Office of Solid Waste, February 5, 1999). Lead regulators for the facility (authorized state or EPA) make the environmental indicator determination (like whether migration of contaminated groundwater is under control), but facilities or their consultants may assist EPA in the evaluation by providing information on the current environmental conditions.

States and regions generate the data and manage data quality related to timeliness and accuracy (i.e., the environmental conditions and determinations are correctly reflected by the data). EPA has provided guidance and training to states and regions to help ensure consistency in those determinations. RCRAInfo documentation, available to all users on-line, provides guidance to facilitate the generation and interpretation of data.

2c. Source Data Reporting:

States: With respect to releases to groundwater controlled, a "yes," "no", or "insufficient information" entry is made in the database. (EPA makes the same kind of entry related to facilities in non-delegated states.)

3a. Relevant Information Systems:

RCRAInfo, the national database which supports EPA's RCRA program, contains information on entities (generically referred to as "handlers") engaged in hazardous waste generation and management activities regulated under the portion of the Resource Conservation and Recovery Act (RCRA) that provides for regulation of hazardous waste.

RCRAInfo has several different modules, and allows for tracking of information on the regulated universe of RCRA hazardous waste handlers, such as facility status, regulated activities, and compliance history. The system also captures detailed data on the generation of hazardous waste by large quantity generators and on waste management practices from treatment, storage, and disposal facilities. Within RCRAInfo, the Corrective Action Module tracks the status of facilities that require, or may require, corrective actions, including the information related to the performance measure.

RCRAInfo is web accessible, providing a convenient user interface for Federal, state and local managers, encouraging development of in-house expertise for controlled cost, and states have the option to use commercial off-the-shelf software to develop reports from database tables.

 RCRAInfo is currently at Version 5 (V5), which was released in March 2010. V5 expanded on V4's capabilities and made updates to the Handler module to support two new rules that went into effect in 2009.

Access to RCRAInfo is open only to EPA Headquarters, Regional, and authorized state personnel. It is not available to the general public because the system contains enforcement sensitive data. The general public is referred to EPA's Envirofacts Data Warehouse to obtain information on RCRA-regulated hazardous waste sites. This non-sensitive information is supplied from RCRAInfo to Envirofacts. For more information, see: http://www.epa.gov/enviro/index.html

Find more information about RCRAInfo at: http://www.epa.gov/enviro/html/rcris/index.html

3b. Data Quality Procedures:

Manual procedures: EPA Corrective Action sites are monitored on a facility-by-facility basis and QA/QC procedures are in place to ensure data validity.

Automated procedures: Within RCRAInfo, the application software enforces structural controls that ensure that high-priority national components of the data are properly entered. Training on use of RCRAInfo is provided on a regular basis, usually annually, depending on the nature of systems changes and user needs. The latest version of RCRAInfo, Version 5 (V5), was released in March 2010 and has many added components that will help the user identify errors in the system.

3c. Data Oversight:

The Information Collection and Analysis Branch (ICAB) maintains a list of the Headquarters, Regional and delegated state/territory users and controls access to the system. Branch members ensure data collection is on track, conduct QA reports, and work with Regional and state partners to resolve issues as they are discovered.

3d. Calculation Methodology:

Annual progress for each measure is found by subtracting the cumulative progress at the end of the previous fiscal year from the cumulative progress at the end of the current fiscal year.

4a. Oversight and Timing of Final Results Reporting:

Program Implementation and Information Division (PIID) data analysts are responsible for the reporting.

4b. Data Limitations/Qualifications:

With emphasis on data quality, EPA Headquarters works with the EPA Regional offices to ensure the national data pulls are consistent with the Regional data pulls.

4c. Third-Party Audits:

US Government Accountability Office (GAO) report: "Hazardous Waste: Early Goals Have Been Met in EPA's Corrective Action Program, but Resource and Technical Challenges Will Constrain Future Progress." (GAO-11-514, August 25, 2011, http://www.gao.gov/assets/330/321743.pdf

Measure Code: CA1 - Cumulative percentage of RCRA facilities with human exposures to toxins under control.

Office of Solid Waste and Emergency Response (OSWER)

Goal Number and Title:
3 - Cleaning Up Communities and Advancing Sustainable Development
Objective Number and Title:
3 - Restore Land
Sub-Objective Number and Title:
3 - Cleanup Contaminated Land
Strategic Target Code and Title:
2 - By 2015, increase the number of Superfund final and deleted NPL sites and RCRA facilities
Managing Office:
Office of Resource Conservation and Recovery
1a. Performance Measure Term Definitions:
The performance measure is used to track the RCRA Corrective Action Program's progress in dealing with immediate threats to human health and groundwater resources. It is meant to summarize and report on facility-wide environmental conditions at RCRA Corrective Action Program's facilities nation-wide. For background information, please visit: http://www.epa.gov/osw/hazard/correctiveaction/index.htm   Cumulative – made up of accumulated parts; increasing by successive additions   RCRA Facilities – facilities subject to restriction or action from the Resource Conservation and Recovery Act;   Human Exposure to Toxins – pathways or means by which toxic substances may come into direct contact with a person   Definition of "under control": Known and suspected facility-wide conditions are evaluated using a series of simple questions and flow-chart logic to arrive at a reasonable, defensible determination. These questions were issued as a memorandum titled: Interim Final Guidance for RCRA Corrective Action Environmental Indicators, Office of Solid Waste, February 5, 1999). Lead regulators (delegated state or EPA Region) for the facility (authorized state or EPA) make the environmental indicator determination, but facilities or their consultants may assist EPA in the evaluation by providing information on the current environmental conditions. The determinations are entered directly into RCRAInfo. EPA collects the determinations as made by the lead regulator, and this total is used for this measure.
2a. Original Data Source:
States and regions enter all data on determinations made.
2b. Source Data Collection:
Known and suspected facility-wide conditions are evaluated using a series of simple questions and flow-chart logic to arrive at a reasonable, defensible determination. These questions were issued as a memorandum titled: Interim Final Guidance for RCRA Corrective Action Environmental Indicators, Office of Solid Waste,

February 5, 1999). Lead regulators for the facility (authorized state or EPA) make the environmental indicator determination (like whether human exposures to toxins are under control), but facilities or their consultants may assist EPA in the evaluation by providing information on the current environmental conditions.

States and regions generate the data and manage data quality related to timeliness and accuracy (i.e., the environmental conditions and determinations are correctly reflected by the data). EPA has provided guidance and training to states and regions to help ensure consistency in those determinations. RCRAInfo documentation, available to all users on-line, provides guidance to facilitate the generation and interpretation of data.

2c. Source Data Reporting:

States: With respect to meeting the human exposures to toxins controlled a "yes," "no", or "insufficient information" entry is made in the database. (EPA makes the same kind of entry related to facilities in non-delegated states.)

3a. Relevant Information Systems:

RCRAInfo, the national database which supports EPA's RCRA program, contains information on entities (generically referred to as "handlers") engaged in hazardous waste generation and management activities regulated under the portion of the Resource Conservation and Recovery Act (RCRA) that provides for regulation of hazardous waste.

RCRAInfo has several different modules, and allows for tracking of information on the regulated universe of RCRA hazardous waste handlers, such as facility status, regulated activities, and compliance history. The system also captures detailed data on the generation of hazardous waste by large quantity generators and on waste management practices from treatment, storage, and disposal facilities. Within RCRAInfo, the Corrective Action Module tracks the status of facilities that require, or may require, corrective actions, including the information related to the performance measure.

RCRAInfo is web accessible, providing a convenient user interface for Federal, state and local managers, encouraging development of in-house expertise for controlled cost, and states have the option to use commercial off-the-shelf software to develop reports from database tables.

RCRAInfo is currently at Version 5 (V5), which was released in March 2010. V5 expanded on V4's capabilities and made updates to the Handler module to support two new rules that went into effect in 2009.

Access to RCRAInfo is open only to EPA Headquarters, Regional, and authorized state personnel. It is not available to the general public because the system contains enforcement sensitive data. The general public is referred to EPA's Envirofacts Data Warehouse to obtain information on RCRA-regulated hazardous waste sites. This non-sensitive information is supplied from RCRAInfo to Envirofacts. For more information, see: http://www.epa.gov/enviro/index.html

Find more information about RCRAInfo at: http://www.epa.gov/enviro/html/rcris/index.html

3b. Data Quality Procedures:

Manual procedures: EPA Corrective Action sites are monitored on a facility-by-facility basis and QA/QC procedures are in place to ensure data validity.
Automated procedures: Within RCRAInfo, the application software enforces structural controls that ensure that high-priority national components of the data are properly entered. Training on use of RCRAInfo is provided on a regular basis, usually annually, depending on the nature of systems changes and user needs. The latest version of RCRAInfo, Version 5 (V5), was released in March 2010 and has many added components that will help the user identify errors in the system.

3c. Data Oversight:

The Information Collection and Analysis Branch (ICAB) maintains a list of the Headquarters, Regional and delegated state/territory users and controls access to the system. Branch members ensure data collection is on track, conduct QA reports, and work with Regional and state partners to resolve issues as they are discovered.

3d. Calculation Methodology:

Annual progress for each measure is found by subtracting the cumulative progress at the end of the previous fiscal year from the cumulative progress at the end of the current fiscal year.

4a. Oversight and Timing of Final Results Reporting:

Program Implementation and Information Division (PIID) data analysts are responsible for the reporting.

4b. Data Limitations/Qualifications:

With emphasis on data quality, EPA Headquarters works with the EPA Regional offices to ensure the national data pulls are consistent with the Regional data pulls.

4c. Third-Party Audits:

US Government Accountability Office (GAO) report: "Hazardous Waste: Early Goals Have Been Met in EPA's Corrective Action Program, but Resource and Technical Challenges Will Constrain Future Progress." (GAO-11-514, August 25, 2011, http://www.gao.gov/assets/330/321743.pdf

Measure Code: ST6 - Increase the percentage of UST facilities that are in significant operational compliance (SOC) with both release detection and release prevention requirements by 0.5% over the previous year's target.

Office of Solid Waste and Emergency Response (OSWER)

Goal Number and Title:
3 - Cleaning Up Communities and Advancing Sustainable Development

Objective Number and Title:
2 - Preserve Land

Sub-Objective Number and Title:
2 - Minimize Releases of Hazardous Waste and Petroleum Products

Strategic Target Code and Title:
2 - Through 2015, increase the percentage of UST facilities in significant operational compliance(SOC)

Managing Office:
Office of Underground Storage Tanks

1a. Performance Measure Term Definitions:
The most current definitions for the EPA's performance measures related to underground storage tanks are available on EPA's website www.epa.gov/oust/cat/camarchv.htm under Definitions. See the definition for the measure number UST-6 in the definitions document. For more information on EPA's Underground Storage Tanks Program, see: http://www.epa.gov/oust/index.htm

2a. Original Data Source:
The data suppliers are the states and territories who are the direct implementers of the program in their respective jurisdictions and the EPA regions who provide assistance to the tribes.

2b. Source Data Collection:
The data is collected by each state and territory using their own systems and databases. They then report this information to OUST using the LUST4 system described under section 3. EPA Quality Assurance Requirements/Guidance under Which Original Data Sources Collect Data: For cooperative agreements: Regional offices include QA Terms and Conditions in their states' assistance agreement. CAs must be current and specify: QA roles and responsibilities for EPA and grantee recipients; and quality requirements including responsibilities for final review and approval. Default quality requirements include: organization-level QA documentation (i.e. QMP) for state agencies and primary contractors; and project-level QAPPs for each CA. In accordance with EPA's Uniform Administrative Requirements for Grants and Cooperative Agreements, 40 CFR Part 31.45, states must develop and implement quality assurance practices. The regulation requires developing and implementing quality assurance practices that will "produce data of quality adequate to meet project objectives and to minimize loss of data to out of control conditions or malfunctions'; see OSWER Directive 9650.10A: www.epa.gov/oust/directiv/d965010a.htm - sec11" For contracts: EPA Regions determine which quality requirements are applicable. Contracts must be current and specify: QA roles and responsibilities for EPA and national LUST contractors; and quality requirements

including responsibilities for final review and approval. Default quality requirements include: organization-level QA documentation (i.e. QMP) for the primary contractors; and project-level QAPPs for each Tribal LUST remedial Work Assignment. Sample EPA contract language: "the Contractor shall comply with the higher-level quality standard selected below: Specifications and Guidelines for Quality Systems for Environmental Data Collection and Environmental Technology Programs (ANSI/ASQC E4, 1994). As authorized by FAR 52.246-11, the higher-level quality standard ANSI/ASQC E4 is tailored as follows: The solicitation and contract require the offerors/contractor to demonstrate conformance to ANSI/ASQC E4 by submitting the quality documentation described below. [Specifically,...] ... The Contractor shall not commence actual field work until until the Government has approved the quality documentation (i.e., QAPP)."

Note: Regions keep copies of individual QAPPs associated with cooperative agreements and contracts. Each EPA regional office manages its own state and tribal assistance agreements.

2c. Source Data Reporting:

Data Submission Instrument:
State-specific databases.

Data Entry Mechanism:
Each state enters their data into the online LUST4 Oracle-based system (see section 3 for more details).

Frequency of Data Transmission to EPA: Twice annually.

Timing of Data Transmission to EPA:
Within 10 days of the end of the reporting period (by April 10 for mid-year, and October 10 for end-of-year).

3a. Relevant Information Systems:

System Description:
LUST4. This database is the master database of all UST program-related data. States, territories and EPA report data for activity and measures directly into LUST4. LUST4 's Oracle Web-based system is accessed through the EPA portal at http://portal.epa.gov/ under the My Communities/Underground Storage Tank Menu Page.

OSWER Performance Assessment Tool (PAT). This tool serves as the primary external servicing resource for organizing and reporting OSWER's performance data. PAT collects information from OSWER program systems, and conforms it for uniform reporting and data provisioning. PAT captures data from LUST4; replicates business logic used by LUST4 for calculating measures; can deliver that data to EPA staff and managers via a business intelligence dashboard interface for analytic and reporting use; enables LUST point of contact to document status and provide explanation for each measure; and transmits data to EPA's Budget Automation System (BAS). No current system specifications document is currently available, but will be provided when available.

BAS. BAS is the final repository of the performance values.

Source/Transformed Data:

LUST4. LUST4 includes both source data and transformed data (e.g., data aggregated into Regional totals).

PAT. PAT includes only transformed data.

BAS. BAS includes only transformed data.

Information System Integrity Standards:

LUST4. LUST4 operates under OSWER's QMP, including the security policy specified in that QMP. LUST4 does not have any stand-alone certifications related to the EPA security policy or the Systems Life Cycle Management policy. The LUST4 system is built upon Oracle Business Intelligence tools provided by the EPA Business Intelligence Analytics Center, which ensures that a stand-alone security certification is not necessary.

PAT. PAT operates under the OSWER Quality Management Plan (QMP). PAT has a security certification confirming that a security policy is not necessary because no sensitive data are handled and PAT is built upon the Oracle-based business intelligence system. PAT's security certification indicates that it follows all security guidelines for EPA's Oracle Portal and that PAT is (1) not defined as a "Major Application" according to NIST Special Publication 800-18, Guide for Developing Security Plans for Information Technology Systems, section 2.3.1; (2) does not store, process, or transmit information that the degree of sensitivity is assessed as high by considering the requirements for availability, integrity, and confidentiality according to NIST Special Publication 800-18, Guide for Developing Security Plans for Information Technology Systems, section 3.7.2. (3) is not covered by EPA Order 2100.2A1 Information Technology Capital Planning and Investment Control (CPIC).

BAS. Not applicable.

3b. Data Quality Procedures:

EPA's regional grants project officers and regional program managers provide first-level data quality reviews and oversight of their recipients' program performance measure results. EPA/OUST reviews, comments and approves each record.

OUST uses a combination of automated validation along with manual QA/QC review.

QA/QC REVIEW BY REGIONS. EPA/OUST oversees the use of the QA/QC checklist, which is incorporated into the LUST4 oracle web-based system. Regions complete the QA/QC checklist, sign it electronically and submit it to EPA/OUST for review, comment and approval of each record.

NOTE: This QA/QC checklist was last updated 10/1/2009 and is accessed through the user interface of LUST4.

Regional QA/QC Evaluation Checklist –

Note: Checklist is to be completed by Regional reviewer and will appear "shaded" to others.

Actions This Reporting Period

Compare this reported percentage to the last three reporting periods for the data submitter. If the current number deviates by more than 10 percentage points from the last period's number or appears otherwise questionable, complete the following actions:

-- Review the state's explanation, if available.

-- If necessary, contact the state to obtain the corrected numbers and/or obtain a sufficient explanation and include the explanation in the "Comments" section for the applicable performance measure.

-- Verify that the numbers being reported were calculated based on the last 12-months of inspections.

AUTOMATED VALIDATION. The LUST4 systems provides an error message when the combined SOC rate (ST6) is higher than either of the individual SOC rates. Also, the system provides an error if the following formula is false:

100% minus UST-4 plus 100% minus UST-5 minus 100% minus UST-6 is equal to or greater than zero (e.g., if UST-4 and UST-5 are 75% and 70%, respectively, UST-6 cannot be lower than 45%. The calculation is ((100%-75%) + (100%-70%))

- (100%-45%) = 0)

EPA/OUST provides second-level data quality reviews of all data.

DATA FLOW:

Step 1. SOC data are entered into LUST4 by state recipients or by Regions (for tribal data).

Step 2. Each Region conducts Regional level review of data from the LUST4 system. Rejected data must be edited by the original data source. Approved data proceed to Step 3.

Step 3. Headquarters' staff perform performs National Program Review, using data from the LUST4 system. Rejected data must be reviewed by the region and, if needed, pushed back to the state for editing(Step 2).

Step 4. PAT pulls data from LUST4. Headquarters staff compare PAT results to LUST4 results. If PAT does not match LUST4 then there was an error with the upload and data is reloaded. Headquarters staff enter into PAT the ACS status information of "Indicator" for each measure and, if desired, explanation.(Note: PAT allows for programs to identify status other than "Indicator." When programs select a status of "no status," "data not available," or "target not met," PAT requires that an explanation be provided. LUST program policy is to resolve all reporting issues prior to ACS reporting, so "Indicator" is the only status chosen and explanations for that status are optional.)

Step 5. Headquarters approves PAT results, and PAT pushes results into BAS/Measures Central.

Step 6. Measures Central aggregates Regional data into a national total. OUST reporting lead reviews and certifies results.

3c. Data Oversight:
Source Data Reporting Oversight Personnel: Regional Program Managers are ultimately responsible for regional-level data.

Source Data Reporting Oversight Responsibilities:
Regional Program Managers conduct their review based upon a national QA/QC checklist, as described in the Data Quality Procedures field.

Information Systems Oversight Personnel:
OUST LUST4 System Manager

Information Systems Oversight Responsibilities:
Maintains a list of the HQ (OUST and OEI), Regional and state/territory primary and backup users; a record of changes to the list is also maintained. Ensures that Regional reporting is on track, conducts QA on LUST performance measures, ensures QA issues are resolved and/or documented, and oversees final reporting to BAS. Works with OUST contractor to resolve any issues with the LUST4 data system.

3d. Calculation Methodology:

Users report SOC by taking the the total number of facilities inspected during the last 12 months that were in significant operational compliance and dividing by the total number of inspections conducted in the last 12 months. For example, 80 facilities inspected in the last 12 months by the state were in compliance out of 100 inspected during this time period, therefore their SOC is 80%.

To calculate the national SOC rate, OUST multiplies each state's SOC rate by their active USTs for the reporting period ("estimated USTs in SOC"). Then OUST sums all of the estimated USTs in SOC, and divides this number by the total number of active USTs in the country.

Unit of Measure: Percent.

Timeframe of Result: Annual, reported twice a year on a rolling 12-month basis. In other words, the April reporting period reflects the SOC of inspections from April through March and the October reporting period reflects the SOC of inspections from Octiber through September.

Documentation of Methodological Changes:
In FY2004 OUST began collecting the combined SOC measure (ST6). From 2001 through 2004 OUST only collected SOC for release detection and SOC for release prevention. From 1997-2000 OUST collected percentages of UST systems equipped to comply with OUST's 1998 regulatory requirements. The changes reflect a move to document the percentage of facilities that were properly operating their required UST equipment after nearly all USTs had the required equipment.

4a. Oversight and Timing of Final Results Reporting:

Final Reporting Oversight Personnel:
Deputy Office Director.

Final Reporting Oversight Responsibilities:
Responsible for final review to ensure LUST 4 System Manager has completed review, and numbers are

accurate.
Final Reporting Timing: Semiannual.

4b. Data Limitations/Qualifications:
Data quality depends on the accuracy and completeness of state records. Also, some states rely on local jurisdictions for their data, which can cause delays for these states. Additionally, the tanks program is primarily run by states, and each state operates their program in a manner that works best for them. Because there are differences between all states, the data from each state can be influenced by the policies and interpretations of each state. This creates limitations when someone compares state-level data.

4c. Third-Party Audits:
None.

Measure Code: 338 - Percent of all SPCC inspected facilities found to be non-compliant which are brought into compliance.

Office of Solid Waste and Emergency Response (OSWER)

Goal Number and Title:
3 - Cleaning Up Communities and Advancing Sustainable Development

Objective Number and Title:
3 - Restore Land

Sub-Objective Number and Title:
2 - Emergency Preparedness and Response

Strategic Target Code and Title:
0 -

Managing Office:
Office of Emergency Management

1a. Performance Measure Term Definitions:
1.   SPCC facility: A facility which must prepare, amend and implement a Spill Prevention, Control and Countermeasure Plan (SPCC) to demonstrate the ability to prevent, prepare and respond to oil discharges to navigable waters and adjoining shorelines.
2.   Initially compliant: The tag used to describe facilities that are not in violation of the requirements in the SPCC regulation of the Oil Pollution Act upon inspection in a given fiscal year.
3.   Non-compliant: The tag used to describe a facility that upon inspection is deemed to be in violation of the SPCC regulation. Non-compliant facilities are brought into compliance through EPA follow-up activity.
4.   Brought into compliance: The tag used to describe a facility that was found non-compliant at some point during the measurement period and then was then found to be in compliance after a follow-up inspection activity.
5.   Not Subject: The tag used to describe a facility that, upon inspection, is not required to comply with the SPCC regulation. Facility may have been subject previously but has made changes to the facility whereby it is no longer subject to the SPCC regulation.
6.   Closed: The tag used to describe a facility that, upon inspection, is closed and no longer subject to the SPCC rule. Facility may have been subject previously.
7.   Carry Over: The tag used to describe facilities that were found non-compliant at some point before the current FY in the measurement period and remain non-compliant as of the beginning of the current FY.
8.   Oil Database Application: The online database that is used to collect and store facility and inspection-related data for SPCC facilities. The Oil Database is not a public database.

2a. Original Data Source:
The primary data used to calculate the oil measures are the number and outcomes of facility inspections and the dates in which non-compliant facilities are brought into compliance. Specifically, calculation of the oil measures involves the recording of the inspection activity and the outcome (in compliance vs. out of compliance) and subsequent tracking of the dates when non-compliant SPCC facilities are brought into compliance. Starting with FY 2011, the reported value of facilities brought into compliance in any given FY must also capture inspections that occurred prior to the current FY (using FY 2010 as the base year).

402

2b. Source Data Collection:

In order to calculate the measure for each FY, the following data must be collected and entered by the regions for each inspection conducted:

1.   Facility Identifier – identifier assigned by the Oil Database Application used to track the activity for each unique facility.

2.   Inspection (Activity) Tracking Number – also assigned by the Oil Database Application, but prior to October 2011, these tracking numbers were Region-specific.

3.   Inspection Date (Activity Start/Schedule Date in the Oil Database Application) – date of the initial inspection or most recent compliance inspection.

4.   Inspection Outcome – either in compliance or not in compliance.

5.   Date of Confirmed Compliance (Brought into Compliance Date in the Oil Database Application) – date the facility is verified to be in compliance overall; could occur in subsequent FYs.

The business rules for entering data into the "Compliance Module" of the Oil Database Application used to calculate the end-of-year measures results are as follows:

·   Inspection Outcomes: When a facility is inspected for SPCC compliance, the outcome of the inspection should be recorded in the application. If the facility is found to be out of compliance, the outcome should be recorded as "out of compliance." If the facility is found to be compliant, then the outcome should be recorded as "in compliance."

·   "Not-Subject Facilities" Reporting: If an inspected facility is found to be not subject to SPCC regulations, the outcome of the inspection should be recorded as "in compliance".

·   Tracking Multiple Inspections: If there are multiple inspections in the same FY for a facility, the SPCC inspection will be counted as a single inspection. Multiple SPCC inspections would not be added together by the Oil Database Application for the same facility. If there are multiple inspections in the same FY for a facility, in order to determine compliance status of the facility, the application will look at the activities to determine if the most recent activity has an "Inspection Outcome" or a "Brought into Compliance" date. If an "Inspection Outcome" or a "Brought into Compliance" date doesn't exist on the most recent activity, the Oil Database Application will look at the previous activity within the FY to determine the facility's compliance status.

·   Tracking Compliance vs. Enforcement: Once a facility is brought into compliance, it should be captured in the "Inspection Outcome" and "Brought into Compliance" fields in the application. If additional enforcement is pursued by the Region after a facility is brought into compliance, the facility record should retain the facility's status as "in compliance."

·   Treatment of Carryover Non-Compliant Facilities: The application automatically tracks the carryover of non-compliant facilities to the next FY, starting at the end of FY 2010. Specifically, the application will review the recorded inspections performed since FY 2010 and for facilities that remain out of compliance as of the date the report is run, these facilities will represent the total population of non-compliant facilities, including carryover and newly-inspected, non-compliant facilities. For example, if you have a facility that was inspected in FY 2010 that was initially found to be "not in compliance" and is still showing as "not in compliance" at the end of FY 2011, the facility will be included in the carryover of non-compliant facilities to the next FY (i.e. FY 2012). This carryover count will be automatically added to the list of facilities found to be non-compliant in the current FY (or FY 2012 in this example).

·   Treatment of Follow-up Inspections: If a non-compliant facility is inspected for SPCC in a subsequent FY, this inspection should be identified as a "Follow-up Inspection". It should be noted that the application will identify this follow-up inspection as tied to the original inspection where the outcome as identified as "out of compliance." This follow-up inspection will not be recorded as new inspection in that FY to avoid double-counting in the "Facilities Inspected" column for the HQ Oil Measures Report. However, for programmatic tracking, this follow-up inspection will be retained in the facility record in the compliance module.

·   Treatment of Inspections with No Outcome: If the inspection does not have an "Inspection Outcome" or a "Brought into Compliance" date populated in the compliance module of the application, the facility will count as "not in compliance" on the HQ Measures Report. When this occurs, the record should be updated with inspection outcome.

·   Tracking of Inspection and Plan Review Outcomes: For inspected facilities that are in compliance at the time of the inspection, but a subsequent Plan review reveals the Plan to be out of compliance, enter the two activities as different activities, field inspection vs. Plan review. Enter the inspection as "out of compliance" (using the date of the inspection), and then enter the Plan review activity as "not in compliance" (using the review date). When the Plan is deemed "in compliance", enter this compliance date in the "Brought into Compliance Field" for the Plan review activity and the inspection activity. Note that the "Brought into Compliance" date for the inspection activity is considered the overall compliance date for oil measures reporting.

2c. Source Data Reporting:

Data Submission Instrument: Data submission and data entry are handled via the Oil Database Application. The Oil Database Application can be accessed at http://emp.epa.gov/

Data Entry Mechanism: Data are entered by EPA personnel in the regions. Facility and inspection results data are entered for each initial inspection and follow-up activity.

Frequency of Data Transmission to EPA: Data are entered on an ongoing-basis as inspections are completed.

The initial measurement period for this measure starts with FY 2010 and concludes in FY 2015. The measure will then be extended to FY 2018 for the next iteration of the Agency's Strategic Plan.

A progressive rate (i.e., cumulative over the measurement period) of non-compliant facilities brought into compliance was selected starting with 15% in the base year leading up to 60% brought into compliance in the final year to account for carryover of non-compliant facilities to subsequent fiscal years. The reasoning behind this approach is based on program experience of the time-lag to bring facilities into compliance. For example, non-compliant facilities may need to install equipment or complete construction activities to achieve compliance. Another example is a facility may be subject to enforcement action as a result of non-compliance. Each or both of these scenarios could result in an extended timeframe to bring the facility into compliance. Thus, a 60% target was selected for FY 2015 to account for the potential time-lag for achieving compliance. The goal over the FY 2016- 2018 measurement period is to maintain a cumulative 60% brought into compliance rate.

3a. Relevant Information Systems:

System Description: The Oil Database Application contains all of the necessary fields and data entry points for a facility inspection. The data are reported via the HQ Measures Report in the Oil Database Application. The Oil Database Application contains data on the following components of a facility and its inspection history:

1. General
2. Address
3. Contacts/Ownership
4. Compliance Module
5. Oil Capacity
6. Discharge History
7. Documents

Source/Transformed Data: The data in the system are source data about the facility and the inspection results.

Information System Integrity Standards: As an application module of the Emergency Management Portal (EMP) the Oil Database application contains Information System Integrity measures to secure its data. EMP has undergone an annual Continuous Monitoring Assessment (CMA) and SAISO Audit of its security posture and continuous monitoring activities. In addition, the CMA was to facilitate ongoing Security Authorization of the system in accordance with Office of Management & Budget Circular A-130, Appendix III, Security of Federal Automated Information Resource; NIST Special Publication 800-37, rev.1, Guide for Applying the Risk Management Framework to Federal Information Systems; and NIST Special Publication 800-53A Rev.1, a subset of the applicable Security Controls of EMP's System Security Plan (SSP) were assessed. EMP's SSP and its Security Controls get reviewed on an annual basis.

3b. Data Quality Procedures:

Facility and Inspection data go through a multi-phase quality check. The Oil Database Application has built in data checks for common data entry problems (e.g., checks for missing data) for each facility and inspection outcome. Regions also engage in periodic (usually quarterly or bi-annual) manual QA/QC procedures to double check the data that are entered in the database.

OEM HQ also assists the regions with their manual data checks and resolves problems with missing or incorrect data. This happens on a bi-annual basis.

3c. Data Oversight:

Source Data Reporting Oversight Personnel:
- HQ Project Manager: Office of Emergency Management –Evaluation and Communications Division & Regulation and Policy Development Division
- HQ Project Support: Office of Emergency Management – Evaluation and Communications Division & Regulation and Policy Development Division
- Regional Oil Program Managers and Inspector Personnel

Source Data Reporting Oversight Responsibilities:
- Ensure Accuracy of source data entered into the Oil Database Application through manual QA/QC and data quality checks built into the application.

Information Systems Oversight Personnel:
- HQ Project Manager: Office of Emergency Management – Evaluation and Communications Division & Regulation and Policy Development Division
- HQ Project Support: Office of Emergency Management – Evaluation and Communications Division & Regulation and Policy Development Division

Information Systems Oversight Responsibilities:
- Maintain information systems and correct errors in report calculations.
- Provide upgrades to database capabilities.

3d. Calculation Methodology:

Explanation of Calculations: Data are subject to the following steps in order to determine the brought into compliance percentage (BIC%) that is the final reporting form for these measures.

1.   First, the Regional BIC% is calculated for each region using the following steps:

a.   For each Fiscal Year of the overall reporting period (initially FYs 2010 to 2015, extended to FY 2018) the number of facilities that are found non-compliant (NC#) is subtracted from the total number of inspections conducted in that FY, which includes those that were found compliant, not subject or closed (Total).   NC# is that region's portion of the total denominator used to calculate the brought into compliance percentage (BIC %).

b.   Facilities that are brought into compliance during the same fiscal (BIC#) year make up that region's portion of the numerator of the BIC %.

c.   So Regional BIC% = BIC# divided by total NC# (facilities found non-compliant in current FY and carryover non-compliant facilities from prior FY)

2.   Then, the National BIC% is calculated by summing the regional BIC# and NC# and dividing them. So, National BIC% = (All Regional BIC# / All Regional NC#)

3.   The Cumulative National BIC% is then calculated by dividing the National BIC# and the National NC# for all FYs from FY 2010 to the current FY. This Cumulative National BIC% accounts for Carry Over facilities from previous FYs, and it is the number that is reported as the final GPRA result for the current FY (i.e., it is not just the National BIC% for the current FY that is reported). So, Cumulative National BIC% = (National BIC# for All FYs / National NC# for All FYs).

The Unit of Measure is percentage of facilities brought into compliance (BIC%).

There are no Assumptions for this calculation.

The Timeframe is from FY 2010 – FY 2018. BIC% represents the percentage of facilities that were initially

inspected and brought into compliance during this timeframe.

4a. Oversight and Timing of Final Results Reporting:

Final Reporting Oversight Personnel:

- HQ Project Manager: Office of Emergency Management – Evaluation and Communications Division & Regulation and Policy Development Division
- HQ Project Support: Office of Emergency Management – Evaluation and Communications Division & Regulation and Policy Development Division
- HQ Results Reporting Support: Office of Emergency Management - Evaluation and Communications Division & Regulation and Policy Development Division

Final Reporting Oversight Responsibilities:

- Ensure Accuracy of source data entered into the Oil Database Application through manual QA/QC and data quality checks built into the application.
- Calculate the Cumulative National BIC% (as outlined in Section 3d of this DQR).
- Enter final performance result into EPA's Budget Automation System (BAS).

Final Reporting Timing: The final cumulative brought into compliance percentage for FY 2010 through the current FY is calculated for SPCC facilities for inclusion in OSWER reporting activities for the Government Performance and Results Act.

4b. Data Limitations/Qualifications:

OEM has not identified any systematic limitations or qualifications to the data beyond typical data entry and user error.

4c. Third-Party Audits:

There are no third-party audits for of this measure.

Measure Code: 112 - Number of LUST cleanups completed that meet risk-based standards for human exposure and groundwater migration.

Office of Solid Waste and Emergency Response (OSWER)

Goal Number and Title:
3 - Cleaning Up Communities and Advancing Sustainable Development
Objective Number and Title:
3 - Restore Land
Sub-Objective Number and Title:
3 - Cleanup Contaminated Land
Strategic Target Code and Title:
5 - Through 2015,reduce the backlog of LUST cleanups
Managing Office:
Office of Underground Storage Tanks (OUST)
1a. Performance Measure Term Definitions:
Cleanups Completed –The number of cleanups completed is the cumulative number of confirmed releases where cleanup has been initiated and where the state has determined that no further actions are currently necessary to protect human health and the environment. This number includes sites where post-closure monitoring as long as site-specific (e.g., risk-based) cleanup goals have been met. Site characterization, monitoring plans, and site-specific cleanup goals must be established and cleanup goals must be attained for sites being remediated by natural attenuation to be counted in this category. Clarification: "Cleanups Completed" is a cumulative category–sites should never be deleted from this category. It is no longer necessary to report separately cleanups completed that are state lead with state money and cleanups completed that are responsible party lead. It is, however, still necessary to report the number of cleanups completed that are state lead with Trust Fund money. A "no further action" determination made by the state that satisfies the "cleanups initiated" measure above, also satisfies this "cleanups completed" measure. This determination will allow a confirmed release that does not require further action to meet the definition of both an initiated and completed cleanup. For complete definition see EPA OUST. UST And LUST Performance Measures Definitions. January 18, 2008. http://www.epa.gov/OUST/cat/PMDefinitions.pdf , which are referenced in the Guidance To Regions For Implementing The LUST Provision Of The American Recovery And Reinvestment Act Of 2009, EPA-510-R-09-003, June 2009, http://www.epa.gov/oust/eparecovery/lustproguide.pdf , p. 7-8. See also: EPA. Environmental Protection Agency Recovery Act Program Plan: Underground Storage Tanks. May 15, 2009. http://www.epa.gov/recovery/plans/oust.pdf Risk-based standards for human exposure and groundwater migration. Reference: Semi-annual Report of UST Performance Measures, End Of Mid Fiscal Year 2011 – as of March 31, 2011 , dated May 2011 ; http://www.epa.gov/OUST/cat/ca_11_12.pdf

2a. Original Data Source:

The original data source is States, DC, and territories who sign Leaking Underground Storage Tank (LUST) agreements with EPA. These entities can delegate reporting to sub-recipients (such as local governments).

Each EPA regional office manages work that occurs within regional boundaries.

For more information:
1. US EPA Office of Underground Storage Tanks. "Guidance to Regions for Implementing the LUST Provision of the American Recovery and Reinvestment Act of 2009." EPA-510-R-09-003. June 2009. http://www.epa.gov/oust/eparecovery/lustproguide.pdf
2. US EPA Office Of Underground Storage Tanks. "Supplemental Guidance on Recovery Act Recipient Reporting (Section 1512) of the American Recovery and Reinvestment Act of 2009." Memo from Carolyn Hoskinson to Regional UST Managers. October 2, 2009. http://www.epa.gov/oust/eparecovery/OUST_1512_Memo_100209.pdf

2b. Source Data Collection:

Determination of cleanup completion requires consideration of environmental data, such as field sampling, which can vary by project. The overall measure requires tabulation of the number LUST clean-ups completed. Spatial Detail: Geographic granularity can vary. Sub-recipient data submissions (when delegated) may be as detailed as the site level, for which granularity is defined in latitude and longitude. Other data are entered by recipients for the entire state/territory (excluding sub-recipient data). Granularity for work in Indian Country is the Regional level.
Spatial Coverage: National

For cooperative agreements: Regional offices include QA Terms and Conditions in their states' assistance agreement. CAs must be current and specify: QA roles and responsibilities for EPA and grantee recipients; and quality requirements including responsibilities for final review and approval. Default quality requirements include: organization-level QA documentation (i.e. QMP) for state agencies and primary contractors; and project-level QAPPs for each CA. In accordance with EPA's Uniform Administrative Requirements for Grants and Cooperative Agreements, 40 CFR Part 31.45, states must develop and implement quality assurance practices. The regulation requires developing and implementing quality assurance practices that will "produce data of quality adequate to meet project objectives and to minimize loss of data to out of control conditions or malfunctions'; see OSWER Directive 9650.10A: www.epa.gov/oust/directiv/d965010a.htm - sec11"

For contracts: EPA Regions determine which quality requirements are applicable. Contracts must be current and specify: QA roles and responsibilities for EPA and national LUST contractors; and quality requirements including responsibilities for final review and approval. Default quality requirements include: organization-level QA documentation (i.e. QMP) for the primary contractors; and project-level QAPPs for each Tribal LUST remedial Work Assignment. Sample EPA contract language: "the Contractor shall comply with the higher-level quality standard selected below: Specifications and Guidelines for Quality Systems for Environmental Data Collection and Environmental Technology Programs (ANSI/ASQC E4, 1994). As authorized by FAR 52.246-11, the higher-level quality standard ANSI/ASQC E4 is tailored as follows: The solicitation and contract require the offerors/contractor to demonstrate conformance to ANSI/ASQC E4 by submitting the quality documentation described below. [Specifically,...] ... The Contractor shall not commence actual field work until until the Government has approved the quality documentation (i.e., QAPP)."

Note: Regions keep copies of individual QAPPs associated with cooperative agreements and contracts. Each EPA regional office manages its own state and tribal assistance agreements.

2c. Source Data Reporting:

Site assessments and cleanup status are recorded as milestones are achieved, in accordance with each site's schedule and each recipient's procedures. Contractors and other recipients individually maintain records for reporting accomplishments into LUST4. Their data systems vary.

States, DC and territories submit location-, funding-, and progress-related data directly into LUST4.

LUST4 also allows for bulk (batch) uploading by states/territories that already have the location & measures-related data captured in a data system or have the technical expertise to create flat files through another method in exactly the format and layout specified. This batch uploading is not supported by OUST; data providers not comfortable with this approach are encouraged to use the interactive online features of the Locations Subsystem and Measures Subsystem. Access to the LUST4 Locations and Measures Subsystems is available online via the EPA portal at http://portal.epa.gov/ under the My Communities/Underground Storage Tank menu page.

3a. Relevant Information Systems:

LUST4. This database is the master database of all LUST program-related data, including but not limited to data supporting Recovery Act measures. Recipients and EPA report data for activity and measures directly into LUST4. LUST4 includes both source data and transformed data (e.g., data aggregated into Regional totals).

The program's Oracle web-based system-- LUST4-- accessed through EPA's portal.

OSWER Performance Assessment Tool (PAT). This tool serves as the primary external servicing resource for organizing and reporting OSWER's performance data. PAT collects information from OSWER program systems, and conforms it for uniform reporting and data provisioning. PAT captures data from LUST4; replicates business logic used by LUST4 for calculating measures; can deliver that data to EPA staff and managers via a business intelligence dashboard interface for analytic and reporting use; enables LUST point of contact to document status and provide explanation for each measure; and transmits data to the Budget Automation System.

Budget Automation System (BAS). BAS is the final repository of the performance values.

3b. Data Quality Procedures:

EPA's regional grants project officers and regional program managers provide first-level data quality reviews and oversight of their recipients' program performance measure results.

OUST uses a combination of automated validation along with manual QA/QC review.

QA/QC REVIEW BY REGIONS. EPA/OUST oversees the use of the QA/QC checklist, which is incorporated into the LUST4 oracle web-based system. Regions complete the QA/QC checklist, sign it electronically and submit it to EPA/OUST for review, comment and approval of each record.
NOTE: This QA/QC checklist was last updated 10/1/2009 and is accessed through the user interface of LUST4.

Regional QA/QC Evaluation Checklist –
Note: Checklist is to be completed by Regional reviewer and will appear "shaded" to others.
1. Previous Totals Column
-- Verify the previous total number is correct by comparing it to the total from the last reporting period. If there is a discrepancy, report the information in the "Correction To Previous Data" column. Please add comments in the "Comments" column for any corrections that are made to the applicable performance measure.
2. Actions This Reporting Period
For each performance measure, if this "Reported" number deviates by more than 10% from the last period's number or appears otherwise questionable, complete the following actions:
-- Review the state's explanation, if available.
-- If necessary, contact the state to obtain the corrected numbers and/or obtain a sufficient explanation and include the explanation in the "Comments" section for the applicable performance measure.
3. Corrections to Previous Data Column
Verify that if any corrections have been listed that an explanation for the correction is provided in the "Comments" column and complete the following actions:
-- Verify and discuss the correction with the state if the correction is >10% or if the correction appears questionable (e.g., database conversions, database cleanup efforts to resolve misclassified data, duplicative records, etc.)
-- Verify if the corrections are anticipated to be a one-time event or occur over multiple years
-- Evaluate if the corrections will impact other performance measures (e.g., if the number of cleanups completed is adjusted downward by a correction, does this also result in a commensurate downward adjustment of cleanups initiated?) Include any additional comments in the "Comments" column as necessary.
4. Totals (Cumulative, if applicable)
-- Verify accuracy of all cumulative totals
-- Include any additional comments in the "Comments" column as necessary

AUTOMATED VALIDATION. For instance upon data entry of any new location, location information is verified automatically by the Facility Registry System (FRS). Location information without latitude and longitude is also geocoded automatically. When entering measure information, the system does not allow a measure value less than the applicable count of locations that are relevant to that measure (the count of location records is automatically generated by the system); a measure value greater than the applicable count of locations requires provision of an explanatory comment.

EPA/OUST provides second-level data quality reviews of all data

LUST4. LUST4 operates under OSWER's QMP, including the security policy specified in that QMP. LUST4 does not have any stand-alone certifications related to the EPA security policy or the Systems Life Cycle Management policy. The LUST4 system is built upon Oracle Business Intelligence tools provided by the EPA Business Intelligence Analytics Center, which ensures that a stand-alone security certification is not necessary.

PAT. PAT operates under the OSWER Quality Management Plan (QMP). PAT has a security certification confirming that a security policy is not necessary because no sensitive data are handled and PAT is built upon the Oracle-based business intelligence system. PAT's security certification indicates that it follows all security guidelines for EPA's Oracle Portal and that PAT is (1) not defined as a "Major Application" according to NIST Special Publication 800-18, Guide for Developing Security Plans for Information Technology Systems, section 2.3.1; (2) does not store, process, or transmit information that the degree of sensitivity is assessed as high by considering the requirements for availability, integrity, and confidentiality according to NIST Special Publication 800-18, Guide for Developing Security Plans for Information Technology Systems, section 3.7.2. (3) is not covered by EPA Order 2100.2A1 Information Technology Capital Planning and Investment Control (CPIC). Data Flow:

Step 1. Performance measure and location data are entered into LUST4 by recipients (or sub-recipients, if delegated) or by Regions (for EPA contractors).Upon entry of any new location, location information is verified automatically by the Facility Registry System (FRS). Location information without latitude and longitude is also geocoded automatically. (FRS data are used solely for data entry QA/QC, not to assist in calculating results.)

Step 2. Each Region conducts Regional level review of data from the LUST4 system. Rejected data must be edited by the original data source. Approved data proceed to Step 3.

Step 3. Headquarters' staff perform performs National Program Review, using data from the LUST4 system. Rejected data must be reviewed by the region and, if needed, pushed back to the state for editing(Step 2).

Step 4. PAT pulls data from LUST4. Headquarters staff compare PAT results to LUST4 results. If PAT does not match LUST4 then there was an error with the upload and data is reloaded. Headquarters staff enter into PAT the ACS status information of "Indicator" for each measure and, if desired, explanation.(Note: PAT allows for programs to identify status other than "Indicator." When programs select a status of "no status," "data not available," or "target not met," PAT requires that an explanation be provided. LUST program policy is to resolve all reporting issues prior to ACS reporting, so "Indicator" is the only status chosen and explanations for that status are optional.)

Step 5. Headquarters approves PAT results, and PAT pushes results into BAS/Measures Central.

Step 6. Measures Central aggregates Regional data into a national total. OUST reporting lead reviews and certifies results.

3c. Data Oversight:

An EPA Headquarters primary contact maintains a list of the HQ (OUST and OEI), Regional and state/territory primary and backup users; a record of changes to the list is also maintained. The primary HQ contact ensures that Regional reporting is on track, conducts QA on LUST performance measures, ensures QA issues are resolved and/or documented, and oversees final reporting to BAS.

Regional Program Managers are ultimately responsible for regional-level data. They conduct their review based upon a national QA/QC checklist.

3d. Calculation Methodology:

The cumulative number of confirmed releases where cleanup has been initiated and where the state or region (for the tribes) has determined that no further actions are currently necessary to protect human health and the environment, includes sites where post-closure monitoring is not necessary as long as site specific (e.g., risk based) cleanup goals have been met. Site characterization, monitoring plans and site-specific cleanup goals must be established and cleanup goals must be attained for sites being remediated by natural attenuation to be counted in this category. (See http://www.epa.gov/OUST/cat/PMDefinitions.pdf.

The unit of analysis is site cleanup

4a. Oversight and Timing of Final Results Reporting:

Semiannual by Deputy Office Director. Responsible for final review to ensure LUST 4 System Manager has completed review, and numbers are accurate.

4b. Data Limitations/Qualifications:

Data quality depends on the accuracy and completeness of state records.

4c. Third-Party Audits:

Not applicable!

Measure Code: 151 - Number of Superfund sites with human exposures under control.

Office of Solid Waste and Emergency Response (OSWER)

Goal Number and Title:
3 - Cleaning Up Communities and Advancing Sustainable Development

Objective Number and Title:
3 - Restore Land

Sub-Objective Number and Title:
3 - Cleanup Contaminated Land

Strategic Target Code and Title:
2 - By 2015, increase the number of Superfund final and deleted NPL sites and RCRA facilities

Managing Office:
OSRTI

1a. Performance Measure Term Definitions:
Definition of Site: "Sites" refers only to National Priorities List (NPL) sites. (See below for definition of NPL.) The term "site" itself is not explicitly defined under Comprehensive Environmental Response, Compensation, and Liability Act (CERCLA) or by the Superfund program; instead "site" is defined indirectly in CERCLA's definition of "facility," as follows: "The term 'facility' means (A) any building, structure, installation, equipment, pipe or pipeline (including any pipe into a sewer or publicly owned treatment works), well, pit, pond, lagoon, impoundment, ditch, landfill, storage container, motor vehicle, rolling stock, or aircraft, or (B) any site or area where a hazardous substance has been deposited, stored, disposed of, or placed, or otherwise come to be located; but does not include any consumer product in consumer use or any vessel." (CERCLA, Title I, Section 101, (9)).   Superfund Alternative Approach (SAA) sites: The program collects and enters into CERCLIS, human exposure determinations at SAA sites, but does not target or report official results at this time.   Definition of National Priorities List (NPL): Sites are listed on the National Priorities List (NPL) upon completion of Hazard Ranking System (HRS) screening, public solicitation of comments about the proposed site, and final placement of the site on the NPL after all comments have been addressed. The NPL primarily serves as an information and management tool. It is a part of the Superfund cleanup process and is updated periodically. Section 105(a)(8)(B) of CERCLA as amended, requires that the statutory criteria provided by the HRS be used to prepare a list of national priorities among the known releases or threatened releases of hazardous substances, pollutants, or contaminants throughout the United States. This list, which is Appendix B of the National Contingency Plan, is the NPL. Visit the HRS Toolbox (http://www.epa.gov/superfund/sites/npl/hrsres/index.htm page for guidance documents that are used to determine if a site is a candidate for inclusion on the NPL. [Source: Superfund website, http://www.epa.gov/superfund/sites/npl/npl_hrs.htm   (Also see Appendix B of the most recent Superfund Program Implementation Manual (SPIM), which is updated each fiscal year and contains definitions and documentation/coding guidance for Superfund measures. The

most current SPIM can be found here: http://epa.gov/superfund/policy/guidance.htm.

Definition of "Current Human Exposure under Control" (HEUC): - Sites are assigned to this category when assessments for human exposures indicate there are no unacceptable human exposure pathways and the Region has determined the site is under control for current conditions site wide.

The human exposure status at a site is reviewed annually by the 10th working day in October, or at any time site conditions change. CERCLIS is to be updated within 10 days of any change in status.

The HEUC documents, for Proposed, Final, and Deleted NPL sites and SAA settlement sites, the progress achieved towards providing long-term human health protection by measuring the incremental progress achieved in controlling unacceptable human exposures at a site. This is also a Government Performance and Results Act (GPRA) performance measure.

Controlling unacceptable human exposures can occur in three ways:
-- Reducing the level of contamination. For purposes of this policy, "contamination"
generally refers to media containing contaminants in concentrations above appropriate protective risk-based levels associated with complete exposure pathways to the point where the exposure is no longer "unacceptable;" and/or
-- Preventing human receptors from contacting contaminants in-place; and/or
-- Controlling human receptor activity patterns (e.g., by reducing the potential frequency or duration of exposure).

Five categories have been created to describe the level of human health protection achieved at a site:
-- Insufficient data to determine human exposure control status;
-- Current human exposures not under control;
-- Current human exposures under control;
-- Current human exposures under control and protective remedy or remedies in place; and
-- Current human exposures under control, and long-term human health protection achieved.

Definition of Accomplishment of "HEUC":
The criteria for determining the Site-Wide Human Exposure status at a site are found in the Superfund Environmental Indicators Guidance Human Exposures Revisions" March 2008
(http://www.epa.gov/superfund/accomp/ei/pdfs/final_ei_guidance_march_2008.pdf [Source: SPIM Appendix B]

(See Appendix B of the most recent SPIM, which is updated each fiscal year and contains definitions and documentation/coding guidance for Superfund measures. The most current SPIM can be found here: http://epa.gov/superfund/policy/guidance.htm.

The Superfund Program's performance measures are used to demonstrate program progress and reflect major site cleanup milestones from start (remedial assessment completion) to finish (number of sites ready for anticipated use sitewide). Each measure marks a significant step in ensuring human health and environment protection at Superfund sites.

References:

 U.S. Environmental Protection Agency, EPA Performance and Accountability Reports, http://www.epa.gov/ocfo/par/index.htm

U.S. Environmental Protection Agency, Superfund Accomplishment and Performance Measures, http://www.epa.gov/superfund/accomplishments.htm

U.S. Environmental Protection Agency, Federal Facilities Restoration and Reuse Office – Performance measures, http://www.epa.gov/fedfac/documents/measures.htm

U.S Environmental Protection Agency, Office of Inspector General, Information Technology - Comprehensive Environmental Response, Compensation, and Liability Information System (CERCLIS) Data Quality, No. 2002-P-00016, http://www.epa.gov/oigearth/eroom.htm

U.S. Government Accountability Office, "Superfund Information on the Status of Sites, GAO/RCED-98-241", http://www.gao.gov/archive/1998/rc98241.pdf

U.S. Environmental Protection Agency, Office of Superfund Remediation and Technology Innovation, Superfund Program Implementation Manuals (SPIM), http://www.epa.gov/superfund/policy/guidance.htm

U.S. Environmental Protection Agency, Office of Solid Waste and Emergency Response, "OSWER Quality Management Plan", http://www.epa.gov/swerffrr/pdf/oswer_qmp.pdf

U.S. Environmental Protection Agency, Office of Environmental Information, EPA System Life Cycle Management Policy Agency Directive 2100.5, http://www.epa.gov/irmpoli8/ciopolicy/2100.5.pdf

U.S. Environmental Protection Agency, Office of Environmental Information, EPA's Information Quality Guidelines, http://www.epa.gov/quality/informationguidelines

2a. Original Data Source:

Original data sources vary, and multiple data sources can be used for each site. Typical data sources are EPA personnel, contractors (directly to EPA or indirectly, through the interagency agreement recipient or cooperative agreement recipient), U.S. Army Corps of Engineers (interagency agreement recipient), and states/tribes/other political subdivisions (cooperative agreement recipients). EPA also collects data via pre-final inspections at sites.

(See item Performance Measure Term Definitions in Tab 1, for more information. Also, detailed information on requirements for source data and completion procedures can be found on the following Superfund website: http://www.epa.gov/superfund/programs/npl_hrs/closeout/index.htm

2b. Source Data Collection:

Collection typically involves some combination of environmental data collection, estimation and/or tabulation of records/activities. Documents such as risk assessments, Record of Decisions (RODs), Action Memoranda, Pollution Reports (POLREPS), Remedial Action (RA Reports), Close-out Reports, Five-year Reviews, NPL Deletion/Partial Deletion Notices are known reliable sources of data and often provide the information necessary for making an HEUC evaluation with reasonable certainty.

Each EPA Region has an information management coordinator (IMC) that oversees reporting.

The Human Exposure Environmental Indicator data were collected beginning in FY 2002.

The collection methods and and guidance for determining HEUC status are found in the Superfund Environmental Indicators Guidance Human Exposures Revisions" March 2008.
(http://www.epa.gov/superfund/accomp/ei/ei.htm

(See item Performance Measure Term Definitions, for more information and references.)

Source data collection frequency: No set interval. Varies by site

Spatial Extent: National

Spatial detail: Site, defined in database by latitude/longitude pair. In cases in which projects work on a smaller part of a site, geography may be defined at a finer grain -- the project-level.

2c. Source Data Reporting:

Varied reporting format for source data EPA uses to make decisions. In many cases, EPA reviews site-specific secondary data or existing EPA-prepared reports. Documents such as risk assessments, RODs, Action Memoranda, POLREPS, RA Reports, Close-out Reports, Five-year Reviews, and NPL Deletion/Partial Deletion Notices are known reliable sources of data and often provide the information necessary for making an HEUC evaluation with reasonable certainty.
EPA's Regional offices and Headquarters enter data into CERCLIS on a rolling basis.
The human exposure status at a site is reviewed annually by the 10th working day in October, or at any time site conditions change. CERCLIS is to be updated within 10 days of any change in status.

The instrument for determining the Site-Wide Human Exposure status at a site is found in the Superfund Environmental Indicators Guidance Human Exposures Revisions" March 2008. Determinations are made by regional staff and management and entered directly into CERCLIS.
http://www.epa.gov/superfund/accomp/ei/ei.htm

See Appendix B of the most recent SPIM, which is updated each fiscal year and contains definitions and documentation/coding guidance for Superfund measures. The most current SPIM can be found here: http://epa.gov/superfund/policy/guidance.htm

3a. Relevant Information Systems:

The HEUC determination is made directly in CERCLIS once it is determined that the site is Under Control and has been approved as such by appropriate regional personnel.

CERCLIS database – The CERCLIS database is used by the Agency to track, store, and report Superfund site information (e.g., NPL sites and non-NPL Superfund sites).

(For more information about CERCLIS, see Appendix E of the most recent SPIM, which is updated each fiscal year and contains definitions and documentation/coding guidance for Superfund measures. The most current SPIM can be found here: http://epa.gov/superfund/policy/guidance.htm.

CERCLIS operation and further development is taking place under the following administrative control quality assurance procedures: 1) Office of Environmental Information Interim Agency Life Cycle Management Policy Agency Directive; 2) the Office of Solid Waste and Emergency Response (OSWER) Quality Management Plan (QMP); 3) EPA IT standards; 4) Quality Assurance Requirements in all contract vehicles under which CERCLIS is being developed and maintained; and 5) EPA IT security policies. In addition, specific controls are in place for system design, data conversion and data capture, as well as CERCLIS outputs.

CERCLIS adherence to the security policy has been audited. Audit findings are attached to this record.

OSWER Performance Assessment Tool (PAT). This tool serves as the primary external servicing resource for organizing and reporting OSWER's performance data, which collects information from OSWER program systems, and conforms it for uniform reporting and data provisioning. PAT captures data from CERCLIS; replicates business logic used by CERCLIS for calculating measures; delivers that data to EPA staff and managers via a business intelligence dashboard interface for analytic and reporting use; and transmits data to the Budget Automated System (BAS). No current system specifications document is currently available for PAT, but will be provided when available. For this measure, PAT transmits Regional-level data to BAS.

PAT operates under the OSWER QMP. PAT has a security certification confirming that a security policy is not necessary because no sensitive data are handled and PAT is built upon the Oracle-based business intelligence system. PAT's security certification indicates that it follows all security guidelines for EPA's Oracle Portal and that PAT is (1) not defined as a "Major Application" according to NIST Special Publication 800-18, Guide for Developing Security Plans for Information Technology Systems, section 2.3.1; (2) does not store, process, or transmit information that the degree of sensitivity is assessed as high by considering the requirements for availability, integrity, and confidentiality according to NIST Special Publication 800-18, Guide for Developing Security Plans for Information Technology Systems, section 3.7.2. (3) is not covered by EPA Order 2100.2A1 Information Technology Capital Planning and Investment Control (CPIC).

EPA Headquarters is now scoping the requirements for an integrated (Superfund Document Management System-) SDMS-CERCLIS system, called the Superfund Enterprise Management System (SEMS). Development work on SEMS began in FY 2007 and will continue through FY 2013.

SEMS represents further re-engineering of the national reporting systems to include additional elements of EPA's Enterprise Architecture. SEMS will provide a common platform for major Superfund systems and future IT development. It will be constructed in part using EPA IT enterprise architecture principles and components. SEMS will provide a Superfund Program user gateway to various IT systems and information collections.

Attached Documents:
19-0585 (CERCLIS QAPP) 2009-0410.doc
CERCLIS July 9 2009 scan High Medium Response.xls
OSWER QMP printed 2010-03-23.pdf

3b. Data Quality Procedures:

The regional SOPs for HEUC data entry, along with review and instructions/guidance for determining the Site-Wide Human Exposure status at a site are found in the Superfund Environmental Indicators Guidance Human Exposures Revisions" March 2008.
http://www.epa.gov/superfund/accomp/ei/ei.htm

A list of all Headquarters-level data sponsors is provided in Exhibit E.2 in SPIM Appendix E, Information Systems. The most current SPIM can be found here:
http://www.epa.gov/superfund/policy/guidance.htm

CERCLIS: To ensure data accuracy and control, the following administrative controls are in place: 1) Superfund Program Implementation Manual (SPIM), the program management manual that details what data must be reported; 2) Report Specifications, which are published for each report detailing how reported data are calculated; 3) Coding Guide, which contains technical instructions to data users including Regional IMCs, program personnel, data owners, and data entry personnel; 4) Quick Reference Guides (QRG), which are available in the CERCLIS Documents Database and provide detailed instructions on data entry for nearly every module in CERCLIS; 5) Superfund Comprehensive Accomplishment (SCAP) Reports within CERCLIS, which serve as a means to track, budget, plan, and evaluate progress towards meeting Superfund targets and measures; 6) a historical lockout feature in CERCLIS so that changes in past fiscal year data can be changed only by approved and designated personnel and are logged to a Change Log report, 7) the OSWER QMP; and 8) Regional Data Entry Control Plans.

EPA Headquarters has developed data quality audit reports and Standard Operating Procedures, which address timeliness, completeness, and accuracy, and has provided these reports to the Regions. In addition, as required by the Office of Management and Budget (OMB), CERCLIS audit logs are reviewed monthly. The system was also re-engineered to bring CERCLIS into alignment with the Agency's mandated Enterprise Architecture. The first steps in this effort involved the migration of all 10 Regional and the Headquarters databases into one single national database at the National Computing Center in Research Triangle Park (RTP) and the migration of SDMS to RTP to improve efficiency and storage capacity. During this process SDMS was linked to CERCLIS which enabled users to easily transition between programmatic accomplishments as reported in CERCLIS and the actual document that defines and describes the accomplishments.

Regional Data Entry Control Plans. Regions have established and published Data Entry Control Plans, which are a key component of CERCLIS verification/validation procedures. The control plans include: (1) regional policies and procedures for entering data into CERCLIS, (2) a review process to ensure that all Superfund accomplishments are supported by source documentation, (3) delegation of authorities for approval of data input into CERCLIS, and (4) procedures to ensure that reported accomplishments meet accomplishment definitions. In addition, regions document in their control plans the roles and responsibilities of key regional employees responsible for CERCLIS data (e.g., regional project manager, information management coordinator, supervisor, etc.), and the processes to assure that CERCLIS data are current, complete, consistent, and accurate. Regions may undertake centralized or decentralized approaches to data management. These plans are collected annually for review by OSRTI/IMB (Information Management Branch). [Source: SPIM FY11, III.J and Appendix E. http://www.epa.gov/superfund/action/process/spim10/pdfs/appe.pdf)

Copies of the 2010 Regional Data Entry Control Plans are provided with this DQR. Current and past year plans are available by contacting the Chief, Information Management Branch, Office of Superfund Remediation and Technology Innovation.

Regions are expected to prepare Data Entry Control Plans consistent with the SPIM and the Headquarters guidance: "CERCLIS Data Entry Control Plan Guidance," June 2009.

Superfund Program Implementation Manual (SPIM). The SPIM should be the first source referred to for additional questions related to program data and reporting. The SPIM is a planning document that defines program management priorities, procedures, and practices for the Superfund program (including response, enforcement, and Federal facilities). The SPIM provides the link between the GPRA, EPA's Strategic Plan, and the Superfund program's internal processes for setting priorities, meeting program goals, and tracking performance. It establishes the process to track overall program progress through program targets and measures.

The SPIM provides standardized and common definitions for the Superfund program, and it is part of EPA's internal control structure. As required by the Comptroller General of the United States, through generally accepted accounting principles (GAAP) and auditing standards, this document defines program scope and schedule in relation to budget, and is used for audits and inspections by the Government Accountability Office (GAO) and the Office of the Inspector General (OIG). The SPIM is developed on an annual basis. Revisions to the SPIM are issued during the annual cycle as needed.

The SPIM contains three chapters and a number of appendices. Chapter 1 provides a brief summary of the Superfund program and summarizes key program priorities and initiatives. Chapter 2 describes the budget process and financial management requirements. Chapter 3 describes program planning and reporting requirements and processes. Appendices A through I highlight program priorities and initiatives and provide detailed programmatic information, including Annual Targets for GPRA performance measures, and targets for Programmatic Measures. [Source: SPIM 2011, Chapter I]

The most current version of the SPIM can be found at: http://epa.gov/superfund/policy/guidance.htm

Data Flow:

Step 1. Original data sources provide information.

Step 2. EPA Region reviews and determines HEUC status at the site and adjusts CERCLIS records as needed.

Step 3. Headquarters' OSRTI data sponsor reviews and approves/disapproves written justifications for Regional determinations of "Not Under Control" and "Insufficient Data," using data from CERCLIS. Data sponsor works with Regional staff to ensure that disapproved justifications comport with Superfund Program guidance.

Step 4. OSWER's PAT pulls data from CERCLIS. Headquarters staff compare PAT results to CERCLIS results. If PAT does not match CERCLIS then there was an error with the upload and data are reloaded. Headquarters staff enter into PAT the Annual Commitment System (ACS) status information for each measure and, if necessary, a status explanation.

Step 5. Headquarters approves PAT results, and PAT pushes results into BAS.

Step 6. BAS aggregates Regional data into a national total. OSRTI reporting lead reviews and certifies results.
Attached Documents:
2009_draft_CERCLIS_DECP_guidance _6-5-09_.pdf

3c. Data Oversight:

The Superfund program has a "data sponsorship" approach to database oversight. Headquarters staff and managers take an active role in improving the quality of data stored in CERCLIS by acting as data sponsors.

Data sponsorship promotes consistency and communication across the Superfund program. Headquarters data sponsors communicate and gain consensus from data owners on data collection and reporting processes. Data sponsors ensure that the data they need to monitor performance and compliance with program requirements is captured and stored properly in CERCLIS. To meet this goal, headquarters data sponsors identify their data needs, develop data field definitions, and distribute guidance requiring submittal of these data. Data owners are normally site managers that need the data in support of site work. Data owners follow the guidance they receive from data sponsors, as they acquire and submit data. Headquarters data sponsors assist data owners in maintaining and improving the quality of Superfund program data. These data are available for data evaluation and reporting. Data sponsorship helps promote consistency in both national and regional reporting. In addition, data sponsorship provides a tool to improve data quality through program evaluation and adjustments in guidance to correct weaknesses detected. Data sponsors may conduct audits to determine if there are systematic data problems (e.g., incorrect use of codes, data gaps, etc.). A list of all Headquarters-level data sponsors is provided in Exhibit E.2 in SPIM Appendix E, Information Systems. The most current SPIM can be found here: http://epa.gov/superfund/policy/guidance.htm [source example for process: Region 2 SOP, page 53, entry for Data-Entry Training/Oversight]

Specific roles and responsibilities of data sponsors:
-- Identify data needs;
-- Oversee the process of entering data into the system;
-- Determine the adequacy of data for reporting purposes;
-- Conduct focus studies of data entered (A focus study is where a data sponsor identifies a potential or existing data issue to a data owner (see below), IMC, or other responsible person to determine if a data quality problem exists, and to solve the problem, if applicable. (IMC responsibilities discussed below.) Focus studies can be informal via electronic messages.);
-- Provide definitions for data elements;
-- Promote consistency across the Superfund program;
-- Initiate changes in CERCLIS as the program changes;
-- Provide guidance requiring submittal of these data;
-- Determine the adequacy of data for reporting purposes;
-- Support the development of requirements for electronic data submission; and
-- Ensure there is "objective" evidence to support the accomplishment data entered in CERCLIS through identifying data requirements and check to assure compliance by performing periodic reviews of a random CERCLIS data sample. [Source: SPIM 2010, III.E and E.A.5]

The primary responsibilities of data owners are (1) to enter and maintain data in CERCLIS and (2) assume responsibility for complete, current, consistent, and accurate data. The data owners for specific data are clearly identified in the system audit tables. Regions annually update region-specific Data Entry Control Plans (DECP). Among other things, Regional data entry control plans identify which Data Sponsors/Data Owners are responsible for different aspects of data entry. (See item 3b., Data Quality Procedures for more information on Data Entry Control Plans.)

Information Management Coordinators (IMC). In each Region, the IMC is a senior position which serves as regional lead for all Superfund program and CERCLIS systems management activities. The following lead responsibilities for regional program planning and management rest with the IMC:
-- Coordinate program planning, budget development, and reporting activities;

-- Ensure regional planning and accomplishments are complete, current, and consistent, and accurately reflected in CERCLIS by working with data sponsors and data owners;

-- Provide liaison to HQ on SCAP process and program evaluation issues;

-- Coordinate regional evaluations by headquarters;

-- Ensure that the quality of CERCLIS data are such that accomplishments and planning data can be accurately retrieved from the system; and

-- Ensure there is "objective" evidence to support accomplishment data entered in CERCLIS. (Objective Evidence Rule: "All transactions must be supported by objective evidence, that is, documentation that a third party could examine and arrive at the same conclusion.") [Source: SPIM 2010, III.E]

The Information Management Officer (IMO) & Director, Information Management and Data Quality Staff. OSWER is the lead point of contact for information about the data from CERCLIS .

The Project Manager for CERCLIS oversees and is the approving authority for quality-related CERCLIS processes, and is closely supported by a Contract Task Manager. (See the CERCLIS QAPP, attached, for more information.) The lead point of contact for information about the data from CERCLIS is the Director, Information Management and Data Quality Staff, Office of Solid Waste and Emergency Response.

PAT Data Entry

The Annual Commitment System (ACS) Coordinator in OSRTI ensures that CERCLIS data for this measure are correctly loaded into PAT. The ACS Coordinator then works with the data sponsor to review uploaded data, edit records as appropriate, and then push data to ACS--part of the Office of Chief Financial Officer's (OCFO) BAS. PAT is maintained by OSWER's System Manager who ensures that the PAT system operates correctly, based on business logic agreed to by OSRTI.

3d. Calculation Methodology:

The performance measure is a specific variable entered into CERCLIS following specific coding guidance and corresponding supporting site-specific documentation.

The unit of measure is number of sites. The calculation only includes NPL sites.

References:

Superfund Data Element Dictionary (DED). The Superfund DED is available online at: http://www.epa.gov/superfund/sites/ded/index.htm The DED provides definitions and descriptions of elements, tables and codes from the CERCLIS database used by the Superfund program. It also provides additional technical information for each entry, such as data type, field length and primary table. Using the DED, you can look up terms by table name or element name, or search the entire dictionary by keyword.

Other additional references that may be useful:

Coding Guide. The Superfund Coding Guide contains technical instructions to data users including Regional IMCs, program personnel, data owners, and data entry personnel. The Remedial component of the Coding Guide is attached to this record.

Quick Reference Guides (QRG). Superfund Quick Reference Guides are available in the CERCLIS Documents Database and provide detailed instructions on data entry for nearly every module in CERCLIS. Sample QRGs are available for entering data related to Remedial Action Starts.

Site Status and Description document: this is a QRG
for CERCLIS users, for filling in information related to site status and
description.

Attached Documents:
Coding Guide - 2009.pdf
Example QRG RA Start.doc
Site Status and Description.doc

4a. Oversight and Timing of Final Results Reporting:

Data Sponsor for HUEC, Annual Commitment System coordinator, and National Program Office (NPO) management.
Progress reporting is done periodically as checks, while official numbers are reported annually.

4b. Data Limitations/Qualifications:

Users of HEUC data should recognize that HEUC status is reviewed at least annually, on a schedule that varies based upon site characteristics. This status review can result in a change in status with regard to this measure, with a site moving from HEUC status to non-HEUC status.

4c. Third-Party Audits:

Three audits, two by the Office Inspector General (OIG) and the other by Government Accountability Office (GAO), assessed the validity of the data in CERCLIS. The OIG audit report, Superfund Construction Completion Reporting (No. E1SGF7_05_0102_ 8100030), dated December 30, 1997, concluded that the Agency "has good management controls to ensure accuracy of the information that is reported," and "Congress and the public can rely upon the information EPA provides regarding construction completions." The GAO report, Superfund: Information on the Status of Sites (GAO/RCED-98-241), dated August 28, 1998, estimated that the cleanup status of National Priority List (NPL) sites reported by CERCLIS as of September 30, 1997, is accurate for 95 percent of the sites. Another OIG audit, Information Technology - Comprehensive Environmental Response, Compensation, and Liability Information System (CERCLIS) Data Quality (Report No. 2002-P-00016), dated

September 30, 2002, evaluated the accuracy, completeness, timeliness, and consistency of the data entered into CERCLIS. The report provided 11 recommendations to improve controls for CERCLIS data quality. EPA has implemented these recommendations and continues to use the monitoring tools for verification.

The IG annually reviews the end-of-year CERCLIS data, in an informal process, to verify data that supports the performance measures. Typically, there are no published results.

EPA received an unqualified audit opinion by the OIG for the annual financial statements and recommends several corrective actions. The Office of the Chief Financial Officer indicates that corrective actions will be taken.

Measure Code: 115 - Number of Superfund remedial site assessments completed.

Office of Solid Waste and Emergency Response (OSWER)

Goal Number and Title:
3 - Cleaning Up Communities and Advancing Sustainable Development

Objective Number and Title:
3 - Restore Land

Sub-Objective Number and Title:
3 - Cleanup Contaminated Land

Strategic Target Code and Title:
1 - By 2015, complete assessments at potential hazardous waste sites

Managing Office:
Office of Site Remediation and Technology Innovation

1a. Performance Measure Term Definitions:

Definition of Assessments: The Superfund site assessment process is used to evaluate potential or confirmed releases of hazardous substances that may pose a threat to human health or the environment. The process is guided by criteria established under the Hazard Ranking System (HRS) and is carried out by EPA, State, Tribal, or other Federal Agency environmental programs. Following notification of a potential site, a series of assessments are carried out until a final decision is reached regarding the need for remedial cleanup attention. [Source: http://www.epa.gov/superfund/programs/npl_hrs/siteasmt.htm

(Also see Chapter V of the most recent Superfund Program Implementation Manual (SPIM), which is updated each fiscal year and contains definitions and documentation/coding guidance for Superfund measures. The most current SPIM can be found here: http://epa.gov/superfund/policy/guidance.htm

Definition of Potential Hazardous Waste Sites: Any site or area where a hazardous substance may have been deposited, stored, disposed of, or otherwise come to be located and is or was assessed by EPA or its State, Tribal, or other Federal partners under the Federal Superfund Program.

Definition of Remedial Response: A remedial response is a long-term action that stops or substantially reduces a release of a hazardous substance that could affect public health or the environment. [Source: Superfund website, http://congressionalresearch.com/97-312/document.php?study=SUPERFUND+FACT+BOOK

Definition of "Other Cleanup Activity": - Sites that are not on EPA's National Priorities List that have completed the Superfund remedial assessment process and determined to need remedial-type cleanup attention may be addressed under a State, Tribal or other Federal Agency environmental cleanup program. EPA refers to these sites as "Other Cleanup Activity (OCA)" sites. Remedial-type work can include comprehensive site investigations in support of making cleanup determinations, interim cleanup actions, removals or final cleanup decisions, including decisions that cleanup is not required. At these sites, there is no continuous and substantive involvement on the part of EPA's site assessment program while remedial-type work is ongoing, such as routinely reviewing work products and other documents and providing comments. Each year, EPA checks in with its State, Tribal and other Federal Agency partners on the status of cleanup work at these sites. Should conditions change such that Federal Superfund involvement becomes necessary, EPA will work with its State, Tribal and other Federal Agency partners to determine an alternative approach for addressing a site. [Source: http://www.epa.gov/superfund/programs/npl_hrs/othercleanup.htm

References:

U.S. Environmental Protection Agency, EPA Performance and Accountability Reports, http://www.epa.gov/ocfo/par/index.htm

U.S. Environmental Protection Agency, Superfund Accomplishment and Performance Measures, http://www.epa.gov/superfund/accomplishments.htm

U.S Environmental Protection Agency, Office of Inspector General, Information Technology - Superfund Enterprise Management System/Comprehensive Environmental Response, Compensation, and Liability Information System (SEMS/CERCLIS) Data Quality, No. 2002-P-00016, http://www.epa.gov/oig/reports/2002/cerlcis.pdf .

U.S. Government Accountability Office, "Superfund Information on the Status of Sites, GAO/RCED-98-241", http://www.gao.gov/archive/1998/rc98241.pdf

U.S. Environmental Protection Agency, Office of Superfund Remediation and Technology Innovation, Superfund Program Implementation Manuals (SPIM), http://www.epa.gov/superfund/policy/guidance.htm

U.S. Environmental Protection Agency, Office of Environmental Information, EPA System Life Cycle Management (SLCM) Requirements Guidance, CIO 2121-G-01.0, http://www.epa.gov/irmpoli8/policies/CIO_2121-G-01.0.pdf

U.S. Environmental Protection Agency, Office of Environmental Information, EPA's Information Quality Guidelines, http://www.epa.gov/quality/informationguidelines

NOTE: Strategic Target Title should read " By 2015, complete 93,400 assessments at potential hazardous waste sites".

2a. Original Data Source:

Original data sources vary, and multiple data sources can be used for each site. Typical data sources are EPA personnel, contractors (directly to EPA or indirectly, through the interagency agreement recipient or cooperative agreement recipient), and states/tribes (cooperative agreement recipients).

(See item Performance Measure Term Definitions in Section 1, for more information.)

2b. Source Data Collection:

Collection typically involves some combination of environmental data collection, estimation and/or tabulation of records/activities. Documents such as Preliminary Assessment Reports, Site Inspection Reports, Expanded

Site Inspection Reports, Pre-SEMS/CERCLIS Screening Reports, Hazardous Ranking Package, and Site Decision Form are known reliable sources of data and provide the information necessary for determining the Assessment is completed.

Each EPA Region has a site assessment manager that oversees reporting.

The completion of site assessment activities has always been required to be entered into SEMS/CERCLIS and tracked on a SCAP-13 report. In the past 2 years, this measure has been added to the strategic plan.

The collection methods and guidance for determining the number of assessments completed are found in the Superfund Program Implementation Manual (SPIM).
(http://epa.gov/superfund/policy/guidance.htm

Source data collection frequency: At the conclusion of each individual site assessment.

Spatial detail: Site, defined in database by latitude/longitude pair.

2c. Source Data Reporting:

SEMS/CERCLIS is used to report the completion of the site assessments. The completion dates come from various site assessment reports such as the Preliminary Assessment report, Site Inspection report, Pre-SEMS/CERCLIS screening report along with other site assessment activity reports and the Site Decision Form. The report date and the Site Decision Form date are known reliable sources of data and provide the information necessary identifying an assessment completion.
EPA's Regional offices and Headquarters enter data into SEMS/CERCLIS as assessments are completed. The site assessment completion is reviewed quarterly by the 5th working day of the start of the following quarter. SEMS/CERCLIS is to be updated prior to the quarterly pull for the quarter in which the event occurs.

See Chapter V of the most recent SPIM, which is updated each fiscal year and contains definitions and documentation/coding guidance for Superfund measures. The most current SPIM can be found here:
http://epa.gov/superfund/policy/guidance.htm

3a. Relevant Information Systems:

The number of sites assessments completed are pulled directly from SEMS/CERCLIS. The assessment completion date is entered into SEMS/CERCLIS when the assessment has been completed and the Site Decision Form is completed and has been approved as such by the appropriate regional personnel.

SEMS/CERCLIS database – The SEMS/CERCLIS database is used by the Agency to track, store, and report Superfund site information (e.g., NPL sites and non-NPL Superfund sites).

(For more information about SEMS/CERCLIS, see Chapter IV of the most recent SPIM, which is updated each fiscal year and contains definitions and documentation/coding guidance for Superfund measures. The most current SPIM can be found here: http://epa.gov/superfund/policy/guidance.htm.

SEMS/CERCLIS operation and further development is taking place under the following administrative control quality assurance procedures: 1) Office of Environmental Information System Life Cycle Management Policy Agency Guidance; 2) the Office of Solid Waste and Emergency Response (OSWER) Quality Management Plan (QMP); 3) EPA IT standards; 4) Quality Assurance Requirements in all contract vehicles under which SEMS/CERCLIS is being developed and maintained; and 5) EPA IT security policies. In addition, specific controls are in place for system design, data conversion and data capture, as well as SEMS/CERCLIS outputs.

See the SEMS/CERCLIS QAPP by going here:
http://www.epa.gov/superfund/sites/phonefax/SEMS/CERCLIS_QAPP.pdf

SEMS/CERCLIS adherence to the security policy has been audited. Audit findings are attached to this record.

OSWER Performance Assessment Tool (PAT). This tool serves as the primary external servicing resource for organizing and reporting OSWER's performance data, which collects information from OSWER program systems, and conforms it for uniform reporting and data provisioning. PAT captures data from SEMS/CERCLIS; replicates business logic used by SEMS/CERCLIS for calculating measures; delivers that data to EPA staff and managers via a business intelligence dashboard interface for analytic and reporting use; and transmits data to the Budget Automated System (BAS). No current system specifications document is currently available for PAT, but will be provided when available. For this measure, PAT transmits Regional-level data to BAS.

PAT operates under the OSWER QMP. PAT has a security certification confirming that a security policy is not necessary because no sensitive data are handled and PAT is built upon the Oracle-based business intelligence system. PAT's security certification indicates that it follows all security guidelines for EPA's Oracle Portal and that PAT is (1) not defined as a "Major Application" according to NIST Special Publication 800-18, Guide for Developing Security Plans for Information Technology Systems, section 2.3.1; (2) does not store, process, or transmit information that the degree of sensitivity is assessed as high by considering the requirements for availability, integrity, and confidentiality according to NIST Special Publication 800-18, Guide for Developing Security Plans for Information Technology Systems, section 3.7.2. (3) is not covered by EPA Order 2100.2A1 Information Technology Capital Planning and Investment Control (CPIC).

EPA Headquarters is now scoping the requirements for an integrated (Superfund Document Management System-) SDMS-SEMS/CERCLIS system, called the Superfund Enterprise Management System (SEMS). Development work on SEMS began in FY 2007 and will continue through FY 2013.

SEMS represents further re-engineering of the national reporting systems to include additional elements of EPA's Enterprise Architecture. SEMS will provide a common platform for major Superfund systems and future IT development. It will be constructed in part using EPA IT enterprise architecture principles and components. SEMS will provide a Superfund Program user gateway to various IT systems and information collections.
Attached Documents:
CERCLIS July 9 2009 scan High Medium Response.xls

3b. Data Quality Procedures:

A list of all data sponsors is provided in Appendix B of the SPIM. The most current SPIM can be found here: http://www.epa.gov/superfund/policy/guidance.htm

SEMS/CERCLIS: To ensure data accuracy and control, the following administrative controls are in place: 1) Superfund Program Implementation Manual (SPIM), the program management manual that details what data must be reported; 2) Report Specifications, which are published for each report detailing how reported data are calculated; 3) Coding Guide, which contains technical instructions to data users including Regional IMCs, program personnel, data owners, and data entry personnel; 4) Quick Reference Guides (QRG), which are available in the SEMS/CERCLIS Documents Database and provide detailed instructions on data entry for nearly every module in SEMS/CERCLIS; 5) Superfund Comprehensive Accomplishment (SCAP) Reports within SEMS/CERCLIS, which serve as a means to track, budget, plan, and evaluate progress towards meeting Superfund targets and measures; 6) a historical lockout feature in SEMS/CERCLIS so that changes in past fiscal year data can be changed only by approved and designated personnel and are logged to a Change Log report, 7) the OSWER QMP; and 8) Regional Data Entry Control Plans.

EPA Headquarters has developed data quality audit reports and Standard Operating Procedures, which address timeliness, completeness, and accuracy, and has provided these reports to the Regions. In addition, as required by the Office of Management and Budget (OMB), SEMS/CERCLIS audit logs are reviewed monthly.

Regional Data Entry Control Plans. Regions have established and published Data Entry Control Plans, which are a key component of SEMS/CERCLIS verification/validation procedures. The control plans include: (1) regional policies and procedures for entering data into SEMS/CERCLIS, (2) a review process to ensure that all Superfund accomplishments are supported by source documentation, (3) delegation of authorities for approval of data input into SEMS/CERCLIS, and (4) procedures to ensure that reported accomplishments meet accomplishment definitions. In addition, regions document in their control plans the roles and responsibilities of key regional employees responsible for SEMS/CERCLIS data (e.g., regional project manager, information management coordinator, supervisor, etc.), and the processes to assure that SEMS/CERCLIS data are current, complete, consistent, and accurate. Regions may undertake centralized or decentralized approaches to data management. These plans are collected annually for review by OSRTI/IMB (Information Management Branch). [Source: SPIM FY12, IV (http://epa.gov/superfund/policy/guidance.htm)

Copies of the Regional Data Entry Control Plans are provided with this DQR. Current and past year plans are available by contacting the Chief, Information Management Branch, Office of Superfund Remediation and Technology Innovation.

Regions are expected to prepare Data Entry Control Plans consistent with the SPIM and the Headquarters guidance: "SEMS/CERCLIS Data Entry Control Plan Guidance."

Superfund Program Implementation Manual (SPIM). The SPIM should be the first source referred to for additional questions related to program data and reporting. The SPIM is a planning document that defines program management priorities, procedures, and practices for the Superfund program (including response, enforcement, and Federal facilities). The SPIM provides the link between the GPRA, EPA's Strategic Plan, and the Superfund program's internal processes for setting priorities, meeting program goals, and tracking performance. It establishes the process to track overall program progress through program targets and measures.

The SPIM provides standardized and common definitions for the Superfund program, and it is part of EPA's internal control structure. As required by the Comptroller General of the United States, through generally accepted accounting principles (GAAP) and auditing standards, this document defines program scope and schedule in relation to budget, and is used for audits and inspections by the Government Accountability Office (GAO) and the Office of the Inspector General (OIG). The SPIM is developed on an annual basis. Revisions to the SPIM are issued during the annual cycle as needed.

The most current version of the SPIM can be found at: http://epa.gov/superfund/policy/guidance.htm

Data Flow:

Step 1. Original data sources provide information.

Step 2. EPA Region enters the assessment completion dates in SEMS/CERCLIS as needed.

Step 3. OSWER's PAT pulls data from SEMS/CERCLIS. Headquarters staff compare PAT results to SEMS/CERCLIS results. If PAT does not match SEMS/CERCLIS then there was an error with the upload and data are reloaded. Headquarters staff enter into PAT the Annual Commitment System (ACS) status information for each measure and, if necessary, a status explanation.

Step 5. Headquarters approves PAT results, and PAT pushes results into BAS.

Step 6. BAS aggregates Regional data into a national total. The OSRTI lead for reporting reviews and certifies results in BAS.

Attached Documents:

2013 DECP guidance10-19-2012.pdf

3c. Data Oversight:

The Superfund program has a "data sponsorship" approach to database oversight. Headquarters staff and managers take an active role in improving the quality of data stored in SEMS/CERCLIS by acting as data sponsors.

HQ managers take an active role in improving the quality of data stored in SEMS/CERCLIS by acting as data sponsors. Data sponsorship promotes consistency and communication across the Superfund program. HQ data sponsors communicate and gain consensus from data owners on data collection and reporting processes. Data sponsors ensure that the data they need to monitor performance and compliance with program requirements are captured and stored properly in SEMS/CERCLIS. To meet this goal, HQ data sponsors identify their data needs, develop data field definitions, and distribute guidance requiring submittal of these data. Data owners are normally site managers that need the data in support of site work. Data owners follow the guidance they receive from data sponsors, as they acquire and submit data.

HQ data sponsors assist data owners in maintaining and improving the quality of Superfund program data. These data are available for data evaluation and reporting. Data sponsorship helps promote consistency in both national and regional reporting. In addition, data sponsorship provides a tool to improve data quality through program evaluation and adjustments in guidance to correct weaknesses detected. Data sponsors may conduct audits to determine if there are systematic data problems (e.g., incorrect use of codes, data gaps, etc.). [Source: XI.A.4 of the FY 2012 SPIM] A list of data sponsors is provided in Appendix B of the SPIM. The latest version of the SPIM can be found here:http://epa.gov/superfund/policy/guidance.htm

Specific roles and responsibilitiesof data sponsors can be found in Chapter IV of the SPIM. http://epa.gov/superfund/policy/guidance.htm

The primary responsibilities of data owners are (1) to enter and maintain data in SEMS/CERCLIS and (2) assume responsibility for complete, current, consistent, and accurate data. The data owners for specific data are clearly identified in the system audit tables. Regions annually update region-specific Data Entry Control Plans (DECP). Among other things, Regional data entry control plans identify which Data Sponsors/Data Owners are responsible for different aspects of data entry. (See item 3b., Data Quality Procedures for more information on Data Entry Control Plans.)

Roles and Responsibilities of the Information Management Coordinators (IMCs). In each Region, the IMC is a senior position which serves as regional lead for all Superfund program and SEMS/CERCLIS systems management activities. Roles and responsibilities of IMCs can be found in IV.B.1a of the FY 2012 SPIM. The latest version can be found here: http://epa.gov/superfund/policy/guidance.htm

The Information Management Officer (IMO) & Director, Information Management and Data Quality Staff. OSWER is the lead point of contact for information about the data from SEMS/CERCLIS .

The Project Manager for SEMS/CERCLIS oversees and is the approving authority for quality-related SEMS/CERCLIS processes, and is closely supported by a Contract Task Manager. (See the SEMS/CERCLIS QAPP here: http://www.epa.gov/superfund/sites/phonefax/SEMS/CERCLIS_QAPP.pdf) The lead point of contact for information about the data from SEMS/CERCLIS is the Director, Information Management and Data Quality Staff, Office of Solid Waste and Emergency Response.

PAT Data Entry

The Annual Commitment System (ACS) Coordinator in OSRTI ensures that SEMS/CERCLIS data for this measure are correctly loaded into PAT. The ACS Coordinator then works with the data sponsor to review uploaded data, edit records as appropriate, and then push data to ACS--part of the Office of Chief Financial Officer's (OCFO) BAS. PAT is maintained by OSWER's System Manager who ensures that the PAT system operates correctly, based on business logic agreed to by OSRTI.

3d. Calculation Methodology:

The performance measure is a specific variable entered into SEMS/CERCLIS following specific coding guidance and corresponding supporting site-specific documentation.

The unit of measure is the number of remedial site assessments completed.

References:

Superfund Data Element Dictionary (DED). The Superfund DED is available online at: http://www.epa.gov/superfund/sites/ded/index.htm The DED provides definitions and descriptions of elements, tables and codes from the SEMS/CERCLIS database used by the Superfund program. It also provides additional technical information for each entry, such as data type, field length and primary table. Using the DED, you can look up terms by table name or element name, or search the entire dictionary by keyword.

Other additional references that may be useful:

Coding Guide. The Superfund Coding Guide contains technical instructions to data users including Regional IMCs, program personnel, data owners, and data entry personnel. The Site Assessment component of the Coding Guide is attached to this record.

[attachment "Coding Guide - 2009.pdf" deleted by Randy Hippen/DC/USEPA/US]

Quick Reference Guides (QRG). Superfund Quick Reference Guides are available in the SEMS/CERCLIS Documents Database and provide detailed instructions on data entry for nearly every module in SEMS/CERCLIS. A sample QRGs is available for entering data related to reporting Non-NPL status.

Site Status and Description document: This QRG describes entering site status and description data into SEMS/CERCLIS.

Attached Documents:

FY 2012 CERCLIS Coding Guide.pdf
ReportingNon-NPLStatus_Feb2008.doc
19-0227 (Site Status Description OUs).doc

4a. Oversight and Timing of Final Results Reporting:

Data Sponsor for Site Assessment Completions, Annual Commitment System coordinator, and National Program Office (NPO) management.

Progress reporting is done periodically, while official numbers are reported annually.

4b. Data Limitations/Qualifications:

The Site Assessment Completions measure is reported at least annually, and data must be entered in SEMS/CERCLIS prior to the annual data pull which is usually the 10th business day following the end of the FYQ4.

4c. Third-Party Audits:

Three audits, two by the Office Inspector General (OIG) and the other by Government Accountability Office (GAO), assessed the validity of the data in SEMS/CERCLIS. The OIG audit report, Superfund Construction Completion Reporting (No. E1SGF7_05_0102_ 8100030), dated December 30, 1997, concluded that the Agency "has good management controls to ensure accuracy of the information that is reported," and "Congress and the public can rely upon the information EPA provides regarding construction completions." The GAO report, Superfund: Information on the Status of Sites (GAO/RCED-98-241), dated August 28, 1998, estimated that the cleanup status of National Priority List (NPL) sites reported by SEMS/CERCLIS as of September 30, 1997, is accurate for 95 percent of the sites. Another OIG audit, Information Technology - Comprehensive Environmental Response, Compensation, and Liability Information System (SEMS/CERCLIS) Data Quality (Report No. 2002-P-00016), dated September 30, 2002, evaluated the accuracy, completeness, timeliness, and consistency of the data entered into SEMS/CERCLIS. The report provided 11 recommendations to improve controls for SEMS/CERCLIS data quality. EPA has implemented these recommendations and continues to use the monitoring tools for verification.

The IG annually reviews the end-of-year SEMS/CERCLIS data, in an informal process, to verify data that supports the performance measures. Typically, there are no published results.

Measure Code: ST1 - Reduce the number of confirmed releases at UST facilities to five percent (5%) fewer than the prior year's target.

Goal Number and Title:
3 - Cleaning Up Communities and Advancing Sustainable Development
Objective Number and Title:
2 - Preserve Land
Sub-Objective Number and Title:
2 - Minimize Releases of Hazardous Waste and Petroleum Products
Strategic Target Code and Title:
3 - Through 2015, minimize the number of confirmed releases at UST facilities
Managing Office:
Office of Underground Storage Tanks
1a. Performance Measure Term Definitions:
The most current definitions for the EPA's performance measures related to underground storage tanks are available on EPA's website www.epa.gov/oust/cat/camarchv.htm under Definitions. See the definition for the measure number LUST-1 in the definitions document. For more information on EPA's Underground Storage Tanks Program, see: http://www.epa.gov/oust/index.htm
2a. Original Data Source:
The data suppliers are the states and territories who are the direct implementers of the program in their respective jurisdictions and the EPA regions who provide assistance to the tribes.
2b. Source Data Collection:
The data is collected by each state and territory using their own systems and databases. They then report this information to OUST using the LUST4 system described under section 3. EPA Quality Assurance Requirements/Guidance under Which Original Data Sources Collect Data: For cooperative agreements: Regional offices include QA Terms and Conditions in their states' assistance agreement. CAs must be current and specify: QA roles and responsibilities for EPA and grantee recipients; and quality requirements including responsibilities for final review and approval. Default quality requirements include: organization-level QA documentation (i.e. QMP) for state agencies and primary contractors; and project-level QAPPs for each CA. In accordance with EPA's Uniform Administrative Requirements for Grants and Cooperative Agreements, 40 CFR Part 31.45, states must develop and implement quality assurance practices. The regulation requires developing and implementing quality assurance practices that will "produce data of quality adequate to meet project objectives and to minimize loss of data to out of control conditions or malfunctions'; see OSWER Directive 9650.10A: www.epa.gov/oust/directiv/d965010a.htm - sec11"

For contracts: EPA Regions determine which quality requirements are applicable. Contracts must be current and specify: QA roles and responsibilities for EPA and national LUST contractors; and quality requirements including responsibilities for final review and approval. Default quality requirements include: organization-level QA documentation (i.e. QMP) for the primary contractors; and project-level QAPPs for each Tribal LUST remedial Work Assignment. Sample EPA contract language: "the Contractor shall comply with the higher-level quality standard selected below: Specifications and Guidelines for Quality Systems for Environmental Data Collection and Environmental Technology Programs (ANSI/ASQC E4, 1994). As authorized by FAR 52.246-11, the higher-level quality standard ANSI/ASQC E4 is tailored as follows: The solicitation and contract require the offerors/contractor to demonstrate conformance to ANSI/ASQC E4 by submitting the quality documentation described below. [Specifically,...] ... The Contractor shall not commence actual field work until until the Government has approved the quality documentation (i.e., QAPP)."

Note: Regions keep copies of individual QAPPs associated with cooperative agreements and contracts. Each EPA regional office manages its own state and tribal assistance agreements.

2c. Source Data Reporting:

Data Submission Instrument:
State-specific databases.

Data Entry Mechanism:
Each state enters their data into the online LUST4 Oracle-based system (see section 3 for more details).

Frequency of Data Transmission to EPA: Twice annually.

Timing of Data Transmission to EPA:
Within 10 days of the end of the reporting period (by April 10 for mid-year, and October 10 for end-of-year).

3a. Relevant Information Systems:

System Description:
LUST4. This database is the master database of all LUST program-related data. States, territories and EPA report data for activity and measures directly into LUST4. LUST4 's Oracle Web-based system is accessed through the EPA portal at http://portal.epa.gov/ under the My Communities/Underground Storage Tank Menu Page.

OSWER Performance Assessment Tool (PAT). This tool serves as the primary external servicing resource for organizing and reporting OSWER's performance data. PAT collects information from OSWER program systems, and conforms it for uniform reporting and data provisioning. PAT captures data from LUST4; replicates business logic used by LUST4 for calculating measures; can deliver that data to EPA staff and managers via a business intelligence dashboard interface for analytic and reporting use; enables LUST point of contact to document status and provide explanation for each measure; and transmits data to EPA's Budget Automation System (BAS). No current system specifications document is currently available, but will be provided when available.

BAS. BAS is the final repository of the performance values.

Source/Transformed Data:
LUST4. LUST4 includes both source data and transformed data (e.g., data aggregated into Regional totals).

PAT. PAT includes only transformed data.

BAS. BAS includes only transformed data.

Information System Integrity Standards:
LUST4. LUST4 operates under OSWER's QMP, including the security policy specified in that QMP. LUST4 does not have any stand-alone certifications related to the EPA security policy or the Systems Life Cycle Management policy. The LUST4 system is built upon Oracle Business Intelligence tools provided by the EPA Business Intelligence Analytics Center, which ensures that a stand-alone security certification is not necessary.

PAT. PAT operates under the OSWER Quality Management Plan (QMP). PAT has a security certification confirming that a security policy is not necessary because no sensitive data are handled and PAT is built upon the Oracle-based business intelligence system. PAT's security certification indicates that it follows all security guidelines for EPA's Oracle Portal and that PAT is (1) not defined as a "Major Application" according to NIST Special Publication 800-18, Guide for Developing Security Plans for Information Technology Systems, section 2.3.1; (2) does not store, process, or transmit information that the degree of sensitivity is assessed as high by considering the requirements for availability, integrity, and confidentiality according to NIST Special Publication 800-18, Guide for Developing Security Plans for Information Technology Systems, section 3.7.2. (3) is not covered by EPA Order 2100.2A1 Information Technology Capital Planning and Investment Control (CPIC).

BAS. Not applicable.

3b. Data Quality Procedures:

EPA's regional grants project officers and regional program managers provide first-level data quality reviews and oversight of their recipients' program performance measure results. EPA/OUST reviews, comments and approves each record.

OUST uses a combination of automated validation along with manual QA/QC review.

QA/QC REVIEW BY REGIONS. EPA/OUST oversees the use of the QA/QC checklist, which is incorporated into the LUST4 oracle web-based system. Regions complete the QA/QC checklist, sign it electronically and submit it to EPA/OUST for review, comment and approval of each record.
NOTE: This QA/QC checklist was last updated 10/1/2009 and is accessed through the user interface of LUST4.

Regional QA/QC Evaluation Checklist –

Note: Checklist is to be completed by Regional reviewer and will appear "shaded" to others.

1. Previous Totals Column

-- Verify the previous total number is correct by comparing it to the total from the last reporting period. If there is a discrepancy, report the information in the "Correction To Previous Data" column. Please add comments in the "Comments" column for any corrections that are made to the applicable performance measure.

2. Actions This Reporting Period

For each performance measure, if this "Reported" number deviates by more than 10% from the last period's number or appears otherwise questionable, complete the following actions:

-- Compare data to additional previous reporting periods to see if this current data deviates by more than 10% from previous reporting periods as well.

-- Review the state's explanation, if available.

-- If necessary, contact the state to obtain the corrected numbers and/or obtain a sufficient explanation and include the explanation in the "Comments" section for the applicable performance measure.

3. Corrections to Previous Data Column

Verify that if any corrections have been listed that an explanation for the correction is provided in the "Comments" column and complete the following actions:

-- Verify and discuss the correction with the state if the correction is >10% or if the correction appears questionable (e.g., database conversions, database cleanup efforts to resolve misclassified data, duplicative records, etc.)

-- Verify if the corrections are anticipated to be a one-time event or occur over multiple years

-- Evaluate if the corrections will impact other performance measures (e.g., if the number of cleanups completed is adjusted downward by a correction, does this also result in a commensurate downward adjustment of cleanups initiated?) Include any additional comments in the "Comments" column as necessary.

4. Totals (Cumulative, if applicable)

-- Verify accuracy of all cumulative totals

-- Include any additional comments in the "Comments" column as necessary

-- Verify that the cumulative total confirmed releases is equal to or greater than the cumulative totals for both cleanups initiated and cleanups completed. The two data elements are subsets of confirmed releases.

AUTOMATED VALIDATION.

LUST4 will show an error message if the user enters values that result in the cumulative total of confirmed releases being less than the cumulative total of either cleanups initiated or cleanups completed.

DATA FLOW:

Step 1. Confirmed releases are entered into LUST4 by state recipients or by Regions (for tribal data).

Step 2. Each Region conducts Regional level review of data from the LUST4 system. Rejected data must be edited by the original data source. Approved data proceed to Step 3.

Step 3. Headquarters' staff perform performs National Program Review, using data from the LUST4 system.

Rejected data must be reviewed by the region and, if needed, pushed back to the state for editing(Step 2).

Step 4. PAT pulls data from LUST4. Headquarters staff compare PAT results to LUST4 results. If PAT does not match LUST4 then there was an error with the upload and data is reloaded. Headquarters staff enter into PAT the ACS status information of "Indicator" for each measure and, if desired, explanation.(Note: PAT allows for programs to identify status other than "Indicator." When programs select a status of "no status," "data not available," or "target not met," PAT requires that an explanation be provided. LUST program policy is to resolve all reporting issues prior to ACS reporting, so "Indicator" is the only status chosen and explanations for that status are optional.)

Step 5. Headquarters approves PAT results, and PAT pushes results into BAS/Measures Central.

Step 6. Measures Central aggregates Regional data into a national total. OUST reporting lead reviews and certifies results.

3c. Data Oversight:

Source Data Reporting Oversight Personnel:
Regional Program Managers are ultimately responsible for regional-level data.

Source Data Reporting Oversight Responsibilities:
Regional Program Managers conduct their review based upon a national QA/QC checklist, as described in the Data Quality Procedures field.

Information Systems Oversight Personnel:
OUST LUST4 System Manager

Information Systems Oversight Responsibilities:
Maintains a list of the HQ (OUST and OEI), Regional and state/territory primary and backup users; a record of changes to the list is also maintained. Ensures that Regional reporting is on track, conducts QA on LUST performance measures, ensures QA issues are resolved and/or documented, and oversees final reporting to BAS.
Works with OUST contractor to resolve any issues with the LUST4 data system.

3d. Calculation Methodology:

At the end of the fiscal year, users report the number of confirmed releases they had over the last six months (July 1 through Sept. 30). The system adds this number to what the user reported in the mid-year report (covering Oct 1 through March 31) to calculate the total number of confirmed releases for the year. The user will also report any corrections to their previous cumulative totals. The corrections and actions during this reporting period are added to the users' previous cumulative total to get the new current cumulative total of confirmed releases in that state.

Unit of Measure: Number of releases from regulated underground storage tanks that are reported and

| verified by the state during the reporting period.

Timeframe of Result: Semi-annual.

Documentation of Methodological Changes: Not applicable. |
|---|

4a. Oversight and Timing of Final Results Reporting:
Final Reporting Oversight Personnel: Deputy Office Director. Final Reporting Oversight Responsibilities: Responsible for final review to ensure LUST 4 System Manager has completed review, and numbers are accurate. Final Reporting Timing: Semiannual.

4b. Data Limitations/Qualifications:
Data quality depends on the accuracy and completeness of state records. Also, some states rely on local jurisdictions for their data, which can cause delays for these states. Additionally, the tanks program is primarily run by states, and each state operates their program in a manner that works best for them. Because there are differences between all states, the data from each state can be influenced by the policies and interpretations of each state. This creates limitations when someone compares state-level data.

4c. Third-Party Audits:
None.

Measure Code: C1 - Score on annual Core NAR.
Office of Solid Waste and Emergency Response (OSWER)

Goal Number and Title:
3 - Cleaning Up Communities and Advancing Sustainable Development

Objective Number and Title:
3 - Restore Land

Sub-Objective Number and Title:
2 - Emergency Preparedness and Response

Strategic Target Code and Title:
1 - By 2015, achieve and maintain percent of the maximum score on the Core National Approach to Response

Managing Office:
Office of Emergency Management

1a. Performance Measure Term Definitions:
The National Approach to Response (NAR) is an Agency-wide mechanism to address effective allocation of resources during an incident and effective implementation of response procedures. To ensure that the goals of the NAR are being met, EPA continues to implement its annual assessment of its response and removal preparedness via the Core National Approach to Response (Core NAR) assessment, which grew out of its Core Emergency Response (ER) program and assessment. The Core NAR evaluation is conducted annually and consists of two parts. The first part, called Core ER, addresses day-to-day preparedness for removal actions for regions, special teams and Headquarters (HQ). The second part addresses national preparedness for chemical, biological, radiological and nuclear (CBRN) incidents. The score on Core NAR, which is reflected as composite percentage of both the ER and CBRN components for the Regions, Special Teams and HQ, reflects Agency performance relative to the evaluation criteria.

2a. Original Data Source:
Information to support Core NAR scoring is collected by the evaluation team. The evaluation team consists of managers and staff from HQ, including contractor support. The data consist of scores (on a scale of 0 to 3 for Core ER and 0 to 5 for Core CBRN) for a number of readiness elements. Scores are collected for each of the 10 EPA regions, HQ and EPA special teams. The original scores are developed by OEM HQ staff and their contract support.

2b. Source Data Collection:
Data are collected through detailed self-evaluation surveys of all regional programs, EPA special teams and HQ offices. Following the self-evaluation surveys, the evaluation team conducts interviews with personnel and managers in order to gather more information and to support their determination of the final score. The evaluation team reviews the data during the data collection and analysis process. The data are collected by a combination of managers and staff to provide consistency across all reviews plus to serve as an important element of objectivity in each review. Standards and evaluation criteria have been developed and reviewed extensively by HQ, and EPA's regional managers and staff and special teams managers and staff. Beginning in FY 2014, the Core NAR evaluation will include Readiness Assessments where response staff will be tested on their ability to use response equipment.

2c. Source Data Reporting:

Scores are entered into a form and sent to the 10 regions and special teams in advance of each Core NAR evaluation conference call (see attached form). The scores are also tabulated by EPA HQ contract support into a database and stored until they are to be revised based on input received during the evaluation call. The scores for Core ER are typically shared with the recipients within three weeks of the completion of individual interviews. The composite score for Core ER and CBRN is provided to all recipients within a month of the completion of all interviews.

Attached Documents:
Core NAR ER Regions FY 2012_FINAL,rcedits.docx

3a. Relevant Information Systems:

System Description: Data from evaluations from each of the 10 regions, special teams, and HQ are tabulated and stored using standard software (e.g., Word, Excel). No specific database has been developed. Currently, there are no plans to develop a dedicated system to manage the data.

Source/Transformed Data: The files identified above contain both source and transformed data (modified and agreed-upon scores from the evaluation interviews).

Information System Integrity Standards: Not Applicable.

3b. Data Quality Procedures:

Data review is conducted after the data have been analyzed by the evaluation team, to ensure the scores are consistent with the data and program information. The scores are developed by a team looking across all 10 regions, special teams, and HQ, allowing for easier cross-checking and ensuring better consistency of data analysis and identification of data quality gaps.

The evaluation team considers any programmatic constraints across the Agency that might call for adjustments to scores for individual Core NAR elements being evaluated. As individual evaluation interviews progress during the year, a particular development or re-occurring instance or theme might emerge for a number of regions or special teams that might warrant revision of scores for groups that had their evaluation earlier in the process. If this instance affects enough of the regions and special teams, then it might, for example, be concluded that no group should receive a score of 3 for the element in question even if a group was given a 3 during an earlier evaluation call.

For the HQ evaluation, two regional managers participate in the process and make the final determination for HQ's scores to ensure fairness and accuracy.

3c. Data Oversight:

Source Data Reporting Oversight Personnel:
- HQ Project Manager: Office of Emergency Management –Evaluation and Communications Division
- HQ Project Support: Office of Emergency Management –Evaluation and Communications Division

Source Data Reporting Oversight Responsibilities:

- Making any necessary changes to the evaluation criteria for Core ER and CBRN
- Evaluating materials submitted by regions, special teams and HQ to determine initial scores
- Schedule and conduct individual evaluation interviews / meetings
- Make revisions to initial scores based on input from interviews / meetings

Information Systems Oversight Personnel:

- HQ Project Manager: Office of Emergency Management –Evaluation and Communications Division
- HQ Project Support: Office of Emergency Management –Evaluation and Communications Division

Information Systems Oversight Responsibilities:

- Tabulate revised scores into spreadsheets to develop final regional, special teams and HQ scores

3d. Calculation Methodology:

Once all of the evaluations are complete, a national average score is calculated by a team looking across all 10 regions, special teams and HQ.

For the two parts of the Core NAR evaluation (Core ER & CBRN), the total score for the region, special team or HQ is divided by the total possible score. This percentage is considered the score for that part of the evaluation for that particular group. For the ER component, percentages are calculated in this manner for all 10 regions, four special teams and EPA HQ. For the CBRN component, one percentage is calculated for the 10 regions, and separate percentages are calculated for each of the four special teams and EPA HQ.

The percentages are given equal weight and averaged to obtain final agency-wide scores for Core ER and CBRN. These two agency-wide scores are averaged to obtain the final, national average Core NAR score for the Agency.

4a. Oversight and Timing of Final Results Reporting:

Final Reporting Oversight Personnel:

- HQ Project Manager: Office of Emergency Management –Evaluation and Communications Division
- HQ Project Support: Office of Emergency Management –Evaluation and Communications Division
- HQ Results Reporting Support: Office of Emergency Management—Evaluation and Communications Division

Final Reporting Oversight Responsibilities:

- Working with EPA HQ contractor support to finalize scores for the regions, special teams and HQ
- Working with EPA HQ contractor support to finalize the composite score for Core ER and CBRN
- Providing the final composite score to OSWER for end-of-year reporting purposes

Final Reporting Timing: The final Core NAR score is reported annually to OSWER for inclusion in end-of-year reporting activities for the Government Performance and Results Act.

4b. Data Limitations/Qualifications:

One key limitation of the data is the lack of a dedicated database system to collect and manage the data.

Standard software packages (word processing, spreadsheets) are used to develop the evaluation criteria, collect the data and develop the accompanying readiness scores. There is also the possibility of subjective interpretation of data.

It is likely that the error for this measure will be small for the following reasons: the standards and evaluation criteria have been developed and reviewed extensively by HQ and EPA's regional managers and staff; the data will be collected by a combination of managers and staff to provide consistency across all reviews plus an important element of objectivity in each review; the scores will be developed by a team looking across all ten regions, special teams, and HQ, allowing for easier cross-checking and ensuring better consistency of data analysis and identification of data quality gaps.

4c. Third-Party Audits:

The Core NAR evaluation has two features that help it achieve validation and verification similar to that of an independent third party audit.
- The final determination on the HQ score is given by two regional managers to help ensure fairness of the evaluation whereas HQ staff gives scores for the regions and special teams.
- Readiness Assessments of EPA's ability to use response equipment will be conducted and overseen by other regions rather than a self evaluation.

Measure Code: CH2 - Number of risk management plan audits and inspections conducted.

Office of Solid Waste and Emergency Response (OSWER)

Goal Number and Title:
3 - Cleaning Up Communities and Advancing Sustainable Development

Objective Number and Title:
1 - Promote Sustainable and Livable Communities

Sub-Objective Number and Title:
3 - Reduce Chemical Risks at Facilities and in Communities

Strategic Target Code and Title:
1 - By 2015, continue to maintain the Risk Management Plan (RMP) prevention program

Managing Office:
The Office of Emergency Management (OEM)

1a. Performance Measure Term Definitions:
Risk Management Plans: Risk Management Plans are documents that are submitted by facilities that store chemicals over a certain threshold quantity. These plans are submitted every five years and document chemical processes, accident history, emergency contact information, etc. Inspections: An inspection is considered "conducted" when the EPA region completes the Inspection Conclusion Data Sheet (ICDS) and enters the information into the Integrated Compliance Information System (ICIS). However this is not always the case. For example, in an ongoing enforcement case, more information or a second site visit might be needed. Audit: Audits are similar to inspections but do not proceed to enforcement. Background: The subobjective's goal is to reduce chemical risks at facilities and in communities. Under the authority of section 112(r) of the Clean Air Act, the Chemical Accident Prevention Provisions require facilities that produce, handle, process, distribute, or store certain chemicals to develop a Risk Management Program, prepare a Risk Management Plan (RMP), and submit the Plan to EPA. The purpose of this performance measure is to ensure that facilities that are required to have risk management plans do indeed have plans and are available in case of an incident. OSWER's Office of Emergency Management implements the Risk Management Program under Clean Air Act section 112(r). Facilities are required to prepare Risk Management Plans (RMPs) and submit them to EPA. In turn, EPA Headquarters (HQ) provides appropriate data to each Region and delegated state so that they have the RMP data for their geographical area. EPA regions and delegated states conduct inspections.

2a. Original Data Source:
Data come from one of two sources: 1) EPA Regions. For most states, EPA regions are the implementing authorities that conduct and make record of inspections.

2) States: Nine states have received delegation to operate the RMP program. These delegated States report audit numbers to the appropriate EPA Regional office so it can maintain composite information on RMP audits.

2b. Source Data Collection:

EPA personnel travel to facilities to conduct inspections, using the Risk Management Plans that the facilities have submitted, as a basis for their inspection. EPA inspects approximately 5 percent of the entire RMP facility universe annually.

2c. Source Data Reporting:

EPA regional staff complete inspections and record information on the ICDS form. Inspections are recorded in the ICIS system as they are completed. EPA headquarters monitors progress of the data collection regularly and reports on the data at mid year and at the end of the fiscal year.

3a. Relevant Information Systems:

The EPA Annual Commitment System (ACS) is the database for the number of risk management plan (RMP) audits. The Integrated Compliance Information System (ICIS) is used for tracking RMP inspection activities. The Risk Management Plan (RMP) database is used to collect RMP information from regulated facilities, and provides essential background information for inspectors. The EPA Annual Commitment System (ACS) is the database for the number of risk management plan (RMP) audits.

3b. Data Quality Procedures:

Facilities submit RMP data via an online system, with extensive validation and quality control measures applied during and after submission to EPA. Regions review RMP data, and compare with information obtained during inspections. Inspection data are collected from states by EPA's Regional offices, and reviewed at the time of Regional data entry. Inspection data are regularly compared to similar data from the past to identify potential errors. Inspection data quality is evaluated by both Regional and Headquarters' personnel. Regions enter data into the Agency's Annual Commitment System, and HQ prepares an annual report.

3c. Data Oversight:

These individuals are Regional Chemical Emergency Preparedness and Prevention managers who are responsible for overseeing the inspections and data entry at the Regional level. Headquarters staff performs QA/QC on the data entered by the Regions and reports data out.

3d. Calculation Methodology:

Regional and National targets for the number of RMP inspections are set based on the FTE and program funding available to the Regions, and our understanding of the resources required to conduct RMP inspections. In prior years, our experience has shown that Regional offices can inspect approximately 5% of the universe of RMP facilities with available resources. However, this percentage is strongly dependent on the size and complexity of facilities inspected. EPA experience indicates that the field portion of RMP facility inspections alone can require anywhere from a single person for one day or less at a simple, single-process facility up to a team of 6-8 inspectors for 1-2 weeks or more at a large chemical plant or refinery. In recent years, EPA has shifted its inspection focus to high-risk RMP facilities by requiring regional offices to conduct a certain percentage of RMP inspections at these facilities. As high-risk facilities generally require the most inspection resources, the agency has reduced the overall RMP inspection target in order to devote additional resources toward high-risk facility inspections. EPA has established criteria for identifying high-risk RMP facilities and provides a list of these facilities at the beginning of each fiscal year to each Regional office. For FY 2013, the overall national RMP inspection target has been reduced from approximately 5% to 4%, while the percentage of high-risk facility inspections has been raised from approximately 25% to 30%.

4a. Oversight and Timing of Final Results Reporting:

These individuals are OEM personnel who work on the Chemical Emergency Preparedness and Prevention programs either in technical expertise or program evaluation.

4b. Data Limitations/Qualifications:

ICIS data quality is dependent on completeness and accuracy of the data provided by state programs and the EPA Regional offices.

Data are count data and not open to interpretation.

RMP data quality is enhanced by system validation, but accuracy is dependent on what the facility submits in their Risk Management Plan.

4c. Third-Party Audits:

There are no third party audits for the RMP measure.

Measure Code: MW8 - Number of tribes covered by an integrated solid waste management plan.

Office of Solid Waste and Emergency Response (OSWER)

Goal Number and Title:
3 - Cleaning Up Communities and Advancing Sustainable Development

Objective Number and Title:
2 - Preserve Land

Sub-Objective Number and Title:
1 - Waste Generation and Recycling

Strategic Target Code and Title:
3 - By 2015, increase the number of tribes covered by an integrated waste management plan

Managing Office:
Office of Resource Conservation and Recovery

1a. Performance Measure Term Definitions:
Tribe: Federally recognized tribes as defined by the Bureau of Indian Affairs. The most recent list can be found here: http://bia.gov/cs/groups/public/documents/text/idc1-023762.pdf (accessed February 11, 2014) Integrated Waste Management Plan: An integrated solid waste management plan provides a tribe with a comprehensive approach to organizing their waste collection and management programs. This is also referred to as an "Integrated Waste Management Plan (IWMP)." The following five elements represent the basic requirements that must be included in a tribe's IWMP for that plan to be considered adequate for GPRA purposes: description of the community service area; description of the tribe's waste management program structure and administration; description of the tribe's current and proposed waste management practices; description of the funding and sustainability and the long-term goals of the tribe's waste management program; and demonstration of approval of the IWMP by an appropriate governing body. Plans that do not meet all five elements may be adequate if a region determines that one or more elements are not applicable to a tribe's waste management program. For more information on the expectations associated with each element, see the following document: U.S. Environmental Protection Agency. "Five Elements of a Tribal Integrated Waste Management Plan". Memorandum from Matt Hale, former Director, Office of Resource Conservation and Recovery. http://yosemite.epa.gov/osw/rcra.nsf/0c994248c239947e85256d090071175f/E7661F353791AD71852573780050876E/$file/14776.pdf (accessed December 17, 2013). For more information on EPA's Tribal Solid Waste Management Program, visit: http://www.epa.gov/wastes/wyl/tribal/index.htm

2a. Original Data Source:
EPA regional personnel evaluate tribal integrated waste management plans, and record the data for this measure in internal Regional data systems and EPA's Budget Automation System (BAS). Regional data is collected by ORCR and maintained in a spreadsheet.

2b. Source Data Collection:

The data set is very small and has only two elements (tribe and fiscal year). The data are reviewed by EPA regional offices for data quality and periodic adjustments are made during these reviews. The minimal data is then provided to ORCR for collection in a spreadsheet.

Because the data sets are small in size on a region-by-region basis, they can be managed efficiently by each regional office.

There is not any geographical or spatial detail of source data. The measure is only determining the number of tribal IWMPs.

2c. Source Data Reporting:

The data is originated by EPA. The data from each EPA Region is entered into BAS and also submitted to the ORCR Project manager.

The instrument that the data is recorded and submitted is a spreadsheet. There is no EPA information system for this data at this time.

EPA staff will manually enter information about each IWMP in a spreadsheet.

The frequency of data transmission to EPA is at least bi-annually. Data may be submitted to the ORCR Project manager on a quarterly basis.

3a. Relevant Information Systems:

EPA Regional offices enter data into their internal data systems and the Annual Commitments System in BAS - see further description of EPA's Annual Commitments System. The internal EPA data set housing the specific integrated waste management plans for each tribe is managed by each regional office and is under the control of each region. DATA is also collected by ORCR in a spreadhseet. As of October 2013, a nationwide total of 173 tribal integrated waste management plans have been counted.

3b. Data Quality Procedures:

The regional data systems are considered to be appropriate for the minimal complexity and small size of the data set.

3c. Data Oversight:

Tribal solid waste program staff in each regional office.

Regional staff ensure that data is entered correctly into BAS.
ORCR/ Federal, State and Tribal Programs Branch, staff reviews data entered into BAS by regional staff to ensure accuracy.

3d. Calculation Methodology:

Units: Number of Tribes
The result is calculated by counting all tribes that EPA Regions have defined as being covered by an Integrated

Waste Management Plan, as defined in field 1a, Performance Measure Term Definitions.

EPA has compiled the regional data into a spreadsheet for national tracking purposes. This spreadsheet is updated quarterly.

4a. Oversight and Timing of Final Results Reporting:
Project Manager/ ORCR/ Federal, State and Tribal Programs Branch. Data is collected from BAS, and confirmed through the informal reporting of data requested from the regions that is placed into a spreadsheet. Official results are reported annually.

4b. Data Limitations/Qualifications:
The data are considered to be accurate on a regional and national scale.

4c. Third-Party Audits:
None.

Measure Code: MW8 - Number of tribes covered by an integrated solid waste management plan.

Office of Solid Waste and Emergency Response (OSWER)

Goal Number and Title:	
3 - Cleaning Up Communities and Advancing Sustainable Development	
Objective Number and Title:	
2 - Preserve Land	
Sub-Objective Number and Title:	
1 - Waste Generation and Recycling	
Strategic Target Code and Title:	
3 - By 2015, increase the number of tribes covered by an integrated waste management plan	
Managing Office:	
Office of Resource Conservation and Recovery	
1a. Performance Measure Term Definitions:	

Tribe: Federally recognized tribes as defined by the Bureau of Indian Affairs. The most recent list can be found here: http://bia.gov/cs/groups/public/documents/text/idc1-023762.pdf (accessed February 11, 2014)

Integrated Waste Management Plan: An integrated solid waste management plan is... []. This is also referred to as an "Integrated Waste Management Plan (IWMP)." The following five elements represent the basic requirements that must be included in a tribe's IWMP for that plan to be considered adequate for GPRA purposes: description of the community service area; description of the tribe's waste management program structure and administration; description of the tribe's current and proposed waste management practices; description of the funding and sustainability and the long-term goals of the tribe's waste management program; and demonstration of approval of the IWMP by an appropriate governing body. Plans that do not meet all five elements may be adequate if a region determines that one or more elements are not applicable to a tribe's waste management program.

For more information on the expectations associated with each element, see the following document: U.S. Environmental Protection Agency. "Five Elements of a Tribal Integrated Waste Management Plan". Memorandum from Matt Hale, former Director, Office of Resource Conservation and Recovery. http://yosemite.epa.gov/osw/rcra.nsf/0c994248c239947e85256d090071175f/E7661F353791AD71852573780050876E/$file/14776.pdf (accessed December 17, 2013).

For more information on EPA's Tribal Solid Waste Management Program, visit: http://www.epa.gov/wastes/wyl/tribal/index.htm

2a. Original Data Source:

EPA Regions evaluate tribal integrated waste management plans, and record the data for this measure in internal Regional data systems and EPA's Budget Automation System (BAS). Regional data is collected by ORCR and housed in a spreadsheet.

2b. Source Data Collection:

The data set is very small and has only two elements (tribe and fiscal year). The data are reviewed by EPA regional offices for data quality and periodic adjustments are made during these reviews. The minimal data is then provided to ORCR for collection in a spreadsheet.

Because the data sets are small in size on a region-by-region basis, they can be managed efficiently by each regional office.

2c. Source Data Reporting:

EPA is the source of the data.
Regions enter the data in the BAS on at least a bi-annual basis (mid-year and end of year) or can report quarterly.

3a. Relevant Information Systems:

EPA Regional offices enter data into their internal data systems and the Annual Commitments System in BAS - see further description of EPA's Annual Commitments System. The internal EPA data set housing the specific integrated waste management plans for each tribe is managed by each regional office and is under the control of each region. DATA is also collected by ORCR in an Excel spreadhseet.

3b. Data Quality Procedures:

The regional data systems are considered to be appropriate for the minimal complexity and small size of the data set.

3c. Data Oversight:

Source Data Reporting Oversight Personnel: [[Staff in each regional office.]]

Source Data Reporting Oversight Responsibilities: [[N/A]]

Information Systems Oversight Personnel: [N/A]]

Information Systems Oversight Responsibilities: [[N/A.]]

3d. Calculation Methodology:

Units: Number of Tribes
The result is calculated by counting all tribes that EPA Regions have defined as being covered by an Integrated Waste Management Plan, as defined in field 1a, Performance Measure Term Definitions.

EPA has compiled the regional data into a spreadsheet for national tracking purposes. This spreadsheet is updated quarterly.

4a. Oversight and Timing of Final Results Reporting:

Final Reporting Oversight Personnel: Wayne Roepe/ Environmental Protection Specialist.

Final Reporting Oversight Responsibilities: [[Data is collected from BAS, and confirmed through the informal reporting of data requested from the regions that is placed into a spreadsheet. Results are reported through budget documents, and the APR.]]

Final Reporting Timing: [Bi-annually according to EPA's BAS reporting deadlines.]]
4b. Data Limitations/Qualifications:
The data are considered to be accurate on a regional and national scale.
4c. Third-Party Audits:
N/A

Measure Code: MW8 - Number of tribes covered by an integrated solid waste management plan.

Office of Solid Waste and Emergency Response (OSWER)

Goal Number and Title:
3 - Cleaning Up Communities and Advancing Sustainable Development
Objective Number and Title:
2 - Preserve Land
Sub-Objective Number and Title:
1 - Waste Generation and Recycling
Strategic Target Code and Title:
3 - By 2015, increase the number of tribes covered by an integrated waste management plan
Managing Office:
Office of Resource Conservation and Recovery
1a. Performance Measure Term Definitions:
Tribe: Federally recognized tribes as defined by the Bureau of Indian Affairs. The most recent list can be found here: http://bia.gov/cs/groups/public/documents/text/idc1-023762.pdf (accessed February 11, 2014) Integrated Waste Management Plan: An integrated solid waste management plan is... []. This is also referred to as an "Integrated Waste Management Plan (IWMP)." The following five elements represent the basic requirements that must be included in a tribe's IWMP for that plan to be considered adequate for GPRA purposes: description of the community service area; description of the tribe's waste management program structure and administration; description of the tribe's current and proposed waste management practices; description of the funding and sustainability and the long-term goals of the tribe's waste management program; and demonstration of approval of the IWMP by an appropriate governing body. Plans that do not meet all five elements may be adequate if a region determines that one or more elements are not applicable to a tribe's waste management program. For more information on the expectations associated with each element, see the following document: U.S. Environmental Protection Agency. "Five Elements of a Tribal Integrated Waste Management Plan". Memorandum from Matt Hale, former Director, Office of Resource Conservation and Recovery. http://yosemite.epa.gov/osw/rcra.nsf/0c994248c239947e85256d090071175f/E7661F353791AD71852573780050876E/$file/14776.pdf (accessed December 17, 2013). For more information on EPA's Tribal Solid Waste Management Program, visit: http://www.epa.gov/wastes/wyl/tribal/index.htm
2a. Original Data Source:
EPA Regions evaluate tribal integrated waste management plans, and record the data for this measure in internal Regional data systems and EPA's Budget Automation System (BAS). Regional data is collected by ORCR and housed in a spreadsheet.
2b. Source Data Collection:

The data set is very small and has only two elements (tribe and fiscal year). The data are reviewed by EPA regional offices for data quality and periodic adjustments are made during these reviews. The minimal data is then provided to ORCR for collection in a spreadsheet.

Because the data sets are small in size on a region-by-region basis, they can be managed efficiently by each regional office.

2c. Source Data Reporting:
EPA is the source of the data. Regions enter the data in the BAS on at least a bi-annual basis (mid-year and end of year) or can report quarterly.

3a. Relevant Information Systems:
EPA Regional offices enter data into their internal data systems and the Annual Commitments System in BAS - see further description of EPA's Annual Commitments System. The internal EPA data set housing the specific integrated waste management plans for each tribe is managed by each regional office and is under the control of each region. DATA is also collected by ORCR in an Excel spreadhseet.

3b. Data Quality Procedures:
The regional data systems are considered to be appropriate for the minimal complexity and small size of the data set.

3c. Data Oversight:
Source Data Reporting Oversight Personnel: [[Staff in each regional office.]] Source Data Reporting Oversight Responsibilities: [[N/A]] Information Systems Oversight Personnel: [N/A]] Information Systems Oversight Responsibilities: [[N/A.]]

3d. Calculation Methodology:
Units: Number of Tribes The result is calculated by counting all tribes that EPA Regions have defined as being covered by an Integrated Waste Management Plan, as defined in field 1a, Performance Measure Term Definitions. EPA has compiled the regional data into a spreadsheet for national tracking purposes. This spreadsheet is updated quarterly.

4a. Oversight and Timing of Final Results Reporting:
Final Reporting Oversight Personnel: Wayne Roepe/ Environmental Protection Specialist. Final Reporting Oversight Responsibilities: [[Data is collected from BAS, and confirmed through the informal reporting of data requested from the regions that is placed into a spreadsheet. Results are reported through budget documents, and the APR.]]

Final Reporting Timing: [Bi-annually according to EPA's BAS reporting deadlines.]]

4b. Data Limitations/Qualifications:
The data are considered to be accurate on a regional and national scale.

4c. Third-Party Audits:
N/A

Measure Code: S10 - Number of Superfund sites ready for anticipated use site-wide.

Office of Solid Waste and Emergency Response (OSWER)

Goal Number and Title:
3 - Cleaning Up Communities and Advancing Sustainable Development

Objective Number and Title:
3 - Restore Land

Sub-Objective Number and Title:
3 - Cleanup Contaminated Land

Strategic Target Code and Title:
7 - By 2015, ensure that 799 Superfund NPL sites are "sitewide ready for anticipated use"

Managing Office:
OSRTI

1a. Performance Measure Term Definitions:
Definition of Site: "Sites" refers only to National Priorities List (NPL) sites. (See below for definition of NPL.) The term "site" itself is not explicitly defined underComprehensive Environmental Response, Compensation, and Liability Act (CERCLA) or by the Superfund program; instead "site" is defined indirectly in CERCLA's definition of "facility," as follows: "The term 'facility' means (A) any building, structure, installation, equipment, pipe or pipeline (including any pipe into a sewer or publicly owned treatment works), well, pit, pond, lagoon, impoundment, ditch, landfill, storage container, motor vehicle, rolling stock, or aircraft, or (B) any site or area where a hazardous substance has been deposited, stored, disposed of, or placed, or otherwise come to be located; but does not include any consumer product in consumer use or any vessel." (CERCLA, Title I, Section 101, (9)). Definition of Sitewide Ready for Anticipated Use (SWRAU): Where for the entire construction complete NPL site: All cleanup goals in the Record(s) of Decision or other remedy decision document(s) have been achieved for media that may affect current and reasonably anticipated future land uses of the site, so that there are no unacceptable risks; and all institutional or other controls required in the Record(s) of Decision or other remedy decision document(s) have been put in place. The Human Exposure determination for sites that qualify for the Sitewide Ready-for-use measure should either be: • "Current Human Exposure Controlled and Protective Remedy in Place"; or • "Long-Term Human Health Protection Achieved In addition, All acreage at the site must be Ready for Anticipated Use (RAU) (i.e., the Superfund Remedial/Federal Facilities Response Universe Acres minus the total RAU Acres must be zero). For more information about the SWRAU performance measure, visit: http://www.epa.gov/superfund/programs/recycle/pdf/sitewide_a.pdf

Also, see Appendix B of the most recent Superfund Program Implementation Manual(SPIM), which is updated each fiscal year and contains definitions and documentation/coding guidance for Superfund measures. The most current SPIM can be found here: http://epa.gov/superfund/policy/guidance.htm

Superfund Alternative Approach (SAA) sites: The program tracks SAA sites that meet Sitewide ready for anticipated use criteria, but does target or report official results at this time.

Definition of National Priorities List (NPL): Sites are listed on the National Priorities List (NPL) upon completion of Hazard Ranking System (HRS) screening, public solicitation of comments about the proposed site, and final placement of the site on the NPL after all comments have been addressed. The NPL primarily serves as an information and management tool. It is a part of the Superfund cleanup process and is updated periodically. Section 105(a)(8)(B) of CERCLA as amended, requires that the statutory criteria provided by the HRS be used to prepare a list of national priorities among the known releases or threatened releases of hazardous substances, pollutants, or contaminants throughout the United States. This list, which is Appendix B of the National Contingency Plan, is the NPL. Visit the HRS Toolbox (http://www.epa.gov/superfund/sites/npl/hrsres/index.htm page for guidance documents that are used to determine if a site is a candidate for inclusion on the NPL. [Source: Superfund website, http://www.epa.gov/superfund/sites/npl/npl_hrs.htm

(Also see Appendix B of the most recent SPIM, which is updated each fiscal year and contains definitions and documentation/coding guidance for Superfund measures. The most current SPIM can be found here: http://epa.gov/superfund/policy/guidance.htm.

The Superfund Program's performance measures are used to demonstrate the agency's progress of site cleanup and reuse. Each measure marks a significant step in ensuring human health and environmental protection at Superfund sites.

References:

U.S. Environmental Protection Agency, EPA Performance and Accountability Reports, http://www.epa.gov/ocfo/par/index.htm

U.S. Environmental Protection Agency, Superfund Accomplishment and Performance Measures, http://www.epa.gov/superfund/accomplishments.htm

U.S. Environmental Protection Agency, Federal Facilities Restoration and Reuse Office – Performance measures, http://www.epa.gov/fedfac/documents/measures.htm

U.S Environmental Protection Agency, Office of Inspector General, Information Technology - Comprehensive Environmental Response, Compensation, and Liability Information System (CERCLIS) Data Quality, No. 2002-P-00016, http://www.epa.gov/oigearth/eroom.htm

U.S. Government Accountability Office, "Superfund Information on the Status of Sites, GAO/RCED-98-241", http://www.gao.gov/archive/1998/rc98241.pdf

U.S. Environmental Protection Agency, Office of Superfund Remediation and Technology Innovation, Superfund Program Implementation Manuals (SPIM), http://www.epa.gov/superfund/policy/guidance.htm (accessed July 30, 2009).

U.S. Environmental Protection Agency, Office of Solid Waste and Emergency Response, "OSWER Quality Management Plan", http://www.epa.gov/swerffrr/pdf/oswer_qmp.pdf

U.S. Environmental Protection Agency, Office of Environmental Information, EPA System Life Cycle Management Policy Agency Directive 2100.5, http://www.epa.gov/oamhpod1/adm_placement/ITS_BISS/slcmmgmt.pdf

U.S. Environmental Protection Agency, Office of Environmental Information, EPA's Information Quality Guidelines, http://www.epa.gov/quality/informationguidelines

2a. Original Data Source:

Original data sources vary, and multiple data sources can be used for each site. Typical data sources are EPA personnel, contractors (directly to EPA or indirectly, through the interagency agreement recipient or cooperative agreement recipient), U.S. Army Corps of Engineers (interagency agreement recipient), and states/tribes/other political subdivisions (cooperative agreement recipients). EPA also collects data via pre-final inspections at sites.

(See item Performance Measure Term Definitions in Tab 1, for more information. Also, detailed information on requirements for source data and completion procedures can be found on the following Superfund website: http://www.epa.gov/superfund/programs/npl_hrs/closeout/index.htm

2b. Source Data Collection:

Collection mode varies, typically with multiple collection modes at each site. Collection typically involves some combination of environmental data collection, estimation and/or tabulation of records/activities. Documents such as risk assessments, Record of Decisions (RODs), Action Memoranda, Pollution Reports (POLREPS), Remedial Action (RA) Reports, Close-out Reports, Five-year Reviews, NPL Deletion/Partial Deletion Notices are known reliable sources of data and often provide the information necessary for making a SWRAU evaluation with reasonable certainty. Regions should also ensure consistency between the SWRAU determination, the All ICs Implemented indicator, and the HEUC environmental indicator.

The SWRAU baseline was established in 2006. Data were from FY 2007 and continues through FY 2012.

The Guidance and Checklist can be found at: http://www.epa.gov/superfund/programs/recycle/pdf/sitewide_a.pdf

(See item Performance Measure Term Definitions, for more information and references.)

Source data collection frequency: On a regular basis with no set schedule, as the data are entered real-time. Varies by site.

Spatial Extent: National

Spatial detail: Site, defined in database by latitude/longitude pair.

2c. Source Data Reporting:

Collection mode varies, typically with multiple collection modes at each site. Collection typically involves some combination of environmental data collection, estimation and/or tabulation of records/activities. Documents such as risk assessments, RODs, Action Memoranda, POLREPS, RA Reports, Close-out Reports, Five-year Reviews, NPL Deletion/Partial Deletion Notices are known reliable sources of data and often provide the information necessary for making a SWRAU evaluation with reasonable certainty. Regions should also ensure consistency between the SWRAU determination, the All ICs Implemented indicator, and the HEUC environmental indicator.

SWRAU source data are entered in CERCLIS on a regular basis with no set schedule, as the data are entered real-time. However, status at a site is reviewed annually by the 10th working day in October, or at any time site conditions change. CERCLIS is to be updated within 10 days of any change in status.

The information is entered in CERCLIS in the Land Reuse Module. This module contains screens for entering and defining acreage data and the checklist, as well as setting the SWRAU status and applicable dates.

In addition to submitting information through CERCLIS, Regions are required to submit a Checklist to Headquarters documenting that the site has met the measure.

3a. Relevant Information Systems:

The SWRAU determination is made directly in Comprehensive Environmental Response, Compensation, and Liability Information System (CERCLIS) once it is determined that the site meets all required criteria and has been approved as such by appropriate regional personnel. The CERCLIS data system meets all relevant EPA QA standards.

CERCLIS database – CERCLIS is EPA's primary database to store and report data for NPL and non-NPL Superfund sites. The Superfund Comprehensive Accomplishment Plan (SCAP) reports in CERCLIS are used to report progress on measures, including SWRAU.

(For more information about CERCLIS, see Appendix E of the most recent SPIM, which is updated each fiscal year and contains definitions and documentation/coding guidance for Superfund measures. The most current SPIM can be found here: http://epa.gov/superfund/policy/guidance.htm.

CERCLIS operation and further development is taking place under the following administrative control quality assurance procedures: 1) Office of Environmental Information Interim Agency Life Cycle Management Policy Agency Directive; 2) the Office of Solid Waste and Emergency Response (OSWER) Quality Management Plan (QMP); 3) EPA IT standards; 4) Quality Assurance Requirements in all contract vehicles under which CERCLIS is being developed and maintained; and 5) EPA IT security policies. In addition, specific controls are in place for system design, data conversion and data capture, as well as CERCLIS outputs.

CERCLIS adherence to the security policy has been audited. Audit findings are attached to this record.

OSWER Performance Assessment Tool (PAT). This tool serves as the primary external servicing resource for organizing and reporting OSWER's performance data, which collects information from OSWER program systems, and conforms it for uniform reporting and data provisioning. PAT captures data from CERCLIS; replicates business logic used by CERCLIS for calculating measures; delivers that data to EPA staff and managers via a business intelligence dashboard interface for analytic and reporting use; ; and transmits data to the Budget Automated System (BAS). No current system specifications document is currently available for PAT, but will be provided when available. For this measure, PAT transmits Regional-level data to BAS.

PAT operates under the OSWER QMP. PAT has a security certification confirming that a security policy is not necessary because no sensitive data are handled and PAT is built upon the Oracle-based business intelligence system. PAT's security certification indicates that it follows all security guidelines for EPA's Oracle Portal and that PAT is (1) not defined as a "Major Application" according to NIST Special Publication 800-18, Guide for Developing Security Plans for Information Technology Systems, section 2.3.1; (2) does not store, process, or transmit information that the degree of sensitivity is assessed as high by considering the requirements for availability, integrity, and confidentiality according to NIST Special Publication 800-18, Guide for Developing Security Plans for Information Technology Systems, section 3.7.2. (3) is not covered by EPA Order 2100.2A1 Information Technology Capital Planning and Investment Control (CPIC).

EPA Headquarters is now scoping the requirements for an integrated (Superfund Document Management System-) SDMS-CERCLIS system, called the Superfund Enterprise Management System (SEMS). Development work on SEMS began in FY 2007 and will continue through FY 2013.

SEMS represents further re-engineering of the national reporting systems to include additional elements of EPA's Enterprise Architecture. SEMS will provide a common platform for major Superfund systems and future IT development. It will be constructed in part using EPA IT enterprise architecture principles and components. SEMS will provide a Superfund Program user gateway to various IT systems and information collections.
Attached Documents:
19-0585 (CERCLIS QAPP) 2009-0410.doc
CERCLIS July 9 2009 scan High Medium Response.xls

3b. Data Quality Procedures:

CERCLIS: To ensure data accuracy and control, the following administrative controls are in place: 1) SPIM, the program management manual that details what data must be reported; 2) Report Specifications, which are published for each report detailing how reported data are calculated; 3) Coding Guide, which contains technical instructions to data users including Regional Information Management Coordinators (IMCs), program personnel, data owners, and data entry personnel; 4) Quick Reference Guides (QRG), which are available in the CERCLIS Documents Database and provide detailed instructions on data entry for nearly every module in CERCLIS; 5) SCAP Reports within CERCLIS, which serve as a means to track, budget, plan, and evaluate progress towards meeting Superfund targets and measures; 6) a historical lockout feature in CERCLIS so that changes in past fiscal year data can be changed only by approved and designated personnel and are logged to a Change Log report, 7) the OSWER QMP; and 8) Regional Data Entry Control Plans.

EPA Headquarters has developed data quality audit reports and Standard Operating Procedures, which address timeliness, completeness, and accuracy, and has provided these reports to the Regions. In addition, as required by the Office of Management and Budget (OMB), CERCLIS audit logs are reviewed monthly. The system was also re-engineered to bring CERCLIS into alignment with the Agency's mandated Enterprise Architecture. The first steps in this effort involved the migration of all 10 Regional and the Headquarters databases into one single national database at the National Computing Center in Research Triangle Park (RTP) and the migration of SDMS to RTP to improve efficiency and storage capacity. During this process SDMS was linked to CERCLIS which enabled users to easily transition between programmatic accomplishments as reported in CERCLIS and the actual document that defines and describes the accomplishments.

Regional Data Entry Control Plans. Regions have established and published Data Entry Control Plans, which are a key component of CERCLIS verification/validation procedures. The control plans include: (1) regional policies and procedures for entering data into CERCLIS, (2) a review process to ensure that all Superfund accomplishments are supported by source documentation, (3) delegation of authorities for approval of data input into CERCLIS, and (4) procedures to ensure that reported accomplishments meet accomplishment definitions. In addition, regions document in their control plans the roles and responsibilities of key regional employees responsible for CERCLIS data (e.g., regional project manager, information management coordinator, supervisor, etc.), and the processes to assure that CERCLIS data are current, complete, consistent, and accurate. Regions may undertake centralized or decentralized approaches to data management. These plans are collected annually for review by OSRTI/IMB. [Source: SPIM FY11, III.J and Appendix E. http://www.epa.gov/superfund/action/process/spim10/pdfs/appe.pdf Copies of the 2010 Regional Data Entry Control Plans are provided with this DQR. Current and past year plans are available by contacting the Chief, Information Management Branch, Office of Superfund Remediation and Technology Innovation.

Regions are expected to prepare Data Entry Control Plans consistent with the SPIM and the Headquarters guidance: "CERCLIS Data Entry Control Plan Guidance," June 2009.

In addition to entering information into CERCLIS, Regions must also submit a Checklist for Reporting the SWRAU Government Performance and Results Act (GPRA) Measure to the headquarters data sponsor. The information provided in the Checklist can be used by Headquarters to ensure sites nominated for the measure conform to the same national best practices.

SWRAU is unique in that sites that cease to meet the measure can be retracted from the national total in the event they no longer meet the criteria. Sites that are retracted must submit a retraction form to Headquarters, explaining the reason for the retractions. For more information, see Appendix A of the Guidance at: http://www.epa.gov/fedfac/sf_ff_final_cprm_guidance.pdf

Superfund Program Implementation Manual (SPIM). The SPIM should be the first source referred to for additional questions related to program data and reporting. The SPIM is a planning document that defines program management priorities, procedures, and practices for the Superfund program (including response, enforcement, and Federal facilities). The SPIM provides the link between the GPRA, EPA's Strategic Plan, and the Superfund program's internal processes for setting priorities, meeting program goals, and tracking performance. It establishes the process to track overall program progress through program targets and measures.

The SPIM provides standardized and common definitions for the Superfund program, and it is part of EPA's internal control structure. As required by the Comptroller General of the United States, through generally accepted accounting principles (GAAP) and auditing standards, this document defines program scope and schedule in relation to budget, and is used for audits and inspections by the Government Accountability Office (GAO) and the Office of the Inspector General (OIG). The SPIM is developed on an annual basis. Revisions to the SPIM are issued during the annual cycle as needed.

The SPIM contains three chapters and a number of appendices. Chapter 1 provides a brief summary of the Superfund program and summarizes key program priorities and initiatives. Chapter 2 describes the budget process and financial management requirements. Chapter 3 describes program planning and reporting requirements and processes. Appendices A through I highlight program priorities and initiatives and provide detailed programmatic information, including Annual Targets for GPRA performance measures, and targets for Programmatic Measures. [Source: SPIM 2011, Chapter I]

The most current version of the SPIM can be found at: http://epa.gov/superfund/policy/guidance.htm

Data Flow:

Step 1. Original data sources provide information.

Step 2. EPA Region reviews and determines SWRAU status at the site and adjusts CERCLIS records as needed.

Step 3. Headquarters' OSRTI data sponsor reviews and approves/disapproves Regional determinations of

SWRAU using data from CERCLIS. Data sponsor works with Regional staff to ensure that determinations follow Superfund Program guidance.

Step 4. The OSWER's PAT pulls data from CERCLIS. Headquarters staff compare PAT results to CERCLIS results. If PAT does not match CERCLIS then there was an error with the upload and data are reloaded. Headquarters staff enter into PAT the Annual Commitment System (ACS) status information for each measure and, if necessary, a status explanation.

Step 5. Headquarters approves PAT results, and PAT pushes results into BAS.

Step 6. BAS aggregates Regional data into a national total. OSRTI reporting lead reviews and certifies results. Attached Documents:
2009_draft_CERCLIS_DECP_guidance _6-5-09_.pdf

3c. Data Oversight:

The Superfund program has a "data sponsorship" approach to database oversight. Headquarters staff and managers take an active role in improving the quality of data stored in CERCLIS by acting as data sponsors. The data sponsor for SWRAU, or "Land Ready for Reuse," is available in Appendix E of SPIM .
http://epa.gov/superfund/action/process/spim11/pdfs/Appendix_E_SPIM_2011_FINAL.pdf

Specific roles and responsibilities of data sponsors:
· Identify data needs;
· Oversee the process of entering data into the system;
· Determine the adequacy of data for reporting purposes;
· Conduct focus studies of data entered (A focus study is where a data sponsor identifies a potential or existing data issue to a data owner (see below), IMC, or other responsible person to determine if a data quality problem exists, and to solve the problem, if applicable. (IMC responsibilities discussed under item 2d, Database Oversight, below.) Focus studies can be informal via electronic messages.);
· Provide definitions for data elements;
· Promote consistency across the Superfund program;
· Initiate changes in CERCLIS as the program changes;
· Provide guidance requiring submittal of these data;
· Support the development of requirements for electronic data submission; and
· Ensure there is "objective" evidence to support the accomplishment data entered in CERCLIS through identifying data requirements and check to assure compliance by performing periodic reviews of a random CERCLIS data sample. [Source: SPIM 2011, III.E and E.A.5]

Measure-specific data sponsor information:

The Headquarter CERCLIS users responsible for the QA/QC of SWRAU data have primary source knowledge of program and site data used in the Superfund Program Implementation Manual (SPIM) and in CERCLIS. The data sponsor is responsible for:

·   ensuring that the correct data enters the system on a real-time basis, as the program/site plans and accomplishments change.

·   assuring procedures for determining that a site's SWRAU eligibility has been accomplished.

·   flipping the special initiative flag in CERCLIS once a site is determined to be SWRAU, and running audit and confirmatory reports from CERCLIS to ensure the information is accurate and up to date.

The Project Manager for CERCLIS oversees and is the approving authority for quality-related CERCLIS processes, and is closely supported by a Contract Task Manager. (See the CERCLIS QAPP, attached, for more information.) The lead point of contact for information about the data from CERCLIS is the Director, Information Management and Data Quality Staff, Office of Solid Waste and Emergency Response.

Information Management Coordinators(IMCs). In each Region, the IMC is a senior position which serves as regional lead for all Superfund program and CERCLIS systems management activities. The following lead responsibilities for regional program planning and management rest with the IMC:

·   Coordinate program planning, budget development, and reporting activities;

·   Ensure regional planning and accomplishments are complete, current, and consistent, and accurately reflected in CERCLIS by working with data sponsors and data owners;

·   Provide liaison to HQ on SCAP process and program evaluation issues;

·   Coordinate regional evaluations by headquarters;

·   Ensure that the quality of CERCLIS data are such that accomplishments and planning data can be accurately retrieved from the system; and

·   Ensure there is "objective" evidence to support accomplishment data entered in CERCLIS. (Objective Evidence Rule: "All transactions must be supported by objective evidence, that is, documentation that a third party could examine and arrive at the same conclusion.") [Source: SPIM 2011, III.E]

The primary responsibilities of data owners are (1) to enter and maintain data in CERCLIS and (2) assume responsibility for complete, current, consistent, and accurate data. The data owners for specific data are clearly identified in the system audit tables. Regions annually update region-specific Data Entry Control Plans (DECP). Among other things, Regional data entry control plans identify which Data Sponsors/Data Owners are responsible for different aspects of data entry. (See item 2e, Regions Have Standard Operating Procedures, for more information on Data Entry Control Plans.)

There is a Project Manager for CERCLIS. S/He oversees and is the approving authority for quality-related CERCLIS processes, and is closely supported by a Contract Task Manager. (See the CERCLIS QAPP, attached, for more information.)

The Information Management Officer (IMO) & Director, Information Management and Data Quality Staff. Office of Solid Waste and Emergency Response is the lead point of contact for information about the data from CERCLIS .

PAT Data Entry

The Annual Commitment System (ACS) Coordinator in OSRTI ensures that CERCLIS data for this measure are correctly loaded into PAT. The ACS Coordinator then works with the data sponsor to review uploaded data, edit records as appropriate, and then push data to ACS--part of the Office of Chief Financial Officer's (OCFO) BAS. PAT is maintained by OSWER's System Manager who ensures that the PAT system operates correctly, based on business logic agreed to by OSRTI.

3d. Calculation Methodology:

The performance measure is a specific variable entered into CERCLIS following specific coding guidance and corresponding supporting site-specific documentation.

The unit of measure is number of sites. The calculation only includes NPL sites

References:

Superfund Data Element Dictionary. The Superfund Data Element Dictionary (DED) is available online at: http://www.epa.gov/superfund/sites/ded/index.htm The DED provides definitions and descriptions of elements, tables and codes from the Comprehensive Environmental Response, Compensation and Liability Information System (CERCLIS) database used by the Superfund program. It also provides additional technical information for each entry, such as data type, field length and primary table. Using the DED, you can look up terms by table name or element name, or search the entire dictionary by keyword.

Other additional references that may be useful:

Coding Guide. The Superfund Coding Guide contains technical instructions to data users including Regional Information Management Coordinators (IMCs), program personnel, data owners, and data entry personnel. The Remedial component of the Coding Guide is attached to this record.

Quick Reference Guides (QRG). Superfund Quick Reference Guides are available in the CERCLIS Documents Database and provide detailed instructions on data entry for nearly every module in CERCLIS. Sample QRGs are available for entering data related to Remedial Action Starts.

Site Status and Description document: this is a Quick Reference Guide
for CERCLIS users, for filling in information related to site status and
description.

Attached Documents: Coding Guide - 2009.pdf Example QRG RA Start.doc Site Status and Description.doc

4a. Oversight and Timing of Final Results Reporting:

Data Sponsor for SWRAU (Land Ready for Reuse), Annual Commitment System (ACS) coordinator, and National Program Office (NPO) management.

Progress reporting is done quarterly as checks, while official numbers are reported annually.

4b. Data Limitations/Qualifications:

Sites that meet the SWRAU performance measure must also meet one of two Human Exposure Under Control (HEUC) performance measures:

• "Current Human Exposure Controlled and Protective Remedy in Place"; or
• "Long-Term Human Health Protection Achieved"

Specific to S10, if the HEUC changes to something other than the two environmental indicators, and will not be restored by the end of the fiscal year, Regions must retract a site from the SWRAU national total, and that retraction counts against the national target. Retractions are made when Regions determine that an entire site no longer meets the SWRAU criteria. Because SWRAU is counted as a "net" number on a yearly basis, a retraction is subtracted from that year's total number of SWRAU sites. If a Region retracts a site, the Region must still achieve its net SWRAU goal. For example, if a Region has a goal of achieving a SWRAU designation for 10 sites, but retracts two that year, the Region must identify 12 sites to meet the SWRAU target that year (12 − 2 = 10).

Based on experience dealing with sites to date, there are three categories of sites that may not meet the SWRAU measure. Even sites that on the surface appear to not meet the measure may have circumstances that require additional consideration.

·   Ground water only sites
·   Sites with overly restrictive ICs
·   Sites that cannot support any use.

NPL sites that have been addressed by state programs for cleanup should NOT be counted as SWRAU if the actions taken are not documented in an EPA CERCLA decision document.

4c. Third-Party Audits:

For CERCLIS data: The GAO report, Superfund: Information on the Status of Sites (GAO/RCED-98-241), dated August 28, 1998, estimated that the cleanup status of National Priority List (NPL) sites reported by CERCLIS as of September 30, 1997, is accurate for 95 percent of the sites.

(www.gao.gov/archive/1998/rc98241.pdf Another OIG audit, Information Technology - Comprehensive Environmental Response, Compensation, and Liability Information System (CERCLIS) Data Quality (Report No.

2002-P-00016), dated September 30, 2002, evaluated the accuracy, completeness, timeliness, and consistency of the data entered into CERCLIS. (See http://www.epa.gov/oig/reports/2002/cerlcis.pdf. The report provided 11 recommendations to improve controls for CERCLIS data quality. EPA has either implemented or continues to implement these recommendations.

The IG also annually reviews the end-of-year CERCLIS data, in an informal process, to verify data that supports the performance measures. Typically, there are no published results.

Annual EPA Office of Inspector General Audit/Report. The EPA OIG provides an annual report to Congress of the results of its audits of the Superfund Program. Those reports are available at: http://www.epa.gov/oig/reports/annual.htm The most recently available report is the FY 2009 report. In that report, EPA received an unqualified audit opinion by the OIG for the annual financial statements, although three material weaknesses and eight significant deficiencies were also identified. The OIG recommended several corrective actions. The Office of the Chief Financial Officer indicated in November 2009 that corrective actions will be taken.

Measure Code: PCB - Number of approvals issued for polychlorinated biphenyl (PCB) cleanup, storage and disposal activities.

Office of Solid Waste and Emergency Response (OSWER)

Goal Number and Title:
3 - Cleaning Up Communities and Advancing Sustainable Development
Objective Number and Title:
2 - Preserve Land
Sub-Objective Number and Title:
2 - Minimize Releases of Hazardous Waste and Petroleum Products
Strategic Target Code and Title:
2 - Prevent exposures at PCB sites
Managing Office:
Office of Resource Conservation and Recovery
1a. Performance Measure Term Definitions:
Approval: The PCB regulations require that many cleanup and disposal activities receive EPA approval. These approvals cover many cleanup, storage, treatment, disposal, and decontamination activities outlined in 40 CFR Part 761. This performance measure captures approvals issued by EPA under the PCB cleanup and disposal program.
2a. Original Data Source:
EPA Regions and Headquarters issue approvals for PCB cleanup and disposal activities. The staff and management responsible for issuing these approvals will provide the data to EPA's Office of Resource Conservation and Recovery (ORCR). The program is not delegated to the States therefore all approvals will be issued from EPA. No external organizations will be responsible for transferring data to EPA.
2b. Source Data Collection:
EPA Regions and Headquarters issue approvals for PCB cleanup and disposal activities. The PCB program is not delegated to the states therefore all approvals will be issued from EPA Regions and HQ. No external organizations will be responsible for transferring data to EPA. Each Regional program and ORCR will compile data for each approval issued for PCB cleanup and disposal activity. The staff and management in the program will confirm the data is accurate and complete before submitting their approval data to ORCR. ORCR will compile all of the Regional and HQ approvals issued and report the total number of approvals issued. The geographical extent of this data could cover all 10 EPA regions, depending on where the approved PCB cleanup and disposal activity is located. There will not be any spatial detail of source data. The measure is only determining the number of approvals issued.
2c. Source Data Reporting:

The data is originated by EPA. The data from each EPA Region will be submitted to ORCR.

The instrument the data will be submitted will be in the form of a spreadsheet. There is no EPA information system for this data at this time.

EPA staff (or management) will manually enter information about each approval issued for PCB cleanup and disposal activity in a spreadsheet.

The frequency of data transmission to EPA is based on when the approvals are issued. ORCR will collect approval data from the Regional programs at the end of each Fiscal Year in order to compile and report the national total number of PCB approvals issued.

3a. Relevant Information Systems:

ORCR is in the process of developing a national database to contain all regional and HQ approval information for tracking and reporting purposes. Until this database is in full operation, the regions will submit spreadsheet databases for their regional accomplishments, to be compiled by ORCR.

Source data will be submitted by EPA Regions and ORCR in a spreadsheet until the national database is completed.

EPA Regions and ORCR will be responsible for submitting accurate and complete data.

3b. Data Quality Procedures:

The Regional data is created and reviewed by the Regional program and then reviewed by ORCR to ensure data quality.

3c. Data Oversight:

EPA Regional PCB program personnel will be responsible for submitting Regional approvals data to ORCR. ORCR personnel will be responsible for submitting HQ approvals data and compiling all national data.

EPA Regional PCB program personnel will review and submit data to ORCR, which will review and compile. When the national approval database is complete, this data will be reported in the national database and quality checked by the Regions and HQ.

ORCR personnel will review and compile data submitted.

ORCR personnel will compile all approvals data to calculate final national totals for reporting purposes.

3d. Calculation Methodology:

The unit of measure is approvals. The results will be calculated by adding the number of approvals issued by both EPA Regions and HQ for PCB cleanup and disposal activities.

4a. Oversight and Timing of Final Results Reporting:

Project Manager/ ORCR/ Cleanup Programs Branch.

Results data are reviewed for accuracy and entered into the EPA's BAS. Official results are reported annually.
4b. Data Limitations/Qualifications:
No limitations noted. The data are considered to be accurate on a regional and national scale.
4c. Third-Party Audits:
None.

Office of Water (OW) Record(s)

Measure Code: dw2 - Percent of person months during which community water systems provide drinking water that meets all applicable health-based standards.

Office of Water (OW)

Goal Number and Title:
2 - Protecting America's Waters
Objective Number and Title:
1 - Protect Human Health
Sub-Objective Number and Title:
1 - Water Safe to Drink
Strategic Target Code and Title:
1 - By 2015,provide drinking water that meets applicable health-based drinking standards for communities
Managing Office:
Office of Ground Water and Drinking Water
1a. Performance Measure Term Definitions:
Community water systems --The U.S. Environmental Protection Agency (EPA) defines a community water system (CWS) as a public water system that serves at least 15 service connections used by year-round residents or regularly serves at least 25 year-round residents. CWSs provide water to more than 280 million persons in the United States. They are a tremendously diverse group. CWSs range from very small, privately owned systems whose primary business is not supplying drinking water (e.g., mobile home parks) to very large publicly owned systems that serve millions of customers. 2006 Community Water System Survey Volume I: Overview http://water.epa.gov/aboutow/ogwdw/upload/cwssreportvolumeI2006.pdf Person months – All persons served by CWSs times 12 months (3,525.1 million for FY2011). This measure is calculated by multiplying the number of months in the most recent four quarter period in which health-based violations overlap by the retail population served. Health-based standards -- exceedances of a maximum contaminant level (MCL) and violations of a treatment technique Effective treatment
2a. Original Data Source:
Data are provided by agencies with primacy (primary enforcement authority) for the Public Water System Supervision (PWSS) program. These agencies are either: States, EPA for non-delegated states or territories, and the Navajo Nation Indian tribe, the only tribe with primacy. Primacy agencies collect the data from the regulated water systems, determine compliance, and report a subset of the data to EPA (a subset of the inventory data and summary violations).
2b. Source Data Collection:

State certified laboratories report contaminant occurrence to states that, in turn, determine exceedances of maximum contaminant levels or non-compliance with treatment techniques and report these violations to EPA. Under the drinking water regulations, water systems must use approved analytical methods for testing for contaminants.

2c. Source Data Reporting:

Public Water Sanitary System (PWSS) Regulation-Specific Reporting Requirements Guidance. Available on the Internet at http://www.epa.gov/safewater/regs.html System, user, and reporting requirements documents can be found on the EPA web site, http://www.epa.gov/safewater/ States may choose to use electronic Data Verification (eDV) tool to help improve data quality.

3a. Relevant Information Systems:

SDWIS/STATE, a software information system jointly designed by states and EPA, to support states as they implement the drinking water program. SDWIS/STATE is an optional data base application available for use by states and EPA regions to support implementation of their drinking water programs. U.S. EPA, Office of Ground Water and Drinking Water. Data and Databases. Drinking Water Data & Databases – SDWIS/STATE, July 2002. Information available on the Internet: http://www.epa.gov/safewater/sdwis_st/current.html SDWIS/FED User and System Guidance Manuals (includes data entry instructions, data On-line Data Element Dictionary-a database application, Error Code Data Base (ECDB) - a database application, users guide, release notes, etc.) Available on the Internet at http://www.epa.gov/safewater/sdwisfed/sdwis.htm System and user documents are accessed via the database link http://www.epa.gov/safewater/databases.html and specific rule reporting requirements documents are accessed via the regulations, guidance, and policy documents link http://www.epa.gov/safewater/regs.html Documentation is also available at the Association of State Drinking Water Administrators web site at www.ASDWA.org SDWIS/Fed does not have a Quality Assurance Project Plan. The SDWIS/FED equivalent is the Data Reliability Action Plan [2006 Drinking Water Data Reliability Analysis and Action Plan, EPA-816-R-07-010 March 2008] The DRAP contains the processes and procedures and major activities to be employed and undertaken for assuring the data in SDWIS meet required data quality standards. This plan has three major components: assurance, assessment, and control. Office of Water Quality Management Plan, available at http://www.epa.gov/water/info.html

3b. Data Quality Procedures:

The Office of Ground Water and Drinking Water is modifying its approach to data quality review based on the

recommendations of the Data Quality Workgroup and on the Drinking Water Strategy for monitoring data.

There are quality assurance manuals for states and Regions, which provide standard operating procedures for conducting routine assessments of the quality of the data, including timely corrective action(s).

Reporting requirements can be found on the EPA web site, http://www.epa.gov/safewater/
SDWIS/FED edit checks built into the software to reject erroneous data.
EPA offers the following to reduce reporting and database errors:
1) training to states on data entry, data retrieval, compliance determination, reporting requirements and error correction, 2) user and system documentation produced with each software release and maintained on EPA's web site, 3) Specific error correction and reconciliation support through a troubleshooter's guide, 4) a system-generated summary with detailed reports documenting the results of each data submission, 5) an error code database for states to use when they have questions on how to enter or correct data, and 6) User support hotline available 5 days a week.

3c. Data Oversight:

The Infrastructure Branch Chief is responsible for overseeing source data reporting.
The Associate Director of Drinking Water Protection is responsible for overseeing information systems utilized in producing performance results.

3d. Calculation Methodology:

Person months – All persons served by CWSs times 12 months (3,525.1 million for FY2011). This measure is calculated by multiplying the number of months in the most recent four quarter period in which health-based violations overlap by the retail population served.

SDWIS contains basic water system information, population served, and detailed records of violations of the Safe Drinking Water Act and the statute's implementing health-based drinking water regulations.

SDWIS/FED data On-line Data Element Dictionary-a database application Available on the Internet at http://www.epa.gov/safewater/sdwisfed/sdwis.htm

Additional information: Several improvements are underway.

First, EPA will continue to work with states to implement the DRAP, which has already improved the completeness, accuracy, timeliness, and consistency of the data in SDWIS/FED through: 1) training courses for specific compliance determination and reporting requirements, 2) state-specific technical assistance, 3) targeted data audits conducted each year to better understand challenges with specific rules and 4) assistance to regions and states in the identification and reconciliation of missing, incomplete, or conflicting data.

Second, more states (as of January 2011, 55 States, Tribes, and territories are using SDWIS/STATE) will use SDWIS/STATE, SDWIS/STATE is an optional data base application available for use by states and EPA regions to support implementation of their drinking water programs.
U.S. EPA, Office of Ground Water and Drinking Water. Data and Databases. Drinking Water Data & Databases

– SDWIS/STATE, July 2002. Information available on the Internet:
http://www.epa.gov/safewater/sdwis_st/current.html a software information system jointly designed by states and EPA, to support states as they implement the drinking water program.

Third, in 2006 EPA modified SDWIS/FED to (1) simplify the database, (2) minimize data entry options resulting in complex software, (3) enforce Agency data standards, and (4) ease the flow of data to EPA through a secure data exchange environment incorporating modern technologies, all of which will improve the accuracy of the data. Data are stored in a data warehouse system that is optimized for analysis, data retrieval, and data integration from other data sources. It has improved the program's ability to more efficiently use information to support decision-making and effectively manage the program.

EPA has also begun a multi-year effort to develop the next generation information system to replace SDWIS/State. In addition to reducing the total cost of ownership to EPA, a high priority goal of this effort is to support improved data quality through the evaluation of all public water system monitoring data.

4a. Oversight and Timing of Final Results Reporting:

The Deputy Director for the Office of Groundwater and Drinking Water and the Evaluation and Accountability Team Leader for the Office of Water are responsible for coordinating the reporting of all measures for the Office of Water.

4b. Data Limitations/Qualifications:

Recent state data verification and other quality assurance analyses indicate that the most significant data quality problem is under-reporting by the states of monitoring and health-based standards violations and inventory characteristics. The most significant under-reporting occurs in monitoring violations. Even though those are not covered in the health based violation category, which is covered by the performance measure, failures to monitor could mask treatment technique and MCL violations. Such under-reporting of violations limits EPA's ability to: 1) accurately portray the percent of people affected by health-based violations, 2) target enforcement oversight, 3) target program assistance to primacy agencies, and 4) provide information to the public on the safety of their drinking water facilities

4c. Third-Party Audits:

N/A

Office of Water (OW)

Goal Number and Title:
2 - Protecting America's Waters
Objective Number and Title:
2 - Protect and Restore Watersheds and Aquatic Ecosystems
Sub-Objective Number and Title:
4 - Improve the Health of the Great Lakes
Strategic Target Code and Title:
1 - Prevent water pollution and protect aquatic systems
Managing Office:
Great Lakes National Program Office
1a. Performance Measure Term Definitions:

Areas of Concern: Great Lakes Areas of Concern (AOCs) are severely degraded geographic areas within the Basin. An AOC is described in the U.S.-Canada Great Lakes Water Quality Agreement (Annex 1 of the 2012 Protocol) as "a geographic area designated by the Parties where significant impairment of beneficial uses has occurred as a result of human activities at the local level. " There were once a total of 43 AOCs: 26 located entirely within the United States; 12 located wholly within Canada; and 5 shared by both countries. There were thus 31 United States or Binational AOCs until the delisting of the Oswego River (NY) AOC in July of 2006 and the Presque Isle Bay (PA) AOC in February of 2013.

Beneficial Use Impairments: This measure tracks the cumulative total number of beneficial use impairments (BUIs) removed within the 26 AOCs located entirely within the United States and the 5 AOCs that are shared by both the United States and Canada. Restoration of U.S. or Binational AOCs will ultimately be measured by the removal of all BUIs. Additional information is available at: http://www.epa.gov/grtlakes/aoc/ An impaired beneficial use means a change in the chemical, physical or biological integrity of the Great Lakes system sufficient to cause any of the following:
- restrictions on fish and wildlife consumption
- tainting of fish and wildlife flavor
- degradation of fish wildlife populations
- fish tumors or other deformities
- bird or animal deformities or reproduction problems
- degradation of benthos
- restrictions on dredging activities
- eutrophication or undesirable algae
- restrictions on drinking water consumption, or taste and odor problems
- beach closings
- degradation of aesthetics
- added costs to agriculture or industry

- degradation of phytoplankton and zooplankton populations
- loss of fish and wildlife habitat

Remedial Action Plans for each of the AOCs address one or up to 14 BUIs associated with these areas.

Removed: A BUI is determined to be removed when:
- A state has established the delisting criteria.
- A state has developed a Stage 2 RAP.
- All management actions necessary for removal of the BUI (determined by the Stage 2 RAP) have been completed and the delisting targets have been met. Also, the state needs to provide documentation that monitoring data indicates that the delisting targets have been met and environmental conditions have improved such that the impairment no longer exists.

2a. Original Data Source:
Great Lakes States, the U.S. Department of State, and the International Joint Commission (IJC).

2b. Source Data Collection:
The designated state environmental office or Office of Great Lakes in the appropriate state e work with the local stakeholders in the AOCs to develop delisting criteria to remove the impaired BUIs. The state offices are the sources of the formal letters (data) for this measure. EPA's AOC program staff lead collects these letters from the state offices. Data is collected and is subject to Quality Assurance procedures established and approved by USEPA. State Quality Programs are reviewed and approved by EPA's Quality Assurance program. State requests to remove BUIs and/or to delist AOCs are submitted to EPA and are reviewed according to the 2001 US Policy Committee document, "Delisting Principles and Guidelines." See: http://www.epa.gov/glnpo/aoc/rapdelistingfinal02.PDF

2c. Source Data Reporting:
Data is being reported via internal tracking and communications with Great Lakes states, the US Department of State, and the International Joint Commission. GLNPO maintains tracking for the removal of U.S. or binational BUIs in office files. Data includes information (such as formal letters) supplied by EPA, the other federal agencies and the state and local agencies involved in AOC work. Data will be reported on a periodic basis or as needed, given changes at the AOC. Results of BUI removals through September 2014 can be reported in October, 2014.

3a. Relevant Information Systems:
System Description: EPA AOC program files. Source/Transformed Data: Official correspondence from applicable states and USEPA. Information System Integrity Standards: Not applicable.

3b. Data Quality Procedures:
GLNPO has an approved Quality Management System in place that conforms to the USEPA Quality Management Order and is audited at least every 3 years in accordance with Federal policy for Quality

Management. GLNPO anticipates submitting an updated version of the Quality Management Plan to OEI for review in December 2014.

3c. Data Oversight:

Source Data Reporting Oversight Personnel: AOC Program Coordinator, located in the GLNPO Technical Assistance and Analysis Branch.

Source Data Reporting Oversight Responsibilities: GLNPO Technical Assistance and Analysis Branch , through the AOC Program Coordinator who is responsible for coordinating amongst federal, state, and tribal agencies; tracking and reporting on progress; and ensuring supporting data and files are maintained.

Information Systems Oversight Personnel: Not applicable.

Information Systems Oversight Responsibilities: Not applicable.

3d. Calculation Methodology:

Decision Rules for Selecting Data: All EPA-approved removals are included. When reasonable and realistic management actions have been completed for a BUI, the appropriate Great Lakes state informs EPA that local environmental conditions are improving and they are on a path to removing a BUI. EPA, State staff and local entities coordinate the information and, once all comments and concerns and documentation that the BUI has met the delisting targets have been addressed, the BUI Removal package is submitted to EPA for approval by the state . When approved by EPA, the information becomes available for reporting.

Definitions of Variables: Not applicable.

Explanation of Calculations: The sum of all approved BUI removals for each fiscal year is added to the cumulative total of BUI removals through the previous fiscal year. Calculations begin with the baseline total of 11 BUIs that had been removed as of the end of FY 2006.

Explanation of Assumptions: See above.

Unit of Measure: Number of BUIs (cumulative) removed.

Timeframe of Result: Through the end of the most recent fiscal year.

Documentation of Methodological Changes: Not applicable.

4a. Oversight and Timing of Final Results Reporting:

Final Reporting Oversight Personnel: GLNPO Technical Assistance and Analysis Branch Chief.

Final Reporting Oversight Responsibilities: Review the reported results for accuracy.

Final Reporting Timing: Annual

4b. Data Limitations/Qualifications:

General Limitations/Qualifications: GLNPO relies on verification of BUI removal by the States to certify that a BUI has been removed. EPA technical staff review such requests, as input to management decisions. Known sources of error include the input of unacceptable data by a state or local partner, data that is incomplete regarding management actions and other data that may be applicable to actions in the AOC but are not relevant to actions that lead to BUI removal. When all BUIs have been removed the site is eligible for the state to formally request delisting as an AOC.

Data Lag Length and Explanation: No more than 3 months. For example, results through September 2014 can generally be reported in October2014; however additional the time could be needed for tabulation/reporting.

Methodological Changes: Not applicable.

4c. Third-Party Audits:

GLNPO's Quality Management System has been given "outstanding" evaluations in previous peer and management reviews. EPA Office of Environmental Information (OEI) conducted a Quality System Assessment (QSA) of GLNPO on June 18 to 20, 2013. There were no findings as a result of the QSA and the report stated "The results of the QSA reflect a quality system that is effective for the organizational structure and environmental data programs. There is senior management commitment to ensuring that quality requirements are met for data and information supporting GLNPO's decisions."

GAO evaluated the EPA Great Lakes program in 2004 and found deficiencies in organizational coordination and information collection. Please see http://www.gao.gov/new.items/d041024.pdf

OMB assessed the EPA Great Lakes program in 2007. Please see
http://www.whitehouse.gov/omb/expectmore/summary/10009010.2007.html

EPA OIG evaluated the Great Lakes' progress in cleaning up AOCs, including recommendations for the data management and reporting of clean-up volume totals and costs. Please see
http://www.epa.gov/oig/reports/2009/20090914-09-P-0231.pdf

Measure Code: 606 - Cubic yards of contaminated sediment remediated (cumulative from 1997) in the Great Lakes.

Office of Water (OW)

Goal Number and Title:
2 - Protecting America's Waters

Objective Number and Title:
2 - Protect and Restore Watersheds and Aquatic Ecosystems

Sub-Objective Number and Title:
4 - Improve the Health of the Great Lakes

Strategic Target Code and Title:
2 - Remediate a cumulative total of 10.2 million cubic yards of contaminated sediment in the Great Lakes

Managing Office:
Great Lakes National Program Office

1a. Performance Measure Term Definitions:
Contaminated sediment: Although many point sources of pollution – discharges from discernible, often end of-pipe conduits – have been reduced, legacy contamination remains. "Legacy contamination" is pollutants largely left over from past practices, but that continue to recirculate through the ecosystem. Such legacy pollutants, often persistent toxic substances (PTS), such as mercury and polychlorinated biphenyls (PCBs), continue to be present at levels above those considered safe for humans and wildlife, warranting fish consumption advisories in the Great Lakes, connecting channels, and Midwestern and New York interior lakes. These contaminated sediments have been created by decades of industrial and municipal discharges, combined sewer overflows, and urban and agricultural non-point source runoff. Buried contaminants posing serious human and ecological health concerns can be resuspended by storms, ship propellers, and bottom-dwelling organisms.

In addition to the well-known toxicants like mercury, PCBs, and banned pesticides, there are chemicals of emerging concern that have been detected in the Great Lakes over the past several years, which may pose threats to the ecosystem. Some such chemicals may include flame retardants, surfactants, pharmaceuticals and personal care product constituents.

Sediments are considered contaminated when they "contain chemical substances in excess of appropriate geochemical, toxicological, or sediment quality criteria or measures, or are otherwise considered to pose a threat to human health or the environment." Source: EPA's Contaminated Sediment Management Strategy, April 1998.

Remediated: An area is considered remediated when sediment is removed, contained, or treated via dredging, capping, in-situ treatment, or natural recovery.

Great Lakes: Sediment remediation information is tracked for harbors and tributaries in the entire Great Lakes basin (AOCs and non-AOCs); not the lakes themselves. |

Background:

Although significant progress over the past 30 years has substantially reduced the discharge of toxic and persistent chemicals to the Great Lakes, persistent high concentrations of contaminants in the bottom sediments of rivers and harbors have raised considerable concern about potential risks to aquatic organisms, wildlife, and humans. EPA's Great Lakes program identifies polluted sediments and air toxics deposition as the largest major sources of contaminants to the Great Lakes food chain. As a result, advisories against fish consumption are in place in most locations around the Great Lakes.

Problem harbor and tributary areas in the Great Lakes basin have been identified and labeled as "Areas of Concern" (AOCs). This measure supports the cleaning up toxics and AOCs, which is the first of five focus areas of the Great Lakes Restoration Initiative (GLRI) being jointly implemented by 11 federal agencies.

GLNPO began tracking sediment remediation actions in the Great Lakes Basin in 1997. At that time, GLNPO's "best guess" of the total number of cubic yards that required remediation in the Great Lakes AOCs was 40 million. In 2004, the U.S. Policy Committee tasked the Great Lakes States with establishing a more comprehensive list of sites requiring remediation in the entire Great Lakes Basin (AOCs and non-AOCs), using best professional judgment to estimate the sediment volumes to be remediated. Using this list of estimated sediment remediation needs created by Great Lakes States in 2004, and sediment remediation estimates reported by Project Managers for calendar years 1997 through 2004, GLNPO estimated the 1997 "universe," for contaminated sediments requiring remediation to be 46.5 million cubic yards. The 2005 "baseline" for this measure for EPA's 2011-2015 Strategic Plan was 3.7 million cubic yards. The 2007 "baseline" for this measure in the GLRI Action Plan was 5.5 million cubic yards.

Efforts to accelerate the rate of sediment remediation in the 29 U.S. Great Lakes AOCs are underway using a variety of funding sources including those under the Great Lakes Legacy Act, Superfund and other programs. For more information, see:

-Great Lakes Restoration Initiative website, http://www.epa.gov/glnpo/glri/

-GLNPO Contaminated Sediments Program website, http://www.epa.gov/glnpo/sediments.html

-Great Lakes Restoration Action Plan, http://greatlakesrestoration.us/action/wp-content/uploads/glri_actionplan.pdf 2/21/2010

-"Indicator 3: Sediment Contamination." Unpublished – in Great Lakes National Program Office files.

2a. Original Data Source:

GLNPO collects sediment remediation data from various State and Federal project managers across the Great Lakes region, who conduct and coordinate contaminated sediments work, including appropriately characterizing and managing navigational dredging of contaminated sediments.

2b. Source Data Collection:

Collection Methodology: The totals for sediment remediation are estimates provided by project managers. Methodologies vary by site. For example, the volume of sediment remediated may be based on either data from depth soundings taken before and after dredging or the weight of sediment (plus possible solidification agents) transported to a landfill or confined disposal facility.

Geographical Extent: U.S. Great Lakes basin

Spatial Detail: City and state

Time Interval Covered by Source Data: 1997-present

Quality Procedures: Quality procedures vary by site. The project manager indicates whether an approved Quality Assurance Project Plan (QAPP) or other quality documentation was in place during remediation of contaminated sediment.

2c. Source Data Reporting:

These data are obtained directly from the project manager via an information fact sheet the project manager completes for any site in the Great Lakes basin that has performed any remedial work on contaminated sediment. The data collected to track sediment remediation in the Great Lakes show the amount of sediment remediated (removed, capped, undergoing natural recovery, or other) for that year and the amount of sediment remediated in prior years for a particular site. This format is suitable for year-to-year comparisons for individual sites.

Project managers report annually and all information about the site must be received by September 30th of the reporting year. The GLNPO project manager is responsible for transferring information from the request forms to the matrix, and generates the associated bar graphs.

More information:
Giancarlo, M.B. "Compilation of Project Managers Informational Sheets". Unpublished - in Great Lakes National Program Office files.

3a. Relevant Information Systems:

System Description: The sediment tracking database houses information on the calculated amount of sediment remediated at individual sites as provided by the project managers. Data tracking sediment remediation are compiled in two different formats. The first is a matrix that shows the annual and cumulative totals of contaminated sediment that were remediated in the Great Lakes basin in the reporting year and from 1997 for each Area of Concern or other non-Areas of Concern with sediment remediation. The second format depicts the yearly and cumulative totals on a calendar year basis graphically.

Please see: Giancarlo, M.B. "Sediment Remediation Matrix". Unpublished - in Great Lakes National Program Office files.
·   Giancarlo, M.B. "Sediment Remediation Graphics." Unpublished - in Great Lakes National Program Office files.

Source/Transformed Data: Individual site data are entered into the matrix as reported by project managers. Occasionally the GLNPO project manager converts the reported estimates into cubic yards before entering the estimates into the matrix (when data is reported in tons, for example).

Information System Integrity Standards: Standards not applicable.

3b. Data Quality Procedures:

It is GLNPO's responsibility to determine if the data are usable based upon the information sheet provided by the project managers. GLNPO does not attempt to verify mass and volume estimates due to the variability in how to calculate them. GLNPO ensures that the estimates provided make sense for the site, and that all estimates are reported in the same units. GLNPO management and Sediment Team members review the data, in the graphic and matrix formats, prior to reporting. GLNPO's Sediment Team works closely with partners and has confidence in those who provide data for the summary statistics. This familiarity with partners and general knowledge of ongoing projects allows GLNPO management to detect mistakes or questionable data. Individual site project managers are also responsible for double checking to ensure that data have been entered properly.

GLNPO does not accept unsolicited data without adequate assurance that quality system documentation was in place and the reporters of the data are not likely to be biased. GLNPO relies on the individual government/agency project managers to provide information on whether an approved QAPP was in place during remediation of contaminated sediment. This information is used to decide if the data provided by the project manager are reliable for GLNPO reporting purposes. If an approved QAPP was not used, sediment data would not likely be reported by GLNPO, unless GLNPO finds that alternative information is available that provides sufficient quality documentation for the project and associated data. This approach allows GLNPO to use best professional judgment and flexibility in reporting data from any cases where there was not a QAPP, but (a) the remedial action is noteworthy and (b) the project was conducted by recognized entities using widely accepted best practices and operating procedures.

The data, in both the graphic and matrix formats, are reviewed by individual project managers, GLNPO's Sediment Team, and management prior to being released. Data quality review procedures are outlined in the QAPP referenced below. See:
Giancarlo Ross, M.B. Quality Assurance Project Plan for "Great Lakes Sediment Remediation Project Summary Support." Unpublished – in Great Lakes National Program Office files, June 2008.

GLNPO has an approved Quality Management System in place that conforms to the USEPA Quality Management Order and is audited at least every 5 years in accordance with Federal policy for Quality Management. See: "Quality Management Plan for the Great Lakes National Program Office." EPA905-R-02-009. Revised and approved May 2008. http://www.epa.gov/glnpo/qmp/

3c. Data Oversight:

Source Data Reporting Oversight Personnel: GLNPO Project Manager
Source Data Reporting Oversight Responsibilities: The GLNPO project manager is responsible for distributing the request form, determining if the data are usable, transferring information from the request forms to the matrix, generating the associated bar graphs, and obtaining final approval from management.

Information Systems Oversight Personnel: Not applicable.

Information Systems Oversight Responsibilities: Not applicable.

3d. Calculation Methodology:

Decision Rules for Selecting Data:All reported volume estimates are included in the sum.

Definitions of Variables: Not applicable.

Explanation of Calculations: GLNPO sums the volume estimates as provided by the individual project managers, but then rounds the totals. For reporting purposes, the yearly volume total is rounded to the nearest one thousand cubic yards and the cumulative volume total is rounded to the nearest one hundred thousand cubic yards. The cumulative total is based off of the previous year's unrounded total.

Explanation of Assumptions: Remedial actions occurred in the Great Lakes basin prior to 1997; however, the GLNPO didn't start tracking the volume of sediment remediated until 1997. GLNPO estimated that as of 1997, the volume, or "universe," of contaminated sediments requiring remediation, was 46.5 million cubic yards.

Unit of Measure: Millions of cubic yards

Timeframe of Result: 1997-present (cumulative)

Documentation of Methodological Changes: In 2008, the yearly and cumulative totals began including large-scale navigation dredging projects that removed a significant amount of contaminated sediment from the environment granted that the site was sufficiently characterized, managed using best practices, and was completed by a recognized entity. No effort has been made to include navigation projects between 1997 and 2008.

4a. Oversight and Timing of Final Results Reporting:

Final Reporting Oversight Personnel: Final Reporting Oversight Personnel: GLNPO Technical Assistance and Analysis Branch Chief

Final Reporting Oversight Responsibilities: Review the list of sites and volume estimates reported by individual project managers for potential mistakes or questionable data based on familiarity with partners and general knowledge of ongoing projects.

Final Reporting Timing: Annual

4b. Data Limitations/Qualifications:

General Limitations/Qualifications:
The data provided in the sediment tracking database should be used as a tool to track sediment remediation progress at sites across the Great Lakes Basin. Many of the totals for sediment remediation are estimates

provided by project managers. For specific data uses, individual project managers should be contacted to provide additional information. The amount of sediment remediated or yet to be addressed should be viewed as qualitative data since a specific error estimate is not able to be calculated.

Data Lag Length and Explanation: One year. For example, the results from calendar year 2011 remediation were reported in FY 2012.

Methodological Changes: See field 3d.

4c. Third-Party Audits:

GLNPO's Quality Management System has been given "outstanding" evaluations in previous peer and management reviews. EPA Office of Environmental Information (OEI) conducted a Quality System Assessment (QSA) of GLNPO on June 18 to 20, 2013. There were no findings as a result of the QSA and the report stated "The results of the QSA reflect a quality system that is effective for the organizational structure and environmental data programs. There is senior management commitment to ensuring that quality requirements are met for data and information supporting GLNPO's decisions."

GLNPO anticipates submitting an updated version of the Quality Management Plan to OEI for review in December 2014.

GAO evaluated the EPA Great Lakes program in 2004 and found deficiencies in organizational coordination and information collection. Please see http://www.gao.gov/new.items/d041024.pdf

OMB assessed the EPA Great Lakes program in 2007. Please see http://www.whitehouse.gov/omb/expectmore/summary/10009010.2007.html

EPA OIG evaluated the Great Lakes' progress in cleaning up AOCs, including recommendations for the data management and reporting of clean-up volume totals and costs. Please see http://www.epa.gov/oig/reports/2009/20090914-09-P-0231.pdf

Office of Water (OW)

Goal Number and Title:
2 - Protecting America's Waters

Objective Number and Title:
2 - Protect and Restore Watersheds and Aquatic Ecosystems

Sub-Objective Number and Title:
1 - Improve Water Quality on a Watershed Basis

Strategic Target Code and Title:
1 - Attain water quality standards for all pollutants and impairments in more than 3,360 water bodies id

Managing Office:
Office of Wetlands Oceans and Watersheds

1a. Performance Measure Term Definitions:

Waterbody segments: A geographically defined portion of navigable waters, waters of the contiguous zone, and ocean waters under the jurisdiction of the United States, including segments of rivers, streams, lakes, wetlands, coastal waters and ocean waters.

Identified by States in 2002 as not attaining standards: In 2002, an estimated 39,503 waterbodies were identified by states or EPA as not meeting water quality standards. These waterbodies and waterbody segments were identified in state-submitted Section 303(d) lists, Section 305(b) reports, and Integrated Reports for the 2002 reporting cycle. (See EPA's guidance for such reporting under "303(d) Listing of Impaired Waters Guidance" at http://www.epa.gov/owow/tmdl/guidance.html. Impairments identified after 2002 are not considered in counting waters under this measure; such impairments may be considered when revising this measure for future updates of the Strategic Plan.

The universe for this measure - the estimated 39,503 waterbodies identified by states or EPA as not meeting water quality standards in 2002 - is sometimes referred to as the "fixed base" or "SP-10 baseline." The universe includes all waters in categories 5, 4a, 4b, and 4c in 2002. Of these waters, 1,703 are impaired by multiple pollutants including mercury, and 6,501 are impaired by mercury alone.

States: All 50 states.

Water quality standards are now fully attained: Attaining water quality standards means the waterbody is no longer impaired for any of the causes identified in 2002, as reflected in subsequent state-submitted assessments and EPA-approved 303(d) lists. Impairment refers to an "impairment cause" in state- or EPA-reported data, stored in ATTAINS (Assessment Total Maximum Daily Load (TMDL) Tracking and Implementation System) or its predecessors NTTS (National TMDL Tracking System) or ADB (Assessment Database). (Any waterbody listed as impaired in these databases must have an impairment cause entered.) There are several reasons why EPA or states may determine that specific waterbodies listed as impaired in

2002 (the baseline year) are no longer impaired in the current reporting year. For example, water quality might improve due to state or EPA actions to reduce point and nonpoint source discharges of pollutants. In other cases, a state or EPA might conduct more robust monitoring studies and use these data to complete more accurate assessments of water quality conditions. In some cases, a state might modify its water quality standards in accordance with EPA's regulations to update scientific criteria or better reflect the highest attainable conditions for its waters. Each of these examples represents a case where an impaired water may no longer exceed water quality standards. Any such removals of waterbody impairments will be recorded based on reports from states scheduled every two years through 2014.

Background:

·   This is a cumulative measure, and it was first tracked in FY 2007. The FY 2007 target was 1,166; the actual result was 1,409. The FY 2008 target was 1,550; the actual result was 2,165. The FY 2009 target for this measure was 2,270; the actual result was 2,505. The FY 2010 target for this measure was 2,809; the actual result was 2,909. The FY 2011 target for this measure was 3,073; the actual result was 3,119.

2a. Original Data Source:

Regional EPA staff, who review and approve states' 303(d) lists.

2b. Source Data Collection:

Approval and Review of 303(d) lists by regional EPA staff: EPA reviews and approves state determinations that water by segments has fully attained standards. The primary data source is states' 303(d) lists of their impaired waterbodies needing development of TMDLs and required submittals of monitoring information pursuant to section 305(b) of the Clean Water Act. These lists/reports are submitted each biennial reporting cycle. EPA regional staffs interact with the states during the process of approval of the lists to ensure the integrity of the data consistent with the Office of Water Quality Management Plan (QMP). EPA review and approval is governed by the Office of Water Quality Management Plan (QMP).

State 303(d) submissions:

States submit 303(d) lists of impaired waterbodies needing development of TMDLs and monitoring information pursuant to section 305(b) of the Clean Water Act. States prepare lists/reports using actual water quality monitoring data, probability-based monitoring information, and other existing and readily available information and knowledge the state has in order to make comprehensive determinations addressing the total extent of the state's waterbody impairments. States exercise considerable discretion in using monitoring data and other available information to make decisions about which waters meet their designated uses in accordance with state water quality standards.

States employ various analytical methods of data collection, compilation, and reporting, including:
1) Direct water samples of chemical, physical, and biological parameters;
2) Predictive models of water quality standards attainment;
3) Probabilistic models of pollutant sources; and
4) Compilation of data from volunteer groups, academic interests and others. EPA-supported models include BASINS, QUAL2E, AQUATOX, and CORMIX. (Descriptions of these models and instructions for their use can be

found at http://www.epa.gov/waterscience/models/

Most states have provided this information in Integrated Reports, pursuant to EPA guidance. An Integrated Report is a biennial state submittal that includes the state's findings on the status of all its assessed waters (as required under Section 305(b) of the Clean Water Act), a listing of its impaired waters and the causes of impairment, and the status of actions being taken to restore impaired waters (as required under Section 303(d)).

QA/QC of data provided by states pursuant to individual state 303(d) lists (under CWA Section 303(d)) and/or Integrated 305(b)/303(d) Reports) is dependent on individual state procedures. EPA enhanced two existing data management tools (STORET and the National Assessment Database) so they include documentation of data quality information.

EPA released the Water Quality Exchange (WQX), which provides data exchange capability to any organization that generates data of documented quality and would like to contribute that data to the national timeframe data warehouse so their data may be used in combination with other sources of data to track improvements in individual watersheds. Currently, data providers must transmit data and required documentation through their own Exchange Network node. EPA has rolled out a web data entry tool called WQXweb for users who have not invested in the node technology.

2c. Source Data Reporting:

Once EPA approves a state's 303(d) list, the information is entered into EPA's Assessment, TMDL Tracking, and Implementation System ATTAINS. After approving a state's 303(d) list, EPA reviews waterbodies in the 2002 universe to determine progress in achieving annual commitments for this measure (coded as SP-10). A waterbody may be counted under this measure when it attains water quality standards for all impairments identified in the 2002 reporting cycle, as reflected in subsequent Integrated Reports. Impairments are identified in later Integrated Reports are not considered for this measure.

Waters delisted for the following reasons can be counted toward meeting this measure:

Results for this performance measure are then entered by the EPA Regional Offices into EPA's Annual Commitment System (ACS).

Guidance Documents:
·The Office of Water has been working with states to improve the guidance under which 303(d) lists are prepared. In 2005, EPA issued listing guidance entitled Guidance for 2006 Assessment, Listing, and Reporting Requirements Pursuant to Sections 303(d), 305(b), and 314 of the Clean Water Act. This document provided a comprehensive compilation of relevant guidance EPA had issued to date regarding the Integrated Report. It included some specific changes from the 2004 guidance. For example, the 2006 Integrated Report Guidance provided greater clarity on the content and format of those components of the Integrated Report that are recommended and required under Clean Water Act sections 303(d), 305(b), and 314. The guidance also gave additional clarity and flexibility on reporting alternatives to TMDLs for attaining water quality standards (e.g.,

utilization of reporting Category 4b). Available at http://www.epa.gov/owow/tmdl/2006IRG

·In 2008, USEPA's Office of Water published Information Concerning 2008 Clean Water Act Sections 303(d), 305(b), and 314 Integrated Reporting and Listing Decisions. Available at http://www.epa.gov/owow/tmdl/2008_ir_memorandum.html

· In May, 2009, EPA released Information Concerning 2010 Clean Water Act Sections 303(d), 305(b), and 314 Integrated Reporting and Listing Decisions. Available at www.epa.gov/owow/tmdl/guidance/final52009.pdf

· EPA issued a 2010 Integrated Report clarification memo (released May 5, 2009 at http://www.epa.gov/owow/tmdl/guidance/final52009.html) which includes suggestions for the use of the rotating basin approach and Category 3, circumstances and expectation for "partial approval/further review pending" determinations, and using and reporting on Statewide Statistical Survey Data in ATTAINS and the National Water Quality Inventory Report to Congress.

· The Consolidated Assessment and Listing Methodology – Toward a Compendium of Best Practices (released on the web July 31, 2002 at www.epa.gov/owow/monitoring/calm.html intended to facilitate increased consistency in monitoring program design and the data and decision criteria used to support water quality assessments.

· The Office of Water (OW) and EPA's Regional Offices have developed the Elements of a State Water Monitoring and Assessment Program (March 2008). This guidance describes ten elements each state water quality monitoring program should contain and directs states to develop monitoring strategies that propose time frames for implementing all ten elements.

· Reporting guidelines for this measure can be found under the water quality sub-objective (SP-10 code) at:http://water.epa.gov/resource_performance/planning/FY-2012-NWPG-Measure-Definitions-Water-Quality.cfm - Measure%20Code_%20WQ_SP10_N11

· Additional information on the use of ATTAINS was furnished in a March 2011 memorandum: "2012 Clean Water Act Sections 303(d), 305(b), and 314 Integrated Reporting and Listing Decisions" issued by the Director, OW/Office of Wetlands, Oceans and Watersheds. This memo is available at http://water.epa.gov/lawsregs/lawsguidance/cwa/tmdl/ir_memo_2012.cfm

3a. Relevant Information Systems:

The Assessment and Total Maximum Daily Load (TMDL) Tracking And ImplementatioN System (ATTAINS) is the database which captures water quality information related to this measure. ATTAINS is an integrated system that documents and manages the connections between state assessment and listing decisions reported under Sections 305(b) and 303(d) (i.e., integrated reporting) and completed TMDLs. This system holds information about assessment decisions and restoration actions across reporting cycles and over time until water quality standards are attained. Annual TMDL totals by state, fiscal year, and pollutant are available at http://iaspub.epa.gov/waters10/attains_nation_cy.control?p_report_type=T - APRTMDLS and TMDL document searches can be conducted at http://www.epa.gov/waters/tmdl/tmdl_document_search.html More information about ATTAINS can be found at http://www.epa.gov/waters/data/prog.html and http://www.epa.gov/waters/ir/about_integrated.html

The Watershed Assessment, Tracking, and Environmental Results System (WATERS) is used to provide water program information and display it spatially using a geographic information system (National Hydrography Dataset (NHD)) integrated with several of EPA's existing databases. These databases include: the STOrage and

RETrieval (STORET) database, the Assessment TMDL Tracking and ImplementatioN System (ATTAINS), the Water Quality Standards Database (WQSDB), and the Grants Tracking and Reporting System (GRTS). This water quality information was previously available only from several independent and unconnected databases. General information about WATERS is available at http://www.epa.gov/waters/ a system architecture diagram is available at http://www.epa.gov/waters/about/arch.html and information about WATERS geographic data is available at http://www.epa.gov/waters/about/geography.html

3b. Data Quality Procedures:

Water Management Divisions in EPA Regional Offices have responsibility for oversight, review and quality assurance of the performance data reported to EPA by the original data source which is the individual states.

3c. Data Oversight:

(1) Source Data Reporting: Water Management Divisions in the EPA Regional Offices. (2) Information Systems Oversight: System Manager for ATTAINS; System Manager for WATERS.

3d. Calculation Methodology:

The calculation methodology is described in Section 2c, "Source Data Reporting". While ATTAINS is the repository for 303(d) lists and 305(b) reports, it is not yet used for tracking performance and success for this measure. EPA is continuing to work to address discrepancies between Regional data records and ATTAINS.

4a. Oversight and Timing of Final Results Reporting:

The Associate Director of the Assessment and Watershed Protection Division is responsible for overseeing final reporting of measure.

4b. Data Limitations/Qualifications:

Delays are often encountered in state 303(d) lists and 305(b) submissions and in EPA's approval of the 303(d) portion of these biennial submissions. EPA encourages states to effectively assess their waters and make all necessary efforts to ensure the timely submittal of required Clean Water Act Section 303(d) impaired waters lists. EPA will continue to work with states to facilitate accurate, comprehensive, and geo-referenced data submissions. Also, EPA is heightening efforts to ensure expeditious review of the 303(d) list submissions with national consistency. Timely submittal and EPA review of integrated reports is important to demonstrate state and EPA success in accomplishing Strategic Plan goals for water quality.

Data may not precisely represent the extent of impaired waters because states do not employ a monitoring design that monitors all their waters. States, territories and tribes collect data and information on only a portion of their waterbodies. States do not use a consistent suite of water quality indicators to assess attainment of water quality standards. For example, indicators of aquatic life use support range from biological community assessments to levels of dissolved oxygen to concentrations of toxic pollutants. These variations in state practices limit how the CWA Sections 305(b) reports and the 303(d) lists provided by states can be used to describe water quality at the national level. There are also differences among sampling techniques and standards.

State assessments of water quality may include uncertainties associated with derived or modeled data. Differences in monitoring designs among and within states prevent the agency from aggregating water quality assessments at the national level with known statistical confidence. States, territories, and authorized tribes monitor to identify problems; typically, lag times between data collection and reporting can vary by state.

Additionally, states exercise considerable discretion in using monitoring data and other available information to make decisions about which waters meet their designated uses in accordance with state water quality standards. EPA then aggregates these various state decisions to generate national performance measures.

Impact of Supplemental Funding. In FY 2010 and CR 2011, the program which this measure supports receives funding from the American Recovery and Reinvestment Act (ARRA). Results from that funding will be reflected in this measure because they cannot easily be separated from results related to other EPA funding.

4c. Third-Party Audits:

Independent reports have cited the ways in which weaknesses in monitoring and reporting of monitoring data undermine EPA's ability to depict the condition of the Nation's waters and to support scientifically sound water program decisions. The most recent reports include the following:

· USEPA, Office of the Inspector General. 2009. EPA Needs to Accelerate Adoption of Numeric Nutrient Water Quality Standards. Available at www.epa.gov/oig/reports/2009/20090826-09-P-0223.pdf

· USEPA, Office of the Inspector General. 2007. Total Maximum Daily Load Program Needs Better Data and Measures to Demonstrate Environmental Results. Available at http://www.epa.gov/oig/reports/2007/20070919-2007-P-00036.pdf

· Government Accountability Office. 2003. Water Quality: Improved EPA Guidance and Support Can Help States Develop Standards That Better Target Cleanup Efforts. GAO-03-308. Washington, DC. Available at www.gao.gov/new.items/d03308.pdf

· Government Accountability Office. 2002. Water Quality: Inconsistent State Approaches Complicate Nation's Efforts to Identify its Most Polluted Waters. GAO-02-186. Washington, DC. Available at: www.epa.gov/waters/doc/gaofeb02.pdf

· Government Accountability Office. 2000. Water Quality: Key EPA and State Decisions Limited by Inconsistent and Incomplete Data. GAO-RCED-00-54. Washington, DC. Available at www.gao.gov/products/RCED-00-54

In response to these evaluations, EPA has been working with states and other stakeholders to improve 1) data coverage, so that state reports reflect the condition of all waters of the state; 2) data consistency to facilitate comparison and aggregation of state data to the national level; and 3) documentation so that data limitations and discrepancies are fully understood by data users. EPA has taken several steps in an effort to make the following improvements:

First, EPA enhanced two existing data management tools (STORET and the National Assessment Database) so that they include documentation of data quality information.

Second, EPA has developed a GIS tool called WATERS that integrates many databases including STORET, ATTAINS, and a water quality standards database. These integrated databases facilitate comparison and understanding of differences among state standards, monitoring activities, and assessment results.

Third, EPA and states have developed guidance. The 2006 Integrated Report Guidance (released August 3,

2005 at http://www.epa.gov/owow/tmdl/2006IRG provides comprehensive direction to states on fulfilling reporting requirements of Clean Water Act Sections 305(b) and 303(d). EPA also issued a 2010 Integrated Report clarification memo (released May 5, 2009 at http://www.epa.gov/owow/tmdl/guidance/final52009.html) which includes suggestions for the use of the rotating basin approach, appropriate use of Category 3, circumstances and expectation for "partial approval/further review pending" determinations, and using and reporting on Statewide Statistical Survey Data in ATTAINS and the National Water Quality Inventory Report to Congress.

Also, the Consolidated Assessment and Listing Methodology – Toward a Compendium of Best Practices (released July 31, 2002 at www.epa.gov/owow/monitoring/calm.html intended to facilitate increased consistency in monitoring program design and the data and decision criteria used to support water quality assessments.

Fourth, the Office of Water (OW) and EPA's Regional Offices have developed the Elements of a State Water Monitoring and Assessment Program(March 2008). This guidance describes ten elements each state water quality monitoring program should contain and directs states to develop monitoring strategies that propose timeframes for implementing all ten elements. (USEPA, Office of Water. 2003. Elements of a State Water Monitoring and Assessment Program. EPA 841-B-03-003. Washington, DC. Available at www.epa.gov/owow/monitoring/elements/.

Measure Code: bps - Number of TMDLs that are established or approved by EPA [Total TMDL] on a schedule consistent with national policy (cumulative). [A TMDL is a technical plan for reducing pollutants in order to attain water quality standards. The terms "approved" and "established" refer to the completion and approval of the TMDL itself.]

Office of Water (OW)

Goal Number and Title:
2 - Protecting America's Waters
Objective Number and Title:
2 - Protect and Restore Watersheds and Aquatic Ecosystems
Sub-Objective Number and Title:
1 - Improve Water Quality on a Watershed Basis
Strategic Target Code and Title:
1 - Attain water quality standards for all pollutants and impairments in more than 3,360 water bodies id
Managing Office:
Office of Wetlands Oceans and Watersheds
1a. Performance Measure Term Definitions:

TMDL: A Total Daily Maximum Load (TMDL) is a calculation of the maximum amount of a pollutant that a waterbody can receive and still safely meet water quality standards. A TMDL is a technical plan for reducing pollutants in order to attain water quality standards. For the purposes of this measure, each individual pollutant for which an allocation has been established/approved is counted as a TMDL. The development of TMDLs for an impaired waterbody is a critical step toward meeting water restoration goals. TMDLs focus on clearly defined environmental goals and establish a pollutant budget, which is then implemented via permit requirements or a wide variety of state, local, and federal programs (which may be regulatory, non-regulatory, or incentive-based, depending on the program), as well as voluntary action by citizens. TMDLs established/approved: The terms "approved" and "established" refer to the completion and approval of the TMDL itself. While the majority of TMDLs are developed by states, territories, or authorized tribes, EPA in some instances may establish a TMDL if:
     ·   EPA disapproves TMDLs submitted by states, territories, or authorized tribes,
     ·   States, territories, or authorized tribes do not submit TMDLs in a timely manner,
     ·   EPA is required to do so pursuant to litigation settlements or judicial orders, or
     ·   States ask EPA to establish TMDLs for particular water bodies.
 	
 Schedule consistent with national policy: National policy states that TMDLs are typically established and approved within 8 to 13 years of the water having been listed as impaired under Clean Water Act Section 303(d). The "state pace" is the number of TMDLs needing to be completed in a given state in a given fiscal year (these TMDLs may eventually be developed by either the state and approved by EPA or established by EPA). State pace is based on state litigation or other schedules or straight-line rates that ensure that national policy is met. Regions collaborate with States to set targets for the number of TMDLs projected to be completed in a given fiscal year. EPA policy has been that targets should be within 80 to 100% of the pace.

Cumulative trend information:

Background:

EPA and States have developed more than 50,000 TMDLs thru FY 2012.

Projecting state TMDL production numbers several months in advance continues to be a challenge as resource constraints and technical and legal challenges still exist. There has also been a notable shift toward the development of more difficult TMDLs that take more time and resources.

As TMDLs and other watershed-related activities are developed and implemented, waterbodies that were once impaired will meet water quality standards. Thus these TMDL measures are closely tied to the program assessment measures WQ-SP10.N11 and WQ-SP-11, "Number of waterbody segments identified by States in 2002 as not attaining standards, where water quality standards are now fully attained," and "remove the specific causes of waterbody impairment identified by states in 2002."

The number of TMDLs needed to address outstanding causes of impairment changes with each 303(d) list cycle; therefore, a baseline as such is not appropriate for these measures.

For more information, please visit http://www.epa.gov/owow/tmdl/

2a. Original Data Source:

State-submitted and EPA-approved TMDLs or EPA-established TMDLs

2b. Source Data Collection:

State-submitted and EPA-approved TMDLs and EPA-established TMDLs are publicly reviewed during their development. Electronic and hard copies of state-submitted and EPA-approved TMDLs are made available by states and often linked to EPA Web sites. The Watershed Assessment, Tracking, and Environmental ResultS system allows search for TMDL documents at http://www.epa.gov/waters/tmdl/tmdl_document_search.html

Explanation:

Office of Water Quality Management Plan. EPA requires that organizations prepare a document called a QMP that: documents the organization's quality policy; describes its quality system; and identifies the environmental programs to which the quality system applies (e.g., those programs involved in the collection or use of environmental data).

2c. Source Data Reporting:

Relevant information from each TMDL is entered into the Assessment and Total Maximum Daily Load (TMDL) Tracking And ImplementatioN System (ATTAINS) data entry system and made available to the public via the web reports. See http://www.epa.gov/waters/ir

3a. Relevant Information Systems:

The Assessment and Total Maximum Daily Load (TMDL) Tracking And ImplementatioN System (ATTAINS) is the database which captures water quality information related to this measure. ATTAINS is an integrated system that documents and manages the connections between state assessment and listing decisions reported under sections 305(b) and 303(d) (i.e., integrated reporting) and completed TMDLs. This system holds information about assessment decisions and restoration actions across reporting cycles and over time until water quality standards are attained. Annual TMDL totals by state, fiscal year, and pollutant are available at http://iaspub.epa.gov/waters10/attains_nation_cy.control?p_report_type=T - APRTMDLSand TMDL document searches can be conducted at http://www.epa.gov/waters/tmdl/tmdl_document_search.html More information about ATTAINS can be found at http://www.epa.gov/waters/data/prog.html and http://www.epa.gov/waters/ir/about_integrated.html

The Watershed Assessment, Tracking, and Environmental Results System (WATERS) is used to provide water program information and display it spatially using a geographic information system (National Hydrography Dataset (NHD)) integrated with several of EPA's existing databases. These databases include the STOrage and RETrieval (STORET) database, the Assessment TMDL Tracking and ImplementatioN System (ATTAINS), the Water Quality Standards Database (WQSDB), and the Grants Tracking and Reporting System (GRTS). This water quality information was previously available only from several independent and unconnected databases. General information about WATERS is available at: http://www.epa.gov/waters/ a system architecture diagram is available at: http://www.epa.gov/waters/about/arch.html and information about WATERS geographic data is available at: http://www.epa.gov/waters/about/geography.html

3b. Data Quality Procedures:

QA/QC of data is provided by EPA Regional staff and through cross-checks of ATTAINS information regarding impaired water listings, consistent with the Office of Water Quality Management Plan (QMP). EPA requires that organizations prepare a document called a QMP that: documents the organization's quality policy; describes its quality system; and identifies the environmental programs to which the quality system applies (e.g., those programs involved in the collection or use of environmental data).

3c. Data Oversight:

The Assessment and Watershed Protection Division Director is responsible for overseeing the source data reporting and information systems.

3d. Calculation Methodology:

Additional information: Internal reviews of data quality revealed some inconsistencies in the methodology of data entry between EPA Regional Offices. In 2005 and 2006, EPA convened a meeting of NTTS users to discuss how to improve the database. As a result, data field definitions were clarified, the users' group was reinstituted, several training sessions were scheduled, and an ATTAINS design made the necessary database upgrades. One of the issues raised included the methodology used to count TMDLs. Previous methodology generated a TMDL "count" based on the causes of impairment removed from the 303(d) impaired waters list as well as the TMDL pollutant. EPA proposed to change the counting methodology to directly reflect only the pollutants given allocations in TMDLs. During a recent EPA Office of the Inspector General review they concurred with this recommendation. This proposed change was vetted during the TMDL Program's annual meeting in March 2007 and implemented in August 2007, resulting in a cumulative net reduction of 1,577 TMDLs.

Guidance:

Detailed measure guidance reporting can be found under the water quality sub-objective (WQ-8a) at http://water.epa.gov/resource_performance/planning/FY-2012-NWPG-Measure-Definitions-Water-Quality.cfm

4a. Oversight and Timing of Final Results Reporting:

The Headquarters point of contact for this measure works with Regions to address any questions and to ensure the TMDL information is correctly entered into and made available to the public in ATTAINS.

Branch Chief for Watershed Branch (WB) is responsible for tracking and reporting on this measure.

4b. Data Limitations/Qualifications:

To meet the increasing need for readily accessible CWA information, EPA continues to improve the database and oversee quality review of existing data. Data quality has been improving and will continue to improve as existing data entry requirements and procedures are being re-evaluated and communicated with data entry practitioners.

4c. Third-Party Audits:

USEPA, Office of the Inspector General. 2007. Total Maximum Daily Load Program Needs Better Data and Measures to Demonstrate Environmental Results. Available at http://www.epa.gov/oig/reports/2007/20070919-2007-P-00036.pdf

USEPA, Office of the Inspector General. 2005. Sustained Commitment Needed to Further Advance the Watershed Approach. Available at http://www.epa.gov/oig/reports/2005/20050921-2005-P-00025.pdf

National Research Council, Committee to Assess the Scientific Basis of the Total Maximum Daily Load Approach to Water Pollution Reduction. 2001. Assessing the TMDL Approach to Water Quality Management. Washington, DC: National Academy Press. http://www.nap.edu/openbook.php?isbn=0309075793

Measure Code: cb6 - Percent of goal achieved for implementing nitrogen reduction actions to achieve the final TMDL allocations, as measured through the phase 5.3 watershed model.

Office of Water (OW)

Goal Number and Title:
2 - Protecting America's Waters

Objective Number and Title:
2 - Protect and Restore Watersheds and Aquatic Ecosystems

Sub-Objective Number and Title:
5 - Improve the Health of the Chesapeake Bay Ecosystem

Strategic Target Code and Title:
1 - Achieve 50 percent of the 185,000 acres of submerged aquatic vegetation necessary to achieve Chesap

Managing Office:
CBPO

1a. Performance Measure Term Definitions:
Percent of goal achieved to achieve final TMDL allocations for nitrogen: In December 2010, the Environmental Protection Agency established a pollution diet for the Chesapeake Bay, formally known as a Total Maximum Daily Load or TMDL. The TMDL is designed to ensure that all nitrogen, phosphorus and sediment pollution control efforts needed to fully restore the Bay and its tidal rivers are in place by 2025, with controls, practices and actions in place by 2017 that would achieve at least 60% of the reductions from 2009 necessary to meet the TMDL. The TMDL sets pollution limits (allocations) necessary to meet applicable water quality standards in the Bay and its tidal rivers. Specifically, the TMDL allocations are 201.63 million pounds of nitrogen, 12.54 million pounds of phosphorus, and 6,453.61 million pounds of sediment per year (note, the nitrogen allocation includes a 15.7 million pound allocation for atmospheric deposition of nitrogen to tidal waters).   As a result of this new Bay-wide "pollution diet," Bay Program partners are implementing and refining Watershed Implementation Plans (WIPs) and improving the accounting of their efforts to reduce nitrogen, phosphorus and sediment pollution. The WIPs developed by Delaware, the District of Columbia, Maryland, New York, Pennsylvania, Virginia and West Virginia identify how the Bay jurisdictions are putting measures in place by 2025 that are needed to restore the Bay, and by 2017 to achieve at least 60 percent of the necessary nitrogen, phosphorus and sediment reductions compared to 2009. Much of this work already is being implemented by the jurisdictions consistent with their Phase I WIP commitments, building on 30 years of Bay restoration efforts.   Planning targets were established August 1, 2011 to assist jurisdictions in developing their Phase II WIPs. Specifically, the planning targets are 207.27 million pounds of nitrogen, 14.55 million pounds of phosphorus and 7,341 million pounds of sediment per year (note, the planning target for nitrogen includes a 15.7 million pound allocation for atmospheric deposition of nitrogen to tidal waters). These planning targets, while slightly higher than the allocations published in the December 2010 TMDL, represent the actions, assumptions, and "level of effort" necessary to meet the TMDL allocations.

The CBP partnership is committed to flexible, transparent, and adaptive approaches towards Bay restoration and will revisit these planning targets in 2017. The partnership will also conduct a comprehensive evaluation of the TMDL and the CBP's computer modeling tools in 2017. Phase III WIPs will be established in 2017 and are expected to address any needed modifications to ensure, by 2025, that controls, practices and actions are in place which would achieve full restoration of the Chesapeake Bay and its tidal tributaries to meet applicable water quality standards.

Annual nitrogen loading, taking into account implementation of nitrogen pollution reduction actions throughout the Chesapeake Bay watershed, will be calculated using the Chesapeake Bay Program phase 5.3.2 Watershed Model. The CBP Watershed Model uses actual wastewater discharge data, which is influenced by annual weather conditions, to estimate wastewater pollution. The influence of weather, rain and snowfall can be quite large and can influence wastewater loads more than the restoration efforts in any single year. However, the indicator does demonstrate long-term progress to reduce wastewater pollution. The Model estimates pollution from other sources such as agriculture or urban runoff using average weather conditions. This allows managers to understand trends in efforts to implement pollution reduction actions.

Data will be from Chesapeake Bay watershed portions of NY, MD, PA, VA, WV, DE, and DC.

This annual loading estimate will be used to identify progress toward the EPA reduction goal, which will be expressed as % of the annual goal achieved. Achieving the Bay TMDL nitrogen allocation is necessary for attaining tidal water quality standards for clarity/submerged aquatic vegetation.

TMDL: is an acronym for Total Daily Maximum Load, a calculation of the maximum amount of a pollutant that a waterbody can receive and still safely meet EPA water quality standards. The Chesapeake Bay TMDL was completed on December 29, 2010. It is the largest and most complex ever developed, involving six states and the District of Columbia and the impacts of pollution sources throughout a 64,000-square-mile watershed. The Bay TMDL is actually a combination of 92 smaller TMDLs for individual Chesapeake Bay tidal segments. It includes limits on nitrogen, phosphorus and sediment sufficient to achieve state clean water standards for dissolved oxygen, water clarity, underwater Bay grasses and chlorophyll-a, an indicator of algae levels. More information about the Chesapeake Bay's TMDL is at
http://www.epa.gov/reg3wapd/tmdl/ChesapeakeBay/tmdlexec.html

Implementing nitrogen pollution reduction actions: Activities by municipalities and state agencies in the Chesapeake Bay watershed to improve stormwater management and wastewater treatment plants (WWTPs), as well as management of septic fields and other nonpoint nitrogen sources, to reduce the amounts of nitrogen that enters the bay.

Phase 5.3 Watershed Model: The CBP Watershed Model uses actual wastewater discharge data, which is influenced by annual weather conditions, to estimate wastewater pollution. The influence of weather, rain and snowfall can be quite large and can influence wastewater loads more than the restoration efforts in any single year. However, the indicator does demonstrate long-term progress to reduce wastewater pollution. The Model estimates pollution from other sources such as agriculture or urban runoff using average weather conditions. This allows managers to understand trends in efforts to implement pollution reduction actions.

Information about the Chesapeake Bay Program Watershed Model can be found in Section 5 of the Bay TMDL (http://www.epa.gov/reg3wapd/pdf/pdf_chesbay/FinalBayTMDL/CBayFinalTMDLSection5_final.pdf Additionally, please see http://ches.communitymodeling.org/models/CBPhase5/index.php for the most recent WSM documentation (December 2010).

Background:
· Nitrogen loads originate from many sources in the Chesapeake Bay watershed. Point sources of nitrogen include municipal wastewater facilities, industrial discharge facilities, combined sewer overflows (CSOs), sanitary sewer overflows (SSOs), NPDES permitted stormwater (MS4s and construction and industrial sites), and confined animal feeding operations (CAFOs). Nonpoint sources include agricultural lands (animal feeding operations (AFOs), cropland, hay land, and pasture), atmospheric deposition, forest lands, on-site treatment systems, nonregulated stormwater runoff, stream banks and tidal shorelines, tidal resuspension, the ocean, wildlife, and natural background.
· The website for EPA's Chesapeake Bay program office is http://www.epa.gov/region3/chesapeake/
For additional information about this indicator, go to
 http://www.chesapeakebay.net/indicators/indicator/reducing_nitrogen_pollution

2a. Original Data Source:

Annual jurisdictional submissions (NY, MD, PA, VA, WV, DE, and DC) of both monitored and estimated wastewater effluent concentrations and flows approved by each jurisdiction as well as best management practice (BMP) data for other sources of pollution tracked by jurisdictions and reported to the Chesapeake Bay Program office. The Phase 5.3.2 watershed model uses many types of data from sources too numerous to describe here. Please see http://ches.communitymodeling.org/models/CBPhase5/index.php for the most recent WSM documentation (December 2010).

2b. Source Data Collection:

Collection Methods:
Jurisdictions from Chesapeake Bay watershed portions of NY, MD, PA, VA, WV, DE, and DC annually submit two kinds of data. One type of data is monitored and estimated wastewater effluent concentrations and flows from WWTPs and industrial facilities. These are approved by each jurisdiction. The second is nonpoint source practice data tracked by jurisdictions based on land uses, animal manure and chemical fertilizer inputs, human population, nonpoint source controls/practices, septic, and atmospheric deposition. This data is reported to the Chesapeake Bay Program office.
For additional information, refer to the Analysis and Methods documentation for this indicator at
http://www.chesapeakebay.net/images/indicators/17662/reducingpollutionamdoc2011_050912.doc

Quality Procedures:
Procedures for compiling and managing wastewater discharge data at the Chesapeake Bay Program Office are documented in the following EPA-approved Quality Assurance Project Plan:

·   "Standard Operating Procedures for Managing Point Source Data – Chesapeake Bay Program" on file for the EPA grant.

Procedures at the Chesapeake Bay Program Office for acquiring and managing data from sources of pollution other than wastewater treatment plants are documented in the following EPA-approved Quality Assurance Project Plan:
·   "Standard Operating Procedures for Managing Nonpoint Source Data – Chesapeake Bay Program" on file for the EPA grant.

Jurisdictions providing wastewater effluent data and BMP data for other sources of pollution to the Bay Program Office have supplied documentation of their quality assurance and quality control policies, procedures, and specifications in the form of Quality Assurance Management Plans and Quality Assurance Project Plans. Jurisdictional documentation can be obtained by contacting the Quality Assurance Coordinator, Mary Ellen Ley, mley@chesapeakebay.net).

Geographical Extent: Jurisdictions from Chesapeake Bay watershed portions of NY, MD, PA, VA, WV, DE, and DC. Refer to map at
http://www.chesapeakebay.net/maps/map/chesapeake_bay_watershed_model_phase_5_modeling_segments

Spatial Detail: Depending on the practice and jurisdiction, data for other sources of pollution are tracked and reported at the following spatial scales:
§   State
§   River Segment
§   State-Segment – intersection of jurisdictional boundary and Watershed Model river segment
§   Major Basin
§   State-Basin – intersection of jurisdictional boundary and Major Basin
§   County
§   County-Segment – intersection of county boundary and Watershed Model river segment

Wastewater: Data can be aggregated to Hydrologic Units (HUC8 and HUC11), counties/cities (FIPS), "state-segments" (the intersection of state boundaries and Phase 5.3.2 Watershed Model river segments), jurisdictional portions of major basins, major basins, jurisdictions, and the Chesapeake Bay watershed as a whole.

Agriculture, Urban/Suburban and Septic, Air:
BMP implementation data to reduce pollution from these sources are aggregated to "state-segments", or the intersection of state boundaries and Phase 5.3.2 Watershed Model river segments, jurisdictional portions of major basins, major basins, major tributaries, jurisdictions, and the Chesapeake Bay watershed as a whole.

2c. Source Data Reporting:
Data Submission/Data Entry: Data are reported to EPA's Chesapeake Bay Program office (CBPO) through the

Chesapeake Bay TMDL Tracking and Accounting System (BayTAS) and the National Environmental Information Exchage Network (NEIEN).

Data Transmission Timing and Frequency: Data are transmitted to CBPO by Dec 31st each year. It takes approximately 3-4 months to process the data and utilize in a watershed model run in order to have final output for use in updating the indicator on an annual basis.

3a. Relevant Information Systems:

System Description: The Chesapeake Bay TMDL Tracking and Accounting System (BayTAS) was developed to inform EPA, the Bay Jurisdictions, and the public on progress in implementing the bay's TMDL. BayTAS stores the TMDL allocations (based on the Watershed Model Phase 5.3.0 and tracks implementation progress (based on the Watershed Model Phase 5.3.2 and the jurisdictions' Phase II Watershed Implementation Plans). For more information about the BayTAS, refer to
http://stat.chesapeakebay.net/sites/all/cstat/tmdl/BayTAS_factsheet.pdf

Source/Transformed Data: Source and transformed data.

Information System Integrity Standards: Refer to Section 5 of the Bay TMDL
(http://www.epa.gov/reg3wapd/pdf/pdf_chesbay/FinalBayTMDL/CBayFinalTMDLSection5_final.pdf Additionally, please see http://ches.communitymodeling.org/models/CBPhase5/index.php for the most recent WSM documentation (December 2010).

3b. Data Quality Procedures:

Procedures for compiling and managing wastewater discharge data at the Chesapeake Bay Program Office are documented in the following EPA-approved Quality Assurance Project Plan:
·   "Standard Operating Procedures for Managing Point Source Data – Chesapeake Bay Program" on file for the EPA grant.

Procedures at the Chesapeake Bay Program Office for acquiring and managing data from sources of pollution other than wastewater treatment plants are documented in the following EPA-approved Quality Assurance Project Plan:
·   "Standard Operating Procedures for Managing Nonpoint Source Data – Chesapeake Bay Program" on file for the EPA grant.

Jurisdictions providing wastewater effluent data and BMP data for other sources of pollution to the Bay Program Office have supplied documentation of their quality assurance and quality control policies, procedures, and specifications in the form of Quality Assurance Management Plans and Quality Assurance Project Plans. Jurisdictional documentation can be obtained by contacting the Quality Assurance Coordinator, Mary Ellen Ley, mley@chesapeakebay.net).

3c. Data Oversight:

Source Data Reporting Oversight Personnel:

Wastewater: Ning Zhou, Wastewater Data Manager, Virginia Polytechnic Institute and State University, Chesapeake Bay Program Office

Best Management Practice and Watershed Model information: Jeff Sweeney, Nonpoint Source Data Manager, USEPA, Chesapeake Bay Program Office.

Source Data Reporting Oversight Responsibilities: Assure quality of data submitted by states.

Information Systems Oversight Personnel:
Wastewater: Ning Zhou, Wastewater Data Manager, Virginia Polytechnic Institute and State University, Chesapeake Bay Program Office

Best Management Practice and Watershed Model information: Jeff Sweeney, Nonpoint Source Data Manager, USEPA, Chesapeake Bay Program Office.

Information Systems Oversight Responsibilities: Assure data has been transferred correctly to watershed model.

3d. Calculation Methodology:

Decision Rules for Selecting Data, Definition of Variables, and Explanation of Calculations: Refer to Section 5 of the Bay TMDL (http://www.epa.gov/reg3wapd/pdf/pdf_chesbay/FinalBayTMDL/CBayFinalTMDLSection5_final.pdf Additionally, please see http://ches.communitymodeling.org/models/CBPhase5/index.php for the most recent WSM documentation (December 2010).

Unit of Measure: Percent of goal achieved

Timeframe of Result: FY 2010 is baseline (based on 2009 progress run) and FY 2026 is the end year (will be based on 2025 progress run. Most recent year of progress will be calculated based on the most current progress run (progress runs are completed on an annual basis).

Documentation of Methodological Changes:Refer to Section 5 of the Bay TMDL (http://www.epa.gov/reg3wapd/pdf/pdf_chesbay/FinalBayTMDL/CBayFinalTMDLSection5_final.pdf Additionally, please see http://ches.communitymodeling.org/models/CBPhase5/index.php for the most recent WSM documentation (December 2010).

4a. Oversight and Timing of Final Results Reporting:

Final Reporting Oversight Personnel:
Jeff Sweeney, Nonpoint Source Data Manager, USEPA, Chesapeake Bay Program Office
Katherine Antos, Water Quality Goal Implementation Team Coordinator, USEPA, Chesapeake Bay Program Office

Final Reporting Oversight Responsibilities: Provide final, approved data for use in indicator.

Final Reporting Timing: Usually by March or April very year.

4b. Data Limitations/Qualifications:

General Limitations/Qualifications: The Chesapeake Bay Program Watershed Model, employed to integrate wastewater technology controls and a large array of BMPs to reduce pollution from other sources, is best utilized when making comparisons among scenarios. For the Reducing Pollution indicators, these comparisons are among the 2009 Bay TMDL baseline, the yearly model assessments of loads, and the Phase II WIP planning targets.

By presenting trends and status at the large scale of the 64,000 square mile watershed over a 20-year period, yearly changes in data tracking mechanisms by particular jurisdictions and changes in methods of data analysis for particular wastewater plants and BMPs are somewhat masked.

The indicators are designed 1) to depict, generally, the degree of progress over the long term toward the implementation goals and 2) to clearly identify pollutant sources where gaps are large and to what extent. The indicators connect efforts (pollutant controls) with results (loading reductions and subsequently, water quality and habitat health).

Data Lag Length and Explanation: Fiscal year end of year results are based on progress run data from the previous year.

Methodological Changes: N/A

4c. Third-Party Audits:

The following evaluation from the National Academies of Sciences assesses Chesapeake Bay nutrient/sediment reduction strategies:
·   The National Research Council (NRC) established the Committee on the Evaluation of Chesapeake Bay Program Implementation for Nutrient Reduction in Improve Water Quality in 2009 in response to a request from the EPA. (Executive Order 13508, Chesapeake Bay Restoration and Protection, called for an independent evaluator to periodically evaluate protection and restoration activities and report on progress toward meeting the goals of the Executive Order.) The committee was charged to assess the framework used by the states and the CBP for tracking nutrient and sediment control practices that are implemented in the Chesapeake Bay watershed and to evaluate the two-year milestone strategy. The committee was also asked to assess existing adaptive management strategies and to recommend improvements that could help CBP to meet its nutrient and sediment reduction goals. See: Committee on the Evaluation of Chesapeake Bay Program Implementation for Nutrient Reduction to Improve Water Quality; National Research Council. Achieving Nutrient and Sediment Reduction Goals in the Chesapeake Bay: An Evaluation of Program Strategies and Implementation (2011). http://dels.nas.edu/Report/Achieving-Nutrient-Sediment-Reduction-Goals/13131

The following reviews are relevant to the Bay Program's modeling approach:
·   An external review of the Bay Program's Phase 5 Watershed Model hydrologic calibrations was

completed in September 2008 and can be found at:

http://www.chesapeakebay.net/content/publications/cbp_51626.pdf

·   In February, 2008, an external panel assembled by the Chesapeake Bay Program's Scientific and Technical Advisory Committee reviewed the Chesapeake Bay Phase 5 Watershed Model assessing (1) work to date, (2) the model's suitability for making management decisions at the Bay Watershed and local scales, and (3) potential enhancements to improve the predictive ability of the next generation of the Chesapeake Bay Watershed Models. A report of the review, with specific recommendations, can be found at the STAC site:

http://www.chesapeake.org/stac/stacpubs.html

·   Another external review of Bay Program modeling efforts "Modeling in the Chesapeake Bay Program: 2010 and Beyond" completed January, 2006 is published by STAC at:

http://www.chesapeake.org/stac/Pubs/ModBay2010Report.pdf

·   In June, 2005, another external review of the Watershed Model addressed the following broad questions: 1) Does the current phase of the model use the most appropriate protocols for simulation of watershed processes and management impacts, based on the current state of the art in the HSPF model development?, and 2) Looking forward to the future refinement of the model, where should the Bay Program look to increase the utility of the Watershed Model? Details of this review and responses can be found at:

http://www.chesapeakebay.net/pubs/subcommittee/mdsc/Watershed_Model_Peer_Review.pdf

The following EPA OIG reports focus on the performance measures or the data underlying the performance measures for the Chesapeake Bay measures discussed in this DQR.

·   Despite Progress, EPA Needs to Improve Oversight of Wastewater Upgrades in the Chesapeake Bay Watershed, Report No. 08-P-0049. Available at http://www.epa.gov/oig/reports/2008/20080108-08-P-0049.pdf

·   Saving the Chesapeake Bay Watershed Requires Better Coordination of Environmental and Agricultural Resources, EPA OIG Report No. 2007-P-00004. Available at http://www.epa.gov/oig/reports/2007/20070910-2007-P-00031.pdf

·   Development Growth Outpacing Progress in Watershed Efforts to Restore the Chesapeake Bay, Report No. 2007-P-00031. Available at http://www.epa.gov/oig/reports/2007/20070910-2007-P-00031.pdf

·   EPA Relying on Existing Clean Air Act Regulations to Reduce Atmospheric Deposition to the Chesapeake Bay and its Watershed, Report No. 2007-P-00009. Available at http://www.epa.gov/oig/reports/2007/20070228-2007-P-00009.pdf

The following GAO reports focus on the performance measures or the data underlying the performance measures for the Chesapeake Bay measures discussed in this DQR.

·   Chesapeake Bay Program: Recent Actions Are Positive Steps Toward More Effectively Guiding the Restoration Effort. Available at http://www.gao.gov/products/GAO-08-1033T (2008)

·   CHESAPEAKE BAY PROGRAM: Improved Strategies Are Needed to Better Assess, Report, and Manage Restoration Progress. Available at http://www.gao.gov/new.items/d0696.pdf (2005)

Measure Code: Opb - Percent of serviceable rural Alaska homes with access to drinking water supply and wastewater disposal.

Office of Water (OW)

Goal Number and Title:
2 - Protecting America's Waters

Objective Number and Title:
2 - Protect and Restore Watersheds and Aquatic Ecosystems

Sub-Objective Number and Title:
1 - Improve Water Quality on a Watershed Basis

Strategic Target Code and Title:
5 - Provide access to basic sanitation for 67,900 American Indian and Alaska Native homes.

Managing Office:
Primary: US. EPA; Office of Water; Office of Wastewater Management; Municipal Support Division; Sustainable Communities Branch; Secondary: EPA Region 10 Office of Water and Watersheds; Grants and Strategic Planning Unit; ANV grant Project Officer (PO)

1a. Performance Measure Term Definitions:
Serviceable rural Alaska home: Roughly 1,500 of the 36,000 Alaskan Native and rural homes are categorized as unserviceable, the remaining are considered serviceable. As defined by the State of Alaska Department of Environmental Conservation, Village Safe Water program, unserviceable homes are those that will likely never receive full scale water and sewer services. General reasons for unserviceable homes may include: the community does not want the services, the home does not have electricity, the home does not have thermostatically controlled heat, the home is structurally unsound, it is located beyond the proper community, there is an extremely high capital cost to make the homes serviceable, and/or it is a seasonally occupied residence. Access to drinking water supply and wastewater disposal: Refers to the reduction in the sanitation deficiency level of a tribal home from a 4 or 5 to a 3 or less. The sanitation deficiency levels definitions are described in Appendix E of the "Indian Health Service Sanitation Deficiency System Guide for Reporting Sanitation Deficiencies for Indian Homes and Communities," working draft, May 2003 (http://www.dsfc.ihs.gov/Documents/SDSWorkingDraft2003.pdf). Sanitation deficiency is an identified need for new or upgraded sanitation facilities for existing homes of American Indians or Alaska Natives. Background: Initiated in FY 2007 through the Program Assessment Rating Tool (PART), this measure was established to track the environmental and human health performance of EPA's Alaska Native Village and Rural Communities Grant Program. This program provides funding to assist Alaska Native Villages (ANVs) and Alaska rural communities with the construction of new or improved drinking water and wastewater systems, and to provide training and technical assistance in the operation and maintenance of these systems. For this program, EPA Headquarters' Office of Wastewater Management allocates funds to the EPA Region 10 Alaska Operations Office, which in turn provides funding to the Alaska Department of Environmental Conservation (DEC). The Alaska DEC administers the funds for projects through its Village Safe Water (VSW) Program.

For more information, please see:

- Website for EPA's ANV and Rural Communities Grant Program (http://www.epa.gov/alaskanativevillages).
- Webpage for Alaska DEC's VSW Program (http://www.dec.state.ak.us/water/vsw/pdfs/vswbrief.pdf).
- Infrastructure Task Force Access Subgroup. "Meeting the Access Goal: Strategies for Increasing Access to Safe Drinking Water and Wastewater Treatment to American Indian and Alaska Native Homes." March 2008. (www.epa.gov/tp/pdf/infra-tribal-access-plan.pdf).

2a. Original Data Source:

Indian Health Service (IHS) through the Alaska Native Tribal Health Consortium and the VSW. For the Alaska Native Villages IHS and VSW enters the original data for this measure into the IHS Sanitation Tracking and Reporting System (STARS). For non-native rural Alaska communities the VSW program is using the U.S. Census Bureau's American Community Survey. This survey collects data annually rather than every ten years and is used to help determine how more than $400 billion in federal and state funding are distributed each year.

2b. Source Data Collection:

The IHS and VSW identify sanitation deficiencies at rural Alaska homes in several ways, the most common of which follow:

- Consultation with Tribal members, community members and other Agencies
- Field visits by engineers, sanitarians or regional health corporation staff
- Public Water System Supervision (PWSS) Sanitary Surveys
- Community Master Plans for Development
- Telephone Surveys
- Feasibility Studies

The most reliable and preferred method is a field visit to each community to identify and obtain accurate numbers of homes with sanitation deficiencies. The number of rural/Native Alaskan homes within the communities must be consistent among the various methods cited above. If a field visit cannot be made, it is highly recommended that more than one method be used to determine sanitation deficiencies to increase the accuracy and establish greater credibility for the data. The identification methods used for each project must be retained and documented in area records.

Sanitation deficiency data entered into STARS undergo a series of highly organized quality control reviews at various levels within the IHS (field, district, and area). Data is also reviewed by the State of Alaska as well as experienced tribal personnel. The data quality review consists of performing a number of established data queries and reports, which identify errors and/or inconsistencies. In addition, the top sanitation deficiency projects and corresponding community deficiency profiles for each area are reviewed against their budgets. Detailed cost estimates are required for the review.

STARS is comprised of several sub-data systems, including the Sanitation Deficiency System (SDS), which contains the data that is used to calculate this measure. The SDS is an inventory of sanitation deficiencies for Indian and rural Alaska homes, ANVs and communities, and is updated on an ongoing basis. The sub-data system called the Housing Inventory Tracking System (HITS) was fully implemented in FY2014 and is a mapped based housing tracking system. The HITS system is more accurate than the previously used tabular housing inventory referred to as the Community Deficiency Profile. The utilization of a map based housing inventory allows for the connection of specific homes to specific projects in SDS. As projects are funded in SDS the deficiency level for the corresponding homes are changed to reflect the change as a result of the funded project.

For more information, please see:
-        Indian Health Service (IHS), Division of Sanitation Facilities (DSFC). Criteria for the Sanitation Facilities Construction Program, June 1999, Version 1.02, 3/13/2003.
http://www.dsfc.ihs.gov/Documents/Criteria_March_2003.cfm .
-        Indian Health Service (IHS), Division of Sanitation Facilities (DSFC). Sanitation Deficiency System (SDS), Working Draft, "Guide for Reporting Sanitation Deficiencies for Indian Homes and Communities", May 2003.
http://www.dsfc.ihs.gov/Documents/SDSWorkingDraft2003.pdf .

2c. Source Data Reporting:

STARS is managed by the IHS Office of Environmental Health and Engineering (OEHE), Division of Sanitation Facilities Construction (DSFC). The State of Alaska retrieves data from STARS via reports generated through STARS dataset queries and through the U.S. Census Bureau's American Community Survey. EPA staff have read only access to STARS. The system is password protected, and IHS maintains ownership and oversight responsibilities. This database is utilized to establish funding priorities for all federal funds identified for water and wastewater infrastructure in rural Native Alaskan communities.

STARS is a web-based application and includes data fields for sanitation deficiencies, Indian homes, and construction projects. The SDS sub-system contains the data that is used to calculate this measure. HITS is a mapped based housing tracking system. As projects are funded in SDS the deficiency level for corresponding homes are changed to reflect the change as a result of the funded project. The HITS image below identifies homes with water service (not circled) and without water service (circled):

HITS – Map Based Housing Inventory

STARS routinely undergoes standard ongoing support and updates to maintain database integrity, efficiency, and accuracy.

For more information, please see:
-        STARS https://wstarstest.ihs.gov/
-        STARS user manual:
http://www.ihs.gov/EHSCT/dsp_folder/dsp_filedownload.cfm?folder=main_resource_docs&filename=STARS_Manual_20080922.pdf&filetype=pdf .

Source Data Reported: Housing information undergoes QA/QC and is compiled annually at the end of the construction year (typically March) in order to capture progress over the previous construction season. For example, housing information collected in March 2014 reflects progress through calendar year 2013. Analysis and data reviews are conducted around April, and the results are made available around May.

Performance Data Reported: annually

Data Lag: Approximately 5 months.

3a. Relevant Information Systems:

System Description: STARS is a web-based application and includes data fields for sanitation deficiencies, American Indian & Alaskan Native homes, and construction projects. This database is utilized to establish funding priorities for all federal funds identified for water and wastewater infrastructure in American Indian and Alaskan Native communities. STARS is comprised of several sub-data systems, including SDS and HITS.

Source/Transformed Data: IHS and VSW staff input data into STARS. The State of Alaska retrieves data for the Alaska Native communities from STARS via an online interface that pulls a STARS dataset query and combines the STARS data from the U.S. Census Bureau's American Community Survey for non-native communities.

Information System Integrity Standards: The combined data undergo a series of quality control reviews at various levels within the IHS and the State of Alaska. IHS Field Engineers, District Engineers and/or the Area Office staff tabulate records and activities. STARS routinely undergoes standard ongoing support and updates to maintain database integrity, efficiency, and accuracy. The State then sorts, filters and QA the data against the previously reported data for reporting on the ANV measure.

3b. Data Quality Procedures:

IHS Area and VSW staff enter data into STARS, the State of Alaska pulls data, sorts, filters and QA the data against the previously reported data in coordination with the EPA Project Officer (PO), results are compiled into a measures monitoring spreadsheet, results are reported as appropriate to: OWM Immediate Office (who imports the data into ACS), OW Immediate Office, OMB, OCFO, etc.

The SDS within STARS data undergo a series of highly organized reviews by experienced tribal, IHS field, IHS district, State of Alaska and IHS area personnel. The data quality review consists of performing a number of established data queries and reports, which identify errors and/or inconsistencies. In addition, the top SDS projects and corresponding community deficiency profiles for each area are reviewed against their budgets. Detailed cost estimates are required for the review.

3c. Data Oversight:

Source Data Reporting Oversight Personnel: EPA ANV HQ Coordinator, Matthew Richardson; EPA ANV Project Officer, EPA Region 10 staff, Dennis Wagner (Reg 10) .

Source Data Reporting Oversight Responsibilities: EPA ANV HQ Coordinator and EPA ANV Project Officer in Region10 store original and transformed data on individual Microsoft Excel files on local EPA networks. State of Alaska duel database querying and QA assurance.

Information Systems Oversight Personnel: IHS staff and State of Alaska Staff.

Information Systems Oversight Responsibilities: IHS staff and State of Alaska staff.

3d. Calculation Methodology:

Unit of Measure: Rural Alaska homes.

4a. Oversight and Timing of Final Results Reporting:

Final Reporting Oversight Personnel: EPA ANV HQ Coordinator, Matthew Richardson; EPA ANV Project Officer, EPA Region 10 staff, Dennis Wagner (Reg 10) .

Final Reporting Oversight Responsibilities: The State of Alaska retrieves data from STARS via reports generated through STARS dataset queries and through the U.S. Census Bureau's American Community Survey. The State then sorts, filters and QA the data against the previously reported data for reporting on the ANV measure. The data is then compiled into a measures monitoring spreadsheet. The data is then provided to the EPA program staff.
OWM Immediate Office (who imports the data into ACS), OW Immediate Office, OMB, OCFO, etc.

Final Reporting Timing: Housing information is undergoes QA/QC and is collected once annually at the end of the construction year (typically March) in order to capture progress over the previous construction season. For example, housing information collected in March 2014 reflects progress through calendar year 2013.

4b. Data Limitations/Qualifications:

The data is limited by the accuracy of reported data in STARS and the quality assurance procedures performed by IHS and the State of Alaska. EPA does not know the precise definitions and data quality procedures used by IHS and the State of Alaska to create the original data. IHS requires that the top 20% of SDS projects are to have cost estimates within 10% of the actual costs.

4c. Third-Party Audits:

No external audits are currently underway (as of February 2014) but EPA in regular communication with IHS about improving STARS data.
In 2011-2012 EPA conducted a detailed internal program evaluation of both the CWISA and the Drinking Water Infrastructure Grant Tribal Set-Aside (DWIG-TSA) Programs, which also use IHS STARS data for reporting their measures. The evaluation assessed the tribal set-aside programs and included program implementation issues, measures and outcomes applicability, and the identification of any needed improvements. The final evaluation report is available online at http://www.epa.gov/evaluate/pdf/water/eval-drinking-water-and-clean-water-infrastructure-tribal-set-aside-grant-programs.pdf

Measure Code: fs1 - Percent of women of childbearing age having mercury levels in blood above the level of concern.

Office of Water (OW)

Goal Number and Title:
2 - Protecting America's Waters

Objective Number and Title:
1 - Protect Human Health

Sub-Objective Number and Title:
2 - Fish and Shellfish Safe to Eat

Strategic Target Code and Title:
0 - By 2015, reduce percentage of women of childbearing age having mercury levels above level of concern

Managing Office:
Office of Science and Technology

1a. Performance Measure Term Definitions:
Women of Childbearing Age: 16 - 49 years of age Mercury Levels in Blood: NHANES collects information about a wide range of health-related behaviors, performs a physical examination and collects samples for laboratory tests. Beginning in 1999, NHANES became a continuous survey, sampling the U.S. population annually and releasing the data in two-year cycles. (Note, however, that the Fourth Report was issued four years after the Third Report.) The sampling plan follows a complex, stratified, multistage, probability-cluster design to select a representative sample of the civilian, noninstitutionalized population in the United States. Additional detailed information on the design and conduct of the NHANES survey is available at http://www.cdc.gov/nchs/nhanes.htm Level of Concern: This measure is the percentage of women of childbearing age with blood mercury concentrations within a factor of 10 of those associated with neurodevelopmental effects. This measure was selected because it provides an indication of levels of exposure in the human population to organic mercury, where the main source is the consumption of fish and shellfish contaminated with methylmercury. As consumers follow fish consumption advice, levels of mercury in blood will decrease. Find out more about EPA's efforts to reduce mercury exposure at http://www.epa.gov/hg/

2a. Original Data Source:
The Centers for Disease Control and Prevention (CDC) National Center for Health Statistics (NCHS). National Center for Health Statistics (NCHS) collects the data for women's blood levels through the National Health and Nutrition Examination Survey (NHANES), and is responsible for releasing the data to the public. NHANES is a survey designed to assess the health and nutritional status of adults and children in the U.S. The NHANES Web site is http://www.cdc.gov/nchs/nhanes.htm Data from NHANES is recognized as the primary database in the United States for national statistics on blood levels of certain chemicals of concern among the general population and selected subpopulation groups.

2b. Source Data Collection:
Collection Methodology: Survey, field sampling. NHANES collects information about a wide range of health-

related behaviors, and includes a physical examination and samples for laboratory tests. The sampling plan follows a complex, stratified, multistage, probability-cluster design to select a representative sample of the civilian, noninstitutionalized population in the United States. The NHANES survey examines a nationally representative sample of approximately 5,000 men, women, and children each year located across the U.S. CDC's National Center for Health Statistics (NCHS) is responsible for the conduct of the survey and the release of the data to the public. The NHANES survey program began in the early 1960s as a periodic study. Beginning in 1999, NHANES became a continuous survey, sampling the U.S. population annually and releasing the data in 2-year cycles. Results are published with a 95% confidence interval.

The NHANES survey contains detailed interview questions covering areas related to demographic, socio-economic, dietary, and health-related subjects. It also includes an extensive medical and dental examination of participants, physiological measurements, and laboratory tests (including blood and urine testing). Specific laboratory measurements of environmental interest include: metals (e.g. lead, cadmium, and mercury), VOCs, phthalates, organophosphates (OPs), pesticides and their metabolites, dioxins/furans, and polyaromatic hydrocarbons (PAHs). NHANES is unique in that it links laboratory-derived biological markers (e.g. blood, urine etc.) to questionnaire responses and results of physical exams. NHANES measures blood levels in the same units (i.e., ug/dL) and at standard detection limits. Additional information on the interview and examination process can be found at the NHANES web site at http://www.cdc.gov/nchs/nhanes.htm

Because NHANES is based on a complex multi-stage sample design, appropriate sampling weights should be used in analyses to produce estimates and associated measures of variation. Analytical guidelines issued by NCHS provide guidance on how many years of data should be combined for an analysis. Details about the methodology, including statistical methods, are reported in the Third and Fourth National Reports on Human Exposure to Environmental Chemicals. The CDC National Center for Health Statistics (NCHS) provides guidelines for the analysis of NHANES data at http://www.cdc.gov/nchs/data/nhanes/nhanes_03_04/nhanes_analytic_guidelines_dec_2005.pdf Assumptions inherent in the CDC's analysis are delineated in the Data Sources and Data Analysis chapter of the national reports. Additional detailed information on the design and conduct of the NHANES survey is available at http://www.cdc.gov/nchs/nhanes.htm

Change in Methodology for Estimating Percentiles, from the CDC: "In the Third National Report on Human Exposure to Environmental Chemicals, weighted percentile estimates for 1999–2000 and 2001–2002 data were calculated using SAS Proc Univariate and a proportions estimation procedure. A percentile estimate may fall on a value that is repeated multiple times in a particular demographic group defined by age, sex and race (e.g., in non-Hispanic white males 12-19 years old, five results that all have a value of 90.1). Since the Third Report, we have improved the procedure for estimating percentiles to better handle this situation. This improved procedure makes each repeated value unique by adding a unique negligibly small number to each repeated value. All data from 1999–2004 have been reanalyzed using this new procedure to handle situations where the percentile falls on a repeating value. Therefore, occasional percentile estimates may differ slightly in the current Fourth Report compared to the Third Report. Appendix A gives the details of the new procedure for estimating percentiles." (http://www.cdc.gov/exposurereport/data_tables/data_sources_analysis.html

Geographical Extent: The sample is selected from the civilian, non-institutionalized U.S. population.

Quality Procedures: The data comes from the NHANES study, which CDC has designed to have a high quality. CDC follows standardized survey instrument procedures to collect data to promote data quality and data are subjected to rigorous QA/QC review. CDC's National Center for Environmental Health (NCEH) and National Center for Health Statistics (NCHS) are responsible for QA/QC of laboratory analysis and NHANES datasets that are made publicly available through CDC/NCEH's website. Background documentation is available at the NHANES Web site at http://www.cdc.gov/nchs/nhanes.htm The following documents provide background information specific to data quality: http://www.cdc.gov/nchs/data/nhanes/nhanes_01_02/lab_b_generaldoc.pdf - search=%22quality%20control%20NHANES%22 and
http://www.cdc.gov/nchs/data/nhanes/nhanes_03_04/lab_c_generaldoc.pdf - search=%22quality%20NHANES%22

Additional information on the interview and examination process can be found at the NHANES web site at http://www.cdc.gov/nchs/nhanes.htm

More information on the CDC's program for improving quality of laboratory testing for mercury can be found at: http://www.cdc.gov/labstandards/lamp.html

2c. Source Data Reporting:

EPA downloads relevant data directly from CDC databases, in an SAS format that does not require conversion. Source data are not entered into an EPA information system. After calculations are conducted, the result is entered into EPA's Annual Commitment System (ACS), prior to publication, by Office of Water budget staff in either the Office of Science and Technology or the immediate Office of Water (HQ).

The data used by EPA for EPA's 2012 result (reflecting 2009-2010 data) are published in Table 5 of the report, Trends in Blood Mercury Concentrations and Fish Consumption Among U.S. Women of Reproductive Age, NHANES, 1999-2010 .The report will be available at
http://water.epa.gov/scitech/swguidance/fishshellfish/fishadvisories/technical.cfm - tabs-4

The CDC National Center for Health Statistics (NCHS) has a policy for release of and access to NHANES data at http://www.cdc.gov/nchs/data/nhanes/nhanes_general_guidelines_june_04.pdf Although the studies are supposed to be on a two-year schedule, they have not always been timely.

Background: The CDC reports the full array of NHANES results in its National Reports on Human Exposure to Environmental Chemicals (and updates to the data tables): http://www.cdc.gov/exposurereport/ The most recent report (published in December 2009), the Fourth National Report on Human Exposure to Environmental Chemicals, is the reporting format for this measure. This report presents exposure data for the U.S. population over the two-year survey period of 2003–2004. The Fourth Report also includes data from 1999–2000 and 2001–2002, as reported in the Second and Third National Reports on Human Exposure to Environmental Chemicals.

In the Fourth Report, CDC presents data on 212 chemicals, including results for 75 chemicals measured for the first time in the U.S. population. The Updated Tables (published in February 2012) provide nationally-representative biomonitoring data from the 2007-2008 survey cycle of the National Health and Nutrition Examination Survey (NHANES) for 51 of the environmental chemicals measured in the Fourth Report on Human Exposure to Environmental Chemicals, as well as results for prior survey cycles.

3a. Relevant Information Systems:

System Description: EPA's Annual Commitment System (ACS) is used to record and transmit the data for performance results for the measure. ACS is a module of the Agency's Budget Formulation system BFS Please see the DQR for BFS for additional information

Source/Transformed Data: ACS contains only transformed data – the final result – for this measure.

Information System Integrity Standards: Please see the DQR for BFS for additional information.

3b. Data Quality Procedures:

Data quality procedures are detailed in the project's Quality Assurance Project Plan titled, "Statistical and Technical Support for Fish Advisory Analyses." The EPA Project Manager maintains the QAPP. NHANES data are evaluated for timeliness, representativeness, comparability, and completeness. The data are downloaded, maintained, and analyzed in SAS files by an EPA contractor – no conversion is necessary. Senior contractor personnel review the work of junior staff and of each other. The contractor examines the NHANES documentation to determine whether the data from the 6 sets of releases were collected in the same manner. In preparing the methodology, the contractor reviewed the NHANES analytic guidelines to ensure use of appropriate statistical methodologies and weights. The EPA Project Manager provides overall management of the project and oversees appropriate internal reviews of the contractor's analysis and reporting.

3c. Data Oversight:

Source Data Reporting Oversight Personnel: Not applicable.

Source Data Reporting Oversight Responsibilities: Not applicable.

Information Systems Oversight Personnel: Please see the DQR for BFS for additional information.

Information Systems Oversight Responsibilities: Please see the DQR for BFS for additional information.

3d. Calculation Methodology:

Decision Rules for Selecting Data: For its analysis, EPA selects all records of women of childbearing age for the sample time frame in question (for 2012, the timeframe was 2009-2010). From those records, EPA captures data on measured blood mercury levels.

Definitions of Variables:
Please see Section 1a, Performance Measure Term Definition.

Explanation of the Calculations:
The percent of women of childbearing age with blood mercury greater than 5.8 µg/L was calculated using SAS

survey procedures. The survey procedures incorporate the sample weights and the stratification and clustering of the design into the analysis, yielding proper estimates and standard errors of estimates. A variable was derived that indicated if a participant had blood mercury greater than 5.8 µg/L, coded as 1 if yes, and 0 if no. Then SAS procedure SurveyMeans was used to determine the proportion of women of childbearing age with levels greater than 5.8 µg/L. Estimating the mean of a 0/1 variable provides a proportion. The standard error of the proportion was also estimated. The calculation used balanced repeated replication weights to account for the survey design. The procedure computes the variance with replication methods by using the variability among replicate estimates to estimate the overall variance.

Explanation of Assumptions: Not applicable.

Unit of Measure: Percent of women of childbearing age

Timeframe of Result. EPA uses the most recent two-year sampling period. For the result reported FY 2012, the most recent available two-year sampling period is 2009-2010.

Documentation of Methodological Changes: Any methodological changes are documented in the project QAPP.

4a. Oversight and Timing of Final Results Reporting:

Final Reporting Oversight Personnel: Branch Chief of the Fish, Shellfish, Beach and Outreach Branch / Standards and Health Protection Division / Office of Science and Technology / Office of Water.

Final Reporting Oversight Responsibilities: Forward result to Office of Water budget staff for entry of data into ACS, and ensure result is accurate prior to publication.

Final Reporting Timing: Every two years, approximately in December, on the following cycle: FY 2012 report using 2009-2010 data; FY 2014 report using 2011-2012 data; etc.

4b. Data Limitations/Qualifications:

General Limitations/Qualifications:

Representativeness:

· 	NHANES is designed to provide estimates for the civilian, non-institutionalized U.S. population. NHANES is a voluntary survey and selected persons may refuse to participate. In addition, the NHANES survey uses two steps, a questionnaire and a physical exam. There are sometimes different numbers of subjects in the interview and examinations because some participants only complete one step of the survey. Participants may answer the questionnaire but not provide the more invasive blood sample. Special weighting techniques are used to adjust for non-response.

· 	The periodic reports from NHANES provide a direct measure of mercury in blood levels in a representative sample of the US population. The current design does not permit examination of exposure levels by locality, state, or region; seasons of the year; proximity to sources of exposure; or use of particular products. For example, it is not possible to extract a subset of the data and examine levels of blood mercury that represent levels in a particular state's population.

Precision: The standard error of the percent over 5.8 µg/L is reported along with the percent. The 95% CI for the percent can be calculated with the equation: % ± (1.96 *SE). This is consistent with the Third and Fourth National Reports on Human Exposure to Environmental Chemicals, which provide 95% confidence intervals for all statistics.

Comparability: The measure can be compared to estimates from other sets of NHANES releases. If doing this, one should note the change in laboratory procedures that occurred between the 2001-2002 and 2003-2004 sets of data.

Data Lag: Data lags may prevent performance results from being determined for every reporting year. Performance results will be updated as NHANES data are published either in the official CDC report on human exposure to environmental chemicals or other journal articles or as the data becomes available. There can be a substantial lag between CDC sampling and publication of data. For instance, the result reported for FY 2012 is based upon data from the sampling period of 2009-2010.

Methodological Changes: Between the Third National Report and the Fourth National Report, the CDC changed its method of estimating percentiles, as described under Source Data Collection Methodology of this DQR. This does not affect the interpretations of the EPA results as the measure is not based off of a percentile.

4c. Third-Party Audits:

The NCHS of CDC appointed a panel to review NHANES. The report is available at:
www.cdc.gov/nchs/data/bsc/NHANESReviewPanelReportrapril09.pdf

Measure Code: 4pg - Loading of biochemical oxygen demand (BOD) removed (million pounds/year) from the U.S.-Mexico border area since 2003.

Office of Water (OW)

Goal Number and Title:
2 - Protecting America's Waters
Objective Number and Title:
2 - Protect and Restore Watersheds and Aquatic Ecosystems
Sub-Objective Number and Title:
9 - Sustain and Restore the U.S.Mexico Border Environmental Health
Strategic Target Code and Title:
1 - Provide safe drinking water or adequate wastewater sanitation to 75 percent of the homes in the U.S.
Managing Office:
Office of Wastewater Management (OWM)
1a. Performance Measure Term Definitions:
U.S.-Mexico Border area: The area 100km North and South of the U.S.-Mexico border.   Loading of biochemical oxygen demand removed since 2003: The amount of pollutant (biochemical oxygen demand [BOD]) from wastewater that has been removed (either through sanitary sewer connections or wastewater treatment plant upgrades) since 2003 as a result of completed Border Environment Infrastructure Fund (BEIF) supported projects. The removal of BOD, which is listed as a conventional pollutant in the US Clean Water Act, can be used as a gauge of the effectiveness of wastewater treatment plants. BOD is released into the environment when homes lack wastewater treatment or when wastewater treatment plants lack adequate treatment processes.   Background: This measure reflects the work of EPA's U.S.-Mexico Border Water Infrastructure Program, a cooperative program that aims to improve human health and environmental quality along the international boundary by improving drinking water quality and wastewater sanitation. For more information, please see http://water.epa.gov/infrastructure/wastewater/mexican/index.cfm
2a. Original Data Source:
Border Environment Cooperation Commission (BECC) and North American Development Bank (NADB). For more information on the BECC and NADB, please visit http://www.cocef.org/ and http://www.nadb.org/ respectively.
2b. Source Data Collection:
Methodology: Projections of BOD removal are based on actual average daily flows at wastewater treatment plants, when available, or incorporate per-capita averages typical of the region. Actual influent and effluent water quality data are used when available and are otherwise based on accepted engineering averages.   Quality Procedures: BECC and NADB are responsible for field verification of project information and progress. EPA Regions are responsible for evaluation of reports from BECC and NADB on drinking water and wastewater

sanitation projects. Regional representatives attend meetings of the certifying and financing entities for border projects (BECC and NADB), review various planning and construction related documents and conduct project oversight visits of projects to confirm information accuracy. EPA Headquarters compiles, reviews and tracks information provided by the EPA Regions.

Geographical Extent: The area 100km North and South of the U.S.-Mexico border.

Spatial Detail: N/A

Dates Covered by Source Data: 2003 to present

2c. Source Data Reporting:

Quarterly reports submitted by BECC and NADB to EPA. The BECC and NADB report on the construction progress of certified drinking water and wastewater projects, as well as homes connected to potable water and wastewater collection and treatment systems, applicable design specifications, and water quality and flow data for removal of biochemical oxygen demand. "Certified" means a project that has completed planning and design and has been approved by the BECC/NADB board for construction funding.

No formal EPA database. Performance is based on construction completion of certified projects, which is tracked and reported quarterly by the Border Environment Cooperation Commission (BECC) and the North American Development Bank (NADB). Data fields are: population served by, and homes connected to, potable water and wastewater collection and treatment systems and, applicable design specifications, water quality and flow data for removal of biological oxygen demand (BOD).

3a. Relevant Information Systems:

System Description: No formal EPA database. Performance is based on construction completion of certified projects, which is tracked and reported quarterly by the Border Environment Cooperation Commission (BECC) and the North American Development Bank (NADB) in an Excel spreadsheet format. Data fields are: population served by, and homes connected to, potable water and wastewater collection and treatment systems; and applicable design specifications, water quality, and flow data for removal of biochemical oxygen demand (BOD).

Source/Transformed Data: Source data is provided by EPA grantees and verified by the EPA project officers.

Information System Integrity Standards: Data quality assurance procedures are articulated in Quality Assurance Project Plans (QAPPs) for each grant. No formal data system (beyond simple spreadsheets and paper files) has been needed to store this information. The Border Program typically completes fewer than 10 projects per year. Thus, there are few data points to track.

3b. Data Quality Procedures:

EPA Regions 6 and 9 hold quarterly meetings with the certifying and financing entities for Border water infrastructure projects, the Border Environment Cooperative Commission (BECC) and the North American Development Bank (NADB). Regional EPA staff review various planning and construction related documents.

These documents include design specifications for each project. Annual BOD targets for this measure are based on these design specifications. Regional staff also conduct oversight visits of project sites to confirm information accuracy and review monthly and quarterly reports from BECC and NADB. The monthly and quarterly reports document project completions. As projects are nearing completion, BOD targets and BOD removal estimates are updated to more accurately reflect actual wastewater treatment volumes, treatment efficiencies, and actual outflows. EPA Headquarters compiles, reviews, and tracks information provided by the EPA Regions. This information can be cross-referenced with the BECC and NADB monthly or quarterly reports as needed.

3c. Data Oversight:

Source Data Reporting Oversight Personnel: Team Leader/Environmental Engineer, Region 6; Team Leader/Environmental Engineer, Region 9; Subobjective Lead/Program Analyst, Headquarters.

Source Data Reporting Oversight Responsibilities: Regional project officers and regional environmental engineers (Regions 6 and 9) gather and verify source data and report annual results to the subobjective lead. The headquarters subobjective lead for the Border Program compiles annual targets and results and reports these results.

Information Systems Oversight Personnel: N/A

Information Systems Oversight Responsibilities: N/A

3d. Calculation Methodology:

Decision Rules for Selecting Data: N/A

Definitions of Variables: N/A

Explanation of Calculations and Assumptions: Concentrations of BOD at wastewater treatment plants (pre- and post- treatment) are multiplied by flow rates to determine the removal of BOD on a mass-basis for each project site. Concentrations are based on actual influent and effluent water quality data when available, or are otherwise based on accepted engineering averages. Flow rates are the actual average daily flows at wastewater treatment plants, when available, or incorporate per-capita averages typical of the region. EPA compiles influent and effluent concentrations (mg/L) of BOD for BEIF-funded wastewater treatment and collection projects from either site-specific water quality data or accepted engineering averages. These influent and effluent concentrations are then multiplied by their site-specific average daily flow rates or per-capita averages typical of the region. The difference between the influent and effluent BOD, when multiplied by the corresponding flow rate, is used as the BOD loading (lbs./yr.) removed from each site. These site-specific reductions are then aggregated annually to determine the total BOD loading removed from the US-Mexico Border resulting from the BEIF-funded wastewater infrastructure projects.

Unit of Measure: Millions of pounds of BOD

Timeframe of Result: 2003-end of most recent fiscal year
Documentation of Methodological Changes: Not applicable.

4a. Oversight and Timing of Final Results Reporting:

Final Reporting Oversight Personnel: Planning and Evaluation Coordinator, OWM/OW
Final Reporting Oversight Responsibilities: Team Leader, Planning and Evaluation Team, OW
Final Reporting Timing: The Planning and Evaluation Coordinator reviews and reports information for the sub-office (OWM). The Planning and Evaluation Team Lead reviews information for the Office of Water.

4b. Data Limitations/Qualifications:

General Limitations/Qualifications: This measure only estimates the amount of waste (BOD) removed from Border area water bodies as a result of EPA-funded wastewater treatment projects. It does not capture the total amount of "BOD removal" from other, non-EPA funded projects, nor does it estimate the total BOD loadings for individual water ways.

Once a project is completed, it's "BOD removal per year" is assumed to be constant. In reality, treatment flows and treatment efficiency can change from year-to-year. The measure is meant to describe the combined impact of multiple projects, but is not meant to track the ongoing performance of each individual project.

Data Lag Length and Explanation: No significant data lag.

Methodological Changes: Not applicable.

4c. Third-Party Audits:

EPA Office of Inspector General (IG) report: http://www.epa.gov/oig/reports/2008/20080331-08-P-0121.pdf

Office of Water (OW)

Goal Number and Title:
2 - Protecting America's Waters
Objective Number and Title:
2 - Protect and Restore Watersheds and Aquatic Ecosystems
Sub-Objective Number and Title:
6 - Restore and Protect the Gulf of Mexico
Strategic Target Code and Title:
1 - Reduce releases of nutrients throughout the Mississippi River Basin
Managing Office:
1a. Performance Measure Term Definitions:
Restore water and habitat quality: 196.45 acres of habitat were restored. Water quality standards: The Gulf of Mexico Program supports the Gulf States' goal to return impaired waters to established designated uses and/or water quality standards. Impaired: The waterbody does not meet water quality standards or is threatened for one or more designated uses by one or more pollutants. Segments: 617 impaired segments removed, 30 restored, and 60 Total Mass Daily Loading (TMDL) determinations established. 13 priority coastal areas: There are 67 coastal watersheds at the 8-digit hydrologic unit code (HUC) scale on the Gulf Coast. The five Gulf States (Florida, Alabama, Mississippi, Louisiana, and Texas) identified 13 priority coastal areas to receive targeted technical and financial assistance for projects that restore impaired water quality. For the current reporting period, the 5 Gulf States have identified 617 specific water segments that are not meeting State water quality standards. Background: The EPA's Gulf of Mexico Program Office (GMPO; see http://www.epa.gov/gmpo/ reports on this performance measure. In FY 2007, this measure replaced measure GM-1, which tracked water and habitat quality in 12 priority coastal areas along the Gulf of Mexico. This measure tracks the number of impaired segments previously listed as not meeting water quality standards for a particular pollutant but are now de-listed from the current 303(d) report and meeting water quality standards. For more information on water quality throughout the United States, visit EPA's "Surf Your Watershed" website at http://cfpub.epa.gov/surf/locate/map2.cfm For more information on water quality in the Gulf of Mexico watersheds, visit EPA GMPO's "Surf Your Gulf Watershed" website at

The tables and maps generated for each water quality monitoring cycle are uploaded to the "Surf Your Gulf Watershed" website, and the website details the impaired segments for the Gulf Program's 13 priority areas, which are the focus of this measure.

2a. Original Data Source:

Gulf States (Texas, Louisiana, Mississippi, Alabama and Florida) provide data on the status of their impaired waterbody segments.

2b. Source Data Collection:

Collection Methodology and Quality Procedures: State-specific Decision Documents provide information on collection methodology and quality procedures, effectively acting as Quality Assurance Project Plans for the state 303(d) data. These decision documents can generally be found on the websites for EPA's Regional offices. A "shapefile" supports the development of maps which depict the status of a stream segment for either impaired, restored or TMDL established. These maps are informational and linked to their respective watershed area.

Geographical Extent: US regional (13 priority coastal areas along the Gulf of Mexico).

Spatial Detail: Gulf-wide; coastal 8-digit HUCs from Texas to Florida.

2c. Source Data Reporting:

Data Submission: One data submission instrument is biannual state 303(d) reports on the status of their impaired waterbody segments, as required under Clean Water Act (CWA) Section 305(b) and as determined by the TMDL schedule. Shapefiles with geospatial data related to 303(d) segments are also acquired from the relevant States and are used in the calculation of this measure.

Data Entry: Data is gathered from public sources (see 3. Information Systems and Data Quality Procedures).

Frequency and Timing of Data Transmission: States submit their 303(d) reports every two years. The EPA's Gulf of Mexico Program does not directly influence the frequency or timing of data transmission. The five Gulf States report data on a schedule coordinated with their respective Regional Office - either Region 4 (FL, MS and AL data) or Region 6 (LA and TX data).

3a. Relevant Information Systems:

System Description: EPA's "Surf Your Watershed" (see http://cfpub.epa.gov/surf/locate/map2.cfm and EPA's WATERS (Watershed Assessment Tracking and Environmental Results) Expert Query Tool (see http://www.epa.gov/waters/tmdl/expert_query.html are the databases for this performance measure.
ATTAINS (see http://www.epa.gov/waters/ir/

Surf Your Watershed: http://cfpub.epa.gov/surf/locate/map2.cfm

WATERS: The Watershed Assessment, Tracking, and Environmental Results System (WATERS) is used to provide water program information and display it spatially using a geographic information system integrated with several of EPA's existing databases. These databases include the STOrage and RETrieval (STORET) database, the Assessment TMDL Tracking and ImplementatioN System (ATTAINS), the Water Quality Standards Database (WQSDB), and the Grants Tracking and Reporting System (GRTS). This water quality information was previously available only from several independent and unconnected databases. Under WATERS, the Water Program databases are connected to a larger framework. This framework is a digital network of surface water features, known as the National Hydrography Dataset (NHD). By linking to the NHD, one Water Program database can reach another, and information can be shared across programs.

General information about WATERS is at http://www.epa.gov/waters/ A system architecture diagram is available at http://www.epa.gov/waters/about/arch.html Information about WATERS geographic data is available at http://www.epa.gov/waters/about/geography.html Information about tools that can be used with WATERS is at http://www.epa.gov/waters/tools/index.html For example, WATERS can be used to view information compiled from states' listings of impaired waters as required by Clean Water Act Section 303(d), which are recorded in the Assessment, TMDL Tracking, and Implementation System (ATTAINS). This information (found at http://iaspub.epa.gov/waters10/attains_nation_cy.control?p_report_type=T is used to generate reports that identify waters that are not meeting water quality standards ("impaired waters") and need one or more TMDLs to be developed.

Source/Transformed Data: Data is not "transformed"; it is extracted from the source material (State-developed Decision Documents, the WATERS Expert Query Tool, and "Surf Your Watershed") in order to monitor 13 priority watersheds around the Gulf of Mexico.

Information System Integrity Standards: The EPA Gulf of Mexico developed a "Quality Assurance Project Plan (QAPP) for the Gulf of Mexico 303(d) Priority Watershed Inventory Mapping", which was approved by EPA Region 4 on April 19, 2007.

3b. Data Quality Procedures:

To create the best report possible, three EPA sources are used to cross-reference the data. Each source is verified with the other two sources. The waterbodies listed as impaired in the Decision Documents for Florida, Alabama, and Mississippi are compared to "Surf Your Watershed" and then to the WATERS Expert Query Tool. Louisiana and Texas have a different form for the Decision Documents in that only the delisted water bodies are listed in the document. For these two states, "Surf Your Watershed" and WATERS Expert Query Tool are used. All the data is cross-referenced; "Surf Your Watershed" is cross referenced with WATERS and the Decision Documents, and WATERS are cross-referenced to the Decision Documents. It is pertinent that each of the sources matches, and no discrepancies in the listed impaired segments are found. No state documents are used in this process, since all state documents have to go through EPA review. Thus, the EPA sources used are a result of EPA reviewing the state documents.

3c. Data Oversight:

Source Data Reporting Oversight Personnel:

Source Data Reporting Oversight Responsibilities: Reference the approved QAPP. Information Systems Oversight Personnel: Chief of Staff. Information Systems Oversight Responsibilities: Reference the approved QAPP.

3d. Calculation Methodology:

Decision Rules for Selecting Data and Definitions of Variables: After all data are cross-referenced against each of the sources, tables are created for each watershed in the Gulf of Mexico Program's Priority Watershed Inventory. In all, 67 tables are created and populated with information obtained from "Surf Your Watershed". These tables include an ID number for the segment to view the location of the segment on the map, the segment ID with a link to "Surf Your Watershed", name of the state basin the segment is located within, the watershed the segment is located in, the name of the waterbody, the number of impairments for that segment, the impairments for that segment, and the year the impairment is listed. Delisting information is also listed in the tables for segments that have that information available. The information available in that table includes the ID number, the segment ID, the waterbody name, what impairment was delisted, the basis for the delisting, and a link to the TMDL document (if it exists). Segments shared among two or more watersheds are highlighted for easier recognition when counting the number of segments duplicated among watersheds.

Shapefiles are acquired from the states that contain the 303(d) segments for that state. Although the segments listed in the shapefile do not always match the documents that EPA provides ("Surf Your Watershed", WATERS Expert Query Tool, and Decision Documents). Therefore, it may be necessary to contact the state for additional shapefiles that contain other segments not available in the shapefile originally obtained from the state. The data is grouped by the watershed with a name to represent the area in the shapefile (ex. 2002_03170009_303d_line). New fields are added to the shapefile to provide meaningful data to the Gulf of Mexico Program Office. New fields include, ID number (which matches the number from the tables), TMDL status (Impaired Water Segment, TMDL Completed, Restored), Number of Impairments for that segment, List of Impairments for that segment, and the waterbody name for that segment. Maps are then generated for each watershed to show the number of impairments in each of the watersheds. Impaired Water Segments are visible with a red cross hatch, while a segment that has a TMDL completed would appear with a yellow cross hatch, and a Restored segment would appear with a blue cross hatch. Each segment is then labeled with the ID number found in the shapefile and table. All maps include the HUC number and name, the map, legend, scale bar, inset map, GMPO logo and a disclaimer for the state (if one was provided), and the date the map was created. In all, 67 maps are created.

Explanation of Calculations: The maps are generated by using the data contained in the shapefiles.

Explanation of Assumptions: Not applicable.

Unit of Measure: Number of impaired segments in each priority watershed.

| Timeframe of Result: Two-year cycle. Current reporting cycle is called "2012" but uses the 2010 reports, as the 2012 reports have yet to be released by the State agencies.

Documentation of Methodological Changes: Not applicable. |
|---|

4a. Oversight and Timing of Final Results Reporting:
Final Reporting Oversight Personnel:

Final Reporting Oversight Responsibilities: Specific duties are specified in the PARS Critical Element for [removed]. Duties include: tracking data and delisting performance for each Gulf State, reporting the data on the "Surf Your Watershed" website maintained by the Gulf of Mexico Program, and creating and posting color maps to visualize the impaired and/or delisted stream segments.

Final Reporting Timing: States report 303(d) data on a 2-year cycle. |

4b. Data Limitations/Qualifications:
General Limitations/Qualifications: No error estimate is available.

Data Lag Length and Explanation: Data is updated every two years on "Surf Your Watershed" and in the WATERS Expert Query Tool due to the fact that states submit a 303(d) report every two years of the status of the impaired segments in each state as required in Clean Water Act (CWA) 305(b) report.

Methodological Changes: Not applicable. |

4c. Third-Party Audits:
There are no outside reviews of the tables and maps used to calculate this performance measure.

Measure Code: wq3 - Improve water quality conditions in impaired watersheds nationwide using the watershed approach (cumulative).

Office of Water (OW)

Goal Number and Title:
2 - Protecting America's Waters

Objective Number and Title:
2 - Protect and Restore Watersheds and Aquatic Ecosystems

Sub-Objective Number and Title:
1 - Improve Water Quality on a Watershed Basis

Strategic Target Code and Title:
2 - Improve water quality conditions in 330 impaired watersheds nationwide using the watershed approach

Managing Office:
Office of Wetlands Oceans and Watersheds

1a. Performance Measure Term Definitions:
This measure demonstrates the capacity for watershed-scale restoration and incremental water quality improvement using the "watershed approach."

Improve water quality conditions: In 2002, 4,737 watersheds were listed as having 1 or more waterbodies impaired. For this measure, one or more of the waterbody/impairment causes identified in 2002 must be removed, as reflected in EPA-approved state assessments, for a) at least 40% of the impaired waterbodies or impaired stream miles/lake acres in the watershed; or b) if there is significant watershed-wide improvement, as demonstrated by valid scientific information, in one or more water quality parameters or related indicators associated with the impairments. Removal of an impairment cause means the original specific impairment cause listed by the state or EPA in 2002 is no longer impairing the waterbody, as reflected in subsequent state-submitted assessments and EPA-approved 303(d) lists.

Impaired watersheds: For purposes of this measure, watershed means (a) a hydrologic unit at the scale of 12-digit hydrologic unit codes, or HUC-12, as determined by the draft or final Watershed Boundary Dataset (WBD); or (b) a regionally-defined hydrologic unit of appropriate scale. Watersheds at this scale average 22 square miles in size. (The second definition includes waters, such as coastal and estuary waters, which fall outside the WBD, and may or may not be hydrologically definable at a scale comparable to inland HUC-12s.) Watersheds or hydrologic units at the 12-digit scale are technically termed "sub-watersheds" by the U.S. Geological Survey (USGS).

An impaired watershed is a watershed containing one or more impaired waterbodies. Impairment refers to an "impairment cause" in state- or EPA-reported data, stored in ATTAINS (Assessment Total Maximum Daily Load (TMDL) Tracking and Implementation System) or its predecessors NTTS (National TMDL Tracking System) or ADB (Assessment Database). (Any waterbody listed as impaired in these databases must have an impairment cause entered.) Any such removals of waterbody impairments will be recorded based on reports from states scheduled every two years through 2012. |

Watershed approach: This term refers to a coordinating process for focusing on priority water resource problems focused on hydrologically-defined areas. This type of approach involves key stakeholders and uses an iterative planning or adaptive management process to address priority water resource goals. It also uses an integrated set of tools and programs. Functionally, the watershed approach is a problem-solving tool for protecting water quality and aquatic resources. It recognizes that factors affecting the health of our nation's waters should be understood within their watershed context. It includes: assessment of relevant watershed hydrological and ecological processes, socioeconomic factors, identification of priority issues and most promising corrective actions, involvement by affected parties throughout the process, and implementation at the required scale. See EPA's website at http://water.epa.gov/type/watersheds/approach.cfm for more information. The watershed approach can be applied at any appropriate scale, including scales smaller or larger than the HUC-12 watersheds described above. Thus, for this measure, one watershed effort could result in improvements in one or in many HUC-12 watersheds, depending on its scale. For consistency, however, all successes under this measure will be reported as numbers of HUC-12 watersheds.

Nationwide: All 50 states.

Background:
This measure is reported as the cumulative number of the 2002-listed watersheds within the 4,737 watersheds; 39,503 water bodies were identified by states or EPA as not meeting water quality standards. The waterbodies and waterbody segments were identified in state-submitted Section 303(d) lists, Section 305(b) reports, and Integrated Reports, for the 2002 reporting cycle. (See EPA's guidance for reporting under "303(d) Listing of Impaired Waters Guidance" at http://www.epa.gov/owow/tmdl/guidance.html. Impairments and/or waterbodies identified after 2002 are not considered under this measure (such changes in scope may be considered when revising this measure for future updates of the Strategic Plan).

This measure is intended to establish and demonstrate a capacity for watershed-scale restoration and protection throughout the country using the "watershed approach." It is not designed to be a measure of what portion of the 12-digit watersheds in the country have improved or meet water quality standards.

2a. Original Data Source:

Regional EPA water quality staff make the determinations about whether an individual watershed meets the criteria for this measure.

2b. Source Data Collection:

For a watershed to be counted under SP-12, the state and Region must demonstrate that the watershed approach was applied and that water quality improved. Results and documentation of evidence that water quality conditions in an impaired watershed have improved, based on the watershed approach, must be reviewed and approved by the Regional office.

EPA's assessment of incremental improvements of water quality conditions utilizes (1) information on impairments from the 2002 303(d) list, described above; (2) 12-digit hydrologic unit code (HUC) boundaries (in 2009, boundaries and data on 12-digit HUC code watersheds were completed, certified and stored on USDA's comprehensive website for HUC watershed information - see http://www.ncgc.nrcs.usda.gov/products/datasets/watershed/index.html); and (3) data and/or information on "watershed-wide water quality improvement" relative to the identified pollutant or response indicator listed as the impairment.

An individual watershed may be counted only once under this measure. That is, a watershed may be counted only when it initially meets the definition. Subsequent actions/restoration efforts resulting in additional impairment causes removed or additional water quality parameters showing water-wide improvements do not enable the watershed to be counted again in subsequent reporting periods.

2c. Source Data Reporting:

When reporting on this measure, the region must provide the following information as demonstration of the watershed approach:
·Applicable HUC-12 or defined watershed(s)
·Key stakeholders involved and role of each
·Description of watershed plan developed and how it was implemented
·Description of restoration activities
·Documentation of results

The guidance for this measure describes two options for meeting the two SP-12 definitions.

Option 1. Reporting Watershed Improvement Based on Impairment Removal. This option corresponds to the first definition of improvement under this measure. The region must demonstrate that the removal of impairment causes meets the 40% threshold. That is, one or more waterbody impairment causes identified in 2002 are removed for at least 40% of the impaired waterbodies or stream miles/lake acres in the watershed. Option 1 is perhaps the most rigorous of the three options.

Option 2. Reporting Watershed-wide Improvement Based on Monitoring. This option corresponds to the second definition of improvement under this measure. It utilizes water quality monitoring data to track improvements occurring across the watershed that have not yet resulted in an impairment being removed. Examples of various monitoring designs are given as part of the guidance. A region may opt to use a statistical approach (option 2a) or multiple-lines-of-evidence approach (option 2b) when documenting the measure.

Results reported under this measure must be provided using a standardized template. When reporting results to ACS, the Region must submit the template immediately or within 45 days after entering results in ACS. Headquarters will provide an electronic storage location for the templates. Currently this location is the EPA Portal at http://portal.epa.gov/ using the Watershed Managers Forum project in the Environmental Science Connector (ESC). This location may change in the future. The regions post their templates to the ESC site and notify Christopher Zabawa at Zabawa.Christopher@epa.gov by e-mail.

Detailed information on meeting the criteria and reporting for this measure is contained in "Guidance for Reporting Watershed Improvement under Measure SP-12 – FY 2009" (December 2008), found at http://water.epa.gov/resource_performance/planning/FY-2012-NWPG-Measure-Definition/cfm - Measure%20Code_%WQ_SP12_N11

3a. Relevant Information Systems:

The Assessment and Total Maximum Daily Load (TMDL) Tracking And ImplementatioN System (ATTAINS) is the database which captures water quality information related to this measure. ATTAINS is an integrated system that documents and manages state assessment and listing/delisting decisions reported under sections 305(b) and 303(d) (i.e., integrated reporting). This system holds information about assessment decisions and restoration actions across reporting cycles and over time until water quality standards are attained More information about ATTAINS can be found at http://www.epa.gov/waters/data/prog.html andhttp://www.epa.gov/waters/ir/about_integrated.html

The Watershed Assessment, Tracking, and Environmental Results System (WATERS) is used to provide water program information and display it spatially using a geographic information system (National Hydrography Dataset (NHD)) integrated with several of EPA's existing databases. These databases include the STOrage and RETrieval (STORET) database, ATTAINS, the Water Quality Standards Database (WQSDB), and the Grants Tracking and Reporting System (GRTS). General information about WATERS is available at: http://www.epa.gov/waters/ , and a system architecture diagram is available at http://www.epa.gov/waters/about/arch.html Information about WATERS geographic data is available at http://www.epa.gov/waters/about/geography.html

3b. Data Quality Procedures:

Managers in the Water Management Divisions in EPA Regional Offices have the responsibility for oversight, review, and quality assurance of the performance data for this measure.

Periodically the results and documentation for at least one submission from each Region will be reviewed by an EPA SP-12 Review Panel. The Panel will consist of at least two reviewers from Regions other than the reporting Region and at least one reviewer from EPA Headquarters. The Review Panel will recommend whether to accept the watershed(s) to be counted and may develop recommendations for improving the submission to ensure consistency.

3c. Data Oversight:

Source data reporting and oversight are the responsibility of staff personnel and managers in the Water Management Divisions of EPA's Regional Offices. Information systems data entry into ATTAINS is performed by the states with EPA regional and headquarters oversight.

3d. Calculation Methodology:

Criteria for selecting which of the 2002-listed watersheds to focus on is at the discretion of the Region water program offices. Results are reported as the cumulative number of 12-digit HUC watersheds that meet the definition of the measure.

4a. Oversight and Timing of Final Results Reporting:

The Associate Division Director of the Assessment and Watershed Protection Division is responsible for overseeing final reporting of this measure. The ADD reports to the Office Director of the Office of Wetlands,

Oceans, and Watersheds.

4b. Data Limitations/Qualifications:

The rate of regional progress varies with some regions making significant strides and others scoring small successes. The main factors in this discrepancy primarily relate to a combination of complex watersheds and the reduction in resources at both the state and federal level (e.g., 319 grant funds). For example, watersheds in the mid-Atlantic states tend to be interstate or otherwise multi-jurisdictional, urban or in highly developed/populated areas, and almost always involve many diverse stakeholders. This coupled with stretched budgets further increases the difficulty of attaining significant water quality improvements. As a result, these regions' annual rate of progress is often less than other regions, where monitoring and active restoration efforts are more straightforward.

4c. Third-Party Audits:

Over the past decade or so, independent reports have cited the ways in which weaknesses in monitoring and reporting of monitoring data undermine EPA's ability to depict the condition of the Nation's waters and adequately track water quality improvements. The most recent report on this subject is: USEPA, Office of the Inspector General. 2007. Total Maximum Daily Load Program Needs Better Data and Measures to Demonstrate Environmental Results. Available at: http://www.epa.gov/oig/reports/2007/20070919-2007-P-00036.pdf .

In response to these evaluations, EPA has been working with states and other stakeholders to improve 1) data coverage, so that state reports reflect the condition of all waters of the state; 2) data consistency to facilitate comparison and aggregation of state data to the national level; and 3) documentation so that data limitations and discrepancies are fully understood by data users. EPA has taken several steps in an effort to make these improvements:

First, EPA enhanced two existing data management tools (STORET and ADB) so that they include documentation of data quality information.

Second, EPA has developed the GIS tool WATERS that integrates many databases including STORET, ATTAINS, and a water quality standards database. These integrated databases facilitate comparison and understanding of differences among state standards, monitoring activities, and assessment results.

Third, EPA has developed several guidance documents. In 2005, EPA issued guidance using and reporting on Statewide Statistical Survey Data in ATTAINS, and in 2008, the Agency issued Elements of a State Water Monitoring and Assessment Program. These guidance documents are available athttp://www.epa.gov/owow/tmdl/guidance/final52009.html and http://www.epa.gov/owow/monitoring/elements/ respectively.

Measure Code: 202 - Acres protected or restored in National Estuary Program study areas.

Office of Water (OW)

Goal Number and Title:
2 - Protecting America's Waters
Objective Number and Title:
2 - Protect and Restore Watersheds and Aquatic Ecosystems
Sub-Objective Number and Title:
2 - Improve Costal and Ocean Waters
Strategic Target Code and Title:
3 - Protect or restore an additional 600,000 acres of habitat
Managing Office:
Office of Wetlands Oceans and Watersheds
1a. Performance Measure Term Definitions:

Acres of habitat: "Habitat" means aquatic and terrestrial areas within the NEP study area. For purposes of this measure, "Habitat Acres Restored and Protected" encompasses a range of activities and is interpreted broadly to include: creation of habitat, acquisition of sites for the purpose of protection, conservation easements and deed restrictions, increasing submerged aquatic vegetation coverage, increasing the number of permanent shellfish bed openings, and increasing the amount of anadromous fish habitat. Habitat acreage serves as an important surrogate and a measure of on-the-ground progress made toward EPA's annual performance goal of habitat protection and restoration in the NEP.

Protected: "Protect" refers to preserving areas through acquisition, conservation easements, deed restrictions, etc.

Restored: "Restore" refers to the return of habitat to a close approximation of its prior condition.

National Estuary Program (NEP) study areas: An estuary is a partially enclosed body of water along the coast where freshwater from rivers and streams meet and mix with salt water from the ocean. The National Estuary Program (NEP) includes 28 estuaries in EPA Regions 1, 2, 3, 4, 6, 9 and 10. EPA provides funding to independent National Estuary Programs for each of those estuaries. For more information about NEP, go to:
http://water.epa.gov/type/oceb/nep/index.cfm

Study areas: The NEP Study Areas include the estuary and adjacent watersheds that could impact the water quality and ecological integrity of the estuary; these are the areas that the NEPs focus on. For a graphical display of the 28 estuaries, visit:
http://water.epa.gov/type/oceb/nep/upload/NatGeo_24x36_final_revised.pdf

Background:
The Office of Wetlands, Oceans, and Watersheds has developed a standardized nomenclature for defining habitat protection and restoration activities

(http://www.epa.gov/owow_keep/estuaries/pivot/habitat/gpra_def.htm) and specifying habitat categories (http://www.epa.gov/owow_keep/estuaries/pivot/habitat/habtype.htm)

2a. Original Data Source:

The 28 National Estuary Programs

2b. Source Data Collection:

Collection Methodology: Primary data are prepared by staff in each NEP based on their own reports and on data provided by partner agencies/organizations that directly engage in habitat protection and restoration activities. NEP documents such as annual work plans, which report on NEP achievements during the previous year, annual progress reports, State of the Bay reports, and implementation tracking materials document the number of acres of habitat restored and protected. EPA has defined and provided examples of protection and restoration activities for purposes of tracking and reporting associated with these measures at the website for the agency's Performance Indicators Visualization and Outreach Tool (PIVOT):

http://www.epa.gov/owow_keep/estuaries/pivot/habitat/hab_fr.htm

Geographical Extent: The study areas of the 28 National Estuary Programs vary from Program to Program. Some are less than 100 square miles, while others are several thousand square miles. For a graphical display of the 28 estuaries, visit:

http://water.epa.gov/type/oceb/nep/upload/NatGeo_24x36_final_revised.pdf

Spatial Detail: NEPs provide latitude and longitude data (where possible) for each protection and restoration project.

Quality Procedures: EPA requests that the NEPs follow EPA guidance to prepare their reports, and to verify the numbers. See "Frequently asked NEPORT Questions" document for more information.

Attached Documents:

Frequently Asked NEPORT Questions 6-21.docx

2c. Source Data Reporting:

Each NEP reports data to the respective EPA regional office. NEPs and EPA track habitat projects using a standardized format for data reporting and compilation, defining habitat protection and restoration activities and specifying habitat categories that the Office of Wetlands Oceans and Watersheds has developed. On or about September 1 each year, the NEPs enter their habitat data into the National Estuary Program On-line Reporting Tool (NEPORT), an online reporting system/database that is managed by EPA. NEPORT is an internal database intended for NEPs use only. Members of the general public do not have access to NEPORT.

Attached Documents:

Frequently Asked NEPORT Questions 6-21.docx

3a. Relevant Information Systems:

System Description:

NEPORT. The National Estuary Program On-Line Reporting Tool (NEPORT) is a web-based database that EPA's Office of Wetlands Oceans and Watersheds developed. NEPORT was developed for National Estuary Programs (NEPs) to submit their annual Habitat and Leveraging reports. http://gispub2.epa.gov/NEPMap/index.html NEPORT was developed by the Office of Wetlands Oceans and Watersheds as a standardized format for data reporting and compilation, defining habitat protection and restoration activities and specifying habitat categories.

NEPORT was intended to reduce the reporting burden on NEPs and the time required for quality assurance and quality control. Starting in FY06, NEPs were required to submit their Habitat and Leveraging reports through NEPORT. NEPORT replaces the prior data reporting protocols in which EPA distributed Habitat and Leveraging forms to NEPs and NEPs completed the forms and submitted them to EPA. Through NEPORT, NEPs are able to download Habitat and Leveraging reports into Microsoft Excel, create pie charts and save them in bitmap format, access data on a secure web site, check report status, and search for NEP staff contact information. At the same time, EPA is able to store NEP data on a centralized database and receive e-mail reports on newly submitted data.

For more information about NEPORT, see http://www.epa.gov/owow_keep/estuaries/neport/index.html

PIVOT. The Performance Indicators Visualization and Outreach Tool (PIVOT) is a reporting tool that visually communicates NEP progress toward protecting and restoring habitat to a wide range of stakeholders and decision makers. It can display aggregate national and regional data for this measurement, as well as data submitted by each NEP. The website highlights habitat loss/alteration, as well as the number of acres protected and restored by habitat type. Data can be displayed numerically, graphically, and by habitat type. PIVOT data are publicly available at http://www.epa.gov/owow_keep/estuaries/pivot/habitat/hab_fr.htm

Source/Transformed Data: Data originates from the NEPs.

Attached Documents:
Frequently Asked NEPORT Questions 6-21.docx

3b. Data Quality Procedures:
Each year, after the data has been entered by the NEPs, the regions complete a QA/QC review within two weeks, to validate the habitat data. For projects where the NEPs provide latitude and longitude data, these data are mapped. Precisely identifying project sites helps to highlight where projects are located in each NEP study area. It also makes it possible for NEPs and EPA to validate NEPORT data, and highlights where different partners may be double counting acreage. This QA/QC may include circling back to a NEP requesting that they redo their submission before the Region "approves" the data.

After Regional review, EPA Headquarters (HQ) conducts a brief examination to finalize and approve all the data 2 weeks after Regional approval. In the process, EPA confirms that the national total accurately reflects the information submitted by each program.

EPA is confident that the annually-reported data are as accurate as possible. EPA actions are consistent with data quality and management policies. The Office of Water Quality Management Plan (July 2002) is available on the Intranet at http://intranet.epa.gov/ow/informationresources/quality/qualitymanage.html

Risk Management Procedures: EPA conducts regular reviews of NEP implementation to help ensure that information provided in NEP documents is accurate, and progress reported is in fact being achieved. EPA's triennial NEP program evaluations include a review of the data reported by the NEPs' over the three year period. Reporting in FY 2007 through FY 2009 did not indicate that any improvements to any of the databases associated with this measure were needed. For information on how the evaluations are conducted, please see EPA's September 28, 2007, National Estuary Program Evaluation Guidance:
http://water.epa.gov/type/oceb/nep/upload/2009_03_26_estuaries_pdf_final_guidance_sept28.pdf

3c. Data Oversight:
Source Data Reporting Oversight Personnel: Headquarters' NEP Coordinator Source Data Reporting Oversight Responsibilities: Reviews the submitted habitat acres data and conducts a QA/QC of the NEP projects. Information Systems Oversight Personnel: Headquarters' NEP Coordinator Information Systems Oversight Responsibilities: Reviews the submitted habitat acres data and conducts a QA/QC of the NEP projects.
3d. Calculation Methodology:
Decision Rules for Selecting Data: The key field used to calculate annual performance is habitat acreage. Definitions of Variables: Not applicable. Explanation of Calculations: After EPA Regional Offices and HQ staff validate individual NEP totals, EPA HQ aggregates the selected acreage data provided by each NEP to arrive at a national total for all 28 estuaries in the NEP. Explanation of Assumptions: Not applicable. Unit of Measure: Acres Timeframe of Result: Regions report the data in early September, a QA/QC is then conducted which takes approximately one month.

Documentation of Methodological Changes: Not applicable.

4a. Oversight and Timing of Final Results Reporting:
Final Reporting Oversight Personnel: Environmental Protection Specialist Final Reporting Oversight Responsibilities: Review the habitat acres reported by each of the Regions and conduct a QA/QC of each of the projects for accuracy. Final Reporting Timing: Annual

4b. Data Limitations/Qualifications:
General Limitations/Qualifications: Current data limitations include: (1) information that may be reported inconsistently across the NEPs because they may interpret the meaning of "protection and restoration" differently; (2) acreage amounts may be miscalculated or incorrectly reported, and (3) acreage may be double-counted (i.e., the same parcel may also be counted by more than one partner, or the same parcel may be counted more than once because it has been restored several times over a period of years). Also habitat restored, improved, and protected may not directly correlate to overall improvements in the health of that habitat (particularly in the year of reporting); rather, habitat acreage protected and restored is only one indicator of habitat health and of on-the-ground progress made by the NEPs. Data Lag Length and Explanation: Data lag time is approximately one month, from the time it is submitted to the time it is approved. Methodological Changes: None

4c. Third-Party Audits:
Not applicable

Measure Code: 4G - Number of acres restored and improved, under the 5-Star, NEP, 319, and great waterbody programs (cumulative).

Office of Water (OW)

Goal Number and Title:
2 - Protecting America's Waters
Objective Number and Title:
2 - Protect and Restore Watersheds and Aquatic Ecosystems
Sub-Objective Number and Title:
3 - Increase Wetlands
Strategic Target Code and Title:
1 - Working with partners, achieve a net increase of wetlands nationwide
Managing Office:
Office of Wetlands Oceans and Watersheds
1a. Performance Measure Term Definitions:
Wetlands: As defined by this measure use the biological definition, Cowardin et al. (1979). This classification system for wetlands became a U.S. Fish and Wildlife Service Standard (1980) as well as the Federal Geographic Data Committee standard for wetlands monitoring and reporting (December 17, 1996). The Cowardin et al definition indicates that wetlands must have one or more of the following three attributes: 1) at least periodically, the land supports predominantly wetland or hydrophytic plants; 2) predominantly undrained hydric soils; and 3) the substrate is non-soil and is saturated with water or covered by shallow water at some time during the growing season of each year. This means that areas that fall into one of the following five categories are considered wetlands for the purpose of this report: 1) areas with hydrophytic plants and with hydric soils, 2) areas without hydrophytic plants but with hydric soils such as mudflats, 3) areas with hydrophytic plants but non-hydric soils which include areas in which hydric soils have not yet developed, 4) areas without soils but with hydrophytic plants such as seaweed covered portions of rocky shores; and 5) areas without soil and without hydrophytic plants such as gravel beaches and rocky shores without vegetation. Restored: "Restore or create" wetlands result in a gain of wetland acres and includes:
a. Creation of wetland that did not previously exist on an upland or deepwater site. These actions are referred to as "establishment" by the White House Wetlands Working Group (WHWWG). "Establishment" is the manipulation of the physical, chemical, or biological characteristics present to develop a wetland on an upland or deepwater site, where a wetland did not previously exist. Establishment results in a gain in wetland acres.
b. Restoration of a former wetland to natural/historic functions and resulting value. Typically, such a former wetland had been drained for some purpose. These actions are known as "re-establishment" by the WHWWG. "Re-establishment" is the manipulation of the physical, chemical, or biological characteristics of a site with the goal of returning natural or historic functions to a former wetland. Re-establishment results in rebuilding a former wetland and results in a gain in wetland acres.
Improved: "Improve" wetlands results in a gain of wetlands function or quality, rather than additional acreage, and includes:
a. Repair of the natural/historic functions and associated values of a degraded wetland. The WHWWG refers

to these actions as "rehabilitation" of wetlands. "Rehabilitation" is the manipulation of the physical, chemical, or biological characteristics of a site with the goal of repairing natural or historic functions of a degraded wetland. Rehabilitation results in a gain in wetland function but does not result in a gain in wetland acres.
 b. Heightening, intensification, or improvement of one or more selected
functions and associated values. The WHWWG called these types of actions "enhancement." Enhancement is undertaken for a purpose such as water quality improvement, flood water retention, or wildlife habitat. "Enhancement" is the manipulation of the physical, chemical, or biological characteristics of a wetland (undisturbed or degraded) site to heighten, intensify, or improve specific function(s) or to change the growth stage or composition of the vegetation present. Enhancement is undertaken for specified purposes such as water quality improvement, flood water retention, or wildlife habitat. Enhancement results in a change in wetland function(s) and can lead to a decline in other wetland functions, but does not result in a gain in wetland acres. This term includes activities commonly associated with enhancement, management, manipulation, and directed alteration.

5-Star: This National Fish and Wildlife Foundation program provides modest financial assistance on a competitive basis to support community-based wetland, riparian, and coastal habitat restoration projects that build diverse partnerships and foster local natural resource stewardship through education, outreach and training activities. NFWF is a 501(c)(3) non-profit that preserves and restores our nation's native wildlife species and habitats. The organization was created by Congress in 1984. In addition to EPA, major funding is provided by FedEx, Pacific Gas & Electric's Nature Restoration Trust and Southern Company. For more information, see EPA's website for the program (http://www.epa.gov/owow/wetlands/restore/5star/ and NFWF's website for the program
(http://www.nfwf.org/AM/Template.cfm?Section=Charter_Programs_List&Template=/TaggedPage/TaggedPageDisplay.cfm&TPLID=60&ContentID=17901

NEP: EPA's National Estuary Program (NEP) was established by Congress in 1987 to improve the quality of estuaries of national importance. The National Estuary Program (NEP) includes 28 estuaries in EPA Regions 1, 2, 3,4,6,9, and 10. For more information, go to: http://water.epa.gov/type/oceb/nep/index.cfm

319: The 1987 amendments to the Clean Water Act (CWA) established the Section 319 Nonpoint Source Management Program. Section 319 addresses the need for greater federal leadership to help focus state and local nonpoint source efforts. Under Section 319, states, territories and tribes receive grant money that supports a wide variety of activities including technical assistance, financial assistance, education, training, technology transfer, demonstration projects and monitoring to assess the success of specific nonpoint source implementation projects. Grant recipients have the option to enter information about whether the project affects wetlands and to indicate the number of acres restored, improved, or protected.

Great Water Body Program: The Great Water Body Programs include: the Chesapeake Bay Program Office located in Region 3, the Great Lakes Program Office located in Region 5, the Gulf of Mexico Program Office located in Region 4.

Cumulative: The baseline for this measure is FY 2006, when EPA reported that 58,777 acres of wetland were restored and improved through the Five Star Restoration Grants, the National Estuary Program, Section 319 Nonpoint Source Grants, Brownfield Grants, and EPA Great Water Body Programs.

Background:

- From 1986-1997, the U.S. had an annual net wetland loss of an estimated 58,500 acres, as measured by the U.S. Fish and Wildlife Service. From 1998-2004, the U.S. achieved a net cumulative increase of 32,000 acres per year of wetlands, as measured by the U.S. Fish and Wildlife Service.

- A number of national programs include efforts to restore and improve wetlands. These acres may include those supported by the Wetland Five Star Restoration Grants, the National Estuary Program, Section 319 Nonpoint Source (NPS) Grants, Brownfield grants, or EPA's Great Water Body Programs. This does not include enforcement or mitigation acres. This measure is shared with other offices including: EPA Office of Wetlands, Oceans, and Watersheds Divisions, EPA Office of Solid Waste and Emergency Response (OSWER) Brownfields Office, EPA Gulf of Mexico Program Office, EPA Great Lakes National Program Office, and the Chesapeake Bay Program Office.

- National Estuary Program (NEP): The Office of Wetlands, Oceans, and Watersheds (OWOW) has developed a standardized nomenclature for defining habitat protection and restoration activities (http://www.epa.gov/owow_keep/estuaries/pivot/habitat/gpra_def.htm and specifying habitat categories (http://www.epa.gov/owow_keep/estuaries/pivot/habitat/habtype.htm Additional information regarding habitat protection is accessible on a web page that highlights habitat loss/alteration, as well as the number of acres protected and restored by habitat type (http://www.epa.gov/owow_keep/estuaries/pivot/habitat/hab_fr.htm The website visually communicates NEP progress toward protecting and restoring habitat to a wide range of stakeholders and decision makers.

2a. Original Data Source:

5-Star: National Fish and Wildlife Foundation (NFWF)

NEP: The 28 National Estuary Programs funded by EPA. The National Estuary Program (NEP) includes 28 estuaries in EPA Regions 1, 2, 3, 4, 6, 9, and 10. For more information about NEP, go to: http://water.epa.gov/type/oceb/nep/index.cfm

319: State agencies that are grant recipients for wetlands projects from State NPS Management Programs and Section 319 funded work programs.

Great Water Body Program: The Great Water Body Programs include and restoration or improvement of wetland resources through: the Chesapeake Bay Program Office located in Region 3, the Great Lakes Program Office located in Region 5, the Gulf of Mexico Program Office located in Region 4. Acreage data from these programs have not been reported under this measure because of their initial inability to provide timely information starting in 2004 when the measure was initiated.

2b. Source Data Collection:

Collection Methodology and Quality Procedures:
5-Star Program: The National Fish and Wildlife Foundation (NFWF), EPA's 5-Star grantee, maintains a subgrant outcome tracking system that tracks the acres of wetlands enhanced, established, or re-established, miles of

riparian buffer restored, and other information such as number of volunteers engaged in restoration activities. 5-Star data entered by grantee, the National Fish and Wildlife Foundation, and the National Association of Counties from annual and final reports from subgrantees into the common grantee managed database. Subgrantees will report the number of acres of wetlands by habitat restoration and improvement activity type from their annual and final reports. EPA has defined and provided examples of protection and restoration activities for purposes of tracking and reporting associated with these measures. Subgrantees determine the number of acres they have restored or improved using hand held GPS units and estimating acreage from those GPS points. Subgrantees provide acres effect and a description of the activities on those acres. EPA then double-checks and determines final restoration or improvement designations for those acres from the description provided for each project.

NEP: Primary data are prepared by staff in each NEP based on their own reports and on data provided by partner agencies/organizations that directly engage in habitat protection and restoration activities. NEP documents such as annual work plans, which report on NEP achievements during the previous year, annual progress reports, State of the Bay reports, and implementation tracking materials document the number of acres of habitat restored and protected. EPA has defined and provided examples of protection and restoration activities for purposes of tracking and reporting associated with these measures at the website for the agency's Performance Indicators Visualization and Outreach Tool (PIVOT): http://www.epa.gov/owow_keep/estuaries/pivot/habitat/hab_fr.htm   EPA requests that the NEPs follow EPA guidance to prepare their reports, and to verify the numbers.

Section 319 Grants:
States have continual access and opportunity to review the information in GRTS to ensure it accurately reflects the data they entered (according to their QA procedures).
·   Nonpoint Source Program and Grants Guidelines for States and Territories. October 23, 2003 (http://www.epa.gov/OWOW/NPS/cwact.html

Great Water Body Program: Acreage data from these programs have not been reported under this measure because of their initial inability to provide timely information starting in 2004 when the measure was initiated.

Geographical Extent: The study areas of the 28 National Estuary Programs. For a graphical display of the 28 estuaries, visit:
http://water.epa.gov/type/oceb/nep/upload/NatGeo_24x36_final_revised.pdf For 5-Star and 319 the study areas are found national-wide. For 5-Star visit: http://water.epa.gov/grants_funding/wetlands/restore/index.cfm

Spatial Detail: NEPs and 5-Star projects provide latitude and longitude data (where possible) for each protection and restoration project. 319 projects provide state, county, township data and will also provide latitude and longitude data.

2c. Source Data Reporting:

5-Star: NFWF provides to EPA annual documentation of acres of wetlands acreage enhanced, established, or re-established and stream miles buffered and/or restored during the life of the cooperative agreement in accordance with OWOW requirements. Data for this measure are kept in the Wetlands Program's Five-Star Restoration Grant Database. For the next four years NFWF will be providing EPA information on or around September 30 through their annual grant report.

NEP: Each NEP reports data to the respective EPA regional office. NEPs and EPA track habitat projects using a standardized format for data reporting and compilation, defining habitat protection and restoration activities and specifying habitat categories that the Office of Wetlands Oceans and Watersheds has developed. On or about September 1 each year, the NEPs enter their habitat data into the National Estuary Program On-line Reporting Tool (NEPORT), an online reporting system/database that is managed by EPA. NEPORT is an internal database intended for NEPs use only. Members of the general public do not have access to NEPORT.

Section 319 Grants: As part of the basic reporting requirements specified by CWA section 319(h), EPA requires reporting through the section 319 Grants Reporting and Tracking System (GRTS). States are encouraged to attach final project reports completed under their grants to the Project Evaluation field in GRTS. States also enter, if applicable, if the project affects wetlands (an optional field) and indicates the number of acres restored, improved, or protected.
·   USEPA. Modifications to Nonpoint Source Reporting Requirements for Section 319 Grants. September 27, 2001.

Great Water Body Program: The Great Water Body programs have not submitted any data for this measure since its inception in 2004 when the measure was initiated.

3a. Relevant Information Systems:
System Description: 5-Star: Five-Star Restoration Grant Database. Data for this measure are kept in the Five-Star Restoration Grant Database. NFWF launched a new, paperless grants management system in 2008 and Five Star subgrants awarded under the current cooperative agreement will be managed using this system. The system allows NFWF to evaluate both the quantitative and qualitative outcomes for individual subgrants and attribute individual projects to the attainment of overall programmatic outcomes. Managing the grants includes overseeing the completion of restoration and training projects and collecting regular financial and programmatic updates from grantees. NFWF also has populated its web-based Grants Library with grant files and subgrant outcomes (final project reports) for all grant programs across the country. Five Star subgrants have been and will continue to be integrated in to this online, browser-based, publically searchable database. NEP: NEPORT. The National Estuary Program On-Line Reporting Tool (NEPORT) is a web-based database that EPA's Office of Wetlands Oceans and Watersheds developed. NEPORT was developed for National Estuary Programs (NEPs) to submit their annual Habitat and Leveraging reports. http://gispub2.epa.gov/NEPMap/index.html NEPORT was developed by the Office of Wetlands Oceans and Watersheds as a standardized format for data reporting and compilation, defining habitat protection and restoration activities and specifying habitat categories.

NEPORT was intended to reduce the reporting burden on NEPs and the time required for quality assurance and quality control. Starting in FY06, NEPs were required to submit their Habitat and Leveraging reports through NEPORT. NEPORT replaces the prior data reporting protocols in which EPA distributed Habitat and Leveraging forms to NEPs and NEPs completed the forms and submitted them to EPA. Through NEPORT, NEPs are able to download Habitat and Leveraging reports into Microsoft Excel, create pie charts and save them in bitmap format, access data on a secure web site, check report status, and search for NEP staff contact information. At the same time, EPA is able to store NEP data on a centralized database and receive e-mail reports on newly submitted data. For more information about NEPORT, see
http://gispub2.epa.gov/NEPMap/index.html

PIVOT. The Performance Indicators Visualization and Outreach Tool (PIVOT) is a reporting tool that visually communicates NEP progress toward protecting and restoring habitat to a wide range of stakeholders and decision makers. It can display aggregate national and regional data for this measurement, as well as data submitted by each NEP. The website highlights habitat loss/alteration, as well as the number of acres protected and restored by habitat type. Data can be displayed numerically, graphically, and by habitat type. PIVOT data are publicly available at http://www.epa.gov/owow_keep/estuaries/pivot/habitat/hab_fr.htm

319: GRTS. The Grants Reporting and Tracking System (GRTS) is the primary tool for management and oversight of the EPA's Nonpoint Source (NPS) Pollution Control Program. GRTS is used by grant recipients (State agencies) to supply information about State NPS Management Programs and annual Section 319 funded work programs, which include wetlands and stream restoration and improvement projects. GRTS pulls grant information from EPA's centralized grants and financial databases and allows grant recipients to enter detailed information on the individual projects or activities funded under each grant.
GRTS also provides EPA and other stakeholders greater and more efficient access to data, information, and program accomplishments than would otherwise be available. GRTS provides detailed georeferencing (i.e., National Hydrography Dataset – or "NHD"-- reach addresses) for 319-funded projects, project cost information, load reduction information, and a host of other elements. For more information:
·   Users Guide: USEPA. GRTS. Grants Tracking and Reporting System. GRTS Web User Guide, Version 1.6 March 15, 2007. USEPA.
·   More information about GRTS is at: http://iaspub.epa.gov/pls/grts/f?p=110:199:3920887085074706

Great Water Body Program: Acreage data from the Great Water Body Programs have not been reported under this measure because of their initial inability to provide timely information starting in 2004 when the measure was initiated. Since then the Great Lakes and Chesapeake Bay programs have developed or in the process of developing databases to collect restoration data under their grant programs.

Source/Transformed Data: All databases listed above contain original source data.

Information System Integrity Standards: The NEW PIVOT and the 319 GRTS data systems are both managed to the relevant EPA standards for information systems integrity including the IT Security policy. The 5-Star data system is managed by an EPA grantee, NFWF, and managed using their data security standards. Acreage data from the Great Water Body Programs have not been reported under this measure because of their initial inability to provide timely information starting in 2004 when the measure was initiated.

3b. Data Quality Procedures:

5-Star:EPA is confident that the annually-reported data are as accurate as possible. Any data collected by NFWF will require all subawards use standard reporting templates and data standards to assist the Foundation in meeting all EPA requirements and to ensure data compatibility with OWOW standards. Five Star projects are generally small restoration projects and do not collect sufficient scientific data warranting extensive QA/QC protocols be employed. Documentation of quality control procedures or any observed QA/QC problems will be included as a component in existing reporting requirements and will serve as the equivalent documentation under the EPA's current QA/QC policy. Specific quality control elements that are to be included in the annual reports include: quantity of data, documentation of how and from whom any data will be obtained, (including secondary data and constraints on the data collection process). In addition NFWF will include in their reporting any specific QA/QC activities that will be conducted during data collection that includes how project data will be analyzed, evaluated and data validation procedures for the reporting period if any data collection has occurred.

NEP: EPA is confident that the annually-reported data are as accurate as possible. Each year, after the data has been entered by the NEPs, the regions complete a QA/QC review within two weeks, to validate the habitat data. For projects where the NEPs provide latitude and longitude data, these data are mapped. Precisely identifying project sites helps to highlight where projects are located in each NEP study area. It also makes it possible for NEPs and EPA to validate NEPORT data, and highlights where different partners may be double counting acreage. This QA/QC may include reporting back to a NEP requesting that they redo their submission before the Region "approves" the data. After Regional review, EPA Headquarters (HQ) conducts a brief examination to finalize and approve all the data 2 weeks after Regional approval. In the process, EPA confirms that the national total accurately reflects the information submitted by each program.

EPA conducts regular reviews of NEP implementation to help ensure that information provided in NEP documents is accurate, and progress reported is in fact being achieved. EPA's triennial NEP program evaluations include a review of the data reported by the NEPs' over the three year period. Reporting in FY 2007 through FY 2009 did not indicate that any improvements to any of the databases associated with this measure were needed. For information on how the evaluations are conducted, please see EPA's September 28, 2007, National Estuary Program Evaluation Guidance:
http://water.epa.gov/type/oceb/nep/upload/2009_03_26_estuaries_pdf_final_guidance_sept28.pdf

319: EPA Regions and Headquarters staff periodically review data entered in GRTS and remind states of the critical importance of their completing mandated data elements in a timely, high-quality manner. Regional personnel also maintain hardcopies of the states work programs, watershed project implementation plans, and Annual Progress Reports. Verification of data in GRTS can be cross-checked with these documents to ensure quality, consistency, and reliability in progress reporting on an incremental (such as, year-to-year) basis, or to note any problems in data quality in GRTS. EPA frequently reviews various aggregation(s) of all the data in GRTS by our use of "ad-hoc" and standard reports available in the GRTS reporting system. The agency sponsors national GRTS-users group meetings each year. These meetings serve not only to meet the training needs of the user community, but also provide a forum for discussing needed enhancements to GRTS. These enhancements range from better capturing environmental results to improving consistency of data entry to facilitate state-by-state comparisons.

State CWA 319 Quality Management Plans (QMPs), are also periodically reviewed and approved by EPA Regions.

Great Water Body Program: Acreage data from the Great Water Body Programs have not been reported under this measure because of their initial inability to provide timely information starting in 2004 when the measure was initiated.

Office of Water: EPA actions are consistent with data quality and management policies. Reporting in FY 2007 through FY 2009 did not indicate that any improvements to any of the databases associated with this measure were needed. The Office of Water Quality Management Plan (July 2002) is available on the Intranet at http://intranet.epa.gov/ow/informationresources/quality/qualitymanage.html

Attached Documents:
OW_QMP.pdf

3c. Data Oversight:
Source Data Reporting Oversight Personnel: 5-Star: 5-Star Grant Project Officer; Headquarters; Office of Water; Office of Wetlands, Oceans, and Watersheds; Wetlands Division; Wetland Strategies and State Programs Branch. NEP: Regional NEP Coordinators; Regions 319: Regional GRTS Coordinators; Regions Great Water Body Program: Acreage data from the Great Water Body Programs have not been reported under this measure because of their initial inability to provide timely information starting in 2004 when the measure was initiated. Source Data Reporting Oversight Responsibilities: All oversight personnel check grantee-reporting data against hardcopies and spot check quality of data entry. Information Systems Oversight Personnel:

5-Star: 5-Star Grant Project Officer; Headquarters; Office of Water; Office of Wetlands, Oceans, and Watersheds; Wetlands Division; Wetland Strategies and State Programs Branch.

NEP: National NEP Coordinator; Headquarters; Office of Water; Office of Wetlands, Oceans, and Watersheds; Oceans and Coastal Protection Division; Coastal Management Branch.

319: National GRTS Coordinator; Headquarters; Office of Water; Office of Wetlands, Oceans, and Watersheds; Assessment and Watershed Protection Division, Nonpoint Source Branch.

Great Water Body Program: Acreage data from the Great Water Body Programs have not been reported under this measure because of their initial inability to provide timely information starting in 2004 when the measure was initiated.

Information Systems Oversight Responsibilities: All information systems oversight personnel manage either grantees or contractors who maintain each of the data systems per contract or grant QA/QC procedures.

3d. Calculation Methodology:

Decision Rules for Selecting Data: Data has to be located in one of the three listed databases after QA/QC procedures have been finalized. Data includes projects that are finalized in the applicable fiscal year; all projects that are not finalized in the applicable fiscal are excluded. Data from all projects that do not address wetlands are excluded (these include upland areas not defined as wetlands on Tab 1 of this database.) All projects for restoration or improvement are added together excluding projects that protect wetlands.

Definitions of Variables: Definitions of all variables are described in Tab 1 of this database.

Explanation of Calculations: The "Wetland Acres Restored or Improved" measure is calculated by adding together wetlands acres from the restoration and improvement projects reported from each of the relevant programs (NEP, 319, and 5-Star) tracking and reporting systems for grants. These databases are as follows: the 319 Grants Reporting and Tracking System (GRTS), NEP's Performance Indicators Visualization and Outreach Tool (PIVOT) and Wetlands Program's Five-Star Restoration Grant Database. Acreage data from the Great Water Body Programs have not been reported under this measure because of their initial inability to provide timely information starting in 2004 when the measure was initialed.

Explanation of Assumptions: All projects are finalized in each applicable fiscal year. Projects do not include routine operations and maintenance of wetlands.

Unit of Measure: Acres of wetlands restored and improved

Timeframe of Result: Annual.

Documentation of Methodological Changes: Not applicable.

4a. Oversight and Timing of Final Results Reporting:
Final Reporting Oversight Personnel: Senior Budget Officer, Headquarters, Office of Water, Office of Wetlands, Oceans, and Watersheds. Final Reporting Oversight Responsibilities: Oversight personnel checks the final numbers provided in the system and checks them for reasonability and approves final number. Final Reporting Timing: Annual by fiscal year.

4b. Data Limitations/Qualifications:
General Limitations/Qualifications: Current data limitations include: (1) information that may be reported inconsistently across the NEPs, CWA 319, and 5-Star projects because they may interpret the meaning of "protection and restoration" differently; (2) acreage amounts may be miscalculated or incorrectly reported, and (3) acreage may be double-counted (i.e., the same parcel may also be counted more than one partner, or the same parcel may be counted more than once because it has been restored several times over a period of years). Data Lag Length and Explanation: No data lag. All data is reported at the end of each fiscal year. Methodological Changes: Not applicable.

4c. Third-Party Audits:
In the past, Nonpoint Source Program reporting under Section 319 had been identified as an Agency-level weakness under the Federal Managers Financial Integrity Act. The Agency's establishment and subsequent enhancements of GRTS has served to mitigate this problem by requiring states to identify the activities and results of projects funded with Section 319(h).

Measure Code: E - Percent of the population in Indian country served by community water systems that receive drinking water that meets all applicable health-based drinking water standards

Office of Water (OW)

Goal Number and Title:
2 - Protecting America's Waters
Objective Number and Title:
1 - Protect Human Health
Sub-Objective Number and Title:
1 - Water Safe to Drink
Strategic Target Code and Title:
2 - By 2015, drinking water that meets health-based drinking water standards for Indian countries
Managing Office:
Office of Groundwater and Drinking Water
1a. Performance Measure Term Definitions:
The definition of Indian country used by the US Department of Justice can be found at this web link:http://www.justice.gov/usao/eousa/foia_reading_room/usam/title9/crm00677.htm Community water systems --The U.S. Environmental Protection Agency (EPA) defines a community water system (CWS) as a public water system that serves at least 15 service connections used by year-round residents or regularly serves at least 25 year-round residents. In FY2011 737 CWSs in Indian country regulated by the EPA and Navajo Nation provided water to more than 918 thousand persons. Health-based drinking water standards-- exceedances of a maximum contaminant level (MCL) and violations of a treatment technique
2a. Original Data Source:
EPA, except for community water systems serving the Navajo Nation, because the Navajo Nation has primacy responsibility for implementing the Safe Drinking Water Act.
2b. Source Data Collection:
The EPA Office of Ground Water and Drinking Water (Headquarters) calculates this measure using data reported in the Safe Drinking Water Information System-Federal (SDWIS-FED) and provides the results to EPA Regions and the Navajo Nation. This measure includes federally-regulated contaminants of the following violation types: Maximum Contaminant Level, Maximum Residual Disinfection Limit, and Treatment Technique violations. It includes any violations from currently open and closed community water systems (CWSs) that overlap any part of the most recent four quarters.
2c. Source Data Reporting:
Public Water Sanitary System (PWSS) Regulation-Specific Reporting Requirements Guidance. Available on the Internet at http://www.epa.gov/safewater/regs.html

System, user, and reporting requirements documents can be found on the EPA web site, http://www.epa.gov/safewater/

3a. Relevant Information Systems:
SDWIS/STATE, a software information system jointly designed by states and EPA, to support states and EPA Regions as they implement the drinking water program. SDWIS/STATE is an optional data base application available for use by states and EPA regions to support implementation of their drinking water programs. EPA Region 9 utilizes an access database system (DIME) to collect and report on tribal community water systems in Region 9.

SDWIS/FED User and System Guidance Manuals (includes data entry instructions, data On-line Data Element Dictionary-a database application, Error Code Data Base (ECDB) - a database application, users guide, release notes, etc.) Available on the Internet at http://www.epa.gov/safewater/sdwisfed/sdwis.htm

System and user documents are accessed via the database link http://www.epa.gov/safewater/databases.html and specific rule reporting requirements documents are accessed via the regulations, guidance, and policy documents link http://www.epa.gov/safewater/regs.html

SDWIS/Fed does not have a Quality Assurance Project Plan. The SDWIS/FED equivalent is the Data Reliability Action Plan [2006 Drinking Water Data Reliability Analysis and Action Plan, EPA-816-R-07-010 March 2008] The DRAP contains the processes and procedures and major activities to be employed and undertaken for assuring the data in SDWIS meet required data quality standards. This plan has three major components: assurance, assessment, and control.

Office of Water Quality Management Plan, available at http://www.epa.gov/water/info.html |

3b. Data Quality Procedures:
The Office of Ground Water and Drinking Water is modifying its approach to data quality review based on the recommendations of the Data Quality Workgroup and on the Drinking Water Strategy for monitoring data.

There are quality assurance manuals for states and Regions, which provide standard operating procedures for conducting routine assessments of the quality of the data, including timely corrective action(s).

Reporting requirements can be found on the EPA web site, http://www.epa.gov/safewater/
SDWIS/FED edit checks built into the software to reject erroneous data.
EPA offers the following to reduce reporting and database errors:
1) training to states on data entry, data retrieval, compliance determination, reporting requirements and error correction, 2) user and system documentation produced with each software release and maintained on EPA's web site, 3) Specific error correction and reconciliation support through a troubleshooter's guide, 4) a system-generated summary with detailed reports documenting the results of each data submission, 5) an error code database for states to use when they have questions on how to enter or correct data, and 6) User support hotline available 5 days a week. |

3c. Data Oversight:

The Drinking Water Protection Division Director oversees the source data reporting and the information systems producing the performance result.
3d. Calculation Methodology:
SDWIS/STATE, a software information system jointly designed by states and EPA, to support states as they implement the drinking water program. SDWIS/STATE is an optional data base application available for use by states and EPA regions to support implementation of their drinking water programs. U.S. EPA, Office of Ground Water and Drinking Water. Data and Databases. Drinking Water Data & Databases – SDWIS/STATE, July 2002. Information available on the Internet: http://www.epa.gov/safewater/sdwis_st/current.html SDWIS/FED User and System Guidance Manuals (includes data entry instructions, data On-line Data Element Dictionary-a database application, Error Code Data Base (ECDB) - a database application, users guide, release notes, etc.) Available on the Internet at http://www.epa.gov/safewater/sdwisfed/sdwis.htm System and user documents are accessed via the database link http://www.epa.gov/safewater/databases.html and specific rule reporting requirements documents are accessed via the regulations, guidance, and policy documents link http://www.epa.gov/safewater/regs.html Documentation is also available at the Association of State Drinking Water Administrators web site at www.ASDWA.org SDWIS/Fed does not have a Quality Assurance Project Plan. The SDWIS/FED equivalent is the Data Reliability Action Plan [2006 Drinking Water Data Reliability Analysis and Action Plan, EPA-816-R-07-010 March 2008] The DRAP contains the processes and procedures and major activities to be employed and undertaken for assuring the data in SDWIS meet required data quality standards. This plan has three major components: assurance, assessment, and control. Office of Water Quality Management Plan, available at http://www.epa.gov/water/info.html
4a. Oversight and Timing of Final Results Reporting:
The Evaluation and Accountability Team Leader is responsible for overseeing the final reporting for the Office of Water
4b. Data Limitations/Qualifications:
Recent state and EPA Regional data verification and other quality assurance analyses indicate that the most significant data quality problem is under-reporting by the states of monitoring and health-based standards violations and inventory characteristics. The most significant under-reporting occurs in monitoring violations. Even though those are not covered in the health based violation category, which is covered by the performance measure, failures to monitor could mask treatment technique and MCL violations. Such under-reporting of violations limits EPA's ability to: 1) accurately portray the percent of people affected by health-based violations, 2) target enforcement oversight, 3) target program assistance to primacy agencies, and 4) provide information to the public on the safety of their drinking water facilities
4c. Third-Party Audits:

Measure Code: co5 - Percent of active dredged material ocean dumping sites that will have achieved environmentally acceptable conditions (as reflected in each site's management plan).

Office of Water (OW)

Goal Number and Title:
2 - Protecting America's Waters
Objective Number and Title:
2 - Protect and Restore Watersheds and Aquatic Ecosystems
Sub-Objective Number and Title:
2 - Improve Costal and Ocean Waters
Strategic Target Code and Title:
2 - Percent of active dredged material ocean dumping sites,will have achieved environmentally acceptable
Managing Office:
Office of Wetlands Oceans and Watersheds
1a. Performance Measure Term Definitions:
Active dredged material ocean dumping site: A dredged material ocean dumping site is a precise geographical area within which ocean dumping of wastes is permitted under conditions specified in permits issued by EPA Regions under section 102 and 103 of the Marine Protection, Research, and Sanctuaries Act (MPRSA). Active refers to a dredged material ocean dumping site that has been used in five years, and/or a site at which there are foreseeable plans for continued use. EPA Regions evaluate whether a site is active at the mid-year and end-of-year periods.

Environmentally acceptable conditions: The responsible EPA Regions determine whether dredged material ocean dumping sites are achieving environmentally acceptable conditions on a case-by-case basis, based on the requirements of the Site Management and Monitoring Plan (SMMP) and site sampling/surveying/monitoring results. On-site monitoring programs are used to collect, test, measure, and analyze data on bathymetry, chemical, biological, and physical conditions (e.g., grain size, current speed) at dredged material ocean dumping sites. Based on the requirements of each SMMP, the responsible Regions may conduct monitoring surveys of the dump sites to determine benthic impacts, spatial distribution of dredged material, characterize physical changes to the seafloor resulting from disposal, pH, turbidity, and other water quality indicators. Monitoring/sampling methodologies and assumptions are site-specific.

Site management plan: Under the MPRSA, each dredged material ocean dumping site must have a Site Management and Monitoring Plan (SMMP). The SMMP includes, but is not limited to, a baseline assessment of the site, a consideration of anticipated use, a monitoring program, and site management conditions or practices that are necessary for protection of the aquatic environment. Each SMMP is unique to the dump site and is developed with the opportunity for stakeholder input.

Background:
·   This performance measure, which is a target in the 2011-2015 Strategic Plan, will be tracked on an

annual basis as a management tool for EPA's ocean dumping program. The baseline year for the measure is 2005. For more information on EPA's ocean dumping program, please visit
http://water.epa.gov/type/oceb/oceandumping/dredgedmaterial/dumpdredged.cfm

2a. Original Data Source:

EPA Regions. EPA Regional offices responsible for management, oversight, and data collection at dredged material ocean dumping sites enter their determinations of sites meeting environmentally acceptable conditions directly into EPA's Annual Commitment System (ACS) database.

2b. Source Data Collection:

Collection Methodology:

EPA Regions determine whether dredged material ocean dumping sites are achieving environmentally acceptable conditions on a case-by-case basis, based on the requirements of the SMMP and site sampling/surveying/monitoring results. For more information on the type of site sampling/surveying/monitoring that is conducted, please see the Performance Measure Term Definitions field. EPA's Oceans and Coastal Protection Division has prepared a template for the Regions to use when preparing survey plans and many oceanographic vessels, such as NOAA vessels, have their own survey plan template. The periodicity of monitoring is determined by the SMMP and is suitable for tracking this measure. Regions collect data per the requirements of the SMMP and based upon site-specific conditions and needs. Regions determine the percentage of active sites meeting environmentally acceptable conditions (as reflected in each site's management plan and measured through on-site monitoring program).

Geographical Extent: Ocean dredged material disposal sites are designated in ocean waters in each of the seven EPA Regions with ocean programs.

Spatial Detail: Ocean dredged material disposal sites are designated in the federal register (lat/log is provided in the federal register for each site) and can vary in shape and size.

Quality Procedures:

Regional OD Coordinators collect and evaluate data to determine if sites are achieving environmentally acceptable conditions on a case-by-case basis, based on the requirements of the Site Management and Monitoring Plan (SMMP) and site sampling/surveying/monitoring results. Regional OD Coordinators collect data/information related to site status (active vs. inactive).

For each survey, the Region is required to submit to EPA Headquarters a survey plan that presents types of sampling techniques, including equipment used, and how data are recorded. Regions must develop a Quality Assurance Project Plan (QAPP), as prescribed by their regional quality assurance procedures, when collecting data at an ocean dumping site. The QAPP outlines the procedures for collection methods, use of analytical equipment, analytical methods, quality control, and documentation and records. Regions must conduct data quality reviews as determined by their quality assurance procedures and included in their QAPPs. If a Region uses a NOAA vessel for the survey, it is expected that the Region will submit a survey plan to NOAA and adhere to NOAA requirements for conducting the survey.

QAPP guidance documents for those Regions responsible for ocean dumping sites may be found at the following internet sites: EPA Region 1 - http://www.epa.gov/ne/lab/qa/pdfs/QAPPProgram.pdf EPA Region 2 - http://www.epa.gov/region2/qa/documents.htm - qag EPA Region 3 – http://www.epa.gov/quality/qmps.html EPA Region 4 - http://www.epa.gov/region4/sesd/oqa/r4qmp.html EPA Region 6 - http://www.epa.gov/earth1r6/6pd/qa/qatools.htm EPA Region 9 - http://www.epa.gov/region9/qa/pdfs/qaprp_guidance3.pdf EPA Region 10 - http://www.epa.gov/quality/qs-docs/g5-final.pdf

2c. Source Data Reporting:

The EPA Regions annually enter their determinations of dredged material ocean dumping sites meeting environmentally acceptable conditions directly into EPA's Annual Commitment System (ACS) database in October. (Regions also provide a report at mid-year).

3a. Relevant Information Systems:

System Description: EPA's Annual Commitment System (ACS) is used to record and transmit the data for performance results for the measure. ACS is a module of the Agency's Budget Formulation system BFS Please see the DQR for BFS for additional information

Source/Transformed Data: The EPA Regions enter data into ACS. The Office of Wetlands Oceans and Watersheds reviews the data to ensure accurate data entry.

Information System Integrity Standards: National Ocean Dumping Program Coordinator, OCPD/OWOW/OW, reviews.

3b. Data Quality Procedures:

National Ocean Dumping Program Coordinator, OCPD/OWOW/OW, reviews.

The data are entered into ACS by EPA Regional OD Coordinators and the HQ National Ocean Dumping Program Coordinator follows up when necessary. HQ maintains a list of designated ocean dredged material disposal sites and works with Regions to verify active or inactive site status and up-to-date site management and monitoring plans (SMMPs). HQ cross-checks data entered into ACS with reporting from years past to ensure consistency and account for irregularities.

Furthermore, Headquarters convenes monthly Ocean Dumping Calls with the Regions, administers a chief scientist certification program, and coordinates, as needed, to address issues associated with ocean dredged material disposal sites, including issues identified through monitoring.

Reporting in FY 2007 through FY 2010 did not indicate that any improvements to the collection and/or evaluation of data to support the measure were needed.

3c. Data Oversight:

Source Data Reporting Oversight Personnel: HQ National Ocean Dumping Program Coordinator,

OCPD/OWOW/OW

Source Data Reporting Oversight Responsibilities: HQ maintains a list of designated ocean dredged material disposal sites and works with Regions to verify active or inactive site status and up-to-date site management and monitoring plans (SMMPs). HQ oversees and reviews final data entry into ACS. HQ cross-checks data entered into ACS with reporting in years past to ensure consistency and account for irregularities.

Information Systems Oversight Personnel: Please see the DQR for BFS for additional information

Information Systems Oversight Responsibilities: Please see the DQR for BFS for additional information.

3d. Calculation Methodology:

Decision Rules for Selecting Data: On a case-by-case basis, active sites are determined to be achieving environmentally acceptable conditions, based on the requirements of the Site Management and Monitoring Plan (SMMP) and site sampling/surveying/monitoring results.

Definitions of Variables: Active dredged material ocean dumping site: A dredged material ocean dumping site is a precise geographical area within which ocean dumping of wastes is permitted under conditions specified in permits issued by EPA Regions under section 102 and 103 of the Marine Protection, Research, and Sanctuaries Act (MPRSA). Active refers to a dredged material ocean dumping site that has been used in five years, and/or a site at which there are foreseeable plans for continued use. EPA Regions evaluate whether a site is active at the mid-year and end-of-year periods.

Environmentally acceptable conditions: The responsible EPA Regions determine whether dredged material ocean dumping sites are achieving environmentally acceptable conditions on a case-by-case basis, based on the requirements of the Site Management and Monitoring Plan (SMMP) and site sampling/surveying/monitoring results. On-site monitoring programs are used to collect, test, measure, and analyze data on bathymetry, chemical, biological, and physical conditions (e.g., grain size, current speed) at dredged material ocean dumping sites. Based on the requirements of each SMMP, the responsible Regions may conduct monitoring surveys of the dump sites to determine benthic impacts, spatial distribution of dredged material, characterize physical changes to the seafloor resulting from disposal, pH, turbidity, and other water quality indicators. Monitoring/sampling methodologies and assumptions are site-specific.

Site management plan: Under the MPRSA, each dredged material ocean dumping site must have a Site Management and Monitoring Plan (SMMP). The SMMP includes, but is not limited to, a baseline assessment of the site, a consideration of anticipated use, a monitoring program, and site management conditions or practices that are necessary for protection of the aquatic environment. Each SMMP is unique to the dump site and is developed with the opportunity for stakeholder input.

Explanation of Calculations: Each EPA Region reports the percent of active sites that are achieving environmentally acceptable conditions, based on the requirements of the Site Management and Monitoring

Plan (SMMP) and site sampling/surveying/monitoring results. The results from the seven EPA regions are averaged.

Explanation of Assumptions: This result does not necessarily reflect monitoring data from that year at all sites. The result reflects several factors such as meeting the requirements of the SMMP as well as site sampling/surveying/monitoring results.

Unit of Measure: Percent of active dredged material ocean dumping sites

Timeframe of Result: annual

Documentation of Methodological Changes: Not applicable

4a. Oversight and Timing of Final Results Reporting:
Final Reporting Oversight Personnel: OCPD/OWOW Division Director
Final Reporting Oversight Responsibilities: Review the data entered into ACS by EPA Regions and national program targets.
Final Reporting Timing: Annual

4b. Data Limitations/Qualifications:
General Limitations/Qualifications:
No error estimate is available for this data. The data collected by the EPA Regions are highly suitable for tracking the performance of this measure, as they are collected for the specific purpose of determining the environmental conditions of the dredged material ocean dump sites.
Data Lag Length and Explanation: Analysis of data collected as part of monitoring surveys may take several months for initial results and even longer for final results.
Methodological Changes: N/A

4c. Third-Party Audits:
N/A

Office of Water (OW)

Goal Number and Title:
2 - Protecting America's Waters

Objective Number and Title:
2 - Protect and Restore Watersheds and Aquatic Ecosystems

Sub-Objective Number and Title:
7 - Restore and Protect the Long Island Sound

Strategic Target Code and Title:
1 - Reduce the maximum area of hypoxia in Long Island Sound

Managing Office:
Long Island Sound Office

1a. Performance Measure Term Definitions:
Goal: Nitrogen waste load allocations (WLA) are specified in the "A Total Maximum Daily Load (TMDL) Analysis to Achieve Water Quality Standards for Dissolved Oxygen in Long Island Sound" (December 2000) that was prepared by the states of New York and Connecticut and approved by EPA in conformance with Section 303(d) of the Clean Water Act. (See: www.longislandsoundstudy.net/wp-content/uploads/2010/03/Tmdl.pdf) The TMDL nitrogen WLAs are included in the NPDES (state-delegated) permits issued by the states for dischargers to Long Island Sound. The baseline for this measure is 211,724 pounds per day of nitrogen or 59,146 TE/lbs-day as calculated in the TMDL. The TMDL established a WLAN of 22,774 TE lbs/day from point sources, to be achieved over a 15-year period beginning in 2000. The TMDL itself does nto establish annual targets for nitrogen reduction. However, EPA developed this measure as a means of tracking annual progress, withe each year's target for the measure equivalent to 1/15 of the overall reduction goal. So the goal is a reduction of 36,372 TE lbs/day (59,146-22,774=36,372). Annualized aggregate reduction = TMDL baseline minus 2014 target (84,474 lbs/day or 22,774 TE/lbs-day) divided by 15 year TMDL time period = 8,487 lbs/day or 2,425 TE lbs/day. The measure will be tracked in lbs/day and Trade Equalized (TE) lbs/day. Trade-equalized (TE): TE lbs/day are pounds of nitrogen adjusted by application of the equivalency factor assigned to each point source based on its proximity to the receiving water body (Long Island Sound) as specified in the TMDL. Trade equalization is a geographical calculation of the effect a pound of nitrogen leaving a point source will eventually have when it reaches western Long Island Sound. The connections among nitrogen, rivers, currents and hypoxia drive this calculation. The calculation takes into account both east-west as well as north-south distance from the western Sound when estimating nitrogen impact on the western Sound. If a coastal wastewater treatment plant is located in the western part of Long Island Sound, 100% of the nitrogen discharged into Long Island Sound could contribute to the hypoxia problem there. However, if a coastal wastewater treatment plant is in the eastern part of the Sound, not all of the nitrogen discharged will end up in the western Sound. Some of the nitrogen will be carried out of the Sound by currents

through the Race, and the calculation of this loss due to currents is called "transfer efficiency." For this reason, an equal amount of nitrogen discharged by these examples will not result in the same amount of nitrogen ending up in the western Sound. Similarly, if an inland wastewater treatment plant discharges nitrogen into a river, some of the nitrogen will be lost before the river waters reach Long Island Sound. This "river attenuation" is also taken into account when calculating nitrogen loads. The combination of transfer efficiency and river attenuation is used to create trade equalization.

Point source nitrogen discharges to Long Island Sound: This measure is the annual aggregate reduction from the TMDL-defined baseline point source nitrogen discharge from 106 sewage treatment plants (STPs) in Connecticut and New York discharging to Long Island Sound waters during the calendar year January-December. Point source pollution is defined in section 502 of the Clean Water Act as "any discernible, confined and discrete conveyance, including but not limited to any pipe, ditch, channel, tunnel, conduit, well, discrete fissure, container, rolling stock, concentrated animal feeding operation, or vessel or other floating craft, from which pollutants are or may be discharged. This term does not include agricultural storm water discharges and return flows from irrigated agriculture.

Background:

Long Island Sound, bounded by New York and Connecticut, has more than 8 million people living within its watershed. It is approximately 110 miles long (east to west) and about 21 miles across at its widest point. Research commissioned by the Long Island Sound Study estimated that more than $5 billion is generated annually in the regional economy from boating, commercial and sport fishing, swimming, and beachgoing in 1990 dollars. In 2012 dollars that value is in excess of $9.5 billion. Congress passed legislation in 1990 establishing an EPA Long Island Sound Office. The office was established in January 1992 with offices in Stamford, CT and Stony Brook, NY. The Long Island Sound Comprehensive Conservation and Management Plan (CCMP), developed under the National Estuary Program, is the result of a strong partnership between EPA Regions 1 and 2 and the states of Connecticut and New York. The CCMP was approved by EPA Administrator Browner and the Governors of Connecticut and New York in September 1994. To address the water quality problems in the Long Island Sound, EPA created the Long Island Sound Study (LISS) in partnership with the Connecticut Department of Energy and Environmental Protection (CTDEEP) and the New York State Department of Environmental Conservation (NYSDEC). The top priority of the LISS is reducing nitrogen loads which contribute to the low levels of oxygen affecting substantial areas of western Long Island Sound in late summer. Other implementation priorities are habitat restoration, watershed management, disposal of dredged materials, and public education and involvement on Long Island Sound issues.

Pollutant sources associated with increased urbanization, including sewage treatment plants and stormwater runoff, have discharged excessive levels of nitrogen to the Sound leading to increased algal blooms and decreased dissolved oxygen (DO) levels, caused when algae die and use up oxygen in the decaying process. As a result of eutrophication and hypoxia, large areas in the western portion of the Sound cannot support aquatic life, recreation, and other important uses. The analysis conducted by the LISS led to the adoption of a 58.5 percent nitrogen reduction target to reduce the extent and duration of hypoxic conditions in the Long Island Sound. Through the TMDL development process, CTDEEP and NYSDEC were able to incorporate the 58.5 percent nitrogen reduction target into a regulatory and legal framework. The Clean

Water Act (CWA) requires implementation of pollutant load reductions through point source permits issued under the National Pollutant Discharge Elimination System (NPDES) Program. The TMDL nitrogen WLAs are included in the NPDES (state-delegated) permits issued by the states for dischargers to Long Island Sound.

Point source nitrogen loads have been reduced over the last 25 years in great part due to the large number of wastewater treatment plant upgrades that have been performed in Connecticut and New York. The relatively flat progress in reducing point source nitrogen to the Sound from 2005-2009 was due to several New York City wastewater treatment plants under construction for nitrogen removal upgrades and their capacity to store and process wastewater has been reduced as a result. Weather and rainfall also affect the ability of wastewater treatment plants to effectively remove nitrogen, i.e., during periods of intense rainfall the capacity of a plant to handle wastewater may be exceeded and excess nitrogen discharged as a result. For more information, please see:

- http://www.ct.gov/dep/cwp/view.asp?a=2719&q=325604&depNav_GID=1654
- http://longislandsoundstudy.net/2010/07/lis-point-source-nitrogen-trade-equalized-loads/
- http://www.longislandsoundstudy.net/pubs/reports/tmdl.pdf
- http://water.epa.gov/lawsregs/lawsguidance/cwa/tmdl/upload/long_island_technical-2.pdf

2a. Original Data Source:

State offices (New York State DEC and Connecticut DEEP).

2b. Source Data Collection:

Collection Methodology: Under NPDES, as part of the discharge monitoring reporting process, the STPs in question must regularly monitor and test effluent for appropriate pollutants, including nitrogen, and annually report pollutant loading data to their respective states (Connecticut and New York). The STPs and the states must follow EPA guidance as part of the DMR process.

Quality Procedures:
- Legal requirements for permittees to self-report data on compliance with effluent parameters in permits generally results in consistent data quality and accuracy.
- Major and selected minor facilities are required to participate in the Discharge Monitoring Report (DMR) Quality Assurance Study Program: http://www.epa.gov/compliance/monitoring/programs/cwa/dmr/

Geographical Extent: Sewage treatment plants in the New York and Connecticut portions of the Long Island Sound watershed. See the TMDL for more information.

Spatial Detail: Facility-level data are provided by the states.

2c. Source Data Reporting:

Data Submission and Data Entry: The EPA Long Island Sound Office (LISO) requests that the states of New York and Connecticut provide information on the pounds of nitrogen discharged by each STP under their jurisdiction for Long Island Sound. The states use the DMR data submitted to them by the STPs. Within each state, the Long Island Sound Coordinator enters annual nitrogen effluent data for each STP into a spreadsheet provided by EPA's Long Island Sound National Estuary Program (NEP) Coordinator. The reporting spreadsheet utilized is a copy of the master spreadsheet used by EPA's Long Island Sound Office store and calculate results (described in more detail in section 3 of this DQR). Upon receiving the state data, the Long Island Sound NEP Coordinator copies relevant data from each state's spreadsheet into EPA's master spreadsheet.

Frequency and Timing of Data Transmission: States annually report nitrogen discharges for the previous calendar year. States provide the data once available, usually by late February or early March.

3a. Relevant Information Systems:
System Description:
The EPA Long Island Sound Office (LISO) stores state-provided data in an Excel spreadsheet stored on the office's network share drive. The filename is a version of the following: "TE WLA 20xx Master File final." The spreadsheet stores current and prior year data for the performance measure. The spreadsheet houses both source data and transformed data, as the spreadsheet conducts calculations on the nitrogen effluent data to determine trade-equalized nitrogen discharge levels. (See the Calculations field for more information.)
Information System Integrity Standards: N/A
3b. Data Quality Procedures:
The Long Island Sound Office assumes the data provided by the states is complete and correct.
3c. Data Oversight:
Source Data and Information Systems Reporting Oversight Personnel: The Long Island Sound NEP Coordinator
Oversight Responsibilities: Coordinate with state personnel to ensure timely and accurate reporting; transfer state data into EPA master spreadsheet; maintain spreadsheet on EPA share drive; and backup data.
3d. Calculation Methodology:
The Long Island Sound NEP Coordinator calculates the result for this measure based on total annual average loads from these 106 STPs discharging to Long Island Sound from Connecticut and New York. LISO uses an Excel spreadsheet that has a column that converts data on pounds of nitrogen discharged by each STP into trade-equalized pounds based on the equivalency factors established and explained in the TMDL referenced above. The STP totals are summed to calculate subtotals for New York and Connecticut and a grand total for both states is calculated as the annual result for this measure.
Unit of Measure: Percent of Goal Achieved
Timeframe of Result: 1999-present (cumulative measure starting with the 1999 baseline.)
Documentation of Methodological Changes: N/A
4a. Oversight and Timing of Final Results Reporting:
Final Reporting Oversight Personnel: Director, EPA Long Island Sound Office

Final Reporting Oversight Responsibilities: Oversight personnel checks the final numbers provided and follows-up with states for confirmation. Director has final review.

Final Reporting Timing: Annually, in February-March

4b. Data Limitations/Qualifications:

General Limitations/Qualifications:
- There may be errors of omission, in classification, documentation or mistakes in the processing of data.
- National trends over the past several years show an average of 94% of DMRs are entered timely and complete.

Data Lag Length and Explanation: There is a lag time of approximately 60-90 days between the end of the reporting year (in this case, a calendar year) and public reporting of the data, given that STPs are required to prepare and provide calendar-year DMR data to the states, who must in turn enter the data and provide to EPA.

Methodological Changes:N/A

4c. Third-Party Audits:

None

Office of Water (OW)

Goal Number and Title:
2 - Protecting America's Waters

Objective Number and Title:
2 - Protect and Restore Watersheds and Aquatic Ecosystems

Sub-Objective Number and Title:
8 - Restore and Protect the Puget Sound Basin

Strategic Target Code and Title:
1 - Improve water quality and enable the lifting of harvest restrictions in shellfish bed growing areas

Managing Office:
Region 10

1a. Performance Measure Term Definitions:

Improve water quality: Measuring the number of acres of shellfish beds with harvest restrictions lifted is not a direct measure of habitat quality, but it is a measure of improvement in water quality with respect to fecal coliform contamination. This measure of recovered shellfish growing acreage serves as an important surrogate for water quality and human health protection in Puget Sound.

Lifting of harvest restrictions: The Washington State Department of Health's (WDOH) Growing Area Classification program is responsible for evaluating all commercially harvested shellfish-growing areas in Washington State to determine their suitability for harvest. The state approves shellfish as safe for harvest if sanitary surveys show that the area is not subject to contamination that presents an actual or potential public health hazard. The sanitary surveys look for the presence of fecal material, pathogenic microorganisms, or poisonous or harmful organisms in concentrations that pose a health risk to shellfish consumers. For more information, see the Washington State Department of Health's Growing Area Classification Program at: http://www.doh.wa.gov/CommunityandEnvironment/Shellfish/GrowingAreas.aspx

Acres of shellfish bed growing areas impacted by degraded or declining water quality: The acreage counted via this measure is based on monitoring and status determinations made by the Washington State Department of Health (WDOH). Commercial shellfish growing areas in Washington State are classified as Approved, Conditionally Approved, Restricted, or Prohibited. (See "Definition of Variables" in section 3-d that follows.) These classifications have specific standards that are derived from the National Shellfish Sanitation Program Guide for the Control of Molluscan Shellfish (Chapter IV, 2009 Revision). Results of these classifications frequently include shellfish harvesting acres affected by National Estuary Program or Section 319 Nonpoint Source grants. The universe of potentially recoverable shellfish areas in the Puget Sound is an estimated 10,000 acres. This estimate is based on a table of potentially recoverable shellfish growing areas in Puget Sound as developed by the WDOH in 2006 The total area of Puget Sound restricted from the safe harvest of shellfish because of the impacts of pollution is approximately 30,000 acres based on baseline data from 2006.

Background:

EPA's Puget Sound webpage is: http://www.epa.gov/pugetsound/index.html

Also see the WDOH annual shellfish reports at:

http://www.doh.wa.gov/CommunityandEnvironment/Shellfish/GrowingAreas/AnnualReports.aspx

2a. Original Data Source:

The Washington State Department of Health (WDOH) determines and tracks the status of shellfish beds. The WDOH does the sampling and analysis, which forms the basis of their shellfish bed status determinations. WDOH provides updates annually and more frequently as requested, to both EPA and Puget Sound Partnership - the lead entity for the Puget Sound National Estuary Program.

2b. Source Data Collection:

Collection Methodology: The Puget Sound Partnership and EPA receive data from the Washington State Department of Health (WDOH), which is the entity that determines and tracks the status of shellfish beds. The WDOH does the sampling and analysis, which forms the basis of their shellfish bed status determinations. WDOH maintains a spreadsheet database of recoverable shellfish growing beds and status of each.

Quality Procedures:

The WDOH's Office of Shellfish and Water Protection (OSWP) has a Quality Management Plan (QMP) on file with EPA's Region 10 Puget Sound Program.

The principal components of OSWP's quality system and corresponding tools for implementing them include:

Quality Management, Accountability, and Performance Policy (Health Policy 02.005)

Public Health Laboratories Quality Management Plan

OSWP Policies/Procedures

Marine Water Sampling (OSWP #3)

Data Flow (OSWP #4)

Shoreline Surveys (OSWP #13)

Voluntary Water Sampling for Growing Area Classifications (OSWP #16)

Establishing Marine Water Sampling Station Locations (OSWP #17)

Closure Zones For Wastewater Treatment Plants (OSWP #19)

Harvest Site Pollution Assessment Policy (OSWP #21)

Recreational Shellfish Beach Classification (OSWP #22)

OSWP Annual Report: Commercial and Recreational Shellfish Areas

OSWP Status and Trends in Fecal Coliform Pollution in Shellfish Growing Areas of Puget Sound

On-site Sewage Systems Recommended Standards and Guidance (RS&G) Document:

OSWP reviews the status of all shellfish growing areas on at least an annual basis, including: review of the past year's water quality sample results, available field inspection reports, and review of available information from other sources. This review is summarized both in individual reports for each growing area, as well as a summary report (http://www.doh.wa.gov/ehp/sf/Pubs/annual-inventory.pdf

Microbiological water quality status and trends are also summarized in a separate report (http://www.doh.wa.gov/ehp/sf/Pubs/fecalreport.pdf

An annual list of Threatened areas identified through these evaluations are published to highlight areas which need attention

(http://www.doh.wa.gov/ehp/sf/Pubs/gareports/threatenlist.pdf

Geographical Extent: US Regional, Washington State, and Puget Sound Basin. Commercial shellfish bed growing areas in the Puget Sound estuary. A .pdf map of the shellfish-growing areas in Washington State, including Puget Sound, can be found

athttp://www.doh.wa.gov/CommunityandEnvironment/HealthyCommunitiesWashington.aspx

Spatial Detail: The spatial extent of growing area classifications is determined by a number of factors, including: marine water quality parameters, proximity to shorelines with sources of potential pollution/contamination (such as outfalls for WWTP or stormwater systems), and dilution and dispersion influences of tidal, wind and current action. Data for the Puget Sound shellfish measure are gathered from growing areas from within the Puget Sound basin, which includes the Strait of Juan de Fuca (but does not include shellfish-growing areas in Washington State located on the Pacific Ocean coast outside of the Puget Sound Basin).

Time Interval Covered by Source Data: WDOH reviews the status of all shellfish-growing areas on at least an annual basis, which includes: review of the past year's water quality sample results, available field inspection reports, and review of available information from other sources. This review is summarized both in individual reports for each growing area as well as a summary report

(http://www.doh.wa.gov/ehp/sf/Pubs/annual-inventory.pdf).

2c. Source Data Reporting:

Data Submission: The Region 10 Puget Sound Program Performance Manager requests the annual shellfish-growing area classification data results from WDOH. Data reported is a single number for total annual results, representing the sum of separate shellfish-growing areas where harvest restrictions were lifted during the reporting period. Region 10's Puget Sound Program Performance Manager receives performance measure data from WDOH either verbally via telephone or electronically in a spreadsheet document. WDOH provides a list containing the number of growing areas with classification upgrades and growing areas with classification downgrades by location. WDOH may report a single net number over the phone, but the official final data is organized by both location and gain/loss for the final net number of acres. The Puget Sound Program Performance Manager reviews submission of data from WDOH in light of previous planning discussions and mid-year status checks on shellfish growing areas targeted for recovery. The Performance Manager compares reported results against the table of potentially recoverable growing areas and uses that as the basis for developing annual targets. The Performance Manager contacts WDOH to ask about any unusual results or uncertainties.

Data Entry: The ACS Coordinator in the Region 10 Office of Water And Watersheds Grants and Strategic Planning Unit enters the data received from the WDOH for this measure into ACS.

Frequency and Timing of Data Transmission: The Region 10 Puget Sound Program Performance Manager requests the annual shellfish-growing area classification data results from WDOH. These data are typically requested in September for end-of-federal fiscal- year reporting. Additional interim data requests may occur for other performance and planning activities including, mid-year results and out-year projections.

3a. Relevant Information Systems:

System Description: EPA's Annual Commitment System (ACS) is used to record and transmit the data for performance results for the measure on the lifting of commercial harvest restrictions for shellfish in Puget Sound. ACS is a module of the Agency's Budget Formulation System (BFS). Please see the DQR for BFS for additional information.

Source/Transformed Data: The EPA Region 10 Office of Water and Watersheds ACS Coordinator enters source data into ACS. The numerical results reported by WDOH are not transformed before entry into ACS.

Information System Integrity Standards: Please see the DQR for BFS for additional information.

3b. Data Quality Procedures:

The Region 10 Puget Sound Program Performance Manager performs a QA/QC on the annual shellfish growing-area classification data results when received from WDOH. Additional QA/QC is achieved during discussions among the Puget Sound Performance Manager, Puget Sound Partnership, and WDOH for annual target setting, mid-year and end-of-year performance reporting, as well as episodic changes in growing area classifications that impact annual results. The source data quality procedures take place at the time monitoring and sampling of marine water quality during the sanitary surveys used to determine the lifting of harvest restrictions.

3c. Data Oversight:

Source Data Reporting Oversight Personnel , Region 10's Puget Sound Program Performance Manager in the Region 10 Office of Water Watersheds is responsible for acquiring the performance data from the WDOH.

Source Data Reporting Oversight Responsibilities: Region 10's Puget Sound Program Performance Manager receives performance measure data from WDOH. The Puget Sound Program Performance Manager reviews submission of data from WDOH in light of previous planning discussions and mid-year status checks on shellfish-growing areas targeted for recovery. The Performance Manager compares the reported results against the table of potentially recoverable growing areas and uses that as the basis for developing annual targets. The Performance Manager contacts WDOH to ask about any unusual results or uncertainties. The Performance Manager verifies that reported results meet the criteria for reporting under the EPA's OW National Water Program Guidance and provides the explanation of results as needed for targets not met or target significantly exceeded. The Puget Sound Performance Manager provides the data to the Region 10 Office of Water and Watersheds ACS Coordinator, who in turn enters the data into the EPA's ACS.

Information Systems Oversight Personnel: Please see the DQR for BFS for additional information.

Information Systems Oversight Responsibilities: Please see the DQR for BFS for additional information.

The Puget Sound Program Performance Manager reviews submission of data from WDOH in light of previous planning discussions and mid-year status checks on shellfish-growing areas targeted for recovery. The Performance Manager compares reported results against the table of potentially recoverable growing areas is the basis for developing annual targets. The Performance Manager contacts WDOH to ask about any unusual results or uncertainties. The Performance Manager verifies that reported results meet the criteria for reporting under the EPA's OW National Water Program Guidance and provides the explanation of results as needed for targets not met or target significantly exceeded.

3d. Calculation Methodology:

The cumulative level of acres where harvest restrictions are lifted is dynamic. The reported performance result is a result of the annual incremental number of shellfish growing acres that have restrictions lifted (and are thus re-classified as Approved or Conditionally Approved), less any acres in growing areas downgraded to Restricted or Prohibited classification during the same period annually. The cumulative measures are a product of the net upgrades minus downgrades for the reporting periods beginning in 2007 from a baseline in 2006.

The universe of 10,000 recoverable harvest acres is static, based on the 2006 baseline. This baseline was developed utilizing a WDOH shellfish-growing area restoration table that identified potentially recoverable growing areas.

Staff working in the WA State Shellfish Growing Area Program continually analyzes marine growing areas to make sure that shellfish in those marine areas are safe to eat. This work involves completing an evaluation of the growing area, assigning a classification to the area based on the results of the evaluation, and monitoring shellfish-growing areas for changes in water quality.

The evaluation process is called a "sanitary survey" and involves:

·A shoreline survey, which identifies pollution sources that may impact water quality. WDOH evaluates sewage treatment plants, onsite sewage systems, animal farms, drainage ways, and wildlife.

·Marine water sampling to determine fecal coliform bacteria levels in the marine water.

·Analysis of how weather conditions, tides, currents, and other factors may affect the distribution of any pollutants in the area.

Decision Rules for Selecting Data: A change in harvest classification is the deciding rule for selecting data. EPA Puget Sound Program's Performance Manager uses the WDOH-reported data for areas that have been upgraded from Prohibited or Restricted and Reclassified to Approved or Conditionally Approved. Additionally, shellfish harvest areas that have been downgraded are selected for inclusion in data to calculate the net change in the cumulative shellfish acreage that have been upgraded to safe harvest conditions. Particular attention is given to data from areas where current or recent restoration and water quality improvement actions have been undertaken, but data is not exclusively limited to these.

Definitions of Variables:

Approved - when the sanitary survey shows that the area is not subject to contamination that presents an actual or potential public health hazard. An Approved classification authorizes commercial shellfish harvest for direct marketing.

Conditionally Approved - when an area meets Approved criteria some of the time, but does not during predictable periods (such as significant rainfall events occurring right after extended dry periods or high boater usage periods during the summer). During these periods, the area is closed. The length of closure is predetermined for each Conditionally Approved area and is based on water sample data that show the amount of time it takes for water quality to recover and again meet Approved criteria.

Restricted - when water quality meets standards for an Approved classification, but the sanitary survey indicates a limited degree of pollution from non-human sources. Shellfish harvested from Restricted growing areas cannot be marketed directly.

Prohibited - when the sanitary survey indicates that fecal material, pathogenic microorganisms, or poisonous or harmful substances may be present in concentrations that pose a health risk to shellfish consumers. Growing areas adjacent to sewage treatment plant outfalls, marinas, and other persistent or unpredictable pollution sources are classified as Prohibited. Growing areas that have not undergone a sanitary survey are also classified as Prohibited. Commercial shellfish harvests are not allowed from Prohibited areas.

Explanation of Calculations:

The calculation of cumulative acres is a result of the annual incremental number of shellfish-growing acres that have restrictions lifted (and thus classified as Approved or Conditionally Approved), less any acres in growing areas downgraded to Restricted or Prohibited during the same period annually. The cumulative measure result is a product of the net upgrades minus downgrades for the reporting periods beginning in 2007 from a baseline in 2006.

Explanation of Assumptions:

Molluscan shellfish such as clams, oysters, and mussels feed by filtering large volumes of seawater. Along with food particles, they can also absorb bacteria, viruses, and other contaminants that are present. If contaminant levels are high enough, shellfish harvested from these areas can make people sick.

Measuring the number of acres of shellfish beds with harvest restrictions lifted is not a direct measure of water quality, but it is a measure of improving water quality with respect to fecal coliform contamination. This acreage serves as an important surrogate for water quality and human health protection in Puget Sound.

Unit of Measure: Acres (e.g., acres of commercial shellfish growing areas within the Puget Sound basin where harvest restrictions have been lifted due to improved water quality parameters).

Time Frame of Result:

The cumulative measure result is a product of the net upgrades minus downgrades for the reporting periods beginning in 2007 from a baseline in 2006.

Documentation of Methodological Changes:
Not applicable.

4a. Oversight and Timing of Final Results Reporting:

Final Reporting Oversight Personnel: The Region 10 Office of Water and Watersheds, Grants and Strategic Planning Unit ACS Coordinator oversees final reporting by the Regional Puget Sound Program. Resource Management Staff in the EPA HQ OW confirm reporting by the Region 10 ACS Coordinator.

Final Reporting Oversight Responsibilities: Resource Management Staff in the EPA HQ OW confirm reporting by the Region 10 ACS Coordinator. Additionally, OW Resource Management Staff will confer with Office of Wetlands Oceans and Watersheds Policy, Communications & Resource Management Staff if questions about the data reported need to be addressed by the NPO.

Final Reporting Timing: Annually, on a fiscal year basis.

4b. Data Limitations/Qualifications:

General Limitations/Qualifications: Data are limited to the commercial shellfish beds which are monitored by the WDOH. Commercial shellfish growing areas are only a part of the potential shellfish harvesting areas in Puget Sound (for example, recreational and non-commercial Tribal shellfish harvesting areas are not included in the data represented in this performance measure). Approximately 30,000 ~ 40,000 acres of potential shellfish harvesting areas in Puget Sound are restricted because of pollution and contamination impacts; the identified recoverable commercial shellfish growing areas are approximately 10, 000 acres. These growing areas have typically been located in more rural areas and downgrades and restrictions to harvest represent the point and nonpoint pollution sources generated in these environments. Currently, the Washington State Department of Health is evaluating some of the shoreline along the more urbanized corridors; however, these areas have not been classified. The lack of data from the more urbanized shoreline limits the evaluation of the more commercial/industrial impacts to the Sound. Even though the monitoring and classification of commercial shellfish growing areas is a limited spatial subset of all the potential shellfish harvest areas in Puget Sound, improvements to the water quality in commercial growing areas indicates a healthier Puget Sound.

Data Lag Length and Explanation: The period of time between changes in the harvest classification of shellfish growing areas and the time in which these changes are reflected in EPA's performance reporting can be as long as 12 months. This data lag can occur because annual performance results reflect changes from the prior year's report. Conceivably, a prior year's report might reflect the status of shellfish growing beds as they were in September of 2011 (i.e., for end-of-year FY 2011), and the end-of-year report for FY 2012 would report the results as of the end of September 2012. Thus, if changes in the shellfish growing beds classification occur very early in the performance period, a longer data lag occurs because reporting won't happen until the end of that performance period. It is also important to recognize that this measure of acres of shellfish growing areas that have harvest restrictions lifted, is a cumulative measure. Consequently, upgrades and downgrades in growing area classifications that occur in prior years still impact the net cumulative results reported in the current year. In this regard, a significant data lag can be embedded in the performance results because any prior years' downgrades are not easily recognized in current years' performance reporting.

Methodological Changes: Not applicable.

4c. Third-Party Audits:

No audits or quality reviews of the primary data have been conducted by EPA. EPA conducted a review of the Puget Sound NEP implementation in spring 2010 to help ensure that information provided is accurate and progress reported is in fact being achieved. EPA Regional staff also met with Washington State Department of Health (WDOH) staff in summer 2010 to review, validate, and update the targets for this performance measure.

www.ingramcontent.com/pod-product-compliance
Lightning Source LLC
Chambersburg PA
CBHW081426170526
45166CB00008B/2113